AF738641

Springer Handbook of Auditory Research

Volume 73

Series Editors

Richard R. Fay, Ph.D., Loyola University Chicago
Arthur N. Popper, Ph.D., University of Maryland

Editorial Board

Karen Avraham, Ph.D., Tel Aviv University, Israel
Andrew Bass, Ph.D., Cornell University
Lisa Cunningham, Ph.D., National Institutes of Health
Bernd Fritzsch, Ph.D., University of Iowa
Andrew Groves, Ph.D., Baylor University
Ronna Hertzano, M.D., Ph.D., School of Medicine University of Maryland
Colleen Le Prell, Ph.D., University of Texas, Dallas
Ruth Litovsky, Ph.D., University of Wisconsin
Paul Manis, Ph.D., University of North Carolina
Geoffrey Manley, Ph.D., University of Oldenburg, Germany
Brian Moore, Ph.D., Cambridge University, UK
Andrea Simmons, Ph.D., Brown University
William Yost, Ph.D., Arizona State University

More information about this series at http://www.springer.com/series/2506

The ASA Press

ASA Press, which represents a collaboration between the Acoustical Society of America and Springer Nature, is dedicated to encouraging the publication of important new books as well as the distribution of classic titles in acoustics. These titles, published under a dual ASA Press/Springer imprint, are intended to reflect the full range of research in acoustics. ASA Press titles can include all types of books that Springer publishes, and may appear in any appropriate Springer book series.

Editorial Board

Mark F. Hamilton (Chair), University of Texas at Austin
James Cottingham, Coe College
Timothy F. Duda, Woods Hole Oceanographic Institution
Robin Glosemeyer Petrone, Threshold Acoustics
William M. Hartmann (Ex Officio), Michigan State University
Darlene R. Ketten, Boston University
James F. Lynch (Ex Officio), Woods Hole Oceanographic Institution
Philip L. Marston, Washington State University
Arthur N. Popper (Ex Officio), University of Maryland
Christine H. Shadle, Haskins Laboratories
G. Christopher Stecker, Boys Town National Research Hospital
Stephen C. Thompson, Pennsylvania State University
Ning Xiang, Rensselaer Polytechnic Institute

ASA PRESS

Ruth Y. Litovsky • Matthew J. Goupell
Richard R. Fay • Arthur N. Popper
Editors

Binaural Hearing

With 93 Illustrations

ASA PRESS

Springer

Editors
Ruth Y. Litovsky
Department of Communication Sciences
and Disorders and Waisman Center
University of Wisconsin-Madison
Madison, WI, USA

Matthew J. Goupell
Department of Hearing and Speech
Sciences
University of Maryland
College Park, MD, USA

Richard R. Fay
Department of Psychology
Loyola University Chicago
Chicago, IL, USA

Arthur N. Popper
Department of Biology
University of Maryland
College Park, MD, USA

ISSN 0947-2657 ISSN 2197-1897 (electronic)
Springer Handbook of Auditory Research
ISBN 978-3-030-57099-6 ISBN 978-3-030-57100-9 (eBook)
https://doi.org/10.1007/978-3-030-57100-9

© Springer Nature Switzerland AG 2021, Corrected Publication 2021
This work is subject to copyright. All rights are reserved by the Publisher, whether the whole or part of the material is concerned, specifically the rights of translation, reprinting, reuse of illustrations, recitation, broadcasting, reproduction on microfilms or in any other physical way, and transmission or information storage and retrieval, electronic adaptation, computer software, or by similar or dissimilar methodology now known or hereafter developed.
The use of general descriptive names, registered names, trademarks, service marks, etc. in this publication does not imply, even in the absence of a specific statement, that such names are exempt from the relevant protective laws and regulations and therefore free for general use.
The publisher, the authors, and the editors are safe to assume that the advice and information in this book are believed to be true and accurate at the date of publication. Neither the publisher nor the authors or the editors give a warranty, expressed or implied, with respect to the material contained herein or for any errors or omissions that may have been made. The publisher remains neutral with regard to jurisdictional claims in published maps and institutional affiliations.

This Springer imprint is published by the registered company Springer Nature Switzerland AG
The registered company address is: Gewerbestrasse 11, 6330 Cham, Switzerland

The Acoustical Society of America

On 27 December 1928 a group of scientists and engineers met at Bell Telephone Laboratories in New York City to discuss organizing a society dedicated to the field of acoustics. Plans developed rapidly, and the Acoustical Society of America (ASA) held its first meeting on 10–11 May 1929 with a charter membership of about 450. Today, ASA has a worldwide membership of about 7000.

The scope of this new society incorporated a broad range of technical areas that continues to be reflected in ASA's present-day endeavors. Today, ASA serves the interests of its members and the acoustics community in all branches of acoustics, both theoretical and applied. To achieve this goal, ASA has established Technical Committees charged with keeping abreast of the developments and needs of membership in specialized fields, as well as identifying new ones as they develop.

The Technical Committees include acoustical oceanography, animal bioacoustics, architectural acoustics, biomedical acoustics, engineering acoustics, musical acoustics, noise, physical acoustics, psychological and physiological acoustics, signal processing in acoustics, speech communication, structural acoustics and vibration, and underwater acoustics. This diversity is one of the Society's unique and strongest assets since it so strongly fosters and encourages cross-disciplinary learning, collaboration, and interactions.

ASA publications and meetings incorporate the diversity of these Technical Committees. In particular, publications play a major role in the Society. *The Journal of the Acoustical Society of America* (JASA) includes contributed papers and patent reviews. *JASA Express Letters* (JASA-EL) and *Proceedings of Meetings on Acoustics* (POMA) are online, open-access publications, offering rapid publication. *Acoustics Today*, published quarterly, is a popular open-access magazine. Other key features of ASA's publishing program include books, reprints of classic acoustics texts, and videos. ASA's biannual meetings offer opportunities for attendees to share information, with strong support throughout the career continuum, from students to retirees. Meetings incorporate many opportunities for professional and social interactions, and attendees find the personal contacts a rewarding experience. These experiences result in building a robust network of fellow scientists and engineers, many of whom become lifelong friends and colleagues.

From the Society's inception, members recognized the importance of developing acoustical standards with a focus on terminology, measurement procedures, and criteria for determining the effects of noise and vibration. The ASA Standards Program serves as the Secretariat for four American National Standards Institute Committees and provides administrative support for several international standards committees.

Throughout its history to present day, ASA's strength resides in attracting the interest and commitment of scholars devoted to promoting the knowledge and practical applications of acoustics. The unselfish activity of these individuals in the development of the Society is largely responsible for ASA's growth and present stature.

Series Preface

Springer Handbook of Auditory Research

The following preface is the one that we published in volume 1 of the *Springer Handbook of Auditory Research* back in 1992. As anyone reading the original preface, or the many users of the series, will note, we have far exceeded our original expectation of eight volumes. Indeed, with books published to date and those in the pipeline, we are now set for over 75 volumes in SHAR, and we are still open to new and exciting ideas for additional books.

We are very proud that there seems to be consensus, at least among our friends and colleagues, that SHAR has become an important and influential part of the auditory literature. While we have worked hard to develop and maintain the quality and value of SHAR, the real value of the books is very much because of the numerous authors who have given their time to write outstanding chapters and to our many co-editors who have provided the intellectual leadership to the individual volumes. We have worked with a remarkable and wonderful group of people, many of whom have become great personal friends of both of us. We also continue to work with a spectacular group of editors at Springer. Indeed, several of our past editors have moved on in the publishing world to become senior executives. To our delight, this includes the current president of Springer US, Dr. William Curtis.

But the truth is that the series would and could not be possible without the support of our families, and we want to take this opportunity to dedicate all of the SHAR books, past and future, to them. Our wives, Catherine Fay and Helen Popper, and our children, Michelle Popper Levit, Melissa Popper Levinsohn, Christian Fay, and Amanda Fay Sierra, have been immensely patient as we developed and worked on this series. We thank them and state, without doubt, that this series could not have happened without them. We also dedicate the future of SHAR to our next generation of (potential) auditory researchers—our grandchildren—Ethan and Sophie Levinsohn, Emma Levit, Nathaniel, Evan, Stella Fay, and Sebastian Sierra.

Preface 1992

The Springer Handbook of Auditory Research presents a series of comprehensive and synthetic reviews of the fundamental topics in modern auditory research. The volumes are aimed at all individuals with interests in hearing research including advanced graduate students, post-doctoral researchers, and clinical investigators. The volumes are intended to introduce new investigators to important aspects of hearing science and to help established investigators to better understand the fundamental theories and data in fields of hearing that they may not normally follow closely.

Each volume presents a particular topic comprehensively, and each serves as a synthetic overview and guide to the literature. As such, the chapters present neither exhaustive data reviews nor original research that has not yet appeared in peer-reviewed journals. The volumes focus on topics that have developed a solid data and conceptual foundation rather than on those for which a literature is only beginning to develop. New research areas will be covered on a timely basis in the series as they begin to mature.

Each volume in the series consists of a few substantial chapters on a particular topic. In some cases, the topics will be ones of traditional interest for which there is a substantial body of data and theory, such as auditory neuroanatomy (Vol. 1) and neurophysiology (Vol. 2). Other volumes in the series deal with topics that have begun to mature more recently, such as development, plasticity, and computational models of neural processing. In many cases, the series editors are joined by a co-editor having special expertise in the topic of the volume.

Richard R. Fay, Chicago, IL, USA
Arthur N. Popper, College Park, MD, USA

SHAR logo by Mark B. Weinberg, Potomac, Maryland, used with permission

Preface 1992

[illegible]

Volume Preface

Binaural hearing is a fundamental component of hearing in all vertebrates. Although the topic has appeared in some earlier SHAR volumes, it has never been dealt with in a comprehensive way. Thus, this volume is intended to serve as a contemporary update on spatial and binaural hearing and also to bring together a cohesive view of the field.

In Chap. 1, Ruth Y. Litovsky and Matthew J. Goupell introduce the volume and provide an overview of the critical concepts related to binaural hearing. In Chap. 2, William M. Hartmann begins with how interaural time and level differences (ITDs and ILDs, respectively) are generated physically by the head and reviews classic studies to give us a historical perspective on the topic. In Chap. 3, William A. Yost, Torben Pastore, and Yi Zhou discuss similar topics but expand from the consideration of static sources to an emerging important area of research using moving sources and/or moving listeners.

After defining the physics of the problems and the signals that are present, Terry Takeshi Takahashi, Lutz Kettler, Clifford Henry Keller, and Avinash Deep Singh Bala discuss in Chap. 4 binaural circuits in avian models. In Chap. 5, Zoe Owrutsky, Victor V. Benichoux, and Daniel J. Tollin focus on mammalian neurons in the nuclei responsible for computation of ITDs and ILDs in the brainstem. These two chapters highlight some of the critical differences in localization circuits between the two model systems.

Then, the focus of the book moves toward binaural processing of complex signals. In Chap. 6, G. Christopher Stecker, Leslie R. Bernstein, and Andrew D. Brown focus on exploiting temporal complexity and the important role of onsets. In Chap. 7, Virginia Best, Matthew J. Goupell, and H. Steven Colburn discuss binaural processing across frequency channels. In Chap. 8, John F. Culling and Mathieu Lavandier cover the increased spectral-temporal complexity of speech signals and the topic of spatial unmasking of speech. In Chap. 9, Pavel Zahorik provides a multidisciplinary overview of stimulus complexity introduced by the environment.

The final chapters tie together the previous material in modeling, how binaural information is processed in the cortex, and binaural hearing in clinical populations. In Chap. 10, Mathias Dietz and Go Ashida describe how models of binaural processing are used for simulating the activity of binaural neurons in animals and for predicting relevant auditory perception in humans. In Chap. 11, Frederick J. Gallun, Nirmal K. Srinivasan, and Anna C. Dietsch discuss auditory dysfunction and the impact of aging, identifying evidence of impaired and preserved binaural function in these listeners. In Chap. 12, Stephen M. Town and Jennifer K. Bizley review what is known about cortical processing of binaural information. In particular, they provide insight into spatial encoding within and beyond auditory cortex support of sound localization. Finally, in Chap. 13, Todd Andrew Ricketts and Alan Kan address the applied issues relevant to patients with hearing loss who wear hearing assistive devices. Because hearing-impaired listeners who are fitted with hearing aids and/or cochlear implants show a gap in performance relative to normal-hearing listeners, the chapter identifies the dire need for improvement in binaural signal processing of assistive devices.

Ruth Y. Litovsky, Madison, WI, USA
Matthew J. Goupell, College Park, MD, USA
Richard R. Fay, Chicago, IL, USA
Arthur N. Popper, College Park, MD, USA

Contents

Contributors

Go Ashida Department of Neuroscience, University of Oldenburg, Oldenburg, Germany

Avinash Deep Singh Bala Institute of Neuroscience, University of Oregon, Eugene, OR, USA

Victor Benichoux Unit of Genetics and Physiology of Hearing, Department of Neuroscience, Institut Pasteur, Paris, France

Leslie R. Bernstein Department of Neuroscience and Department of Surgery (Otolaryngology), University of Connecticut Health Center, Farmington, CT, USA

Virginia Best Department of Speech, Language and Hearing Sciences, Boston University, Boston, MA, USA

Jennifer K. Bizley University College London Ear Institute, London, UK

Andrew D. Brown Department of Speech and Hearing Sciences, University of Washington, Seattle, WA, USA

H. Steven Colburn Department of Biomedical Engineering, Boston University, Boston, MA, USA

John F. Culling School of Psychology, Cardiff University, Cardiff, Wales, UK

Anna C. Diedesch Department of Communication Sciences and Disorders, Western Washington University, Bellingham, WA, USA

Mathias Dietz Department of Medical Physics and Acoustics, University of Oldenburg, Oldenburg, Germany

Frederick J. Gallun VA RR&D National Center for Rehabilitative Auditory Research, VA Portland Health Care System, Portland, OR, USA

Department of Otolaryngology, Head and Neck Surgery, Oregon Health and Science University, Portland, OR, USA

Matthew J. Goupell Department of Hearing and Speech Sciences, University of Maryland, College Park, MD, USA

Department of Biomedical Engineering, Boston University, Boston, MA, USA

William M. Hartmann Department of Physics and Astronomy, Michigan State University, East Lansing, MI, USA

Alan Kan School of Engineering, Macquarie University, Macquarie Park, NSW, Australia

Clifford Henry Keller II Institute of Neuroscience, University of Oregon, Eugene, OR, USA

Lutz Kettler Technical University of Munich (TUM), Freising-Weihenstephan, Germany

Mathieu Lavandier Laboratoire Génie Civil et Bâtiment, École Nationale des Travaux Publics de l'État, Université de Lyon, Vaulx-en-Velin Cedex, France

Ruth Y. Litovsky Department of Communication Disorders and The Waisman Center, University of Wisconsin-Madison, Madison, WI, USA

Zoe L. Owrutsky Department of Physiology and Biophysics, University of Colorado Anschutz Medical Campus, Aurora, CO, USA

M. Torben Pastore Spatial Hearing Laboratory, College of Health Solutions, Arizona State University, Tempe, AZ, USA

Todd Andrew Ricketts Department of Hearing and Speech Sciences, Vanderbilt University Medical Center, Nashville, TN, USA

Nirmal K. Srinivasan Department of Speech-Language Pathology & Audiology, Towson University, Towson, MD, USA

G. Christopher Stecker Center for Hearing Research, Boys Town National Research Hospital, Omaha, NE, USA

Department of Hearing and Speech Sciences and Vanderbilt Brain Institute, Vanderbilt University School of Medicine, Nashville, TN, USA

Terry Takeshi Takahashi Institute of Neuroscience, University of Oregon, Eugene, OR, USA

Daniel J. Tollin Department of Physiology and Biophysics, University of Colorado Anschutz Medical Campus, Aurora, CO, USA

Stephen M. Town University College London Ear Institute, London, UK

William A. Yost Spatial Hearing Laboratory, College of Health Solutions, Arizona State University, Tempe, AZ, USA

Pavel Zahorik Department of Otolaryngology-Head and Neck Surgery and Communicative Disorders and Department of Psychological and Brain Sciences, University of Louisville, Louisville, KY, USA

Yi Zhou Spatial Hearing Laboratory, College of Health Solutions, Arizona State University, Tempe, AZ, USA

Chapter 1
Binaural Processing of Sounds

Ruth Y. Litovsky and Matthew J. Goupell

1.1 Introduction

Think about navigating the real world on an everyday basis, such as visiting a new city and trying to find your way among the noise generated by traffic, construction, and pedestrians communicating loudly as they traverse busy streets to their respective destinations. As a pedestrian heading to a restaurant to meet a friend, your task is to pay attention to a multitude of moving objects, oncoming vehicles at crosswalks, sirens, and the voices of people surrounding you. You are able to turn your head toward immediately relevant sounds, ignore irrelevant sounds, and safely navigate across the street until you arrive at your destination. The restaurant is popular, bustling with people; conversations are lively and exuberant while music plays in the background. Not only can you focus on your friend's voice but you can also locate the source of the music, despite the complexity of the auditory world.

In the restaurant scenario, the binaural auditory system plays a critical role in facilitating sound localization and in the segregation of speech from background noise. Binaural hearing arises from the integration of sounds arriving at the two ears. These sounds often differ because of the acoustical transformation imposed by the features of the head with one ear on either side. This arrangement provides an avenue for sounds arriving from each location around the listener to produce a unique set of interaural differences (i.e., differences between the ears in the time of

---

R. Y. Litovsky (✉)
Department of Communication Disorders and The Waisman Center,
University of Wisconsin-Madison, Madison, WI, USA
e-mail: litovsky@waisman.wisc.edu

M. J. Goupell
Department of Hearing and Speech Sciences, University of Maryland,
College Park, MD, USA
e-mail: goupell@umd.edu

© Springer Nature Switzerland AG 2021

R. Y. Litovsky et al. (eds.), *Binaural Hearing*, Springer Handbook of Auditory Research 73, https://doi.org/10.1007/978-3-030-57100-9_1

arrival and the intensity of a sound). Specific neural circuits and mechanisms in the brain process that information from the two ears to help the localization of sounds and the organization of complex auditory scenes.

1.1.1 The Roles of Binaural Hearing

Through binaural hearing, humans and animals can localize sound sources in space without a priori knowledge of the location of a sound source or without other sensory information such as visual cues (Masterton et al. 1969). For humans, examples of sounds whose location is important to identify include those that occur in everyday listening environments, such as the voice of someone calling one's name or the source of potentially dangerous traffic at a busy intersection. For animals in their habitat, significant ecologically important sounds can emerge from a potential food source, potential mate, or an approaching predator.

Moreover, in addition to the role of binaural hearing in facilitating precise sound localization abilities, the brain has the amazing ability to organize complicated auditory scenes with multiple sound sources. This ability allows someone to hear a target source of interest in the context of a cacophony of competing background sounds. Such a skill is critical for communication and is typically utilized in work, school, or crowded social situations.

1.1.2 The Importance of Binaural Hearing

Investigations of binaural hearing encompass studies of normal and abnormal aspects of auditory perception and physiology as well as of modeling. Over the past few decades, binaural processes involved in both sound localization and the ability to perceive speech in the presence of other sounds have gained a broader interest and have received growing attention in the published literature. Moreover, the subject has undergone some significant changes so that there is now a much richer understanding of the many aspects comprising binaural processing, its role in development, and in the success and limitations of hearing-assistive-device use. The topic of binaural hearing is also particularly unique in generating the interest and dedication of researchers with varying backgrounds and expertise, including psychoacousticians, physiologists, engineers, modelers, and animal behaviorists. Because of the immediate and significant relevance in aiding people with hearing loss, binaural hearing is also a critical topic for clinicians in audiology and otolaryngology.

1.1.3 Physiological Substrate of Binaural Hearing

Unlike most auditory perceptions, the physiological substrate where binaural processing first occurs, the superior olivary complex, is anatomically unique. A few small nuclei that make up this brain region are responsible for integrating input from the two ears and transmitting information to higher brain regions. These nuclei are distinguished by being identifiable both physiologically and anatomically. However, as in all auditory processing, when binaural information advances to higher auditory centers (i.e., the midbrain, thalamus, and cortex), the manner in which individual neurons and neural networks handle this information becomes increasingly complex. Thus, at more central brain regions involved in auditory processing, binaural information is preserved, although the role of these centers in binaural processing is enigmatic. What we do know is that neurons at all levels at and above the superior olivary complex and auditory midbrain show preferences for binaural cues associated with the locations of sound sources.

Overall, the field of binaural hearing has benefitted from an intricate marriage of three main scientific approaches: perception, physiology, and modeling. Together, studies using these approaches have been woven into a fascinating narrative about encoding of the binaural cues associated with locations and how it enables perception of objects in the world.

Psychoacousticians have focused on understanding the relationship between the perception of sounds in space and numerous stimulus parameters including frequency, amplitude, and duration, to name a few. At the heart of research on binaural hearing is the understanding of the relative importance of interaural timing difference (ITD) and interaural level difference (ILD) cues at low and high frequencies, respectively.

This work intricately connects to the research of physiologists, who have devoted decades of effort to understanding whether binaural processing of ITDs can be explained through a simple model that assumes the existence of neurons in the brainstem that code for coincident arrival of neural inputs from the left and right auditory nerves (Jeffress 1948). That model assumed that coding of ITDs relies on excitatory inputs from the two ears. More recently (e.g., Brand et al. 2002), understanding the role that inhibition plays in modulating binaural sensitivity has altered and expanded on the simple excitatory coincidence model. The role of ILDs was, for many years, downplayed relative to that of ITDs; however, in recent years, there has been growing evidence for joint coding of ILD and ITD cues, a departure from the simple physiological understanding of ITDs and ILDs being processed separately.

1.2 Definition of Critical Concepts and Processing of Binaural Cues

As described in Chap. 2 by William M. Hartmann, it is well-known that binaural cues arise from physical acoustics resulting from the manner in which sounds in space reach the ears of the listeners. These acoustics cues consist of differences between the ears, ITDs and ILDs, both of which are coded by the auditory system and provide information regarding sound location in the horizontal plane. Chapter 2 provides an in-depth understanding of the physics of how the head generates interaural differences as well as how these cues relate to perception of static sound location. The head can be modeled as a simple sphere that generates different ITDs and ILDs depending on the location of the sound source. Perception of these cues can be measured in a variety of tasks, such as free-field sound localization or headphone lateralization experiments.

Historically, it was first determined that ITDs and ILDs operated primarily at low and high frequencies, respectively, for human sound localization (Strutt 1907), and this has been called the "duplex theory" of sound localization. More recent work suggests that access to both ITDs and ILDs are processed at all frequencies (e.g., Henning 1974; Yost and Dye 1988), but it is a matter of how these cues are weighted across frequency. With both cues for complex sounds like speech, humans primarily use low-frequency ITDs for sound localization (Wightman and Kistler 1992; Macpherson and Middlebrooks 2002).

Not all sound sources and receivers are stationary. Chapter 3 by William A. Yost, Torben Pastore, and Yi Zhou expand the information from Chap. 2 to dynamic localization situations that people experience daily. Critical to understanding sound localization in these dynamic situations are the changes in auditory-spatial and head-position cues, and how multisensory information from the visual and vestibular systems is integrated into these perceptions. This area of study was first investigated and probed around the 1940s (e.g., Wallach 1940) but only recently has been energized by renewed research interest.

The duplex theory (the low- vs high-frequency dichotomy for processing ITDs and ILDs, respectively), as thought about in humans, may not be the same in animals with relatively large or small heads. There is a general shift from ILD to ITD processing as head size increases, allowing for low-frequency cues to be more effectively captured. For example, mice (*Mus musculus*) and rats (e.g., *Rattus rattus* or *Rattus norveticus*) primarily utilize high frequencies from about 15 to 80 kHz for sound localization (Ehret and Dreyer 1984; Heffner and Heffner 2016). Animals with relatively larger heads, such as humans or cats, utilize ITDs at low frequencies.

ITDs and ILDs are primarily processed in separate brainstem nuclei, the medial and lateral superior olivary complexes, respectively. Much of the research over the years has been on ITDs, perhaps because Jeffress (1948) proposed a simple and compelling model that focused on coincidence detection in brainstem neurons, and both psychoacousticians and physiologists were driven to either confirm or deny the utility of such a model. Early physiology studies confirmed the existence of binaural coincidence neurons (Goldberg and Brown 1968, 1969). Neurophysiological and anatomical studies showed that correlated brainstem nuclei in the owl brainstem have clear spatial maps. Chapter 4 by Terry T. Takahashi, Lutz Kettler, Clifford H. Keller, and Avinash D. S. Bala describes the ascending anatomy and physiology of the monaural pathways in birds and the binaural computations that produce the receptive fields found in the midbrain nucleus, the inferior colliculus.

Over the years, particularly with the finding that inhibitory inputs were also important for ITD encoding (Brand et al. 2002) and findings of critical differences between avian and mammalian binaural processing centers, it has become more apparent that applicability of the Jeffress framework has limitations, particularly across species. Chapter 5 by Zoe L. Owrutsky, Victor Benichoux, and Daniel J. Tollin discusses how ITDs and ILDs are processed in mammals, with a particular focus on ILD encoding and envelope processing in the lateral superior olive. The specializations in the processing of the inputs to the lateral superior olive allow for joint coding of the ILD and ITD, a departure from the simple physiological understanding of the duplex theory that suggests primarily ILD processing at relatively high frequencies.

To summarize, although evidence for the Jeffress model remains somewhat elusive and of considerable debate, as discussed in Chaps. 2, 4, 5, 6, 7, and 10, there has been great progress and refocusing in recent years on understanding the roles of ITDs and ILDs. A more nuanced appreciation of ITD and ILD processing and the duplex theory has emerged, namely that ITDs in the envelope of high-frequency sounds and ILDs can be processed at low frequencies, which has been confirmed in perception and neurophysiological recordings. Overall, an increasingly important role of ILDs in binaural hearing has emerged (see Chaps. 4 and 5).

Another important factor to consider is that ITD and ILD cues for sound localization change as signals become complex. One aspect of signal complexity comes in the temporal domain through temporal modulations. As discussed in Chap. 6 by G. Christopher Stecker, Leslie R. Bernstein, and Andrew D. Brown, temporal onsets (i.e., positive-going envelope fluctuations) appear to provide critically important localization cues (see also Hartmann, Chap. 2, and Zahorik, Chap. 9, concerning the precedence effect). Processing of modulated signals at high frequencies allow for the conveyance of ITD information, another departure from the duplex theory. Signal factors related to processing of this information, for example, are the envelope modulation rate and envelope sharpness.

Another aspect of stimulus complexity arises in the spectral domain. Across-frequency processing of binaural information is also important, which is reviewed in Chap. 7 by Virginia Best, Matthew J. Goupell, and H. Stephen Colburn. Although a relatively small area of research to review compared with temporal complexity, basic models of across-frequency binaural processing consider aspects of similarity of the cues across frequencies, how the interaural differences match those naturally produced by the head, and how those frequencies might be weighted differently. For across-frequency ITD processing, a dominant region of frequency dependence is revealed around 600–700 Hz, which is a common feature in most models in this area.

1.3 Spatial Unmasking and the Effects of Rooms

Humans are social beings and communication is critical to them. The ability to communicate in acoustically complex and challenging environments that are embedded with background noise and reverberation is often necessary. Chapter 8 by John F. Culling and Mathieu Lavandier reviews the basic findings of unmasking of tones (i.e., binaural masking level differences) and speech. The tonal conditions are thought to be simplifications of more complex spatial situations where there is masking of speech signals. For signals such as speech, which are both temporally and spectrally complex, there is a robust literature examining the spatial release from masking. Many studies have shown that spatial separation between the target speech and interfering sound sources provides speech understanding benefits, which can be partially understood by binaural and monaural cues arising from the physics of the head and partially by the neural computations of interaural differences. The authors review the relationship between binaural masking level differences and the spatial release from masking.

Chapter 9 by Pavel Zahorik reviews how rooms and reverberation affect the physical properties of the sounds that reach the ears, which then impact communication and spatial hearing. Room acousticians have quantified a number of metrics relevant to the perception of sounds in reverberant spaces, those that affect the intelligibility of speech (which is important for classrooms and lecture halls) and the quality of sound (which is important for concert halls and music venues). Localization abilities are altered by rooms, and features of binaural hearing like the precedence effect are clearly critical for successful communication in such environments.

1.4 Relationship Between Perception, Physiology, and Modeling

In the area of binaural hearing, there have always been strong ties between perception, physiology, anatomy, and modeling. In Chap. 10, Mathias Dietz and Go Ashida bring together many of the ideas in the previous chapters. Basic model concepts in binaural hearing are reviewed as well as the state of the art in the understanding of this topic, starting with a toolkit of basic principles on which to build these models. The types of possible models range in complexity, from simple signal-processing models to highly complex biophysical models of neural processing, all of which are necessary for the deepest understanding of binaural hearing the perception of sound source location.

1.5 Aging and Hearing Impairment

A hallmark of hearing and communication needs is the known fact that hearing disorders occur in a significant proportion of the population. Hearing impairment through noise exposure and damage, genetic reasons, or the natural aging process (i.e., presbycusis) affects spatial hearing abilities and is particularly detrimental to communication in noisy and reverberant environments. In Chap. 11, Frederick J. Gallun, Nirmal K. Srinivasan, and Anna C. Diedesch review the multiple factors that can impair binaural hearing. Of particular importance for comparing impaired and typical hearing systems is understanding the role of the physical aspects of the encoded sound, such as the sensation level of sound presentation.

Issues of spatial encoding can occur at the highest levels of auditory encoding, a topic relevant to clinical populations and discussed in Chap. 12 by Stephen M. Town and Jennifer K. Bizley. As binaural information ascends to the cortex and that information is transformed and integrated with other aspects of the neural encoding of extracted sound features, more complicated neural response patterns occur. By the time that processing reaches the level of the auditory cortex, the mechanisms of sound source location extraction are obscured by the complexity of the neural system. Chapter 12 reviews the brain regions that support spatial information encoding, much of this information learned from clinical populations. The authors also discuss issues related to how hearing impairment might be compensated for by plasticity in the cortex.

Finally, hearing interventions for people with hearing loss, such as hearing aids and cochlear implants, also affect spatial hearing. This topic is reviewed by Todd A. Ricketts and Alan Kan in Chap. 13. Although bilateral devices often improve hearing compared with unilateral devices, they are not currently optimized to convey binaural information. Part of the reason this occurs is that these devices were originally designed for unilateral use. Numerous factors that currently stymie improved processing with binaural hearing with hearing-assistive devices are reviewed in this chapter.

1.6 Concluding Remarks

In conclusion, binaural hearing is one of the most intriguing aspects of the auditory system because it is relevant to a wide range of species and impacts nearly every aspect of communication, learning, and safety in naturalistic environments. Unique to binaural hearing is the obvious physiological substrates, which have created a synergy between psychoacousticians, physiologists, and modelers. The attraction to binaural hearing extends broadly and deeply into clinical fields including audiology, otolaryngology, and neurology because it is recognized that having access to sound in both ears is extremely important for children and adults alike.

The area of binaural hearing has undergone many advances in recent years, thus motivating the current comprehensive review of state-of-the-art studies and recent progress. The purpose of this book is to build on the pillars of the seminal binaural literature and launch into a discussion of recent advances in the area of binaural hearing. Many of these advances have been made possible by the emergence of novel biological and imaging techniques and greater computing power. This book aims to be a desktop resource for researchers and students, in this research area and more broadly, with the hopes of providing enthusiasm for future directions and interdisciplinary collaborations.

Acknowledgments The time to collect the ideas reported here was supported by the National Institute on Deafness and Other Communication Disorders of the National Institutes of Health under Grants R01-DC-003083 and R01-DC-008365 to Ruth Y. Litovsky and R01-DC-014948 to Matthew J. Goupell. The content is solely the responsibility of the authors and does not necessarily represent the official views of the National Institutes of Health.

Compliance with Ethics Requirements
Ruth Y. Litovsky declares that she has no conflict of interest.
Matthew J. Goupell declared that he has no conflict of interest.

References

Brand A, Behrend O, Marquardt T, McAlpine D, Grothe B (2002) Precise inhibition is essential for microsecond interaural time difference coding. Nature 417(6888):543–547. https://doi.org/10.1038/417543a

Ehret G, Dreyer A (1984) Localization of tones and noise in the horizontal plane by unrestrained house mice (Mus musculus). J Exp Biol 109:163–174

Goldberg JM, Brown PB (1968) Functional organization of the dog superior olivary complex: an anatomical and electrophysiological study. J Neurophysiol 31(4):639–656

Goldberg JM, Brown PB (1969) Response of binaural neurons of dog superior olivary complex to dichotic tonal stimuli: some physiological mechanisms of sound localization. J Neurophysiol 32(4):613–636

Heffner HE, Heffner RS (2016) The evolution of mammalian sound localization. Acoust Today 12(1):20–27

Henning GB (1974) Detectability of interaural delay in high-frequency complex waveforms. J Acoust Soc Am 55(1):84–90

Jeffress LA (1948) A place theory of sound localization. J Comp Physiol Psychol 41(1):35–39

Macpherson EA, Middlebrooks JC (2002) Listener weighting of cues for lateral angle: the duplex theory of sound localization revisited. J Acoust Soc Am 111(5 Pt 1):2219–2236

Masterton B, Heffner H, Ravizza R (1969) The evolution of human hearing. J Acoust Soc Am 45(4):966–985. https://doi.org/10.1121/1.1911574

Strutt JW (1907) On our perception of sound direction. Phil Mag 13:214–232

Wallach H (1940) The role of head movements and vestibular and visual cues in sound localization. J Exp Psychol 27:339–368

Wightman FL, Kistler DJ (1992) The dominant role of low-frequency interaural time differences in sound localization. J Acoust Soc Am 91(3):1648–1661

Yost WA, Dye RH Jr (1988) Discrimination of interaural differences of level as a function of frequency. J Acoust Soc Am 83(5):1846–1851

Chapter 2
Localization and Lateralization of Sound

William M. Hartmann

2.1 Introduction

The ability of human (and other) listeners to use their ears to locate the sources of sound in space is a remarkable capability. Unlike the sensations of vision or touch, the sensory system that localizes sound is not spatially distributed as on the retina or surface of the skin. Instead, sound location needs to be computed by the nervous system. Listeners are able to localize sound sources in three dimensions: in polar coordinates, the dimensions of azimuth, elevation, and distance. Of these, the azimuthal dimension is most related to binaural hearing. The azimuthal dimension is readily associated with the locations in the horizontal plane, a plane that approximately passes through the two ears and the nose. Localization in this plane has been the most studied, and most of this chapter is devoted to it.

The chapter begins with physical modeling of the interaural differences that are used by listeners to localize sound sources. It describes the several different coordinate systems regularly used to quantify sound source location in three dimensions and shows how to convert from one system to another. The chapter continues with a review of human sensitivity to interaural differences and raises questions about how the results of headphone experiments can be compared with sound localization in the free field or rooms, especially the matter of externalization of sound images.

For the most part, this review of perceptual aspects of binaural hearing is confined to simple sounds, continuously ongoing sounds that are readily described mathematically. The perceptual aspects are followed by elementary descriptions of important neurally based signal-processing models for the binaural system. Viable neurophysiological models require a complicated mix of frequency-dependent effects. The range of hearing is so broad that there is a range of relationships between anatomical parameters and signal wavelength. A similarly complicated relationship

W. M. Hartmann (✉)
Department of Physics and Astronomy, Michigan State University, East Lansing, MI, USA
e-mail: wmh@msu.edu

© Springer Nature Switzerland AG 2021

R. Y. Litovsky et al. (eds.), *Binaural Hearing*, Springer Handbook of Auditory Research 73, https://doi.org/10.1007/978-3-030-57100-9_2

applies to audible waveform frequencies and the timing limitations of auditory neurons. Consequently, the different interaural differences are seen to play different roles in different circumstances. The chapter concludes by describing how binaural perception changes when continuously ongoing sounds are replaced by sounds with onsets and other temporal envelope features. Most important is the precedence effect, and the several different manifestations of precedence are described.

2.2 Interaural Differences

The interaural time difference (ITD; see Table 2.1 for a list of abbreviations) is the difference in arrival time of waveform features at the two ears. For a sine tone, the ITD is related to the interaural phase difference (IPD) by IPD $= f \times$ ITD, where f is the frequency in Hertz and the IPD is in number of cycles. The right-hand side of this equation can be multiplied by 360 to obtain the IPD in degrees. For a complex tone, the temporal differences become more complicated where the cycle-to-cycle

Table 2.1 List of abbreviations

Abbreviation	Definition
a	Head radius
c	Speed of sound or cross-correlation function
d	Source distance from a listener
f	Frequency
h	Hankel function
H	Transfer function to an ear
k	Wavenumber
p	Wave pressure with respect to ambient atmospheric
P_n	Legendre polynomial
θ	Azimuth re: listener's forward direction
θ_E	Ear angle re: Forward direction
τ	Model interaural delay-line delay
AZ	Source azimuth
EL	Source elevation
HRTF	Head-related transfer function
ICI	Interaural click interval (of time)
ILD	Interaural level difference
IPD	Interaural phase difference
ITD	Interaural time difference
JND	Just-noticeable difference
LA	Source lateral angle
MSO	Medial superior olive
PA	Source polar angle

variation, as present in a sine tone, is known as the fine structure, and there are larger scale variations to consider such as onsets, envelopes, and modulation.

The interaural level difference (ILD) is the difference in level between the two ears, expressed in decibels. It is related to the intensities of waves at right and left ears (I_R and I_L, respectively) by

$$\mathrm{ILD} = 10\log_{10}(I_R/I_L) \tag{2.1}$$

For sine waves with amplitudes (A_R and A_L), it is

$$\mathrm{ILD} = 20\log_{10}(A_R/A_L) \tag{2.2}$$

2.2.1 *Airhead Model*

The simplest model for binaural differences consists of two point-like receivers, representing the ears, with only air between them. This is the airhead model, and it was used by Koenig (1950), replacing the two points by microphones. It is an approach widely used in stereophonic sound recording even today. The model is shown in Fig. 2.1, where a is half the separation between the ears and d is the distance to the source. The azimuth of the source with respect to the forward direction is angle θ.

From the cosine law of trigonometry, the ITD is given by

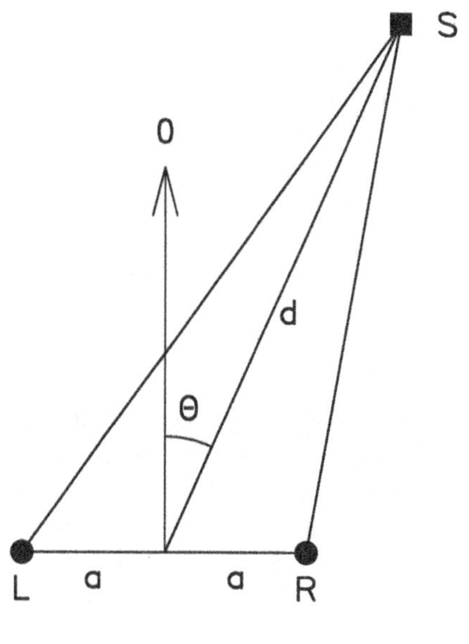

Fig. 2.1 The geometry of a source of sound (S) and two ears (L and R) separated by distance 2a. The source is distance d away from the midpoint between the ears and at an angle θ to the reference direction. For horizontal plane geometry, the reference direction is the forward direction (established by the listener's nose) and θ is an azimuth within the plane

$$\mathrm{ITD} = (d/c)\left[\sqrt{1+x^2+2x\sin\theta} - \sqrt{1+x^2-2x\sin\theta}\right] \tag{2.3}$$

where $x = a/d$ and c is the speed of sound.

It is usual for x to be small because the source distance is usually much larger than the distance between the ears. Expanding Eq. 2.3 in powers of x leads to the simple first-order approximation ITD = $(2a/c)\sin(\theta)$. The approximation is exact in both limits: $\theta = 0°$ and $\theta = 90°$. Taking the human "head" radius to be $a = 8.75$ cm (Hartley and Fry 1921) and the speed of sound to be $c = 34{,}400$ cm/s, the maximum ITD ($\theta = 90°$) is 509 μs. The first nonzero correction to this approximation is order x-cubed, and it is negative. Therefore, the first-order approximation overestimates the ITD in the airhead model, but it is nevertheless too small to be a realistic model for an animal with an actual head between the ears.

In the airhead model, the ILD depends on the inverse square law. Consequently, it becomes a less efficient sensor as the source distance increases, ultimately resulting in no ILD information at all for large distances. Nevertheless, the airhead model is useful because many terrestrial mammals have their ears on the tops of their heads. The airhead model is a reasonable first approximation for such animals and also for some robots (Baumann et al. 2015).

2.2.2 *Woodworth Model*

In contrast to top-of-head animals, squirrel monkeys (*Saimiri sciureus*), meerkats (*Suricata suricatta*), and humans have heads between their ears. The heads lead to increased interaural differences which, in principle, should lead to more accurate localization, especially for distant sources. For humans, the obstructing-head advantage serves as compensation for a very limited ability to move the ears to enhance the interaural differences.

The Woodworth model (1938) is a simple model for computing the ITD. It assumes a spherical head where waves from a source an infinite distance away creep around the segments of the head that are occluded from the source, as shown in Fig. 2.2. The ears are assumed to be antipodal (diametrically opposite on the sphere). According to the Woodworth formula, the ITD is given by

$$\mathrm{ITD} = (a/c)\left[\theta + \sin\theta\right] \tag{2.4}$$

where azimuth θ is given in radians for a source in front ($-\pi/2 \le \theta \le \pi/2$). The creeping wave approximation is accurate when the frequency is very high, and the Woodworth formula becomes monotonically increasingly accurate as the frequency increases. The largest ITD from the formula occurs for a source at 90°, and for a 8.75-cm head radius the value is $(a/c)[\pi/2 + 1] = 654$ μs. Aaronson and Hartmann (2014) generalized the creeping wave approximation to allow for arbitrary ear

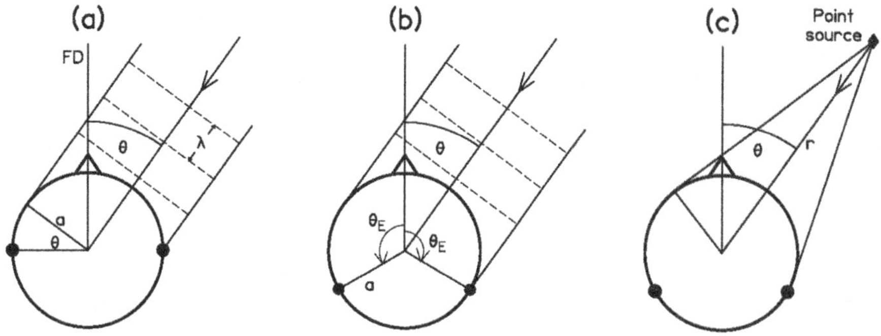

Fig. 2.2 Creeping-wave models. (**a**) Geometry for the Woodworth model with antipodal ears (solid circles) and a plane wave (wavelength λ), incident from direction θ with respect to the forward (nose) direction (FD). (**b**) Extended spherical model with ear angles (θ_E) greater than 90°. (**c**) The same but with a finite source distance ($r=d$). (Reprinted from Aaronson and Hartmann (2014), with the permission of the Acoustical Society of America)

angles (θ_E; Fig. 2.2b) and a finite source distance (*d*; Fig. 2.2c). The Woodworth formula is simple to implement and often used. Its main problem is that the predicted ITD is independent of frequency, but in reality, the phase shift caused by diffraction of waves around a sphere depends on frequency (Strutt 1907).

The Woodworth model (1938) is applied to the ITD. There is no creeping wave parallel for the ILD in popular use. The ILDs can be calculated with a complete spherical-head model, which features both shadows on the far side and enhancements on the near side.

2.2.3 Spherical-Head Model

The spherical-head model begins with the solution to the wave equation in air where the wave encounters a sphere. Because the sphere is rigid compared with the air, it prevents the sound wave from forcing air particles into the region of space occupied by the sphere, although air particles can graze the sphere surface. Mathematically, that means that the normal component of the particle velocity field is zero at the surface of the sphere. The "normal component" refers to a component of a vector quantity that is perpendicular to a surface. Because the surface varies in space, the direction of the normal component varies with it. The surface of the sphere thus imposes a *boundary condition* on the solution to the equation for the wave propagation. Because of the boundary condition, the general solution for a wave traveling in a space becomes specialized to the geometry of the object.

As for previous models, the two ears and the source location define a plane, and only two variables are needed to describe the geometry. Points in space at which the sound field is calculated can be defined in terms of the distance from the center of the sphere (*r*) and the angle between the reference direction and a line drawn from the center of the sphere to the point (θ). The reference direction is the direction of any perpendicular bisector of the interaural axis. The model does not yet have a nose that would establish a forward direction.

The wave equation then has a separable solution, meaning that the pressure dependence on r and θ can be written as a product of two functions that depend on r and θ uniquely,

$$p(r,\theta) = R(r)\Theta(\theta) \tag{2.5}$$

For hearing, the most important part of the sound field is the pressure on the surface of the head ($r = a$). The solution to the angular part of the solution is a Legendre polynomial $\Theta(\theta) = P_n(\cos\theta)$. The nature of Legendre polynomials is well-known because they are the functions that famously describe atomic orbitals, as studied in chemistry class. For $n = 0$, there is no θ dependence and the function P_0 represents a sphere like the *s*-orbital of an atom. For $n = 1$, the θ dependence produces a pattern like a figure eight, as for a *p*-orbital. Higher values of n lead to an increasing number of lobes in the function. In fact, n is the number of times that the Legendre polynomial equals zero as angle θ runs from zero to 180°. Overall, the Legendre polynomials form a complete set that can produce a unique description of angular dependence for any arbitrarily complicated pattern.

The pressure at the surface of a sphere with radius a can be written as the sum of incident and scattered waves. A solution satisfying the boundary condition at the surface of the sphere can be expressed as a sum of partial waves, where each partial wave has a unique angular dependence from its Legendre polynomial (Rayleigh and Lodge 1904). There are two especially useful solutions, depending on assumptions about the source distance (d). If the source is infinitely far away, the incoming wavefront is a plane wave and the solution involves mainly the angle of incidence or source azimuth (Kuhn 1977)

$$H(\theta_a) = \left(\frac{1}{ka}\right)^2 \sum_{n=0}^{N} \frac{i^{n+1}(2n+1)P_n(\cos\theta_a)}{h_n'(ka)} \tag{2.6}$$

The solution of Eq. 2.6 is in the form of a transfer function (H). Like every transfer function, H is a frequency-dependent relationship between one signal and another (e.g., the input and the output of a device or process). Equation 2.6 relates the signal on the surface of a sphere to the signal that would exist at the location of the center of the sphere if the sphere were absent. With the sphere as an approximation for a head, function H might be called a "head-related transfer function" (HRTF). Given a free-field environment and a loudspeaker with an ideally flat frequency response, function H can be used to relate interaural differences to an electrical signal sent to the loudspeaker. The frequency dependence is given by parameter k. Parameter k is the wavenumber, equal to 2π divided by the wavelength or $k = 2\pi f/c$, where f is the frequency and c is the speed of sound. Angle θ_a is the angle between a line drawn from the center of the sphere to the observation point on the surface (most usefully the location of an ear) and a line drawn from the center of the sphere to the source. Function h is a spherical Hankel function, a complex function (with real and imaginary parts) that is readily available in computational libraries. Function h' is the derivative of h with respect to its argument. In principle, there are an infinite number of terms in this sum, but in practice, the value of N for adequate accuracy depends on the frequency because of the Hankel functions. Up to 1000 Hz, $N = 20$ terms is more than adequate.

A more general transfer function that applies to a finite source distance d was developed by Rabinowitz et al. (1993) and explicated by Duda and Martens (1998)

$$H(\theta_a) = \left(\frac{d}{ka^2}\right)\exp(-ikd)\sum_{n=0}^{N}\frac{i^{n+1}(2n+1)P_n(\cos\theta_a)h_n(kd)}{h_n'(ka)} \tag{2.7}$$

Here, the numerator involves the Hankel function depending on the source distance and the denominator again involves the derivative depending on the head radius.

The transfer functions in Eqs. 2.6 and 2.7 can be used to compute the interaural differences. Because the transfer function for an ear at the computed point is $H = |H| \exp i\varphi$, the phase of the waveform at the ear, namely φ, is given by the imaginary part of the natural log of H.

Up to this point, the model head has spherical symmetry and no nose. Adding a nose establishes the horizontal plane where IPDs are calculated by finding the phase at each ear and taking the difference. The first step is to relate both the source azimuth θ and the ear angle (θ_E) to the forward direction (nose direction). That can be done by replacing θ_a in Eqs.2.6 and 2.7 by $\theta - \theta_E$ for the ear nearer the source and by $\theta + \theta_E$ for the ear farther from the source. By definition, θ_E is a positive number, the same for both ears. It is the angular distance between the forward direction and the location of the ears on the surface of the head. For antipodal ears, $\theta_E = 90°$. Although $\theta_E = 90°$ is a common assumption, it is known that the ears are somewhat further back on the head. Treeby et al. (2007) used $\theta_E = 100°$.

2.2.3.1 Interaural Time Difference

The spherical-head model leads to a useful calculation of the ITD in seconds by dividing the phase difference in radians by the angular frequency in radians per second

$$\text{ITD} = \text{Im}\left\{\ln\left[H(\theta_E + \theta)/H(\theta_E - \theta)\right]\right\}/\omega \tag{2.8}$$

where ω is the angular frequency ($\omega = 2\pi f$).

The ITD calculated from the spherical-head model is shown in Fig. 2.3, *red dashed lines*, where the calculation can be compared with measurements made by Cai et al. (2015) on a manikin in an anechoic room. The low-frequency limit of the

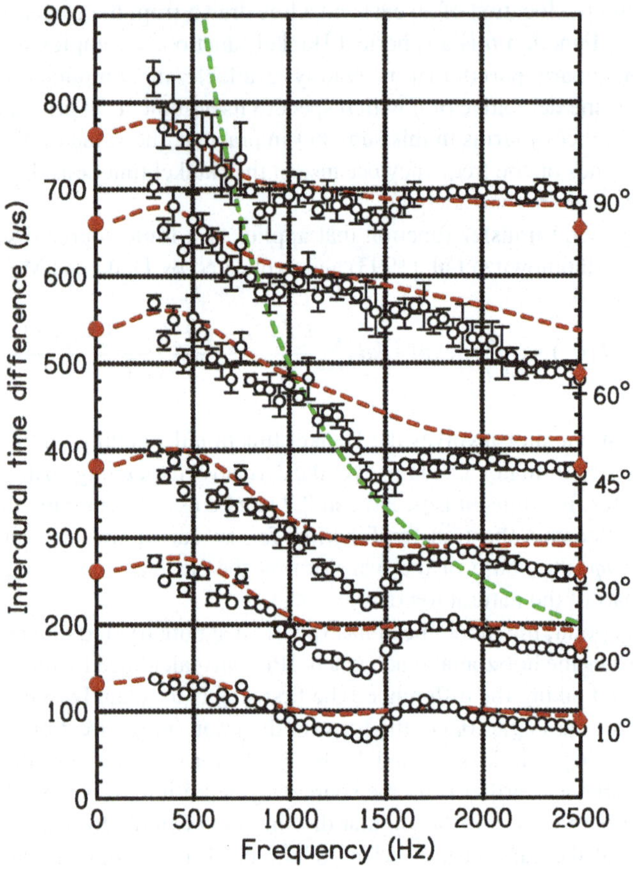

Fig. 2.3 Interaural time differences for six source azimuths with respect to the forward direction. The source distance was 3 meters. *Open circles*, measured values ±1 SD in length for a manikin in the free field. Error bars smaller than the circles are not visible. *Green dashed line*, locus of half-period interaural time differences, where the interaural phase difference is 180°. *Red diamonds*, predictions from the Woodworth model; *solid red circles*, values from the low-frequency limit of the model. *Red dashed line* was computed from the spherical-head model (Eq. 2.7). Adapted from Cai et al. (2015), with the permission of the Acoustical Society of America

model formula turns out to involve the first two terms of the series and a small-*k* approximation for the Hankel functions. It is given by

$$\text{ITD} = 3(a/c)\sin\theta_E \sin\theta \tag{2.9}$$

For antipodal ears (θ_E = 90°), *a* = 8.75 cm, and *c* = 34,400 cm/s; the ITD in microseconds becomes ITD = 763 sin *θ*. The maximum value is 763 μs, which is not an unreasonable estimate of the maximum value for human heads.

Figure 2.3 shows that the spherical model calculation for the ITD exhibits a strong frequency dependence (dispersion). It is even stronger for the manikin:

Fig. 2.3, *circles*, have a more dramatic frequency dependence than Fig. 2.3, *red dashed lines*. Dispersion can be expected to lead to binaural incoherence in a broadband signal because different frequency regions of the signal experience different interaural delays. The dispersion causes the waves in the two ears to have different shapes as functions of time. Incoherence appears as a reduction in the peak of the cross-correlation function (CCF) because the two waveforms cannot match exactly for any value of lag. Different waveform shapes in the two ears can lead to different neural firing patterns in a single-auditory frequency channel. Alternatively, dispersion can lead to different ITDs in different auditory channels. Either way, dispersion tends to reduce the information available in binaural comparison. Nevertheless, Constan and Hartmann (2003) found that listeners are insensitive to the dispersion seen in the spherical-head model. Listeners might be sensitive to the larger dispersion observed for the manikin, but it mainly occurs above 1000 Hz where listeners are less sensitive to ITD itself.

Cai et al. (2015) found that replacing a model sphere with a finite-element model ellipsoid considerably improved the agreement between physical model calculations and ITDs measured on the manikin, reducing the root-mean-square (RMS) deviation by 50%. The measured HRTFs show traces of oscillations, also reported by Algazi et al. (2001). Cai et al. (2015) had some success in attributing this structure to reflections from shoulders and torso.

2.2.3.2 Interaural Level Difference

For human listeners, the ILD is mainly attributable to the head itself as a sound-obstructing object. Unlike the airhead model, where ILDs disappear for distant sources, the spherical-head model leads to ILDs that remain large and distance independent for large source distances. They depend on distance only for nearby sources. The ILD is given by

$$\mathrm{ILD} = 20\log_{10}\left|H(\theta_E+\theta)/H(\theta_E-\theta)\right| \tag{2.10}$$

Sound waves diffract around the head, and the physics of diffraction makes the ILD strongly frequency dependent. According to the spherical-head model, the ILD grows as the fourth power of frequency for frequencies of 200 Hz and below. This power law represents Rayleigh scattering, famously the origin of the blue color of the sky. Above 200 Hz, the growth with frequency is less steep than the fourth power. Rayleigh scattering occurs when the scattering object (e.g., the head) is small compared with the wavelength. Consequently, the Rayleigh limit is proportionately higher than 200 Hz for heads smaller than human.

Figure 2.4 shows the ILD as measured in the free field compared with a calculation from the spherical-head model. The figure shows a complicated structure. There is a strong non-monotonic frequency dependence, especially for small azimuths such as 30°. This effect has appeared in free-field measurements made in

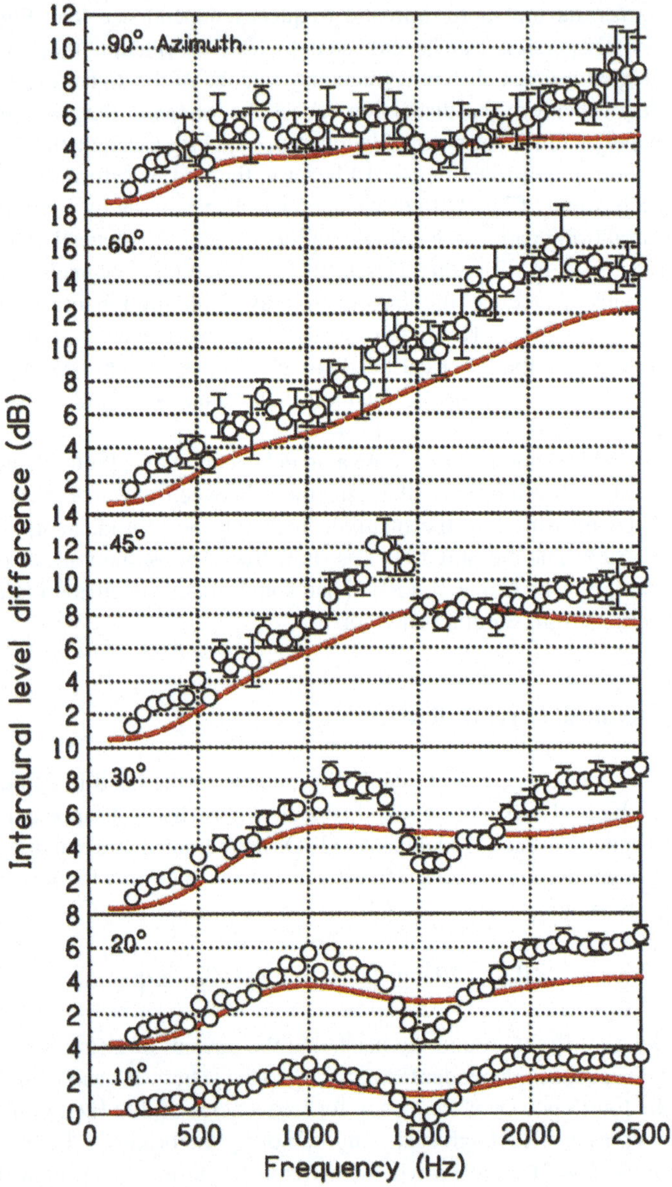

Fig. 2.4 Interaural level differences for six source azimuths. Measurements and calculations are the same as in Fig. 2.3. Adapted from Cai et al. (2015), with the permission of the Acoustical Society of America

several laboratories (Kayser et al. 2009; Cai et al. 2015) and it is clearly underestimated by the spherical-head model. An ellipsoidal head model led to considerably improved agreement with measurements, mainly caused by a realistic increase in the height of the head. The improvement was especially notable for azimuths greater than 30°, where the spherical-head model can be in error by as much as 5 dB (Cai et al. 2015).

Although there is a clear tendency for the ILD to grow with an increasing source azimuth, the ILD is not a monotonic function of azimuth, as shown by spherical-head model calculations in Fig. 2.5. For a 1500-Hz tone, the ILD decreases from 8 dB to 4 dB as the azimuth increases from 50° to 90°. This decrease is caused by the "bright spot." The bright spot occurs for a source near a 90° azimuth, where the far ear is directly opposite the source. Sound waves diffracting around the head arrive in phase at this opposite point and add constructively. The constructive interference leads to the bright spot at the opposite ear and a consequent reduction in ILD. As the frequency increases, the reduction in the ILD becomes more dramatic. For 4000 Hz, the ILD decreases by more than 10 dB as the azimuth increases from 70° to 90°. This high-frequency decrease is far greater than the 4 dB decrease seen for 1500 Hz, but the high-frequency decrease occurs for a smaller range in azimuth. The bright spot has pronounced perceptual effects. The *1500-Hz curve* in Fig. 2.5 would predict that listeners in an anechoic room cannot discriminate between a source at 30° and a source at 75°. That turned out to be true (Macaulay et al. 2010).

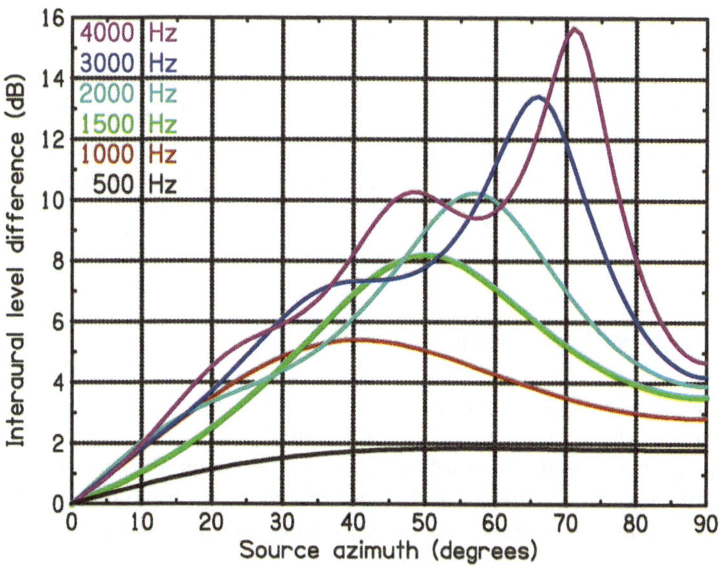

Fig. 2.5 Interaural level differences as a function of azimuth for six frequencies computed from the spherical-head model. Adapted from Macaulay et al. (2010), with the permission of the Acoustical Society of America

2.2.4 Cones of Confusion and Coordinate Systems

For a spherical head with antipodal ears, the geometrical relationship between any point in space and each of the two ears is unchanged by a rotation of the point about an axis passing through the two ears. This rotation defines a circle in a sagittal plane on the flat end surface of a cone with its apex at one of the ears. The cone itself is a figure of rotation about the interaural axis. Calculations within the spherical-head model show that for all practical purposes, the ITD is insensitive to distance along the cone (i.e., distance from the ear at the apex). For distances of 1 or 2 meters from the head, the ILD is similarly insensitive to the distance along the surface of the cone. Consequently, the cone is known as a "cone of confusion" in that any point on the surface of the cone leads to the same values of ITD and ILD. However, if the source is closer to the head than 1 or 2 meters, the ILD also depends on distance (Brungart and Rabinowitz 1999; Shinn-Cunningham et al. 2000).

The importance of the cone of confusion has motivated a special coordinate system for the study of sound localization, the "lateral-polar" system. This section relates the lateral-polar system to a psychophysical version of the more common spherical-polar coordinate system. Both systems are related to the cartesian system (x, y, and z). The listener is at the origin of the coordinates $x = y = z = 0$. The positive y-axis is the listener's forward direction, and the positive x-axis extends to the listener's right. The x-and y-axes define the horizontal plane. The z-axis is straight up.

2.2.4.1 Spherical-Polar System

The angles are azimuth (AZ) and elevation (EL). The *azimuth* is measured positive to the right of the forward direction (y) in the horizontal plane. Therefore, the positive x-axis corresponds to AZ = 90°. The range of the azimuth extends over the full 360° around the listener. The *elevation* is measured upward from the horizontal plane. The range of the elevation extends from −90° (straight down) to +90° (straight up).

2.2.4.2 Lateral-Polar System

The angles are the lateral angle (LA) and the polar angle (PA), as seen in Fig. 2.6. The *lateral angle* defines the location of a sagittal plane. It is an angle in the horizontal plane between the forward direction and the sagittal plane containing the point of interest. The *polar angle* is measured along a circle within that sagittal plane; the circle is the intersection of that plane with the sphere having a radius equal to the distance of the point of interest from the origin.

The LA spans the range $-90° \leq \text{LA} \leq 90°$. The PA spans the range $-90° < \text{PA} \leq 270°$. Positive LAs are to the listener's right; PAs are zero on the

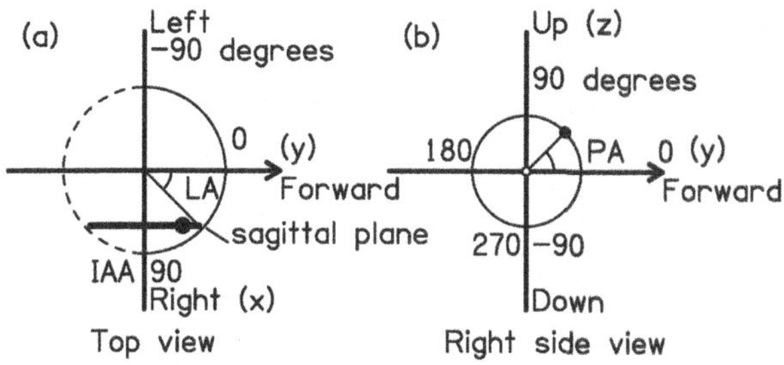

Fig. 2.6 The position of the solid dot is given in terms of lateral angle (LA) and polar angle (PA). (**a**) View from the top shows how the LA defines a sagittal plane (*thick line*) perpendicular to the interaural axis (IAA). *Dashed line part of the circle* shows that lateral angles are not defined over that part. (**b**) View from a position to the right of the listener shows the intersection of the sagittal plane with the sphere of radius *r* and the definition of the PA. The IAA is seen end-on (*open circle at center*)

horizontal plane and become positive above that plane. PAs between −90° and +90° indicate positions in front of the listener. PAs between 90° and 270° indicate positions behind the listener. It takes a little time for a user to become comfortable with the idea that the distinction between front and back all lies in the PA.

The advantage of this lateral-polar coordinate system is that the surface defined by constant lateral angle corresponds to the surface of the cone of confusion for a spherical head with antipodal ears (i.e., the lateral angle is the cone angle; Searle et al. 1976). This coordinate system has been widely adopted (e.g., Morimoto and Aokata 1984; Algazi et al. 2001), although the precise definitions of the angles has differed among different authors. Figure 2.6 adopts the notation used by Macpherson and Middlebrooks (2000). Equations 2.11, 2.12, and 2.13 give the cartesian coordinates of a point in terms of the spherical-polar system and then in terms of the lateral-polar system.

$$x = r\cos(\mathrm{EL})\sin(\mathrm{AZ}) = r\sin(\mathrm{LA}) \tag{2.11}$$

$$y = r\cos(\mathrm{EL})\cos(\mathrm{AZ}) = r\cos(\mathrm{LA})\cos(\mathrm{PA}) \tag{2.12}$$

$$z = r\sin(\mathrm{EL}) = r\cos(\mathrm{LA})\sin(\mathrm{PA}). \tag{2.13}$$

The transformation between the spherical-polar system and the lateral-polar system can be made by equating the expressions for *x*, *y*, and *z*, as given in Eqs. 2.11, 2.12, and 2.13. Therefore, if the AZ and the EL are known, there are three equations for the two unknowns LA and PA. Initially, this appears to be problematic, but the unknowns are not really overdetermined by the equations because the equations are redundant, meaning that any one of them can be derived from the other two.

2.2.4.3 Solving for Lateral and Polar Angles

The LA is given in terms of azimuth and elevation by

$$\mathrm{LA} = \arcsin\left[\cos(\mathrm{EL})\sin(\mathrm{AZ})\right] \tag{2.14}$$

The principal value range of the arcsin function is −90° to +90° and that agrees perfectly with the allowed range of the LA. Within the horizontal plane, the elevation angle EL is zero and the LA and the azimuth become equal; both are then equivalent to angle θ as it appears in equations for the horizontal plane only.

The PA is given in terms of azimuth and elevation by

$$\mathrm{PA} = \arctan\left[\tan(\mathrm{EL})/\cos(\mathrm{AZ})\right] \tag{2.15}$$

with the adjustment that whenever cos(AZ) is negative, one should add 180° to the value of PA computed from the arctan in Eq. 2.15.

Whenever the azimuth is ±90°, Eq. 2.15 says that the PA is ±90°, with the sign chosen to match the sign of the elevation. If the azimuth is ±90° and the elevation is zero, the disc in Fig. 2.6b, around which the PA is measured, shrinks to a point and the PA does not exist.

2.2.4.4 Solving for Azimuth and Elevation

It is possible to solve the equations in reverse to obtain the azimuth and elevation in terms of the LAs and PAs.

The elevation is given in terms of the LAs and PAs by

$$\mathrm{EL} = \arcsin\left[\cos(\mathrm{LA})\sin(\mathrm{PA})\right] \tag{2.16}$$

The principal value of the arcsin function agrees with the allowed range of the elevation and no adjustment to this equation for the elevation is ever needed. The azimuth is given in terms of the LAs and PAs by

$$\mathrm{AZ} = \arctan\left[\tan(\mathrm{LA})/\cos(\mathrm{PA})\right] \tag{2.17}$$

with the adjustment that whenever cos(PA) is negative, one should add 180° to the AZ computed from the arctan in Eq. 2.17.

When the PA is −90° or 90° or 270°, cos(AZ) = 0° and the azimuth is ±90°, with the sign chosen to agree with the sign of the LA. If cos(AZ) is zero and the LA is zero, the point in question is directly above or below and the azimuth is undefined, but it is convenient to call it "zero."

2.3 Interaural Sensitivity

Interaural sensitivity, as it relates to sound source location, has been measured using two general classes of measurement. The first class is through sound localization, where a listener discriminates between two different source locations in a three-dimensional space. The results of such measurements can be expressed geometrically in terms of minimum audible angles, with no reference to the signals in the ear canals. Ear canal measurements and/or posthoc computations may make the connection to ear canal signals. The second class of measurements exploits the control of individual ear signals that is available through headphone listening. Sound image location measurements made with headphones are called lateralization to distinguish them from localization. Lateralization measurements show that listeners are amazingly sensitive to small changes in ITD (Klumpp and Eady 1956; Brughera et al. 2013; see also Chap. 6 by Stecker, Bernstein, and Brown). The just-noticeable difference (JND) in the forward direction is less than 20 μs, meaning that listeners can discriminate between +10 μs and −10 μs for a frequency of 800 Hz. For an ITD given by $\mathrm{ITD}_{max} \sin \theta$, with the ITD_{max} reduced to 711 μs at 800 Hz because of dispersion, the ITD corresponds to the difference between +0.8° and −0.8° of azimuth. Sensitivity to a change in ITD decreases with increasing ITD. By differentiating the inverse sine function, the ITD in azimuth (in degrees) is found to be related to the ITD JND by

$$\Delta\theta = \frac{180/\pi}{\sqrt{\mathrm{ITD}_{max}^2 - \mathrm{ITD}^2}} \Delta ITD \tag{2.18}$$

or

$$\Delta\theta = \frac{180/\pi}{\cos\theta} \frac{\Delta ITD}{\mathrm{ITD}_{max}} \tag{2.19}$$

indicating that the discriminability of the ITD decreases markedly as the ITD or the source azimuth increases. Mills (1958) reported that the difference limen in azimuth grew as an exponential function of the azimuth itself.

There is a strong frequency dependence in the sensitivity to the ITD. Although the sensitivity appears to be greatest at about 800 Hz, an octave higher, there is no sensitivity at all. The dramatic decrease in ITD sensitivity as the frequency approaches 1500 Hz was attributed by Brughera et al. (2013) to the membrane properties of coincidence cells in the medial superior olive (MSO).

Human listeners also become insensitive to the ITD for low frequencies, especially below 100 Hz. Experiments in this low-frequency range are difficult, with a large variability between and within individual listeners. Over the range from about 500 to 100 Hz, the ITD JND becomes inversely proportional to the frequency. Such an inverse first-power law is exactly what is expected if discrimination is based on a constant threshold IPD. A threshold phase shift of about 0.02 cycle fits the average data over the range from 500 to 100 Hz

$$\Delta ITD = 0.02/f \tag{2.20}$$

As the frequency decreases below 100 Hz, the threshold ITD appears to increase somewhat faster than the inverse first power of the frequency. This behavior might be understood in terms of a decreasing number of neurons sensitive to very long ITDs (see Sect. 2.3.4).

2.3.1 Interaural Phase Difference

The IPD for sine tones is simply related to the ITD thorough the tone frequency, IPD = *f*ITD, where the IPD is given in number of cycles. As noted by Yost and Hafter (1987), the IPD JND is roughly independent of frequency, but the ITD JND is not, as seen in Eq. 2.20. It appears as though the binaural system might be essentially an IPD detector. It has been argued that the proximal parameter for azimuth perception is indeed the IPD. Experimental evidence in favor of maximum perceived laterality for an IPD of 90° comes from headphone experiments by Young (1976) and Yost (1981), although Mills (1958) reported that the apparent azimuth in the low-frequency range is proportional to the ITD up to a limit that corresponds to an IPD of 180°.Experiments specifically designed to measure the perceived laterality of tones do not find that IPD is the determining parameter. Instead, the data are consistent with a model in which laterality is determined by the ITD. Schiano et al. (1986) performed a headphone experiment in which listeners adjusted the ILD of a noise pointer to match the lateral position of tones (or narrowband noise) having various interaural delays. This acoustical lateralization pointing task suggested that constant laterality occurred for constant ITD, although experimental values of the IPD were limited to 90°.

Hartmann et al. (2016) used transaural synthesis with probe microphones in the ear canals to reproduce individualized ITDs from 25 locations in a 180° arc in a free field, with the ILD set equal to zero. The listener responses are plotted two ways in Fig. 2.7, as a function of IPD and as a function of ITD. There were three different frequencies. The overlapping data in the ITD plot indicate that for localization (as previously found for lateralization), the perceptually relevant variable is the ITD.

2.3.2 Lateralization

In everyday listening in air, sounds are perceived to be localized in a three-dimensional space in the external world. A sound image may be punctate or spatially distributed, but it is perceived as external. There is normally some sense of

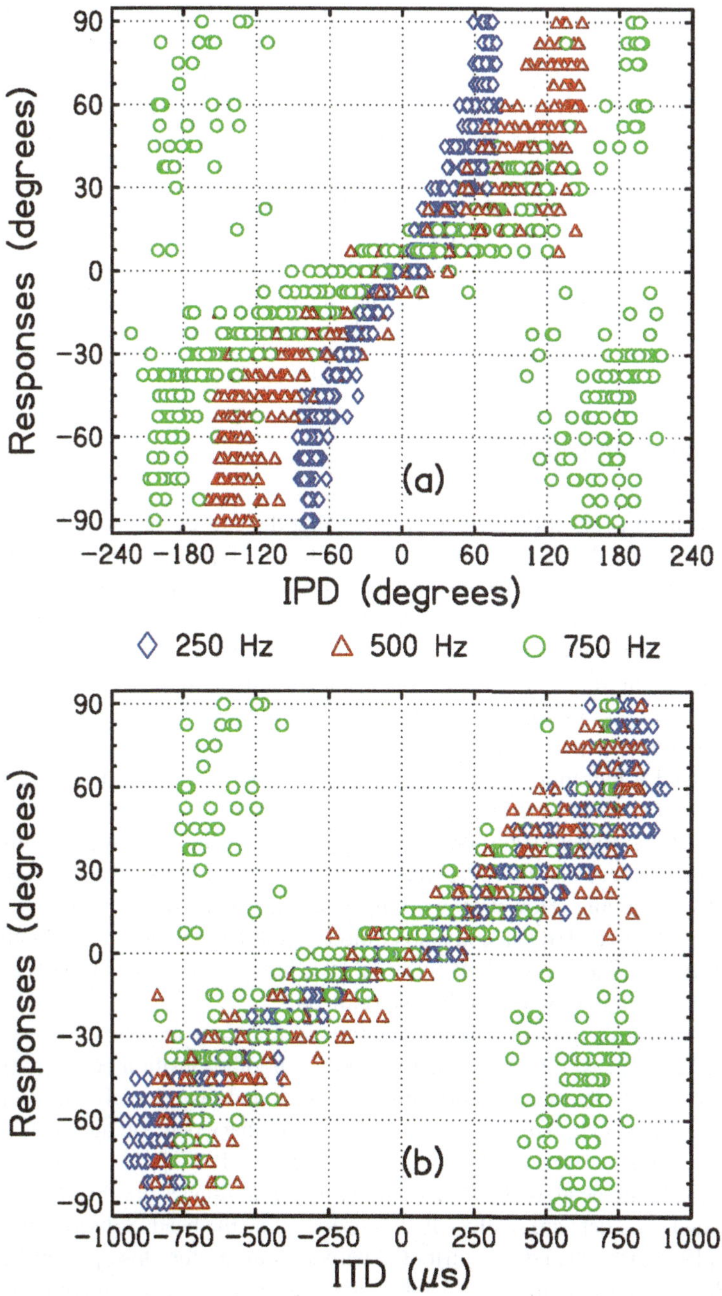

Fig. 2.7 *Symbols* show the accumulated localization responses across five listeners to pure-tone targets of three frequencies as indicated. The same data are plotted as a function of the interaural phase difference (IPD; **a**) and of the interaural time difference (ITD; **b**). Reprinted from Hartmann et al. (2016), with the permission of the Acoustical Society of America

distance, although that sense may be greatly in error (Mershon and King 1975). When sounds are presented by headphones, the impression is normally internal to the head and may be more or less diffuse (Blauert and Lindemann 1986a,b).

Mills (1958, 1960) showed that JNDs in azimuth agreed with ITD and ILD JNDs, as linked by anatomical measurements made on a manikin. Since that time, localization and lateralization have been regarded as fundamentally equivalent. Experiments on lateralization using headphones are easier to do than experiments on localization in a free field because the anechoic conditions required by free-field experiments are generally less available than headphones. The research literature displays little doubt that azimuthal plane localization and lateralization involve functionally the same process. However, there is the matter of externalization.

2.3.3 Externalization

Externalization is the subjective impression that the source of a sound is out in the world and not inside the head. Inside the head is the typical experience with headphone listening (or while listening through rubber tubes; Thompson 1877). Toole (1970) reviewed an array of often fantastical theories explaining why the sound is heard inside the head: static pressure on the head caused by wearing earphones, slightly different transmission paths to the ears, nonlinear distortion in the transducers, lack of body irradiation with earphones, abnormal acoustic loading of the middle ear, and abnormal interaction of the head with the sound field. Blauert (1997, p.133) added a few more: overmodulation of the nervous system or an unnatural proportion of bone conducted sound.

Interestingly, Bauer (1965) observed that a sound from a loudspeaker directly in front of the listener can be perceived inside the head if the listener's head is prevented from moving. However, working with stationary listeners in a free field, using individualized HRTFs and also compensating headphones for individual differences in a virtual acoustics experiment, Wightman and Kistler (1989a,b) found that listeners reported images to be externalized.

As a subjective attribute, externalization is not easy to study. Plenge (1972) attempted loudspeaker experiments to create a smooth transition between externalization and inside-the-head localization. He concluded that no such transition could be created. Thus, externalization appeared to be a binary percept; an image is either internalized or it is externalized. By contrast, Blauert (1997, p. 131 ff) treated sound images inside the head as part of a continuum; some images are distant, others are closer, and still others are so close that they are inside the head. Subsequent work by Plenge (1974) introduced the attribute "verged cranial" for close proximity to the head. Hartmann and Wittenberg (1996) combined free-field experimental trials with virtual-reality trials based on simultaneous ear canal recordings. They parameterized the virtual-reality trials to be more or less realistic. At one end of the parameter space, sounds were internalized. At the other end, they were indistinguishable from an actual source in the space. The experiments supported the continuum concept, including images that were neither externalized nor internalized but were pressing on the face. Mixing experiments that varied the preservation of the HRTF also found a continuum (Boyd et al. 2012). Cross-talk cancellation experiments with loudspeakers show that a synthesis that accurately maintains the HRTF from the source to the individual ears, including the effects of the room if there is one, leads to an image that is externalized (e.g., Zhang and Hartmann 2010).

However, externalization can involve more than just getting the details right for a static signal. Sakamoto et al. (1976) observed that even a poorly synthesized binaural signal can be externalized if enough artificial reverberation is added. Data from Volk et al. (2008) agreed. Durlach et al. (1992) reviewed unpublished experiments, indicating an important role for reverberation as well as realistic interaural changes with head motion and monaural cues from spectral features above 5000 Hz. The literature points to varying roles for the multiple cues to externalization. Brimijoin et al. (2013) emphasized the effect of binaural presentation that faithfully tracked head motion, with interaural parameters varying with motion in a realistic way. Failure to track immediately resulted in an internalized image. Using nonindividualized HRTFs, Hendrickx et al. (2017) found that an image, once externalized by tracking of head motion, retained its externalization when the head motion stopped. An unresolved problem is that a sine tone (e.g., 500 Hz) presented with headphones is internalized. With such an impoverished stimulus, that result could be expected only if internalization is a default percept and externalization requires additional information. More surprising is that listeners claim that a sine tone is externalized when presented with a loudspeaker in a free field. The paradox could be resolved if there are inevitable micromovements by the listener that the listener associates with adequate auditory changes. That idea would be consistent with the emphasis on clamping the head observed by Bauer (1965) in the internalization of an image produced by a loudspeaker. However, Brimijoin et al. (2013) and Hendrickx et al. (2017) found no role for micromovements. Additional information on listener motion is in Chap. 3 by Yost, Pastore, and Zhou. In addition to the qualitative impression of externalization, Simon et al. (2016) identified other perceptual attributes such as *immersion, realism* and *relief/depth* that are associated with differences in HRTFs.

At this point, one must conclude that externalization is difficult to study. It is subjective and, unlike source localization, there is no correct answer. Furthermore, it is not an attribute that listeners are generally skilled at evaluating. It is likely that further progress in this area can only be made through virtual-reality experiments in which the listener evaluates both well-controlled transaural synthesis and real sources presented in random order (e.g., Hartmann et al. 2016). Obviously, this eliminates headphone presentation as a procedure. The virtual-reality approach combining transaural trials with real source trials has the benefit of the baseline synthesis in which the transaural synthesis procedure is applied in such a way as to simulate a real source. If a listener can distinguish the baseline synthesis from a real source, then the synthesis has failed and the system needs to be recalibrated before continuing with the experiment.

2.3.4 Interaural Time Difference Encoding

Encoding the ITD requires that neural spikes from the left and right peripheries be transmitted to a central location where they can be compared. The comparison process was modeled as a cross-correlation by Jeffress (1948). Cross-correlation could be realized through cells with thresholds so high that spikes arriving simultaneously from both sides were required to activate. The cells were distributed along model left-right delay lines. Given a relative delay in the input spikes from the periphery, caused by an ITD in the stimulus, a cell could fire only if its location on the internal delay line canceled the original stimulus ITD. Even for heads as large as the human head, it is plausible to imagine neural delays large enough to cancel. The Jeffress model is therefore a form of "place" model for ITDs, with specific cells tuned to specific values of ITD and hence for specific source azimuths.

The mammalian brainstem includes the MSO where spikes arrive from the anteroventral cochlear nucleus on each side. Cells of a mammalian MSO have been found to operate just as the Jeffress model imagined (Yin and Chan 1990). That model is especially applicable to the barn owl (*Tyto alba*; Carr and Konishi 1990). In addition to excitation by arriving spikes, the MSO also appears to include inhibition (Grothe 2000). The cross-correlation models consider only the excitation.

In 1973, Colburn reported computational modeling aimed at understanding binaural observations on human listeners on the basis of the interactions of realistic spike trains. His models found it necessary for spikes to be compared only within left and right channels tuned to the same stimulus frequency. Most models of binaural interaction now operate within frequency-tuned channels, although some employ cochlear delays between best frequency places (Schroeder 1977; Shamma et al. 1989) and experiments by Joris et al. (2006) showed ITD sensitivity to excitation at peripheral sites with different best frequencies.

In 1977, Colburn studied a model delay line with the important feature that the characteristic delays (τ) for cross-correlation cells were not uniformly distributed. Instead, the density of cells favored small interaural delays, tailing off exponentially for large delays. The density function, as shown in Fig. 2.8, was called $p(\tau)$.

The density is highest at a delay of 0 and remains high out to 200 μs. That choice is reasonable considering that ITD JNDs for human listeners are smallest at small ITD values, changing little as the ITD increases to 200 μs. For an ITD of 700 μs, characteristic of maximum human head delays, the compensating internal delay of $\tau = 700$ μs brings $p(\tau)$ down to only 4% of its maximum value. An alternative $p(\tau)$ used by Shackleton et al. (1992) is considerably wider.

Stern and Colburn (1978) chose to predict the lateralization of a tone by the centroid of a model CCF as weighted by $p(\tau)$

$$\bar{\tau}(\Delta t) = \frac{\int d\tau\, p(\tau)\tau\, c(\tau - \Delta t)}{\int d\tau\, p(\tau) c(\tau - \Delta t)} \tag{2.21}$$

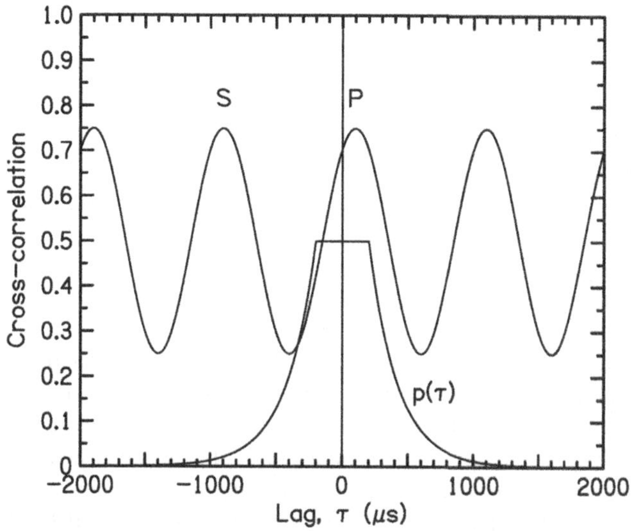

Fig. 2.8 The cross-correlation function (CCF) for model neural spikes generated by a 1000-Hz tone and the density of coincidence cells in the model medial superior olive [MSO; $p(\tau)$]. The CCF has a primary peak (P), indicating that the tone has an ITD of 100 μs. The secondary peak (S) is one period (1 ms) removed. The CCF and $p(\tau)$ are multiplied together to generate an effective pattern. In the model by Stern and Colburn (1978), the centroid of that effective pattern is the prediction for the lateralization of the tone

Here the $\bar{\tau}$ is the centroid (center of mass) when the ITD is Δt. For a periodic signal, the CCF for neural spikes is an oscillating function, as shown in Fig. 2.8. The function has the period of the tone.

The centroid model has attractive features that become evident by mentally morphing Fig. 2.8. First, the displacement of the primary peak (P) of the CCF shows that the centroid will be positive after the multiplication is done. Next, it is clear that in the limit of very low frequency, the CCF will approach a flat line and the value of the centroid will approach zero regardless of the ITD. In the limit of high frequency, the secondary peak (S) will enter the $p(\tau)$ window and reduce the centroid. The low- and high-frequency limits of the model are approximate representations of the frequency effects seen in ITD threshold experiments. It is clear too, that as the frequency increases, with point P remaining at 100 μs, eventually point S will reach the −100-μs location. This will occur when the period of the tone becomes as small as twice the ITD (i.e., the IPD becomes as large as 180°). Then the centroid will again be zero. That result agrees with the fundamental ambiguity of a 180° IPD.

Stern and Colburn (1978) extended their model to incorporate the ILD by introducing a third multiplicative function that weighted the pattern by a bell-shaped curve. The curve was displaced to emphasize the pattern on the side where the level is larger. The multiplicative ILD function removes the symmetry of the pattern and removes the ambiguity of a 180° IPD. If the ILD has the same sign as the ITD (i.e., both favor the same side), the IPD can grow well beyond 180°, and yet the predicted laterality will remain on the same side. That prediction too agrees with experimental observation (Domnitz and Colburn 1977).

If the ILD has a sign opposite to the ITD, the results can be complicated. The free-field virtual-reality experiments by Hartmann et al. (2016) are consistent with the following model for an opposing ILD. For a low frequency (e.g., 250 Hz), the secondary peak is 4000 μs away from the primary peak. The model interaural delay line does not have many cells at such long delays, and the opposing ILD can only enhance the near tail of the pattern. The result is to displace the centroid in the direction of the opposing ILD, as per the model by Stern and Colburn (1978). The displacement increases with increasing ILD. For a higher frequency (e.g., 750 Hz), the behavior depends on the range of the ITD. If the ITD is small, the secondary peak occurs near the 1300-μs point on the delay line. Again, the centroid is displaced, an exaggerated version of the displacement at low frequency. However, if the ITD is as large as 500 μs, the secondary peak appears near the 800-μs point. At that point, the model has a significant cell density, and the centroid is completely dominated by the secondary peak. The interpretation is that the ILD acts mainly as a switch because the perceived image switches sides, appearing at the location cued by the secondary ITD peak.

2.3.5 *Duplex Theory*

The ITD and ILD are effective cues for sound localization/lateralization. The duplex theory attempts to describe relative roles of these two cues. In the broadest terms, the theory says that the ITD is dominant for low frequencies and the ILD is dominant for high frequencies. The theory originated in observations by Lord Rayleigh (Strutt 1907), although Rayleigh put the boundary between ITD dominance and ILD dominance at a frequency that was far too low. In 1936, Stevens and Newman performed a classic experiment measuring the localization error rate for tones as a function of frequency. The function showed a peak near 3000 Hz, which they interpreted as the frequency at which neither the ITD or the ILD was effective. However, setting the boundary between ITD and ILD dominance at 3000 Hz is far too high. Subsequent experiments by Mills (1958) got the boundary about right, namely, 1500 Hz. The boundary is caused by an ILD cue that becomes gradually more effective as the frequency increases, coupled with an ITD cue that becomes precipitously less effective as frequency increases. For pure tones above about 1500 Hz, no human fine-structure ITD sensitivity could be detected at all (Brughera et al. 2013). There is also evidence to support a rule of thumb that the ITD will be dominant whenever it can be detected. In headphone experiments using broadband noise and individualized HRTFs, Wightman and Kistler (1992) showed that the ITD strongly dominated contradictory ILD- and HRTF-spectral cues. Similarly, by preserving individual HRTFs, Macpherson and Middlebrooks (2002) found that ILDs were ineffective at low frequencies even though the ILDs were physically quite large enough to be effective.

The role of ILDs in localizing low-frequency sound appears to depend on method. In headphone experiments, Yost (1981) found that the perceived laterality of tones depended on the ILD according to a function that was essentially identical for all frequencies from 200 to 5000 Hz. For instance, an ILD of 2 dB always led to an image located 20% of the way from the center of the head to the extreme right or left. This observation does not comport with ILDs in the physical world. An ILD of 2 dB corresponds to an azimuth of less than 10° for a 2000-Hz tone but to an azimuth of 90° for a 200-Hz tone. At the same time, in a free-field experiment with a substantial ITD, imposing a 2 dB ILD on a 200-Hz tone might not even be noticed, unless the sign of the ILD pointed in a direction opposite to the ITD.

Opposing ITDs and ILDs are at the basis of time-intensity trading experiments, begun at least as early as 1932 (Shaxby and Gage). The experiments determine the opposing ITD that is required to offset a standard ILD. Understandably, the experiments are normally done using headphones, and the results are trading ratios expressed in microseconds per decibel. Depending on the stimulus, the answers range from less than 2 μs/dB to hundreds of microseconds per dB. As noted by a number of authors (e.g., Ignaz et al. 2014), the answers depend sensitively on method, especially, which interaural parameter is used as the standard. Another problem with the trading experiments is that listeners may not perceive a single image. Instead, they may lateralize both a "time" image and an "intensity" image (Whitworth and Jeffress 1961). Even if an image can be lateralized based on trading, the trading was found to be incomplete by Hafter and Carrier (1972), in that the image can be distinguished from a tone with no conflicting cues but having the same laterality.

The observation of split images or incomplete trading is contrary to the Stern and Colburn model (1978) that incorporates the ILD into the delay line to form a single image. The observation is consistent with the physiological fact that the ITD and ILD are processed by different segments of the superior olivary complex as described in Chap. 4 by Takahashi, Kettler, Keller, and Bala and in Chap. 5 by Owrutsky, Benichoux, and Tollin. It is possible that split images occur only in uncompensated headphone listening and that the externalization perceived in virtual-reality experiments serves to fuse an image made with contradictory cues.

2.3.6 Opponent-Hemifield Model

Within the past 20 years, the Jeffress model has been challenged by the opponent-hemifield model (see Chap. 10 by Dietz and Ashida). The challenge began with the observation that for small mammals, such as the gerbil, the distribution of best ITDs in the MSO does not represent all the ITDs nor does it peak at a best delay of zero. In fact, the distribution peaks for best delays that are longer than the longest delay afforded by the head diameter, the limit of the "physiological range" (McAlpine et al. 1996; Brand et al. 2002). The advantage of a cell with such a long best delay is that the slope of its excitation function is steep for delays within the physiological

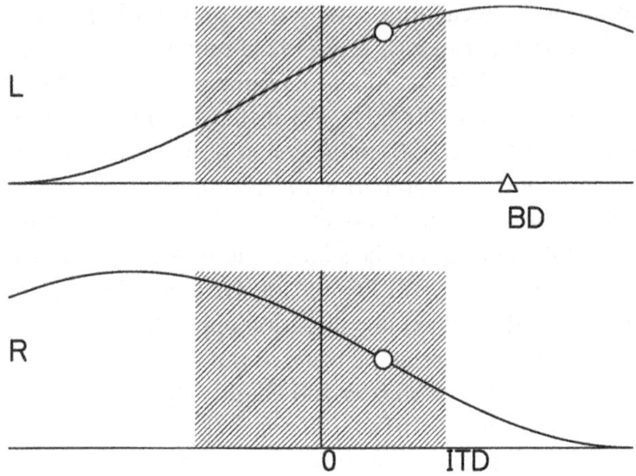

Fig. 2.9 Cartoon for the opponent-hemifield model. The firing rates for corresponding cells on the left and right sides are shown as a function of the ITD. As is typical for small mammals, the peak rate occurs at a best delay (BD) outside the physiological range of ITDs (*shaded box*). For an ITD (*open circle*), the excitation is greater on the left side because the source of sound is on the listener's right side. The difference in rate encodes the localization of the source

range. That feature affords maximum sensitivity to changes within that range. However, a cell like that does not exhibit tuning within that range. Instead, it exhibits only rate sensitivity. To display localization, there needs to be a higher center that compares the rates from equivalent cells on both sides (opponency), as indicated in Fig. 2.9. A survey of different animals by Harper et al. (2014) concluded that although rate coding (opponent-hemifield model) may suffice for small mammals, human localization can best be understood as a combination of tuned (or place) models and rate coding models. The combined system features a tuned, Jeffress-type model for the higher frequencies and a rate model for lower frequencies. The distinction between higher and lower depends on head size. For human heads, the division occurs near 400 Hz (Harper and McAlpine 2004).

One can make several observations. Although the neural delays for MSO cells on one side of the head are partly attributable to the longer path length from the opposite side, there is no necessary reason within the Jeffress model to have coincidence cells on both sides of the brain. A single MSO would serve. By contrast, the opponent-hemifield model clearly needs to be bilateral. Although practical ITDs can be expected to fall within the range of human head delays, expected to be less than 800 μs, listeners are sensitive to interaural delays that are much longer, extreme delays as long as 10,000 μs (Mossop and Culling 1998). There is no measurable ITD discrimination for such long delays, but a lateral sensation is there. A differential excitation rate for neurons tuned to much shorter ITDs on opposite sides of the head is an attractive way to account for sensitivity to such extreme delays. That argues for the opponent-hemifield model.

2.4 Precedence Effect

The ITD and ILD provide the information that listeners use to localize the azimuth of the source of a sound. The effects of these cues are readily observed in headphone experiments or in a free field. But listeners are not normally in a free field. People in developed societies spend most of their time in reflective environments and have done so for centuries. Reflective environments lead to standing waves that transform the well-ordered relationships between interaural cues and source azimuth into wildly varying, misinformative functions. As a listener moves about in this space, the functions verge on chaotic for frequencies greater than a few hundred hertz. Figure 2.10 for a 1000-Hz tone shows that the IPD experienced by a listener varies over the complete 360° range as the listener moves a few inches. Further discussion of spatial hearing in rooms appears in Chap. 9 by Zahorik.

Given the physical distortions of steady-state interaural cues in everyday reflective environments, it is perhaps surprising that listeners continue to find it profitable to retain their maps relating interaural cues to source azimuths. Possibly, the mapping is reinforced by experience with nearby sound sources where the direct wave dominates the reflections. Possibly, it is reinforced by successful experiences out of doors. The most potent of the effects, however, is that listeners have subconsciously learned to accord overwhelming perceptual weight to the interaural cues that arrive first, leading to a dominance of the direct sound over reflections. Exactly what is meant by "arrive

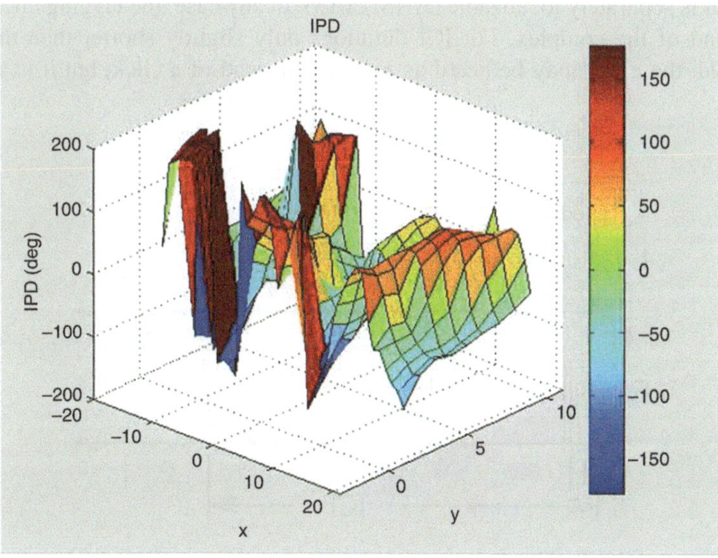

Fig. 2.10 The IPD for a 1000-Hz tone, as measured by Mr. Zane Crawford, in an office using a model head. The head location was moved in one-inch steps around the horizontal (*x*-*y*) plane. Axis labels show inches (1 inch = 2.54 centimeters). The nose always faced the source, 2.5 m away

first" is left for discussion later. Since 1949 (Wallach, Newman, and Rosensweig), this first-arrival effect has been known as the *precedence effect* (Hartmann 1997).

2.4.1 Precedence Attributes

The precedence effect is most easily approached with a stimulus consisting of two clicks. Each click consists of a left ear pulse and a right ear pulse, as shown in Fig. 2.11. The first click, with ITD1, represents a direct sound; the second click, with ITD2, represents a reflection. The direct click and the reflection click are separated by the reflection delay, here called the "interclick interval (ICI)." The special character of this stimulus is that none of the four sounds overlap in time and that makes it straightforward to study. The review by Litovsky et al. (1999) deals specifically with this four-pulse stimulus and summarizes three essential aspects of the precedence effect: fusion, localization dominance, and lag discrimination suppression.

2.4.1.1 Fusion

Although Fig. 2.11 shows four pulses, the entire complex is heard by the listener as a single event if the ICI duration is shorter than the echo threshold. An echo is defined as a reflected (lagging) sound that is heard as a distinct repetition of a direct sound; it is separately localizable (ANSI 2013). In any case, the lagging click alters the sound of the complex. For ICI durations only slightly shorter than the echo threshold, the event may be heard as a scratch instead of a click, but it is a *single* scratch.

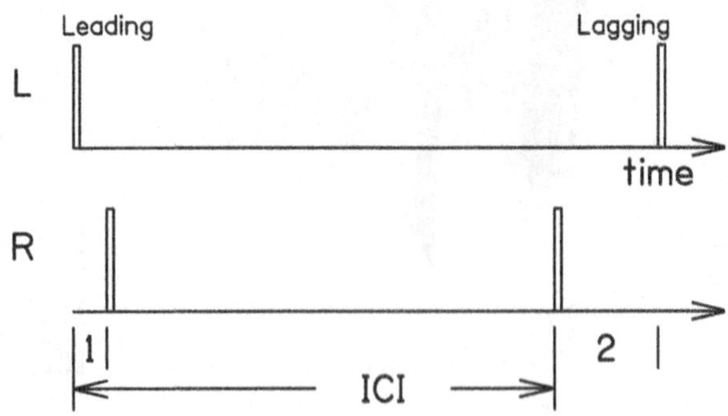

Fig. 2.11 Model pulses as they arrive at a listener's ears. The leading click arrives first at the left ear (ITD1). The lagging click (reflection) arrives first at the right ear (ITD2). The lead and lag are separated by the interclick interval (ICI). If the ICI is long enough, the lagging click will become an echo. If not, the entire sequence of four pulses is heard as a single click

2.4.1.2 Localization Dominance

Figure 2.11 shows a direct sound arriving from an azimuth somewhat to the left of the listener's forward direction and a reflection that comes from an azimuth considerably to the right. Because the leading sound dominates the perceived localization, the listener hears the sound on the left, even though the evidence for a source on the right (occurring in the reflected click) appears to be the stronger because ITD2 is greater than ITD1. The precedence effect can be studied quantitatively by finding the interaural delay for the reflected click (lagging click) that would cancel the leftward influence of the leading click. Localization dominance for the four-pulse stimulus is strongest with an ICI of about 2 ms (Litovsky et al. 1999). Using 1-ms noise bursts and headphone listening, Shinn-Cunningham et al. (1993) modeled their localization data with the formula

$$\mathrm{ITD}_{\mathrm{Effective}} = c\,\mathrm{ITD}_1 + (1-c)\mathrm{ITD}_2 \tag{2.22}$$

where c usually fell between 80% and 100%.

Localization dominance also depends on the ITD values of leading and lagging clicks. Using a stimulus with ITD1 = 0 and an ICI of 2000 μs, Wallach et al. (1949) found a curious localization reversal, an anomalous lateralization. For modest values of ITD2, the localization was biased in the direction indicated by ITD2, as expected. However, as the ITD2 increased to 700 μs, the direction of localization reversed. An explanation for this curious effect is that the frequency tuning in the cochlea causes the neurons of the peripheral auditory system to ring and the resulting neural pulse trains overlap in time, as pointed out by McFadden (1973a,b). Because of the ringing, the signs of the relevant interaural differences in lagging clicks become ambiguous. Given an appropriate ring time, the directional cues in the lagging click could be reversed. For instance, if the second click leads in the right ear by ITD2 = 600 μs and the ring time is 700 μs, there is a neural pulse on the left side that precedes a pulse on the right side by 100 μs, reversing the direction of the cues. Wallach et al. (1949) found maximum ambiguity for ITD2 = 675 μs. According to the ringing hypothesis, maximum ambiguity occurs for a ring time that is twice that value, 1350 μs, corresponding to a filter with center frequency of 741 Hz. Similarly, Tollin and Henning (1999) found anomalous localization that could be attributed to band-pass filtering centered on 750 Hz. Just why the binaural system should pay so much attention to that particular filter is unclear. It does correspond to the frequency of the minimum ITD threshold (Brughera et al. 2013). Experiments on dichotic pitch suggested a binaural weighting, with a peak near 500 or 600 Hz (Bilsen and Raatgever 2000; see also Chap. 7 by Best, Goupell, and Colburn).

2.4.1.3 Lag Discrimination Suppression

The fusion and localization dominance aspects of the precedence effect imply a suppression of the lagging sound. Lag discrimination experiments probe a listener's ability to gain localization information from a lagging sound. Zurek (1980) compared ITD discrimination for a leading burst and a lagging burst as a function of ICI. For the shortest ICI values, shorter than 1 ms, there is summing localization, where leading and lagging sounds both contribute to the perceived location. For very long values of ICI, beyond the echo threshold where fusion clearly fails, leading and lagging clicks are comparably discriminable. At ICI values near 2 ms, where localization dominance is maximum, the discrimination of the lagging burst compared with that for the leading burst is found to be the worst.

The four-pulse experiment leaves open questions about the precedence effect as it applies to stimuli of longer duration. To bridge the gap between lateralization with headphones and localization in rooms, Rakerd and Hartmann (1985) put a heavy, reflective wall (or floor or ceiling) in an anechoic room to create an acoustical condition with a single reflection and no reverberation. Using 50-ms 500-Hz tones with rectangular envelopes, they found that precedence from the sharp onset led to source localization that was mostly accurate but biased. The bias was not necessarily toward the wall. Instead, it was in the direction indicated by the interaural differences in the summed wave, the ongoing direct wave plus the reflected wave. That observation was a variation on lag discrimination suppression. The reflection from the wall was not influential on its own.

2.4.2 Modeling the Precedence Effect

The observed nature of the four-pulse experiment was successfully explained by Hartung and Trahiotis (2001) using a model that was entirely peripheral. Key to the model was the interaction of clicks in cochlear filters followed by hair cell compression. Bianchi et al. (2013) found support for a significant cochlear contribution to the suppression of a lagging click by comparing click-evoked otoacoustic emissions with auditory brainstem responses. However, precedence cannot reside entirely within the cochlea because it is experienced by listeners with cochlear implants (Brown et al. 2015). Cochlear implants bypass the basilar membrane with its delays and directly excite the auditory nerve.

Similarly, a peripheral (monaural) candidate for the precedence effect was discovered by Wickesberg and Oertel (1990) in slices of the mouse cochlear nucleus. Excitation of cells in the dorsal cochlear nucleus leads to inhibitory postsynaptic potentials in the anteroventral cochlear nucleus. Both pathways, ITD and ILD, were inhibited. Because the inhibition requires an extra synapse, there is an opportunity for a leading sound to be transmitted to the superior olivary complex for binaural interaction while a lagging sound is inhibited. Models of the precedence effect, starting as early as the thesis by Franssen (1960), tend to be inhibitory additions to a Jeffress delay line model. Lindemann (1986a) inserted attenuators, activated by contralateral excitation, within the delay line. The effect of that addition was to immediately incorporate the ILD into the ITD process. Lindemann (1986b) applied the same model to a dynamic situation. With the right inhibitory time constants, a lagging click was eliminated from the delay line, thereby simulating precedence. Subsequent work tried to extend the model to cope with ongoing stimuli. Braasch and Blauert (2003) found that models depending on specifically tuned channels were too sensitive to account for precedence in continuous sounds with different bandwidths. A Lindemann-type model, but with the ITD sensitivity removed, remained successful.

Greater realism in precedence effect modeling would preserve a role for reflected sounds in enhancing perceived level and generating a sense of spaciousness. A model by Braasch (2016) retained information about reflected sounds, including incident direction. The model was based on multiple stages of cross-correlation and applied to running speech in reverberant environments. Directions were cued by the ITD.

2.4.3 The Central Precedence Effect

Although the peripheral and brainstem observations and the neural models of precedence may capture important elements of the precedence effect, it is generally acknowledged that poking around in the precedence arena uncovers complicated behaviors that are difficult to square with simple neural signal-processing models. It is an ongoing challenge to sort out the aspects attributable to central processes, including listener expectations (Clifton et al. 2002).

2.4.3.1 Stimulus Type

Essential parameters of the precedence effect depend on the nature of the stimulus. The echo threshold, which is about 5 ms for clicks, becomes 50 ms if the sound source is speech (Haas 1949). The echo threshold is even longer for music, and for musical performances that benefit from long reverberation times (e.g., Gregorian chant), it may not even exist. The 50-ms threshold is commonly used by audio engineers and acoustical designers. The problem for binaural modeling is that simple signal processors do not change their time constants for different signal types.

2.4.3.2 Development

Infants are sensitive to the left-right directions of sounds. Left-right distinction for single sounds is one of the infant screening tests done at birth. However, the precedence effect develops more slowly. Clifton et al. (1981) showed that leading-source dominance does not occur in infants. It does appear, though incompletely, at 6 months. Such a delayed onset suggests a precedence effect with a complex central involvement developed through experience. Litovsky (1997) showed that for discrimination, the precedence effect is not fully developed at 5 years of age. Litovsky and Godar (2010) showed that for localization dominance, the precedence effect is weaker in 5-year-old children than in adults, extending out to tens of milliseconds.

2.4.3.3 Breakdown and Buildup

The breakdown effect (Clifton 1987) is optimally observed in the free field, with two spatially separated loudspeakers reproducing the four-pulse stimulus in Fig. 2.11. An interclick delay, not much smaller than the 5-ms threshold, occurs between the clicks sent to the loudspeakers and the interaural differences are caused by the listener's head. The listener hears a single click event located at the leading loudspeaker. If the order of leading and lagging loudspeakers is suddenly reversed, fusion breaks down. The listener hears two clicks, one from each loudspeaker. The listener continues to hear two clicks for some seconds, until the precedence effect builds up again and a single click is heard, located at the other (now-leading) loudspeaker. The long time constants observed in these effects are difficult for simple neural signal-processing models to explain.

2.4.3.4 Franssen Illusion

The Franssen illusion (Franssen 1960) occurs with pure tones in a room environment. There are two loudspeakers, placed to the left and right of the listener. A tone is turned on abruptly at the first loudspeaker and then faded out. As the tone is faded out, it is similarly faded in in the second loudspeaker. The second loudspeaker continues to sound indefinitely. The illusion is that the listener continues to locate the source of the tone at the *first* loudspeaker, even for tens of seconds after the transition. The illusion appears to depend on a central competition between the strengths of localization cues. Because of the precedence effect, the first speaker is strongly localized. Because of standing waves in the room, the second speaker can be localized only with great uncertainty. Hence, the only good spatial evidence is for a tone originating from the first loudspeaker. That explanation was easily validated by moving Franssen's experiment into an anechoic room, where it was found that the illusion failed completely (Hartmann and Rakerd 1989). Without the confusion of room reflections, listeners localized the second loudspeaker reliably. Even in a room environment, the Franssen illusion fails for a broadband noise stimulus. That result might have been predicted from the experiment by Tobias and Schubert (1959), which showed that the precedence effect has very limited influence for broadband noise as heard in a room through loudspeakers.

2.4.3.5 Median Sagittal Plane

The precedence effect is typically studied using sources in the horizontal plane or otherwise with interaural differences produced by headphone signals. Models for the effect are essentially binaural. However, it is not necessary for there to be an interaural difference at all. Litovsky et al. (1997) found precedence-like effects for sources at different elevations in the median sagittal plane where interaural differences are minimal or nonexistent. Different elevations are distinguished by different spectral signatures (Blauert 1997, p. 97 ff). With minimal (or zero) interaural differences, vertical-direction click-train localization experiments on human listeners showed leading source localization dominance. The dominance was similar in ICI dependence for both the sagittal plane and horizontal plane, although it was stronger in the horizontal plane (Litovsky et al. 1997). With similar stimuli, psychophysical recordings in the inferior colliculus of cats showed lagging click suppression (Litovsky et al. 1997).

2.4.3.6 Back to the Periphery

The buildup and breakdown effects are regarded as evidence for central processes in the precedence effect (Clifton et al. 2002) because of the long times over which these effects occur. Also, the appearance of an interpretative component to the fusion effect for click trains makes this aspect of the precedence effect seem to resemble auditory stream segregation. At the central level, the essential feature for clicks is localization (lateralization) left or right. Whether the location is cued by the ITD or by the ILD would not seem to be of much importance; the two cues are to a great extent tradeable. Therefore, it was a considerable surprise when headphone experiments by Krumbholz and Nobbe (2002) showed that the breakdown effect is not symmetrical in the two cues. It was found that for most listeners, switching the ILD caused fusion to break down, as would be expected from Clifton's 1987 experiments, but switching the ITD did not. Brown and Stecker (2013) confirmed the more robust nature of the ITD and suggested that fusion and localization dominance might be dissociated. The asymmetry implies a role for processes at a lower level where the effects of the two interaural cues are different.

2.4.4 Ongoing Precedence Effect

The four-pulse stimulus is especially straightforward because it avoids temporal overlap of any parts of the stimulus. However, the precedence effect is a powerful effect in everyday listening where overlap is the rule. Increasingly, the precedence effect is considered to be one of an ensemble of perceptual reweightings of temporally evolving events. For example, Dietz et al. (2013) studied the role of interaural differences presented at different phases of an amplitude-modulated signal. They found a strong dominance for interaural differences presented during the rising part of the envelope. Analogs to the human psychophysics results were observed in the MSO and inferior colliculus of anesthetized guinea pigs (Dietz et al. 2014). More about the ongoing precedence effect can be found in other chapters of this volume.

2.5 Summary

Localization of sounds in the horizontal plane depends on interaural differences established by the sound source location, the environment, and the listener's anatomy. Complex environmental effects can be simplified by experiments in the free field, where the listener receives only direct sounds and there are no reflections.

The important interaural differences are the ILD, as a decibel measure of relative intensities at the two ears, and the ITD. The ITD may be a difference in arrival time of the fastest details of the pressure waveform at the head or it may be a difference in arrival time for onsets or other more slowly modulated waveform features such as the envelope. The interaural differences can be calculated from various models that simplify the relevant anatomy. The spherical-head model captures all the variety of physical effects that occur for human heads, including the frequency dependence of the interaural differences as well as the azimuth dependence. It captures these effects with varying degrees of accuracy, sometimes contrary to physical intuition.

Extending sound localization to three dimensions requires an extension of coordinate systems. The cones of confusion characteristic of the spherical-head model have important applicability to real-world human sound localization, and they are best represented by the lateral-polar system of angular coordinates. This chapter gives transformations between that system and the more common spherical-polar system.

The sensitivity of human listeners to interaural differences can be studied directly using headphone presentation or practically using spatially distributed real sound sources. Gross anatomical features are involved in the latter. The difference between these two experimental modes corresponds to the perceptual difference between lateralization, where headphone-presented sounds are perceived to be within the head, and localization for real sources, where sounds are perceived to be external. There are likely to be other perceptual differences too, a topic that has been very little studied. Both lateralization and localization experiments indicate that the best representation of ITD effects is through the ITD itself. Although a transformation to the IPD offers some attractive simplifications, it is less related to perceived location than is the ITD.

There are different models for the neural processing of interaural differences. According to the duplex model, ITDs are the dominant cues for localization of low-frequency sounds, whereas ILDs are dominant for high. There is a role for the ITD at a high frequency in comparing arrival times of envelope features of the sound. Models for ITD processing are based on a CCF. The Jeffress model encodes location through the variation in the lag value of the peak of the function. The opponent-hemifield model encodes through the variation of magnitudes of the function compared binaurally for a reduced range of lags. Models such as the position-variable model attempt to account for the combined effects of the ITD and ILD.

Although sound wave reflections play havoc with the physical localization cues available to listeners, it is found that listeners localize most real-world sounds with little difficulty in a room environment, even in an acoustically lively room. This apparent paradox is explained by the precedence effect, by which perceived spatial attributes of sound are dominated by the first arriving wavefront. Although there are features of precedence that accord with peripheral processing of sounds, an explanation of the large variety of precedence-related effects and time constants requires a system with considerable central processing. Precedence is now understood as one manifestation of the temporal weighting of binaural cues.

Acknowledgments I acknowledge my Michigan State colleagues, especially Brad Rakerd and Eric Macaulay, for years of collaboration. The work summarized here was supported by the National Institutes of Health and the Air Force Office of Scientific Research.

Compliance with Ethics Requirements

William M. Hartmann declares that he has no conflict of interest.

References

Aaronson NL, Hartmann WM (2014) Testing, correcting, and extending the Woodworth model for interaural time differences. J Acoust Soc Am 135:817–823

Algazi VR, Avendano C, Duda RO (2001) Elevation localization and head-related transfer function analysis at low frequencies. J Acoust Soc Am 109:1110–1122

ANSI (2013) Acoustical terminology, ANSI/ASA S1.1-2013. American National Standards Institute, Washington, D.C.

Bauer B (1965) Transmission of directional perception. IEEE Trans Audio 13:5–8

Baumann C, Rogers C, Massen F (2015) Dynamic binaural sound localization based on variations of interaural time delays and system rotations. J Acoust Soc Am 138:635–650

Bianchi F, Verhulst S, Dau T (2013) Experimental evidence for a cochlear source of the precedence effect. J Assoc Res Otolaryngol 14:769–779

Bilsen FA, Raatgever J (2000) On the dichotic pitch of simultaneously presented interaurally delayed white noises. Implications for binaural theory. J Acoust Soc Am 108:272–284

Blauert J (1997) Spatial hearing. MIT Press, Cambridge, MA

Blauert J, Lindemann W (1986a) Auditory spaciousness: Some further psychoacoustic analyses. J Acoust Soc Am 80:533–542

Blauert J, Lindemann W (1986b) Spatial mapping of intracranial auditory events for various degrees of interaural coherence. J Acoust Soc Am 79:806–813

Boyd AW, Whitmer WM, Soraghan JJ, Akeroyd MA (2012) Auditory externalization in hearing impaired listeners: The effect of pinna cues and number of talkers. J Acoust Soc Am 131:EL268–EL274

Braasch J (2016) Sound localization in the presence of multiple reflections using a binaurally integrated cross-correlation/autocorrelation mechanism. J Acoust Soc Am 140:EL143–EL148

Braasch J, Blauert J (2003) The precedence effect for noise bursts of different bandwidths, II. Comparison of model algorithms. Acoust Sci Tech 24:293–303

Brand A, Behrend O, Marquardt T, McAlpine D, Grothe B (2002) Precise inhibition is essential for microsecond interaural time difference coding. Nature 417:543–547

Brimijoin WO, Boyd AW, Akeroyd MA (2013) The contribution of head movement to the externalization and internalization of sounds. PLOS One 8(12):e83068.1–e8306812

Brown AD, Stecker GC (2013) The precedence effect: Fusion and lateralization measures for headphone stimuli lateralized by interaural time and level differences. J Acoust Soc Am 133:2883–2898

Brown AD, Jones HG, Kan A, Thakkar T, Stecker GC, Goupell MJ, Litovsky RY (2015) Evidence for a neural source of the precedence effect in sound localization. J Neurophysiol 114:2991–3001

Brughera A, Dunai L, Hartmann WM (2013) Human interaural time difference thresholds for sine tones. The high-frequency limit. J Acoust Soc Am 133:2839–2855

Brungart DS, Rabinowitz WM (1999) Auditory localization of nearby sources. Head-related transfer functions. J Acoust Soc Am 106:1465–1479

Cai T, Rakerd B, Hartmann WM (2015) Computing interaural differences through finite element modeling of idealized human heads. J Acoust Soc Am 138:1549–1560

Carr CE, Konishi M (1990) A circuit for detection of interaural time difference in the brain stem of the barn owl. J Neurosci 10:3227–3246

Clifton RK (1987) Breakdown of echo suppression in the precedence effect. J Acoust Soc Am 82:1834–1835

Clifton RK, Morrongiello BA, Kulig JW, Dowd JM (1981) Newborns' orientation toward sound: Possible implications for cortical development. Child Dev 52:833–838

Clifton RK, Freyman RL, Meo J (2002) What the precedence effect tells us about room acoustics. Percept Psychophys 64:180–188

Colburn HS (1973) Theory of binaural interaction based on auditory-nerve data I. General strategy and preliminary results on interaural discrimination. J Acoust Soc Am 54:1458–1470

Colburn HS (1977) Theory of binaural interaction based on auditory-nerve data II. Detection of tones in noise. J Acoust Soc Am 61:525–533

Constan ZA, Hartmann WM (2003) On the detection of dispersion in the head-related transfer function. J Acoust Soc Am 114:998–1008

Dietz M, Marquardt T, Salminen NH, McAlpine D (2013) Emphasis of spatial cues in the temporal fine structure during the rising segments of amplitude-modulated sounds. Proc Natl Acad Sci U S A 110:15151–15156

Dietz M, Marquardt T, Stange A, Pecka M, Grothe B, McAlpine D (2014) Emphasis of spatial cues in the temporal fine structure during the rising segments of amplitude-modulated sounds II: single-neuron recordings. J Neurophysiol 111:1973–1985

Domnitz RH, Colburn HS (1977) Lateral position and interaural discrimination. J Acoust Soc Am 61:1586–1598

Duda RO, Martens WL (1998) Range dependence of the response of a spherical-head model. J Acoust Soc Am 104:3048–3058

Durlach NI, Rigopulos A, Pang XD, Woods WS, Kulkarni A, Colburn HS, Wenzel EM (1992) On the externalization of auditory images. Presence 1:251–257

Franssen NV (1960) Some considerations on the mechanism of directional hearing. Thesis. Technische Hogeschool, Delft

Grothe B (2000) The evolution of temporal processing in the medial superior olive, an auditory brainstem structure. Prog Neurobiol 61:581–610

Haas H (1949) Uber den Einfluss des Einfachechos auf die Horsamkeit von Sprache. Acustica 1:49–58, tr. (1972) The influence of a single echo on the audibility of speech. J Audio Engr Soc 20, 146–159

Hafter ER, Carrier SC (1972) Binaural interactions in low-frequency stimuli: The inability to trade time and intensity completely. J Acoust Soc Am 51:1852–1862

Harper NS, McAlpine D (2004) Optimal neural population coding of an auditory spatial cue. Nature 430:682–686

Harper NS, Scott BH, Semple MN, McAlpine D (2014) The neural code for auditory space depends on sound frequency and head size in an optimal manner. PLoS One 9(11):e108154
Hartley RVL, Fry TC (1921) The binaural localization of pure tones. Phys Rev 18:431–442
Hartmann WM (1997) Listening in a room and the precedence effect. In: Gilkey RH, Anderson TR (eds) Binaural and spatial hearing in real and virtual environments. Lawrence Erlbaum and Associates, Maywah, pp 191–210
Hartmann WM, Rakerd B (1989) Localization of sound in rooms IV – The Franssen effect. J Acoust Soc Am 86:1366–1373
Hartmann WM, Wittenberg A (1996) On the externalization of sound images. J Acoust Soc Am 99:3678–3688
Hartmann WM, Rakerd B, Crawford ZD, Zhang PX (2016) Transaural experiments and a revised duplex theory for the localization of low-frequency tones. J Acoust Soc Am 139:968–985
Hartung K, Trahiotis C (2001) Peripheral auditory processing and investigations of the precedence effect which utilize successive transient stimuli. J Acoust Soc Am 110:1505–1513
Hendrickx E, Stitt P, Messonnier J-C, Lyzwa J-M, Katz BF, de Boishéraud C (2017) Influence of head tracking on the externalization of speech stimuli for non-individualized binaural synthesis. J Acoust Soc Am 141:2011–2023
Ignaz A, Lang A-G, Buchner A (2014) The impact of reference tones on the adjustment of interaural cues. J Acoust Soc Am 135:1896–1992
Jeffress LA (1948) A place theory of sound localization. J Comp Physiol Psychol 41:35–39
Joris PX, Van de Sande B, Louage DH, van der Heijden M (2006) Binaural and cochlear disparities. Proc Natl Acad Sci 103:12917–12922
Kayser H, Ewart SD, Anemüller J, Rohdenburg T, Hohmann V, Kollmeier B (2009) Database of multichannel in-ear and behind-the-ear head-related and binaural room impulse responses. EURASP J Adv Signal Process. https://doi.org/10.1155/2009/298605
Klumpp RB, Eady HR (1956) Some measurements of interaural time difference thresholds. J Acoust Soc Am 28:859–860
Koenig W (1950) Subjective effects in binaural hearing. J Acoust Soc Am 22:61–62
Krumbholz K, Nobbe A (2002) Buildup and breakdown of echo suppression for stimuli presented over headphones - the effects of interaural time and level difference. J Acoust Soc Am 112:654–663
Kuhn GF (1977) Model for the interaural time differences in the azimuthal plane. J Acoust Soc Am 62:157–167
Lindemann W (1986a) Extension of a binaural cross-correlation model by contralateral inhibition I. Simulation of lateralization for stationary signals. J Acoust Soc Am 80:1608–1622
Lindemann W (1986b) Extension of a binaural cross-correlation model by contralateral inhibition II. The law of the first wavefront. J Acoust Soc Am 80:1623–1630
Litovsky RY (1997) Developmental changes in the precedence effect: Estimates of minimum audible angle. J Acoust Soc Am 102:1739–1745
Litovsky RY, Godar SP (2010) Difference in precedence effect between children and adults signifies development of sound localization abilities in complex listening tasks. J Acoust Soc Am 128:1979–1991
Litovsky RY, Rakerd B, Yin TCT, Hartmann WM (1997) Psychophysical and physiological evidence for a precedence effect in the median sagittal plane. J Neurophysiol 77:2223–2226
Litovsky R, Colburn HS, Yost WA, Guzman S (1999) The precedence effect. J Acoust Soc Am 106:1633–1654
Macaulay EJ, Hartmann WM, Rakerd B (2010) The acoustical bright spot and mislocalization of tones by human listeners. J Acoust Soc Am 127:1440–1449
Macpherson EA, Middlebrooks JC (2000) Localization of brief sounds: Effects of level and background noise. J Acoust Soc Am 108:1834–1849
Macpherson EA, Middlebrooks JC (2002) Listener weighting of cues for lateral angle: The duplex theory of sound localization revisited. J Acoust Soc Am 111:2219–2236

McAlpine D, Jiang D, Palmer AR (1996) Interaural delay sensitivity and the classification of low best-frequency binaural responses in the inferior colliculus of the guinea pig. Hear Res 97:136–152
McFadden D (1973a) Precedence effects and auditory cells with long characteristic delays. J Acoust Soc Am 54:528–530
McFadden D (1973b) A note on auditory neurons having periodic response functions to time-delayed, binaural stimuli. Physiol Psychol 1:265–266
Mershon DH, King LE (1975) Intensity and reverberation as factors in the auditory perception of egocentric distance. Percept Psychophys 18:409–415
Mills AW (1958) On the minimum audible angle. J Acoust Soc Am 30:237–246
Mills AW (1960) Lateralization of high frequency tones. J Acoust Soc Am 32:132–134
Morimoto M, Aokata H (1984) Localization cues of sound sources, in the upper hemisphere. J Acoust Soc Japan 5:165–173
Mossop JE, Culling JF (1998) Lateralization of large interaural delays. J Acoust Soc Am 104:1574–1579
Plenge G (1972) Über das Problem der Im-Kopf-Lokalisation. Acust 26:241–252
Plenge G (1974) On the differences between localization and lateralization. J Acoust Soc Am 56:944–951
Rabinowitz WM, Maxwell J, Shao Y, Wei M (1993) Sound localization cues for a magnified head: Implications from sound diffraction about a rigid sphere. Presence 2:125–129
Rakerd B, Hartmann WM (1985) Localization of sound in rooms II: The effects of a single reflecting surface. J Acoust Soc Am 78:524–533
Rayleigh, Lord (Strutt, J.W.S.) and Lodge, A. (1904). On the acoustic shadow of a sphere. Phi Trans Roy Soc London 203, 87–110 https://doi.org/10.1098/rsta.1904.0016
Sakamoto N, Gotoh T, Kimura Y (1976) On 'out-of-head localization' in headphone listening. J Audio Engr Soc 24:710–716
Schiano JL, Trahiotis C, Bernstein LR (1986) Lateralization of low-frequency tones and narrow bands of noise. J Acoust Soc Am 79:1563–1570
Schroeder MR (1977) New viewpoints in binaural interaction. In: Evans EF, Wilson JP (eds) Psychophysics and physiology of hearing. Academic, New York, pp 455–456
Searle CL, Braida LD, Davis MF, Colburn HS (1976) Model for auditory localization. J Acoust Soc Am 60:1164–1175
Shackleton TM, Meddis R, Hewitt MJ (1992) Across frequency integration in a model of lateralization. J Acoust Soc Am 91:2276–2289
Shamma SA, Shen N, Gopalaswamy P (1989) Stereausis: Binaural processing without neural delays. J Acoust Soc Am 86:989–1006
Shaxby JH, Gage FH (1932) The localization of sounds in the median plane. Special report 166, Medical Research Council, reports of the Committee upon the Physiology of Hearing, H.M Stationery Office Code 45-5-66, Universal Decimal Classification 612-858-751+535-76, pp 1–32
Shinn-Cunningham BG, Zurek PM, Durlach NI (1993) Adjustment and discrimination measurement of the precedence effect. J Acoust Soc Am 93:2923–2932
Shinn-Cunningham BG, Santarelli S, Kopco N (2000) Tori of confusion: Binaural localization cues for sources within reach of a listener. J Acoust Soc Am 107:1627–1636
Simon LSR, Zacharov N, Katz BFG (2016) Perceptual attributes for the comparison of head-related transfer functions. J Acoust Soc Am 140:3623–3632
Stern RM, Colburn HS (1978) Theory of binaural interaction based on auditory-nerve data IV. A model for subjective lateral position. J Acoust Soc Am 64:127–140
Stevens SS, Newman EB (1936) The localization of actual sources of sound. Am J Psych 48:297–306
Strutt JW (1907) On our perception of sound direction. Phil Mag 13:214–232
Thompson SP (1877) On binaural audition. Phil Mag XXXVI:274–276

Tobias JV, Schubert ER (1959) Effective onset duration of auditory stimuli. J Acoust Soc Am 31:1595–1605
Tollin DJ, Henning GB (1999) Some aspects of the lateralization of echoed sound in man. II. The role of the stimulus spectrum. J Acoust Soc Am 105:838–849
Toole FE (1970) In-head localization of acoustic images. J Acoust Soc Am 48:943–949
Treeby BE, Paurobally RM, Pan J (2007) The effect of impedance on interaural azimuth cues derived from a spherical-head model. J Acoust Soc Am 121:2217–2226
Volk F, Heinemann F, Fastl H (2008). Externalization in binaural synthesis: effects of recording environment and measurement procedure. Proceedings of the International Congress on Acoustics (Paris) pp 6419–6424
Wallach H, Newman EB, Rosensweig MR (1949) The precedence effect in sound localization. Am J Psych 62:315–337
Whitworth RH, Jeffress LA (1961) Time vs intensity in the localization of tones. J Acoust Soc Am 33:925–929
Wickesberg RE, Oertel D (1990) Delayed, frequency-specific inhibition in the cochlear nuclei of mice: A mechanism for monaural echo suppression. J Neurosci 10:1762–1768
Wightman FL, Kistler DJ (1989a) Headphone simulation of free field listening I: Stimulus synthesis. J Acoust Soc Am 85:858–867
Wightman FL, Kistler DJ (1989b) Headphone simulation of free field listening II: Psychophysical validation. J Acoust Soc Am 85:867–878
Wightman FL, Kistler DJ (1992) The dominant role of low-frequency interaural time differences in sound localization. J Acoust Soc Am 91:1648–1661
Woodworth RS (1938) Experimental psychology. Holt, New York, pp 520–523
Yin TCT, Chan JCK (1990) Interaural time sensitivity in medial superior olive of cat. J Neurophysiol 64:465–488
Yost WA (1981) Lateral position of sinusoids presented with interaural intensive and temporal differences. J Acoust Soc Am 70:397–409
Yost WA, Hafter ER (1987) Lateralization. In: Yost WA, Gourevitch G (eds) Directional hearing. Springer, New York, pp 49–84
Young LL (1976) Time-intensity trading functions for selected pure tones. J Speech Hear Res 19:55–67
Zhang PX, Hartmann WM (2010) On the ability of human listeners to distinguish between front and back. Hear Res 260:30–46
Zurek PM (1980) The precedence effect and its possible role in the avoidance of interaural ambiguities. J Acoust Soc Am 67:952–964

Chapter 3
Sound Source Localization Is a Multisystem Process

William A. Yost, M. Torben Pastore, and Yi Zhou

3.1 Introduction

This chapter reviews the literature concerning human sound source localization in a sound field. Other aspects of spatial hearing, such as studies of lateralization of sounds presented over headphones (Blauert 1997), spatial unmasking (Litovsky 2012), the cocktail party effect (Middlebrooks et al. 2017), localization in reverberation (Litovsky et al. 1999; Brown et al. 2014), and anatomical/physiological investigations of sound source localization and studies of sound source localization in nonhuman species, are not reviewed in any detail. Many of these topics have been reviewed in other Springer Handbook of Auditory Research (SHAR) books (e.g., Yost et al. 1993, 2008; Popper and Fay 2005; Middlebrooks et al. 2017).

There are several examples in the literature of how information from other sensory and neural systems bias sound source localization perceptions (see Yost et al. 2015; Van Opstal 2016). However, the motivating theme of this chapter is the insight that Wallach first presented in a series of papers (1938, 1939, 1940) regarding the fact that localizing sound sources in the actual world is not possible without both auditory-spatial cues and information about the location of the head. Although it may appear obvious that auditory-spatial processing must refer to the world around the listener, it is not obvious how this is achieved by the brain. Ultimately, obtaining information about the position of the head probably involves several neural systems (i.e., is a multisystem process). These neural estimates of head position must be integrated in some way with neural estimates of the location of sound sources relative to the head that are derived from auditory-spatial cues. This chapter reviews the literature describing the interaction of auditory-spatial and head-position cues in sound source localization.

W. A. Yost (✉) · M. T. Pastore · Y. Zhou
Spatial Hearing Laboratory, College of Health Solutions, Arizona State University, Tempe, AZ, USA
e-mail: william.yost@asu.edu; yizhou@asu.edu

© Springer Nature Switzerland AG 2021
R. Y. Litovsky et al. (eds.), *Binaural Hearing*, Springer Handbook of Auditory Research 73, https://doi.org/10.1007/978-3-030-57100-9_3

3.2 A Brief History of Sound Source Localization (1825–1942)

The earliest empirical studies of sound source localization, in the mid-nineteenth century, were tied to a major question in science and philosophy of whether the mind is different from the body in kind (the mind is distinct and separate from the body) or only in degree (Cartesian mind-body dualism; Boring 1942; Crane and Patterson 2002). If the mind is different from the body in degree only (i.e., the mind is not distinct and separate), then the mind should be able to be studied in ways similar to the study of the body (Fechner 1860). Many argued that the mind represents the external world through sensations (Boring and Gardener 1918; Boring 1942). Those arguments suggested that sensations can be measured; they have attributes such as quality, intensity, duration, and extension, which provide a basis for perception. During the half century between approximately 1825 and 1875 when the field of experimental psychology emerged, many debated the exact definitions and means of measuring sensations and their attributes as well as of their perception. At this same time, some scholars addressed the issue of sound source localization (i.e., the perception of the location of a sound source). Most suggested that sound has no attributes of extension (e.g., size and shape) and that there are no auditory spatial receptors (Boring 1942). With sound having no attributes of extension and there being no spatial receptors, it was postulated by some that sound could not, by itself, form a spatial perception of a sound source. Nevertheless, many early scholars indicated that listeners appeared to localize sounds to their sources using only the sound. Empiricists like Wundt (reviewed by Pierce 1901; Boring 1942) argued that sound source location was mediated by other senses such as vision, touch, and the vestibular sense for which there were spatial receptors that sensed extension. Early psychologists such as Berkeley (reviewed by Pierce 1901; Boring 1942) argued that experience helps sound source localization. For example, experience indicates that if the characteristics of the intensity of a sound are known, then when the sound is perceived as soft, its source is likely to be further away than when it is perceived as loud. During the quarter of a century between approximately 1850 and 1875, the ability to determine sound source location was assumed by many to be largely based on other sensory systems and/or experience. At the end of the nineteenth century, it became accepted that the sensations of sound had attributes related to extension when considering sound source localization. The question regarding sound source localization, as stated by Boring (1942), became, "Can the organism discriminate the relative positions of *sounds*, and, if so, how?" This approach, exemplified by the work of Lord Rayleigh (1876, 1907), shifted the view of sound source localization from that of a multisystem process to one based on the ability of the mind (brain) to use inputs from the two ears to extract sound source localization cues and thereby estimate sound source location based entirely on the sound produced by the source.

Most of the literature from the end of the nineteenth century through the first third of the twentieth century is characterized by studies of the possible binaural auditory cues used for sound source localization (Boring 1942; Yost 2017a). In his first lecture on sound source localization, Lord Rayleigh (1876) suggested that the ratio of sound intensity at the two ears could explain his ability to locate a person speaking in a circle of people. Others (e.g., Steinhauser 1879) also suggested that a sound-intensity ratio could be a basis for sound source localization. Following the invention of the vacuum tube in about 1910 (Yost 2017a), this binaural ratio of sound intensity at the two ears was expressed as an interaural level difference (ILD; see Table 3.1 for a list of abbreviations) in decibels. Rayleigh (1876) was less sure of the reason for his limited ability to locate the position of a tuning fork producing a low-frequency sound. Two years later, Thompson (1878) made several observations suggesting that the interaural phase might be a basis for sound source localization. However, the observations of Thompson (and others; e.g., Boring 1942) that the interaural phase might play a role in sound source localization were counter to Helmholtz's (1863) claims that the "ear is phase deaf." Thus, not much attention was paid to the interaural phase as a cue for sound source localization for almost 30 years. Having won the Nobel Prize in 1904 for his work on gases, Rayleigh's (1907) second lecture on sound source localization, "On Our Perception of Sound Direction," received considerable attention. Although Rayleigh never proposed a "duplex theory" of sound source localization, he did acknowledge in that lecture that tuning forks producing very low frequency sounds (less than 256 Hz) were most likely localized on the basis of interaural phase differences, whereas sources with sounds of higher frequencies were almost certainly localized using a binaural ratio. In the 1920s, in conjunction with the discovery of neural action potentials

Table 3.1 List of abbreviations

Abbreviation	Definition
CI	Cochlear implant
FBR	Front-back reversal
G	Movement gain
HRTF	Head-related transfer function
ILD	Interaural level difference
ITD	Interaural time difference
MAMA	Minimum audible movement angle
MMAA	Minimal movement audible angle
PSE	Point of subjective equality
T	Time
WAI	Wallach azimuth illusion
WEI	Wallach elevation illusion
α	Angular acceleration
β	Azimuth angle
δ	Elevation angle
θ	Actual angle
θ'	Estimated angle
$\theta(t)$ or $\theta'(t)$	Angle changing as a function of time
θ_{Abc_0}	Subscripted angles: A, sound source (*S*) or head (*H*); b, reference system (world-centric [*w*] or head-centric [*h*]); c, reverse angle (*r*); subscript 0 indicates an initial fixed angle
ω_0	Initial constant velocity
Ψ	Lateral angle

(Lord Adrian 1928), scholars such as von Hornbostel and Wertheimer (1920) and Klemm (1920) proposed that interaural phase differences could be converted to interaural time differences (ITDs).

Lord Rayleigh (1876, 1907) further observed that sources of speech (complex) sounds were more easily localized and less prone to front-back reversals (FBRs) than tuning fork (simple) sounds. Lord Rayleigh (1876) also commented that he could resolve FBRs when he rotated his head. Thirty years after Lord Rayleigh's widely cited paper in 1907, Stevens and Newman (1936) firmly established a duplex account of sound source localization in the azimuth plane (ITDs for low frequencies and ILDs for high frequencies; also see Hartmann et al. 2016). Stevens and Newman (1936) showed greater accuracy in localizing sources of complex sounds than those of tonal sounds. They also showed the existence of FBRs for sound source localization of tonal sound sources located in the azimuth plane. Stevens and Newman also suggested a possible role for the pinnae in resolving FBRs using loudness differences at high frequencies.

By the late 1930s, the binaural cues used for azimuthal sound source localization were fairly well established and any multisystem interactions were seldom mentioned. Then Wallach (1938) calculated cones of confusion (although he did not use that term) and showed how head motion might resolve cones of confusion errors in judging both the azimuthal and elevation location of sound sources (Wallach 1939, 1940).

Wallach's (1938, 1939, 1940) work was often cited at the time (Boring 1929, 1942; de Boer and Van Urk 1941; Kohlrausch and Altosaar 2011). In Boring's (1929, 1942) widely read historical reviews of experimental psychology and sensation and perception, he covered the topic of sound source localization, and concluded his 1942 review with the following observation: "Wallach has made it quite clear that localization is not purely auditory, but the product of an integration of auditory, kinesthetic and, when the eyes are open, visual factors." Thus, by the early 1940s, there was a growing literature supporting the notion that sound source localization was a multisystem process (i.e., involving binaural cues and other neural system cues that indicate head position), and one might have thought that this would have been studied extensively after World War II. But it was not, until recently. For this reason, this chapter explains Wallach's (1940) arguments and describes the work that followed after Wallach's experiments, with a focus on a resurgence of interest in sound source localization as a multisystem process.

3.3 Wallach's Multisystem Proposal for Sound Source Localization

Wallach's (1938, 1940) derivation of cones of confusion was in terms of the lateral angle (ψ; see also Mills 1972) where the lateral angle is defined by:

$$\sin(\psi) = \sin(\theta)\cos(\delta) \tag{3.1}$$

where δ is the elevation angle and θ is the azimuth angle (see Chap. 2).

Note that sin(θ) and sin(180° − θ) are the same value, as are cos(δ) and cos(− δ). Thus, azimuth angles that differ by 180° produce the same lateral angle as are elevation angles above and below an azimuth plane, resulting in front-back and up-down reversals, and these are loci of locations that are part of cones of confusion. In other words, listeners should not be able to differentiate the location on an azimuth plane of sound sources in the front hemifield from those in the back hemifield when each sound source produces the same binaural cues. Nor should listeners be able to use binaural cues to locate sound sources above a location on the azimuth plane from those below that azimuthal location.

Wallach (1939, 1940) asked how could binaural cues be used for sound source localization in light of the cones of confusion. Wallach's (1940) answer was, "Two sets of sensory data enter into the perceptual process of localization, (1) the changing binaural cues and (2) the data representing the changing position of the head." That is, integration of information about auditory-spatial and head-position cues should enable the brain to determine the location of sound sources in the actual world, even when FBRs and up-down reversals exist.

Sections 3.3, 3.4, and 3.5 summarize what is currently known about auditory-spatial cues (Sect. 3.3), head-position cues (Sect. 3.4), and how the integration of the two cues might be the basis for sound source localization (Sect. 3.5).

3.4 Auditory-Spatial Cues

Separate sets of cues are used to localize sources of sounds in each of the three spatial dimensions: azimuth, elevation, and range/distance.

3.4.1 Sound Source Localization in Azimuth

ITDs are used for sound source localization of low-frequency sounds (less than 1400 Hz; Hartmann et al. 2016). ITDs for low frequencies (less than approximately 1500 Hz) are related to azimuth angle by ITD $\approx 3k\cos(\theta)$ and at high frequencies by ITD $\approx 2k\cos(\theta)$, where θ is azimuth angle and k is the ratio of head radius (r) to the speed of sound (c; $k = r/c$; Kuhn 1987; Aaronson and Hartmann 2014). Fine-structure ITD cues are not usable for sound source localization of high-frequency stimuli due to, among other things, the time it takes sound to travel from one side of the head to the other, which is approximately 0.8 ms (Kuhn 1987). This is the period of a 1250-Hz tone. If the frequency of a tonal sound is less than approximately 1250 Hz, then the time lag between the ears will be less than one cycle, leading to an unambiguous estimate of which ear received the sound first. If the tonal frequency is higher than 1250 Hz, there may be an ambiguity as to which ear received the leading sound (see Hartmann, Chap. 2).

A large literature (e.g., see Blauert 1997 for a review) on sound source lateralization has shown that high-frequency envelope ITDs can affect the perceived lateral position when sounds are presented over headphones (see Owrutsky, Benichoux, and Tollin, Chap. 5; Stecker, Bernstein, and Brown, Chap. 6). However, papers by Macaulay et al. (2017) and Yost (2017b) suggest that envelope ITD cues might not be used for sound source localization in a sound field.

ILDs in a sound field are mainly based on reflection and diffraction of sound off and around the head. The threshold for discriminating a change in ILDs is approximately 1 dB (e.g., Middlebrooks and Green 1991; Goupell and Stakhovskaya 2018). Although sound source distance, azimuth angle, and sound frequency affect ILD magnitude, ILDs of more than 1 dB are typically measured for higher frequencies and are very common above 1500 Hz for even small azimuthal angles (Kuhn 1987; see Hartmann, Chap. 2, Fig. 2.4; Owrutsky, Benichoux, and Tollin, Chap. 5). Thus, it is assumed that ILD cues are useable for sound source localization at frequencies higher than approximately 1500 Hz (see Hartmann et al. 2016 for a discussion of the frequency region where sound source localization accuracy is poor due to a poor resolution of ITDs and ILDs). Because ILDs depend on sound frequency and sound source azimuth, the relationship between the magnitude of the ILD and the azimuth is often not monotonic as it is with ITDs. Above approximately 1000 Hz, because of constructive interference, the greatest ILD does not occur at a 90° azimuth (i.e., a sound source directly opposite either of the two ears). Rather, the greatest ILD is between approximately 45°and 70°, depending on frequency; this is called the "bright spot" (Kuhn 1987; Macaulay et al. 2010).

3.4.2 Sound Source Localization in Elevation

Although Wallach (1939, 1940) described how head motion (rotation and tilt) could be used to estimate sound source elevation (see Sect. 3.6.4), the role of head motion in the perception of sound source elevation has not been studied in much detail. Today, it is known that spectral features of complex sounds caused by the filtering of sound as it travels from a sound source and passes over the torso, head, and pinna (head-related transfer function [HRTF]) on its path to the ear canal can provide cues that help determine the perceived elevation of sound sources (see reviews by Middlebrooks and Green 1991; Blauert 1997). These HRTF spectral features at each ear depend on the position of the sound source relative to the ears. Monaural HRTF spectral features aid elevation estimations only if the listeners have prior information about the spectrum of the sound (e.g., Wightman and Kistler 1997). An exact description of the HRTF spectral features responsible for elevation judgments has not been agreed on at this time (Blauert 2013), partly because HRTFs vary considerably across listeners and different listeners may use different spectral features based on the uniqueness of their HRTF (the "individualized HRTF" challenge; e.g., Wenzel et al. 1993, but see Middlebrooks 1992). That is, listeners seem to localize sound sources with the highest accuracy when they are provided HRTF information based on their own pinnae. There is, however, evidence that when listeners move their heads, they might be able to locate sound sources based on HRTF features other than those produced by their own pinnae (Hendrickx et al. 2017).

3.4.3 *Cones of Confusion*

Wallach (1938) calculated cones of confusion, as did Woodworth (1938; see also Mills 1972; Aaronson and Hartmann 2014). Although FBRs based on ITDs occur in the azimuth plane, ILDs might not always produce FBRs due to the constructive interference of the bright spot. That is, ILDs may not be the same for sources at θ and at $180° - \theta$ in that one angle may be in the bright spot and another not. The literature (see Blauert 1997 for a review) suggests that HRTF cues can help resolve FBRs. HRTF spectral features may provide cues as to the azimuthal and, especially, the elevation angle of a sound source. In addition, HRTF spectral features may indicate in which hemifield a sound source is located, whereas ITD and, perhaps, ILD cues determine the sound source location within the hemifield (Blauert 1997). Such HRTF features are most prominent for broadband, high-frequency (greater than approximately 3000 Hz) sounds. Head motion has also been shown to resolve FBRs (e.g., Wallach 1940; Wightman and Kistler 1999; McAnally and Martin 2014). The role of head motion in sound source localization will be described in Sect. 3.5.

Nonauditory factors can also affect FBRs. Wightman and Kistler (1999) showed that when the experimenter rotated a sound source, the rotation had no effect on FBRs. However, if the listener rotated the sound source, FBRs were often substantially reduced, leading Wightman and Kistler (1999) to conclude, "It appears that if the listener is aware of the direction of movement, the necessary information can be extracted." By necessary information, they were referring to information about sound source location. Thus, expectation regarding where a sound source might be appears to provide additional opportunities for reducing FBRs. Finally, when considering multisensory opportunities for influencing FBRs, Montagne and Zhou (2018) showed that providing a frontal visual cue increased FBRs for 15-ms noise bursts coming from the rear hemifield, indicating that forms of visual capture can influence FBRs.

3.4.4 *Sound Source Localization in Distance/Range*

Due to the inverse-square law, sound intensity decreases in proportion to an increase in distance between a sound source and a listener. Thus, if a listener has prior knowledge of the expected level of a sound from a source, then the source of a softer sound is likely to be perceived as further away than that of a louder sound.

A sound source that is less than approximately 1 m, depending on frequency, from the listener is in the near field. Within this range, the sound does not travel as a plane wave and its nonlinear propagation produces low-frequency ILD changes that might indicate distance (Brungart et al. 1999). For sound sources that are further away (i.e., at a distance equal to at least the equivalent of two wavelengths) from a listener, the sound source is in the far field. Several cues have been implicated for judging distance based on far-field conditions (Kolarik et al. 2016). Sounds propagating over very far distances are low-pass filtered by the atmosphere, providing possible spectral cues for relative long-distance sound source estimations. In reverberant environments, the ratio of the levels of the direct sound and associated reflections (the direct-to-reverberant ratio; see Zahorik, Chap. 9) can also provide a cue for judging relative distance. The greater the ratio, the closer the sound source is likely to be (i.e., the direct sound level will dominate). Exact models and/or physiological descriptions of distance perception based on a direct-to-reflected sound level ratio have yet to be proposed partly because direct-to-reverberant ratios depend on the specific space one is in, and so using them to infer the distance of a sound source probably requires some knowledge of the acoustics of that space. The Doppler effect (Middlebrooks and Green 1991), in which the motion of a sound source toward or away from a listener, produces a change in the pitch/timbre of a sound that can also be used to determine if a sound source is traveling toward or away from a listener.

The auditory motion parallax may provide a cue for discerning relative sound source distance. The motion parallax is a powerful visual cue used to judge relative distance (e.g., Steinman and Garzia 2000). Genzel et al. (2018) presented evidence that an auditory motion parallax could assist in determining the relative distance of sound sources. They used a virtual panning process to present two sounds (a low- and a high-frequency sound), each at a different panned (phantom) distance. With no rotation of the sound sources and listeners, the listeners could not determine if one sound source was further away than the other because all known distance cues were eliminated. However, when the listeners moved, they were much better at discriminating the differences in panned distances than when the sound source moved and the listener was stationary. That is, the listeners could use the fact that the near panned source appeared to move faster than the far one as the head moved to infer that one sound source was closer than the other. This is consistent with the visual analogy in which the motion parallax assists in relative distance judgments.

Although there is considerable evidence supporting the auditory-spatial cues cited in Sect. 3.4 for judging sound source location in any one dimension (especially azimuth and elevation), there is very little information about the weighting and integration of the various cues when an actual sound source is located in the three-dimensional world, especially if sound sources and/or listeners move.

3.5 Head-Position Cues in Sound Source Localization

In contrast to the richly documented literature of almost 150 years on auditory-spatial cues, much less is known about head-position cues as they might relate to localizing sound sources relative to the surrounding environment, as suggested by

Wallach (1940). The possible head-position cues for sound source localization have been largely derived from the visual literature (see any textbook on visual perception; e.g., see Goldstein 2013 for a summary) because the visual system must cope with some of the same challenges as the auditory system in localizing objects when the object and/or the observer moves (e.g., determining whether an object and/or the observer moved when the visual-spatial cues change). Retinal rods and cones act as spatial filters, but because the eye constantly moves, the spatial output of the rods and cones would provide a very blurry view of the world without some compensating mechanisms. A major means of spatial retinal compensation is via vestibular feedback systems. The vestibulo-ocular reflex (e.g., Straka and Dieringer 2004), in combination with the optokinetic reflex (e.g., Mustari and Ono 2010), provides one means for the retinal output to enable the visual object location to be corrected (compensated) for retinal movement, leading to a stabilized perception of stationary visual objects.

The head can move independently of the eyes; thus, in the absence of stabilization through other mechanisms, head motion would produce a blurred retinal output. Again, there are compensatory mechanisms involving head-position cues provided by the vestibular and proprioceptive systems. The term "proprioception" is used for the sense of the relative position of the body parts and the strength of effort being employed in movement. Neuro-motor circuits control the position of body parts. Kinesthesia has also been used to describe the movement senses, but it has been used inconsistently in the literature, so the term will not be used in this chapter.

In the visual system, proprioceptive compensation involves efferent-copy (efference-copy or corollary-discharge) feedback mechanisms (e.g., Teuber 1960; Bridgeman and Stark 1991; see Goldstein 2013 for a general description). The general idea is that when a neuromotor signal that controls head position is generated, an efferent copy is also made. This efferent-copy signal is then integrated with the processed retinal spatial signal to yield a stable perception of the world (see Freeman et al. 2017 for an explanation involving sound source localization). For example, if there is a stationary light source and the head moves, the processed retinal output would change, signaling a possible change in the position of the light source (that did not change position). The neuromotor control signal that causes a head movement also generates an efferent-copy signal about that head motion, and this signal effectively "cancels" the processed retinal-change signal, yielding an estimate of the fixed (stable) visual object. In the visual literature, there are several well-established examples of such efferent cancellation based on head movements (e.g., Bridgeman and Stark 1991; see Goldstein 2013 for a review).

Based on the work in vision, Wallach (1940) assumed that three systems provide head-position information for integration with the binaural cues: "proprioceptive stimulation from the muscles engaged in active motion, stimulation of the eyes, and stimulation of the vestibular apparatus." Yost et al. (2015) indicated that cues in addition to the three suggested by Wallach (1940) might also provide head-position cues: somatosensory cues, additional auditory cues, and cues based on established spatial maps. At any moment in time, any or all of these multisystem cues could indicate that the head moved in a particular direction at a relative velocity (accelerated, decelerated, or constant velocity) and/or where the head is located in each of the three spatial planes.

At time t, the angle of a rotating object [$\theta(t)$], can be determined from information about the rotation of the object (Yost et al. 2015, 2019) and the time at which an angle is to be determined

$$\theta(t) = \theta_0 + \omega_0 t + \frac{\alpha}{2} t^2 \tag{3.2}$$

where θ_0 is the initial angle, ω_0 is an initial angular rotation velocity, and α is angular rotation acceleration. Thus, at a given time, the head-position angle could be computed from rotation kinetics based on acceleration, initial constant velocity, and the initial angle of the listener. It is not clear if the head-position angle is required to integrate the head position with auditory-spatial cues or if information about just the direction and/or rate of head movement is all that is required. For example, Freeman et al. (2017) suggest that rotation velocity is required, whereas Yost et al. (2015) suggest that head rotation velocity and direction may not be sufficient. Some of what is known about these head-position cues as they relate to sound source localization is summarized.

3.5.1 *Visual Head-Position Cues*

There is little doubt that vision provides a major, heavily weighted estimate of head position (e.g., Wallach 1940). Vision provides information about the angle of the head at each moment in time and, therefore, the direction and relative rate of rotation of the source. Of course, for visual cues to indicate head position in terms of providing the required information for sound source localization, stationary visual objects must be perceived as stable. This requires the use of the other processes involved in head and/or eye motion, and stable visual perception described in Sect. 3.5.

3.5.2 *Vestibular Head-Position Cues*

Another head-position cue is provided by the vestibular system (e.g., Britton and Arshad 2019). In most experiments, listeners rotate in the azimuth plane so that the vestibular cue is the angular acceleration or deceleration of the head, triggering semicircular canal output that provides head-rotation cues (Lackner and DiZio 2004; Britton and Arshad 2019). The semicircular canals act as accelerometers so there is no change in semicircular canal vestibular outputs when the head is stationary or rotating at constant velocity. In most situations, listeners rotate themselves. Thus, the head rotates in an accelerating and then decelerating manner and is rarely, if ever, at a constant velocity. When a rotating chair or platform is used for self-rotation, the rotations also usually involve head acceleration and deceleration, especially for rotations over small arcs. Yost et al. (2015, 2019) performed what appear to be the only studies in which sound source localization judgments were made when listeners rotated in an accelerating, decelerating, and constant velocity manner.

The vestibular system, when the head angularly accelerates or decelerates, provides information that the head is rotating, the direction of rotation, and the relative velocity of rotation, but the vestibular system itself does not provide an estimate of head angle. As Eq. 3.2 indicates, head angle, using vestibular information about angular acceleration, could be computed if the time over which the rotation had occurred and the starting head location were known. This would require memory. This idea has not been explored in terms of establishing head-position cues for sound source localization.

3.5.3 Proprioceptive Head-Position Cues

Another head-position cue suggested in several sound source localization studies (Wallach 1940; Brimijoin and Akeroyd 2012; Genzel et al. 2016) is related to proprioception or neuro-motor control of head rotation. When listeners rotate themselves, neuro-motor control signals initiate and control the rotation. These neuro-motor control signals could indicate head angle. And, as mentioned in Sects. 3.5 and 3.5.3, it is possible that neuro-motor control signals also trigger efferent-copy signals that could be used to "cancel" auditory-spatial signals that might otherwise indicate that a sound source had moved when, in fact, it was the head that moved (Freeman et al. 2017). Although there is no direct physiological evidence for such efferent-copy processes in the mammalian auditory system, several authors (e.g., Genzel et al. 2016; Freeman et al. 2017) have found indirect evidence for such processes for sound source localization.

In addition, it is possible that when listeners are rotated by some external means and must keep their heads still, the resistance to the rotation would stimulate muscles (e.g., neck muscles) that could trigger neural signals as a means of indicating head rotation, but it is not clear how such resistance information would inform head-position angle.

3.5.4 Other Possible Head-Position Cues

Although Yost et al. (2015) suggested that somatosensory cues, auditory cues from sound sources not associated with the source to be localized, and listener experience might provide head-position information, these processes appear not to have been examined. Some investigators have argued that the contributions of head-position cues may be weighted differently in determining head position (e.g., Brimijoin and

Akeroyd 2012; Genzel et al. 2016; Freeman et al. 2017). This weighting may change based on the sound source localization scenario (e.g., Genzel et al. 2016).

There is an intriguing literature related to head-direction (HD) cells found in several brain areas of rodents and a few other animals and are probably part of the spatial-map system involving the hippocampus and entorhinal cortex (e.g., Taube 2007). HD cells record the orientation of an animal's head, even in the dark. Such HD cells could provide valuable information about head position, although there appears to be no research involving HD cells in the context of sound source localization.

3.6 Sound Source Localization as the Integration of Head-Position and Auditory-Spatial Cues

Spatial location is a relative measure so it is crucial to note which spatial reference system when specifying sound source location. A "world-centric" reference system (also called "allocentric" or "room-centric") refers to sound source location relative to the actual environment (e.g., room). For example, noting which loudspeaker presented a sound is specifying a world-centric location. A "head-centric" reference system (also called "ego-centric" or "cranial-centric") specifies sound source location relative to the head (i.e., relative to an interaural axis going from one ear to the other or to an axis perpendicular to the interaural axis). The head-centric location is determined by auditory-spatial cues. Whereas in vision there may be an "eye-centric" reference system because the eyes can move independently of the head, the perceived direction of a sound source appears in most situations to be unaffected by shifts in gaze position through eye movement (Warren 1970; Vliegen 2004).

Wallach (1940) appears to be the first to infer that binaural cues cannot, by themselves, indicate the world-centric location of a sound source. In his study, he used listener and sound source rotation to investigate the integration of the binaural cues (head-centric cues) and head-position cues to account for world-centric sound source localization perceptions. The issue of such cue integration has received very little mention in the literature.

Figures 3.1 and 3.2 display scenarios concerning sound source localization in the azimuth plane. All angles shown in Figs. 3.1 and 3.2 are expressed relative to the frontal midline position in a world-centric coordinate system. The nomenclature described by (Yost et al. 2019) and provided in the legend to Fig. 3.1 is used to define the various angles.

3.6.1 *Head Rotation Resolves Front-Back Reversals*

Figure 3.1a displays the basic relationship among the five angles and how the world-centric sound source angle might be determined by the sum of the world-centric head-position angle and the head-centric sound source angle. Comparing Fig. 3.1b to Fig. 3.1a illustrates how head rotation could resolve FBRs.

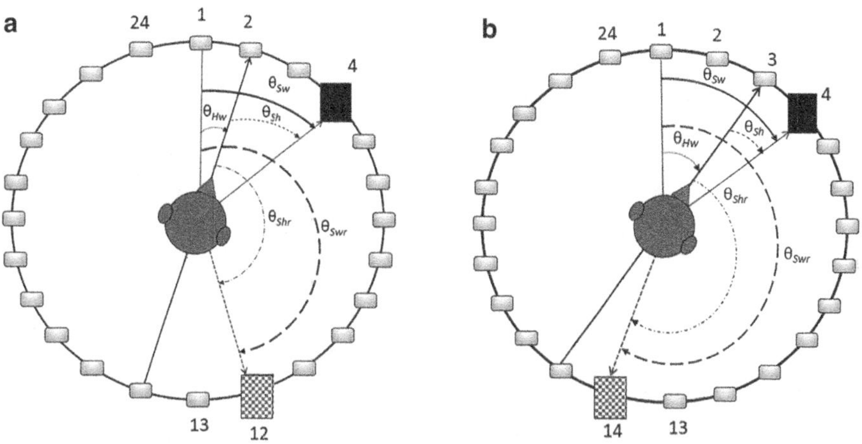

Fig. 3.1 Definitions of angles: θ, azimuth angle; θ', estimated angle. The first subscript, in capitals, is the object of the measured angle: *S*, sound source; *H*, the head. The second subscript, in lowercase, is the coordinate system for the measurement: *w*, world-centric; *h*, head-centric. If there is a third subscript, it is for a front-back reverse location (*r*). For example, $(\theta')_{Swr}$ indicates the estimated angle of the sound source (*S*) in world-centric coordinates (*w*) at a front-back reversed angle (*r*). A subscript (0) is used to indicate an initial condition (i.e., before a head turn and/or sound source location change). For example, $\theta(_{Hw0})$ indicates the initial world-centric angle of the head before a head rotation occurred. (**a**) listener facing loudspeaker 2 at the world-centric head-position angle (θ_{Hw}) and the actual sound presented from loudspeaker 4 at the world-centric angle (θ_{Sw}), resulting in the sound source head-centric angle (θ_{Sh}). The listener has direct information about, θ_{Sh} and θ_{Hw} but not θ_{Sw}. θ'_{Sw} could be estimated as the sum of the estimates of θ'_{Sh} and θ'_{Hw}. The reverse world-centric sound source angle (θ_{Swr}), occurring with a reverse head-centric angle (θ_{Shr}), generates the same interaural differences as a sound source with angle θ_{Sh} (i.e., sound at loudspeaker 12). (**b**) the head rotates from loudspeaker 2–3 with sound remaining at loudspeaker 4, generating a reverse world-centric sound source location at loudspeaker 14

In Fig. 3.1a, the listener faces loudspeaker 2 while the sound is presented from loudspeaker 4. In Fig. 3.1b, the listener rotates to facing loudspeaker 3 while the sound is still presented from loudspeaker 4. The position of a reverse sound source location is shown to be at loudspeaker 12 in Fig. 3.1a and loudspeaker 14 in Fig. 3.1b. The estimated head-centric angle of the sound source (θ'_{Sh}) can be obtained from the auditory-spatial cues, and the world-centric head-position angle (θ'_{Hw}) can be estimated from multisystem head-position cues. However, there is no direct estimate of the world-centric angle of the sound source (θ'_{Sw}; i.e., the angle of the source-producing sound). Yost et al. (2015) suggested that addition of the estimated head-centric sound source and world-centric head-position angles could estimate the world-centric sound source angle (see also Mills 1972) at some moment in time (*t*)

$$\theta'_{Sw}(t) = \theta'_{Hw}(t) + \theta'_{Sh}(t). \tag{3.3}$$

There are other ways (e.g., Mills 1972; Braasch et al. 2015) to integrate the head-centric and head-position angles to determine the world-centric sound source angle

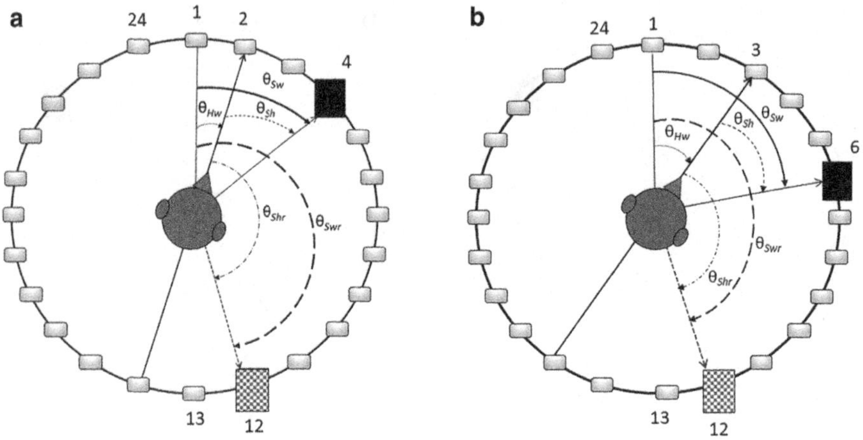

Fig. 3.2 Similar format as in Fig. 3.1, but when the head and sound rotations are in the 2:1 rotation relationship leading to the Wallach azimuth illusion (WAI). (**a**) Same as Fig. 3.1a. (**b**) head rotates from loudspeaker 2–3 and the sound rotates from loudspeaker 4–6 (the 2:1 rotation scenario). The world-centric location of the actual sound source changes to loudspeaker 6. The location of the reversed sound source remains at loudspeaker 12

besides addition, but the necessity of the integration of the two estimated angles was one of Wallach's (1940) main insights.

The reverse world-centric sound source angle in Fig. 3.1, A and B, is computed as

$$\theta'_{Swr}(t) = \theta'_{Hw}(t) + \theta'_{Shr} = \theta'_{Hw}(t) + \left[180^{\circ} - \theta'_{Sh}(t)\right] \tag{3.4}$$

Thus, in Fig. 3.1, the sound source could be at a location associated with the head-centric sound source angle, θ_{Sh} (i.e., at loudspeaker 4), or at the location associated with the reverse head-centric sound source angle, $180^{\circ} - \theta_{Sh}$, (i.e., at loudspeaker 12 or 14) because both angles (θ_{Sh} and $180^{\circ} - \theta_{Sh}$) are the result of the same interaural difference.

Assume that a head rotation is some increment ($\Delta\theta_{Hw}$) of the angle $\theta_{Hw}(t)$ where

$$\theta_{Hw}(t) = \theta_{Hw_0} + \Delta\theta_{Hw} \tag{3.5}$$

The increment might be derived from Eq. 3.2: $\Delta\theta_{Hw} = \omega_{0\,Hw}t + \frac{\alpha Hw}{2}t^2$.

From Fig. 3.1, the estimated head-centric sound source angle after head rotation would be

$$\theta'_{Sh}(t) = \theta'_{Sh_0} - \Delta\theta_{Hw} \tag{3.6}$$

and the estimated reverse head-centric sound source angle would be

$$\theta'_{\mathrm{Shr}}(t) = 180^{\circ} - \theta'_{Sh}(t) = 180^{\circ} - \theta'_{Sh_0} + \Delta\theta_{Hw} \tag{3.7}$$

The actual head-centric sound source angle changes in the opposite direction as the direction of head rotation, whereas the reverse head-centric sound source angle changes in the same direction. Experience could lead one to assume that the actual sound source angle is that which causes the auditory spatial cues to rotate in the opposite direction as the head rotated, thus avoiding a FBR. Then, this head-centric sound source angle would have to be integrated with the world-centric head-position angle to determine an estimate of the world-centric sound source angle.

The other possibility for how head rotation might resolve FBRs is that the "decision" as to which angle represents the actual world-centric sound source location is based on the computation of the two world-centric angles (the actual angle and the reverse angle). The estimated actual world-centric sound source angle, given the head-centric angle shown in Eq. 3.6, is

$$\theta'_{Sw}(t) = \theta'_{Hw_0} + \Delta\theta'_{Hw} + \theta'_{Sh_0} - \Delta\theta'_{Hw} = \theta'_{Hw_0} + \theta'_{Sh_0} = \theta'_{Sw_0} \tag{3.8}$$

and the estimated reverse world-centric sound source angle, given the reverse head-centric angle shown in Eq. 3.7, is

$$\begin{aligned} \theta'_{Swr}(t) &= \theta'_{Hw_0} + \Delta\theta'_{Hw} + \left[180^{\circ} - \left(\theta'_{Sh_0} - \Delta\theta'_{Hw}\right)\right] \\ &= \theta'_{Hw_0} + \left(180^{\circ} - \theta'_{Sh_0}\right) + 2\Delta\theta'_{Hw} \end{aligned} \tag{3.9}$$

Thus, the actual world-centric sound source angle does not change with head rotation, but the reverse world-centric sound source angle changes by twice the head-rotation angle, in the same direction as the head turn (compare Fig. 3.1a with Fig. 3.1b). Wallach (1940) noted that listeners nearly always perceive a stationary sound source as stationary during and after head movements. Wallach wondered why would the listener choose the stationary estimate of sound source location (θ'_{Sw}) over the moving estimate (θ'_{Swr})? Wallach developed the "selective principle of rest" to answer his question, "Of all the directions which realize the given sequence of lateral angles, that one is perceived which is covariant with the general content of the surrounding space." In other words, if objects around the listener are generally not moving, then the listener will choose the estimate that does not change as a result of head movements. This approach gives the listener a strategy for distinguishing his/her own movements from the movement of sound sources.

Although Wallach (1940) is often cited as the person who first showed how head motion could resolve FBRs, he did not test this directly. As a result, there is not a clear resolution as to which set of cues (head- or world-centric) are used to resolve FBRs. Wallach was interested in how elevation could be determined given up-down reversals (see Sect. 3.6.4). He noted that there was symmetry in the changes in the head-centric and world-centric sound source angles when the head rotated and the sound source did not (Fig. 3.1) compared with a scenario in which the sound source rotated at twice the rate of head rotation (Fig. 3.2). This 2:1 rotation scenario became the main condition Wallach (1939, 1940) used to explore the integration of dynamic auditory-spatial and head-position cues to account for world-centric sound source localization despite cones of confusion. As Sect. 3.6.2. explains, the 2:1 rotation scenario can result in an illusionary perception of a stationary sound source in the azimuth plane, referred to as the Wallach azimuth illusion (WAI; Brimijoin and Akeroyd 2012).

3.6.2 *The 2:1 Rotation Scenario and the Wallach Azimuth Illusion*

Figure 3.2a and b, displays the 2:1 rotation in the azimuth plane. Figure 3.2a is the same as Fig. 3.1a but is shown to facilitate comparisons between Fig. 3.2a and b as well as between Figs. 3.1 and 3.2. Figure 3.2b indicates a rotation of the head (from facing loudspeaker 2 in Fig. 3.2a to facing loudspeaker 3 in Fig. 3.2b), and the 2:1 rotation of the sound source is from loudspeaker 4 in Fig. 3.2a to loudspeaker 6 in Fig. 3.2b.

The 2:1 rotation scenario yields the estimate of the head-centric sound source angle of the actual sound source as

$$\theta'_{Sh}(t) = \theta'_{Sh_0} + \Delta\theta_{Hw} \tag{3.10}$$

and the head-centric sound source angle of the reverse sound source as

$$\theta'_{Shr}(t) = 180^\circ - \left[\theta'_{Sh_0} + \Delta\theta_{Hw}\right] = 180^\circ - \theta'_{Sh_0} - \Delta\theta_{Hw} \tag{3.11}$$

Note that the head-centric sound source angle is in the same direction of head rotation, whereas the reverse head-centric sound source angle is in the opposite direction. This is opposite to what occurred when the head rotated and the sound source remained stationary as in Fig. 3.1 and Eqs. 3.7 and 3.8. Summing the head-centric angle (Eq. 3.3) with the head-position angle (Eq. 3.10) yields the estimated world-centric sound source angle of the actual sound source as

$$\begin{aligned}\theta'_{Sw}(t) &= \theta'_{Hw}(t) + \theta'_{Sh}(t) = \left[\theta'_{Hw_0} + \Delta\theta_{Hw}\right] + \left[\theta'_{Sh_o} + \Delta\theta_{Hw}\right] \\ &= \left[\theta'_{Hw_0} + \theta'_{Sh_0}\right] + \left[2\Delta\theta_{Hw}\right] = \theta'_{Sw_0} + 2\Delta\theta_{Hw}\end{aligned} \tag{3.12}$$

Likewise, summing Eqs. 3.4 and 3.11 yields the estimated world-centric sound source angle of the reverse sound source as

$$\theta'_{Swr}(t) = \theta'_{Hw}(t) + \theta'_{Shr}(t) = \left[\theta'_{Hw_0} + \Delta\theta_{Hw}\right] + \left[180^\circ - \theta'_{Sh_0} - \Delta\theta_{Hw}\right]$$
$$= \left[\theta'_{Hw_0} + 180^\circ - \theta'_{Sh_0}\right] = \theta'_{Swr_0} \quad (3.13)$$

There are several aspects of the world-centric sound source locations shown in Fig. 3.2 and the angles in Eqs. 3.12 and 3.13. First, the world-centric sound source angle changes for the actual sound source (Eq. 3.12) in the 2:1 rotation scenario but does not for the reverse world-centric sound source angle (Eq. 3.13). Second, this is the opposite of the result shown in Fig. 3.1 and Eqs. 3.7 and 3.8 when the listener rotated but the sound source did not. Third, in the 2:1 rotation scenario, the angle of the stationary reverse world-centric sound source is that of the reverse sound source angle that exists at the start of rotation. Fourth, if the selective principle of rest is employed, the perceived world-centric sound source location in the 2:1 rotation scenario would be that of a stationary sound source at the reverse location when rotation began, even though the actual sound source is rotating. These aspects of the perception of sound sources prone to FBRs in the 2:1 rotation scenario define the WAI (Brimijoin and Akeroyd 2012; Yost et al. 2019).

In cases where sounds are not prone to FBRs, the perceived world-centric location of the sound source in the 2:1 rotation scenario would likely be $\theta'_{Sw}(t)$ (Eq. 3.12) because there is no $\theta_{Swr}(t)$. As such, the perceived world-centric sound source location would be veridical, rotating twice as fast as the listener and in the same direction. Thus, in addition to the four aspects that define the WAI cited above, listeners should perceive rotating sound sources for sounds not prone to FBRs.

3.6.3 Studies of the Wallach Azimuth Illusion

Wallach (1940) tested five listeners on the WAI with musical sounds produced by a Victrola record. Listeners were manually rotated quickly in about the same way in a chair over an arc from approximately 30° left to 30° right, back and forth several times, with their eyes closed and head fixed in a head holder. Wallach (1939, 1940) used a device that connected head motion to which loudspeaker presented a sound (see Blauert 1997 for a schematic description of the apparatus). For the WAI conditions, a head rotation of x degrees resulted in a $2x$ degree change in which loudspeaker presented the music, thus generating the 2:1 rotation scenario. When sound source and listener rotations started, the loudspeaker presenting music was either directly in front of or behind the listener.

In the 2:1 rotation WAI condition when the musical sound source started in front, Wallach (1940) argued that listeners should perceive a stationary sound source at the world-centric location behind the listener. Likewise, in the back condition, listeners should perceive a stationary sound source in the front position. All five listeners reported these perceptions with no exceptions. Because the listeners' eyes were closed and the body or head was passively rotated, Wallach (1940) assumed that the vestibular system indicated head position.

Wallach (1940) also used the illusionary perception of self-rotation vection to investigate the WAI. In self-rotation vection (e.g., Britton and Arshad 2019), an observer views only a surrounding screen with clear markings rotating around their head in one direction. After several seconds, the rotating screen is perceived to rotate in the opposite direction. Then, a few seconds later, observers experience the sensation that they are rotating in a direction opposite to the original screen rotation and the actual rotating screen is perceived as stationary. That is, the rotating screen induces the illusionary perception of self-motion at a rate equal to the screen rotation but in a direction opposite to the actual screen rotation. When Wallach (1940) presented his musical sounds rotating around the azimuth loudspeaker array at twice the rate of the induced self-rotation vection and in the same direction as the illusionary rotation, listeners experienced the WAI. It appears as if actual head rotation is not required for one to experience the WAI. This implies that only a sense of where the head might be is needed for head-position information to be integrated with auditory-spatial cues for world-centric sound source localization.

It is important to note that Wallach (1940) was unaware of the use of spectral (HRTF) features in sound source localization (i.e., these features were not revealed until several decades after Wallach's work). He did recognize that the pinnae might play a role in sound source localization but thought this would occur only in unusual, rare cases. Thus, he argued that the pinnae were a "secondary factor" for sound source localization and that his calculations applied to all stimuli. Wallach (1940) articulated that all sounds would be affected by head motion through his "principle of least displacement." That is, the smallest relative angular change that a perceived sound source may undergo when the head moves is the probable basis for determining perceived world-centric sound source location for any sound.

The WAI was not studied again until Macpherson (2011) tested how dynamic cues affected FBRs for noise stimuli with different high- and low-frequency content. He was particularly interested whether Wallach's (1940) principle of least displacement applies across different sounds. Macpherson (2011) used virtual sound source locations produced by HRTFs to produce spatially perceived virtual sound sources over headphones, and he provided data for one listener. One of the conditions he tested was like the WAI with the 2:1 rotation of the listener and virtual sound sources. The listener perceived a stationary sound source in back when it was originally presented in front, as Wallach (1940) had reported, but only when the stimulus was a low-pass noise (0.5–1 kHz), not when it was wideband (0.5–16 kHz) or narrowband, high-frequency (6.0–6.5 kHz) and especially when this stimulus was short (less than approximately 75 ms). Based on the illusion being frequency dependent, Macpherson (2011) suggested that low-frequency ITDs, but not high-frequency ILDs, may be a basis for generating the WAI. Another possibility is that because listeners tend to localize 4–6-kHz narrowband stimuli to the front, regardless of stimulus location (e.g., so-called directional bands; Middlebrooks 1992; Blauert 1997), FBRs may not have occurred for this listener when the stimuli were presented from the front. Without FBRs, the WAI cannot exist.

Brimijoin and Akeroyd (2012) investigated the WAI as listeners moved their heads back and forth between ±15° of midline. A camera system recorded head motion, and the system's output controlled the amplitude panning of the sound so that the "phantom" location of the sound rotated at twice the listener's head angle, thereby generating the 2:1 rotation scenario. A speech signal was low-pass filtered in octave steps from 0.5 to 16 kHz across different blocks of the experiment. The listeners started their head rotations while facing the forward, centered loudspeaker. The speech sound was either initially presented from the front or back center loudspeaker. As a control condition, a sound was presented from a stationary source, either in front or in back, for each filter condition. Figure 3.3 shows the proportion of "front" responses as a function of the cutoff frequency of the filter. When the sound contained only very low frequencies (therefore FBRs were highly likely), the front sound was perceived in back and the back sound in front, entirely consistent with the WAI. As the filter cutoff was increased, allowing for more high-frequency content, the percent of "front" judgments tends toward an asymptote at 50% (e.g., perhaps chance; Yost et al. 2019). That is, the WAI "weakens" as the stimulus contains more high-frequency content (fewer FBRs). Judgments of the position of the stationary front and back stimuli were not affected by the filter conditions. Brimijoin and Akeroyd (2012) did not provide a clear explanation as to why the proportion of front responses reached an asymptote at 50% (Yost et al. 2019).

Following the work of Wallach (1940), Macpherson (2011), and Brimijoin and Akeroyd (2012), the WAI appeared to be a robust phenomenon that might be useful

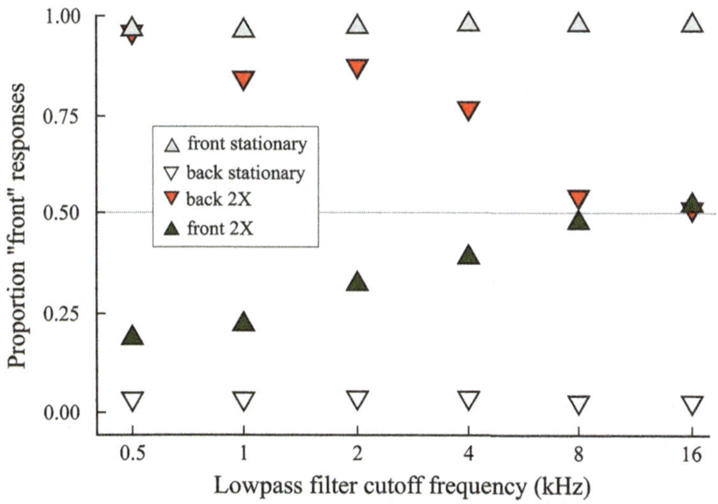

Fig. 3.3 A sound was panned from loudspeaker to loudspeaker at twice the rate the listeners rotated themselves. The starting location of the sound was either directly in front (up; *green triangles*) or directly in back (down; *red triangles*) of the listener, and the listeners indicated if the perceived position of the sound was in front as they rotated. Data are mean proportions of front responses as a function of the low-pass cutoff of the filtered-speech signal (seven listeners). A control condition included a front stationary sound source (up; *gray triangles*) or a back stationary sound source (down; *open triangles*) as the listeners rotated. Adapted from Brimijoin and Akeroyd (2012)

for further exploration of the interaction of head-position and auditory-spatial cues. However, the previous literature suggested that several aspects of the illusion needed to be more fully explained. Yost et al. (2019) conducted a study of the WAI using arcs and durations of rotation much larger and longer than had previously been used. This allowed for a more thorough investigation of the full range of possible responses in the 2:1 rotation scenario. In Experiment 1, they measured FBRs for five 75-ms filtered-noise bursts: two with a wideband high-frequency spectral content that yielded a low proportion of FBRs (HF1: 2–8 kHz; HF2: 0.125–8 kHz); two with only a low-frequency spectral content that yielded large proportions of FBRs (LF1: 0.125–0.5 kHz; LF2: 0.244–0.263 kHz); and a narrowband, high-frequency filtered noise that yielded an intermediate proportion of FBRs (MF: 3.9–4.2 kHz). Then, in Experiment 2, using the same listeners and filtered noises as were used in Experiment 1, they tested a 2:1 rotation scenario where listener and sound source rotation started as acceleration (listener 5°/s^2; sound 10°/s^2), then led to a constant-velocity period of rotation (listener 60°/s, sound 120°/s) and ended with rotation deceleration (listener −5°/s^2; sound −10°/s^2). Both visual and vestibular information about head motion were available during the acceleration/deceleration stages of rotation, but only visual information would have been available during the constant velocity stage (see Sect. 3.5.2). Each rotation stage lasted 12 s, although the noise bursts were presented only for the last 6 s of acceleration, the middle 6 s of constant velocity, and the first 6 s of deceleration to ensure that the rate of listener rotation was not near zero and to allow time for listener responses. For each stage of rotation and for each filtered noise, there were four possible starting head-centric sound source locations. At the end of each of the three stages of rotation, listeners were asked to indicate if the preceding repeating 75-ms filtered-noise bursts appeared to be stationary and, if so, at which of 24 possible azimuth loudspeaker locations (see Fig. 3.1 for the loudspeaker layout). If the listeners did not perceive the repeating filtered-noise bursts to be stationary, they were to indicate if the repeating noise bursts were perceived as rotating around the azimuth array in a clockwise or counterclockwise direction.

Figure 3.4 shows the results averaged over the three stages of rotation and four starting head-centric sound source locations because these two variables did not have a significant influence on listeners' WAI and rotation responses. The results plotted in the positive direction (Fig. 3.4, *solid bars*) represent the proportion of stationary WAI responses (i.e., stationary responses at or near the reverse loudspeaker location when the sound started). The results plotted in the negative direction (Fig. 3.4, *horizontal-hashed bars*) represent the proportion of clockwise rotation responses. All listeners in Experiment 2 reported almost 100% WAI responses when the filtered noises produced a large proportion of FBRs in Experiment 1 (the green LF1 and LF2 bars), and many listeners had a large proportion of clockwise-rotating responses for filtered noises producing a low proportion of FBRs (red HF1 and HF2 bars). For most listeners, the proportion of responses when the filtered noise produced an intermediate proportion of FBRs (MF blue bars) was between that obtained for the low-frequency and high-frequency filtered noises. Almost all of the responses were either WAI responses or clockwise-rotating

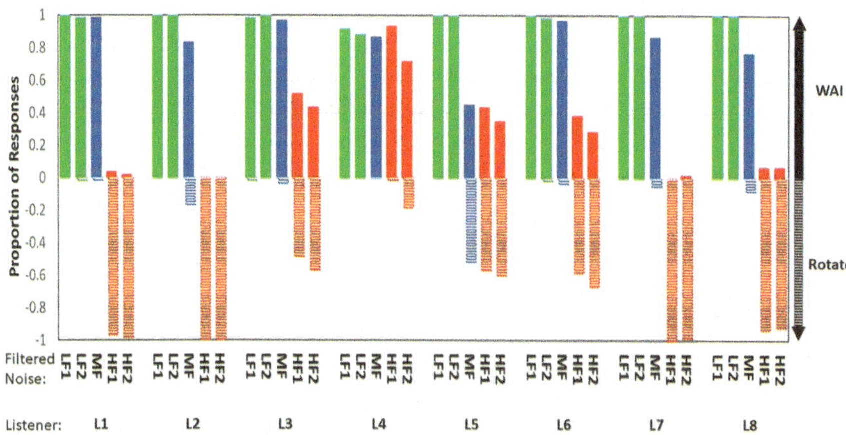

Fig. 3.4 The proportion of WAI and rotation responses averaged over the four starting head-centric loudspeaker locations and the three stages of rotation for each of the five filtered-noise conditions (LF1, LF2, MF, HF1, and HF2) for each of the eight listeners. *Solid bars*, WAI responses; *horizontal hashed bars*, clockwise rotating responses; *green bars*, low-frequency filtered-noise conditions with a high proportion of front-back reversals (FBRs) measured in Experiment 1 (LF1 and LF2); *red bars*, filtered-noise conditions with broadband high-frequency content generating a low proportion of FBRs (HF1 and HF2); *blue bars*, narrowband high-frequency filtered-noise condition with an intermediate proportion of FBRs (MF). Adapted from Yost et al. (2019)

responses. If listeners did not indicate a rotating sound source, they almost always indicated it was at or near the predicted WAI location. There was a strong positive correlation (0.81) between the listeners' proportion of WAI stationary responses measured in Experiment 2 and their proportion of FBRs measured in Experiment 1 across all listeners and filtered noises, indicating a strong relationship between the occurrence of FBRs and the existence of the WAI.

Thus, when listeners have ample time to judge if a sound is rotating or stationary and if rotation is in the 2:1 rotation scenario, listeners almost always perceive a rotating sound that is prone to FBRs as being stationary, at a location based on the reverse sound source location of the sound when sound rotation begins (i.e., the WAI). When a sound is not prone to FBRs, listeners usually indicate that the sound source is rotating and only occasionally indicate that the WAI occurred. The WAI in the Yost et al. (2019) study appeared to be based entirely on visual input about head position because changes in vestibular input (stage of rotation) did not interact with the visual input to affect the listeners' responses. The WAI does appear to be a robust phenomenon that could allow for further examination of the integration of auditory-spatial and head-position cues in sound source localization.

Although these experiments reveal that the WAI mostly occurs for sounds prone to FBRs, Wallach (1940) suggested, based on his principle of least displacement, that the illusion applied to all sounds. It is possible (as Yost et al. 2019 suggested) that Wallach's (1940) findings were a result of using 1940s Victrola records, record player, and loudspeakers that probably produced low-frequency filtered sounds that would be prone to FBRs. Today, it is known that sounds with high frequencies and wide bandwidths are not prone to FBRs. Thus, Wallach's (1940) principle of least displacement probably does not apply to all sounds as he proposed.

3.6.4 Wallach Elevation Illusion

Wallach used the 2:1 scenario to motivate his arguments for the role of head motion in sound source elevation judgments, which was the main focus of his three papers (1938, 1939, 1940). Wallach's description of how head motion might yield elevation estimates has received very little attention in the literature, probably because of the large literature showing the influence of HRTF features (particularly those from pinna filtering) on sound source elevation judgments (see Middlebrooks and Green 1991; Blauert 1997, 2013 for reviews). Because head motion might play a role in sound source elevation judgments when HRTF features are not available and because there is not a full explanation of what HRTF features provide cues for sound source elevation judgments, Wallach's (1940) arguments concerning head rotation and elevation judgments seem worth considering.

Wallach (1940) developed his elevation calculations based on cones of confusion; Eq. 3.1 indicates that sound source rotation through a lateral angle (ψ) will rotate around different azimuth planes (θ) with different angular velocities (Mills 1972; Pastore et al. 2020). Binaural disparities for a sound source directly above a listener will not change as the listener rotates their head. At the other extreme, binaural disparities change by the full amount of the head turn for a sound source on the δ° elevation azimuth plane.

Wallach (1940) did not have listeners judge the location of sound sources placed at different elevations. Rather, he used arguments similar to those used for the WAI to show how azimuthal head rotation coupled to azimuthal rotating sound sources (both on the horizontal plane as for the WAI) could lead to the illusory perception of elevated sound sources. In this chapter, these elevation illusions are called the Wallach elevation illusion (WEI).

In his main WEI experiment, Wallach (1940) tried to simulate a sound source at an elevation of 60° above the horizontal plane by rotating the listener and the sound source at different rates. According to a simplification of Eq. 3.1 (Wallach 1940; Pastore et al. 2020), if a sound stimulus that is likely to produce FBRs is rotated azimuthally at 1.5 times the rate of head rotation, a listener should perceive the world-centric location of a stationary azimuthally reversed sound source at an elevation of +60° above the horizontal plane because that spatial auditory estimate remains unchanged with head rotation (i.e., the selective principle of rest). Five of 15 listeners reported the sound source at an elevation near +60° when they were blindfolded. The rest of the listeners reported the elevation to be less (sometimes much less) than +60°. All but three listeners indicated a sound source elevation near or somewhat less than +60° when the blindfold was removed. Note that because

Wallach's (1940) calculations only indicate a change in elevation relative to the horizontal plane and not the direction (above or below), a response of a sound source elevation below the horizontal plane near −60° would also be consistent with his calculations. Wallach (1940) made two somewhat inconsistent assumptions about these calculations. First, he assumed that listeners are naturally biased by experience to perceive sound sources above rather than below them. He also argued that the underestimation of sound source elevation could be due to the fact that the calculated locations indicate locations below the listener and that this influenced listeners, leading to underestimations.

It appears as if only Perrett and Noble (1997) have attempted to replicate Wallach's (1940) WEI findings. They found that the WEI appeared to exist only for low-frequency sounds, and even then, the correspondence between the predicted sound source elevation and the judged elevations was weak. Thus, head rotation is probably not used to determine elevation for sources of sounds for which there are not cues associated with HRTF features.

Zhong et al. (2016) used the WEI concept to show that machine learning algorithms (e.g., Kalman filters) could use simulated head-rotation cues to determine the location of three different sound sources located in different azimuthal and vertical locations. Thus, there might be some utility in using head rotation to determine sound source elevation.

3.6.5 Other Studies of the Integration of Head-Position and Auditory-Spatial Cues in Sound Source Localization

Brimijoin and Akeroyd (2014) studied the moving minimum audible angle (MMAA), i.e., the minimum perceivable angle between two sound sources when both sound sources rotate relative to a listener's head. The MMAA is different from the minimal audible movement angle (MAMA), which is the smallest perceivable angle between a moving sound source and a stationary sound source required to detect if a sound source was moving (Chandler and Grantham 1992). Brimijoin and Akeroyd (2014) showed that the MMAA was smaller when the listeners' heads rotated and the sound sources were stationary in world-centric coordinates than when the listeners' heads were stationary and the sound sources rotated. Thus, multisystem information about self-motion assisted in localizing moving sound sources.

Pettorossi et al. (2005) rotated listeners in an oscillating manner on a platform while they were blindfolded. Their listeners used a laser pointer to indicate sound source location. As the listeners were blindfolded and passively rotated, the authors assumed that the cues indicating listener position were provided only by the vestibular system. The pointer responses were consistent with the listeners integrating information about their head location with the head-centric location of the sound source to indicate the world-centric location of the sound sources.

Genzel et al. (2016) used listener rotation and sound source localization measures to study "spatial updating" (i.e., the "updating" of the perceived world-centric location of a sound source based on integration of head-position and auditory-spatial cues). This concept is consistent with the use of updating in the vision literature (Klier and Angelaki 2008; Medendorp 2011). Three conditions were tested by Genzel et al. (2016): listeners moved their heads (active condition), were rotated on a platform (passive condition), or moved their heads in one direction and rotated in the other direction (counter-rotation condition) such that head-centric auditory cues essentially did not change as the head rotated. Listeners performed a two-alternative, forced-choice sound source localization task, determining whether a test sound source was to the left or right of a preceding reference sound source in world-centric coordinates. The results by Genzel et al. (2016) showed that the percentage of "right" judgments of the second sound relative to the first sound depended on the way the listeners moved in the active condition, in the passive condition, and especially in the counter-rotation condition. Genzel et al. (2016) assumed that the three means of listener rotation involved different weighting of head-motion cues for determining head position and interpreted their data as indicating that both proprioceptive/efferent-copy and vestibular cues play a role in determining head position, but vestibular cues are weighted more heavily. Although there are several relatively untested assumptions underlying these interpretations, the data by Genzel et al. (2016) clearly support the basic idea that sound source localization depends on the integration of head-motion and auditory-spatial cues, that the two cues might be linearly combined, and that vestibular function and proprioception/efferent copy are possible indicators of head position.

Freeman et al. (2017) observed that listeners with their eyes closed and head rotating could perceive a rotating world-centric sound source as rotating in the same or opposite direction of head rotation or not moving at all depending on the relative rates of head and sound rotation (Yost et al. 2015). The listeners were trained to move their heads back and forth (±30° arc) at different rates. For each rate, the listeners were asked to determine if a sound source rotating back and forth was perceived as moving in the same or opposite direction of their head rotation. The sound rotation was a version of panning, and the rate of sound rotation was proportional to the head-rotation rate and referred to as movement gain ("g"; the proportion of listener-rotation rate that equals the head-rotation rate). A psychometric function indicating the proportion of times the listeners reported that sound source rotation was in the direction of head rotation was determined as a function of g (where g less than 0 indicates a perceived sound source rotating in the direction opposite to head rotation). The experiments also included measurements when the sound source was at different azimuth eccentricities as it was panned. The point of subjective equality (PSE) on psychometric functions was assumed to equal the g required for the listeners to perceive the sound as not moving (compensation). The PSE occurred for small and positive g values and did not change with eccentricity. That is, as in vision, the sound source had to rotate about 15% of head rotation speed (g equals 0.15) for listeners to perceive a stationary sound source. This result is consistent with an integration of head-motion and auditory-spatial cues similar to the way head-motion and visual cues interact (called the Filehne illusion in vision; Mack and Herman

1973). The study underscores that the perceptual error (the Filehne illusion) exists in relating sound source rotation to head rotation, suggesting that a simple summation of head-position and auditory-spatial cues may not be entirely appropriate.

When sound stimuli are emitted from a visible sound source, both auditory and visual cues exist for identifying sound source location. Because the eyes and the head can move independently of each other in the direction of a visual source, an object in space can have an eye-centric or a head-centric location. There is a growing literature concerning the interaction of eye movements (visual gaze), head movements, and auditory-spatial cues affecting sound source localization (for a thorough review, see Van Opstal 2016). Also, because the eyes tend to move in the direction of a sound source, eye movement can be used to measure perceived sound source location. As a result, how are head and eye movements coordinated when a listener "looks at" the location of a sound source? Several studies (see Van Opstal 2016 for a review) have used the "double-step" paradigm to investigate spatial updating of eye movements and head movements in determining visual and auditory object location. The double-step paradigm can involve a stationary head or a moving head. For simplicity, assume a head-fixed scenario in which the observer first uses saccadic eye movements to orient the eyes toward a light source located in two dimensions (left-right, up-down). A second stimulus is then presented later at a different location and observers must move their eyes to the new location. The stimulus for the second location can be a light or a sound source. To successfully locate the second stimulus, there must be some form of updating of eye position after the eye movement to the first stimulus. That is, the reference position changed after the first light was presented and to find the position of the second source requires information about this first position. Several experiments indicate that observers are successful in locating the second-stimulus location (updating) if the second stimulus is either a light or a sound source. A displacement feedback/efferent copy model fits the data and has physiological data supporting it. Several of these studies have shown that such oculomotor efferent-copy feedback could also explain how the eyes locate an auditory stimulus. That is, there could be oculomotor and auditory-spatial cue integration. Again, these eye-movement studies support a multisystem interaction for sound source location, in this case when eye position indicates sound source location.

3.7 Studies of Sound Source Localization When Head-Position or Auditory-Spatial Cues May Be Degraded

3.7.1 Auditory-Spatial Cue Degradation

Brimijoin and Akeroyd (2016) investigated the WAI for listeners with hearing loss, including those who use hearing aids. Listeners with hearing impairment, with and without hearing aids, participated in the same paradigm used by Brimijoin and Akeroyd (2012; see Sect. 3.6.3; Fig. 3.3). Hearing-impaired listeners were less accurate at discriminating front from back and were less affected by high-frequency content than normal-hearing listeners. There was a large listener variability when the listeners wore hearing aids, but the average data suggested a spectrally dependent increase in front responses; the more high-frequency energy in the signal, the more likely listeners were to report its source as in front. Brimijoin and Akeroyd (2012) concluded that "hearing impairment was associated with a decrease in the accuracy of self-motion processing for both static and moving signals. Hearing aids may not always reproduce dynamic self-motion cues with sufficient fidelity to allow reliable front/back discrimination."

Pastore et al. (2018) measured sound source localization in bilateral cochlear-implant (CI) listeners using sound processors with behind-the-ear microphones. They were asked to localize high-frequency filtered-noise bursts that were 3 s long. In the first experiment, the CI and normal-hearing control listeners were instructed not to move their heads. In the second experiment, they were told to move their heads and were instructed in how to do so (e.g., move only the head and not the body). With the head stationary, normal-hearing listeners had fewer than 5% FBRs for the high-frequency noise burst, but the CI listeners had over 40% FBRs, most likely because with the microphone of the sound processor behind the ear, the pinna is unable to provide HRTF features that can be used as cues for sound source localization (Majdak et al. 2011). However, when the CI listeners rotated their heads over the 3 s, they substantially reduced the proportion of FBRs to just under 7%. Thus, head rotation reduces FBRs for some bilateral CI listeners who have a high proportion of FBRs when the head is fixed due to a lack of HRTF spectral cues.

Measures of stability are used to diagnose vestibular disorders, and vision has a clear impact on stability. If a patient is asked to stand or walk in particularly difficult scenarios, they can do so if their eyes are open, but they sway (or even fall) if their eyes are closed. The degree and kind of sway provide diagnostic information about vestibular function (Fukuda 1959; McCaslin et al. 2008). It is assumed that vision provides reference locations that help maintain stability. Zhong and Yost (2012) showed that if listeners localized a sound source and their eyes were closed, they tended to sway less than when the eyes were closed and no sound source was present, suggesting that localizable sound sources can be used as spatial references similar to visual references. Such auditory-spatial references would not work to reduce sway if the perceived location of the sound source perceptually moved as the listener swayed (i.e., based on changing head-centric auditory-spatial cues). Listeners appeared to perceive a stationary sound source, likely based on integration of auditory-spatial and head-position cues.

3.7.2 Head-Position Cue Degradation

Several studies cited previously (e.g., Wallach 1939, 1940; Freeman et al. 2017) have investigated sound source localization when listeners were blindfolded, usually to eliminate vision for providing head-position information. If listeners rotate themselves (e.g., Brimijoin and Akeroyd 2012, 2016) and are blindfolded, then the proprioceptive system could provide head-position cues. If listeners are rotated in a chair and are blindfolded, it is often assumed that neither vision nor proprioception provide head-position cues (Wallach 1939, 1940; Pettorossi et al. 2005; Genzel et al. 2016; Freeman et al. 2017); however, vestibular function could still indicate head position. Thus, all of these studies assume that even when some head-position cue information is degraded, other systems provide head-position information. As Wallach (1940) and now others have argued (Yost et al. 2015), this could mean that listeners localize sound sources in world-centric coordinates when some but not all head-position cues are degraded. Providing different forms of head-position information allowed several authors to estimate the relative weight these cues might provide for world-centric sound source localization (e.g., Genzel et al. 2016; Freeman et al. 2017).

Yost et al.'s (2015) investigation appears to be the first study that attempted to deny listeners access to any and all information about head position. The authors argued that if all head-position cues are denied (or severely degraded), then listeners would be able to determine a sound source location only in head-centric coordinates. In this study, naive listeners were rotated (no proprioceptive cues) at slow rates (no opportunity for muscular resistance to rotation or somatosensory cues from something like wind motion), at a constant velocity (no vestibular cues), and with their eyes closed (no visual cues) and were provided no feedback (no cognitive cues). Several rotation scenarios were used to show that listeners made sound source location judgments in world-centric coordinates when they had access to head-position cues via vision, but judgments were made in head-centric coordinates when head-position cues (vision) were not available. For instance, when head-position cues were not available, a broadband sound source that rotated at the same rate and direction as the listener was perceived at fixed, head-centric locations (consistent with Eqs. 3.6 and 3.7) when they were denied information about head position.

Similar outcomes occurred when the sounds were presented in an accelerating and decelerating manner compared with the constant-velocity condition. However, the responses in the eyes-closed condition, when the sound rotation was accelerating or decelerating, were not as different from the eyes-open condition as they were in the constant-velocity condition. Yost et al. (2015) suggested that some aspects of the experiment (e.g., listeners were moving so slowly at the start of acceleration and end of deceleration that rotation may not have been sensed) might have influenced the eyes-closed data during acceleration and deceleration, so caution is required in interpreting data comparing acceleration/deceleration to constant velocity rotation.

It has been assumed that if listeners are able to determine a sound source location in a world-centric coordinate system, then some form of head-position information must have been available. Conversely, if sound source location judgments are made in only a head-centric reference system, then it has been assumed that all forms of head-position information have been denied or significantly degraded (Yost et al. 2015). One difficulty with these assumptions is that listeners can make both head-centric and world-centric sound source localization judgments when head-position cues are readily available. Everyday experience suggests that listeners almost always make world-centric sound source location judgments when they can. However, this appears to have not been studied. Thus, the specific question asked of listeners may influence the type of sound source location judgment they make. If a listener is asked what loudspeaker presented a sound, the response is likely to represent a world-centric judgment that would require information about head position. If the question is vague, such as "where is the sound coming from," then an answer such as "from my right" might be hard to interpret. For instance, is this a head-centric response (from the right of my head), a world-centric response (from the loudspeaker on my right), or some form of both types of response? Furthermore, a head-centric-type response does not mean that the listener is not capable of making a world-centric-type response (i.e., has the requisite head-position information). To obtain good information about what head-position cues are integrated with auditory-spatial cues and how this integration occurs, it will be necessary to carefully design experiments that allow for clear interpretation of listeners' responses concerning the coordinate system their responses represent.

3.8 Additional Observations Concerning the Integration of Auditory-Spatial and Head-Position Cues

It seems logical that an evolving species would survive much better if it could determine the world-centric location of sound sources rather than only their head-centric locations. If so, then it might be that the spatial auditory system evolved to take advantage of any head-position information it could to allow for world-centric sound source localization. Thus, one might have to be careful in making assumptions about the extent to which a particular experimental condition may or may not controlled for information about one or more head-position cue. For instance, Wallach (1940) assumed that because he rotated blindfolded listeners and they perceived the WAI, the vestibular system was providing head-position information. Wallach did not directly manipulate vestibular cues. It is possible that Wallach's brief repeated rotation of the chair over a short arc and a brief period of time did not fully eliminate his experienced listeners' information about the position of their heads as they were rotated blindfolded.

This chapter has referred to spatial maps (e.g., Leutgeb et al. 2005; LaChance et al. 2019) as a basis for providing head-position information. Spatial maps are based on memory. Attention of one sort or another is used to "find" locations in a spatial map that are required for some sort of behavioral outcome (e.g., where is an object or where am I?). Using a spatial map effectively for sound source localization might require some neural form of accurate estimates of elapsed time (i.e., time between events). Consider the example above about the possibility that Wallach's (1940) experienced listeners might have known where they were in the room even though they were blindfolded. This could have occurred because the listeners had developed spatial maps of the room. They then remembered where they were in the room when rotation started (e.g., the initial angle in Eq. 3.2). Being able to measure elapsed time over a short time period combined with vestibular information about the relative rate and direction of rotation might allow for the estimation head-position angle via Eq. 3.2.

Although it seems clear that auditory-spatial and head-position cues are integrated for world-centric sound source localization, it is not well understood how the various cues are weighted and integrated (see Pastore et al. 2020 for a computational framework for studying the weights and integration). There might be a benefit in studying existing models of multisensory cue integration (e.g., Stein and Meredith 1993; Ernst and Bulthoff 2004). A large amount of work has been devoted to the study of visual capture ("ventriloquism effect") in which there can be an integration of visual and auditory-spatial cues in determining the perceived location of a frontal source that generates both sound and light (or reflected light) at about the same time (see Montange and Zhou 2018 for a review). Growing evidence suggests that visual and auditory-spatial information can be combined nearly optimally to form a fused percept of source location (Alais and Burr 2004). However, these multisensory models of cue integration might not be directly applicable to world-centric sound source localization perceptions because head and head-centric sound source positions provide partial estimations of the world-centric sound source position variable as opposed to two independent estimations of the same position variable as in visual capture. On the other hand, models used in the visual capture literature (e.g., Alais and Burr 2004) could be useful in suggesting ways to integrate head-position and head-centric variables.

This chapter has dealt primarily with location judgments along the azimuth plane. Presumably, similar computations would be made for elevation and distance and then these computations would be combined to account for three-dimensional sound source localization. Alternatively, the head-centric location of a sound source in three-dimensional space could be computed and then integrated with a representation of head position to form a three-dimensional prediction of world-centric sound source location. Wigderson et al. (2016) provided physiological evidence that the output of the cochlear nucleus (primarily the dorsal cochlear nucleus) generates neural information about the integration of spectral auditory-spatial and head-position cues. This work supports the idea that spatial-spectral information is first integrated with head-position information before a three-dimensional determination of world-centric sound source location is formed, which occurs higher somewhere in the ascending auditory pathway. Clearly, additional research is required to better understand not only how the various cues are weighted and integrated but also what neural processes perform these tasks. These processes are probably adaptable, as is the process of auditory sound source localization itself.

3.9 Summary

Wallach's (1938, 1939, 1940) observations that sound source localization is a multisystem process that involves an integration of auditory-spatial and head-position cues has received very little attention until recently. This chapter described some of the history of the study of sound source localization as a multisystem process, and it reviewed investigations of the multisystem processes that integrate auditory-spatial and head-position cues for humans to localize sound sources in the real world.

Acknowledgments We were supported by grants to William A. Yost and M. Torben Pastore from the National Institute of Deafness and Other Communication Diseases, National Institutes of Health, and from the Facebook Reality Laboratories; and to Yi Zhou from the National Science Foundation. We appreciate the assistance of Kathryn Pulling and Christopher Montagne.

Compliance with Ethics Requirements

William A. Yost declares that he has no conflict of interest.

M. Torben Pastore declares that he has no conflict of interest.

Yi Zhou declares that she has no conflict of interest.

References

Aaronson NL, Hartmann WM (2014) Testing, correcting, and extending the Woodworth model for interaural time difference. J Acoust Soc Am 135:817–823

Adrian L (1928) The basis of sensation: the action of the sense organs. Hafner, New York

Alais D, Burr D (2004) The ventriloquist effect results from near-optimal bimodal integration. Curr Biol 14:257–262

Blauert J (1997) Spatial hearing. The MIT Press, Cambridge, MA

Blauert J (ed) (2013) The technology of binaural listening. Springer-Verlag, Berlin

Boring EG (1929) A history of experimental psychology. Appleton-Century-Crofts, New York. (Reprinted in 1950)

Boring EG (1942) Sensation and perception in the history of experimental psychology. Appleton-Century-Crofts, New York

Boring EG, Gardener L (eds) (1918) A history of psychology in autobiography. Appleton-Century-Crofts, New York

Braasch J, Clapp S, Parks A, Pastore MT (2015) A binaural model that analyses aural spaces and stereophonic reproduction systems by utilizing head movements. In: Blauert J (ed) The technology of binaural listening. Springer-Verlag, Berlin-Heidelberg, pp 201–224

Bridgeman B, Stark L (1991) Ocular proprioception and efference copy in registering visual direction. Vis Res 31:1903–1913

Brimijoin WO, Akeroyd MA (2012) The role of head movements and signal spectrum in auditory front/back illusion. Iperception 3:179–181

Brimijoin WO, Akeroyd MA (2014) The moving minimum audible angle is smaller during self-motion than during source motion. Front Neurosci 8:1–8

Brimijoin WO, Akeroyd MA (2016) The effects of hearing impairment, age, and hearing aids on the use of self motion for determining front/back location. J Am Acad Audiol 27:588–600

Britton Z, Arshad Q (2019) Vestibular and mulitsensory influences upon self-motion perception and the consequences for human behavior. Front Neurosci 7:1–10

Brown AD, Stecker GC, Tollin DJ (2014) The precedence effect in sound localization. J Assoc Res Otolaryngol 13:1–28

Brungart DS, Durlach NI, Rabinowitz WM (1999) Auditory localization of nearby sources. II. Localization of a broadband source. J Acoust Soc Am 106:1956–1968

Chandler DW, Grantham DW (1992) Minimum audible movement angle in the horizontal plane as a function of stimulus frequency, bandwidth, source azimuth, and velocity. J Acoust Soc Am 91:1624–1632

Crane T, Patterson S (2002) History of the mind-body problem. Taylor & Francis, New York
de Boer K, Van Urk AT (1941) Some particulars of directional hearing. Philips Tech Rev 6:359–364
Ernst MO, Bülthoff HH (2004) Merging the senses into a robust percept. Trends Cogn Sci 8:162–169
Fechner G (1860) Elements of psychophysics. Holt, Rinehart, and Winston Inc, New York. (Reprinted in 1942)
Freeman TCA, Culling JF, Akeroyd MA, Brimijoin WO (2017) Auditory compensation for head rotation is incomplete. J Exp Psychol Hum Percept Perform 43:371–380
Fukuda T (1959) The stepping test: two phases of the labyrinthine reflex. Acta Otolaryngol 50:95–108
Genzel D, Firzla U, Wiegrebe L, MacNeilage PR (2016) Dependence of auditory spatial updating on vestibular, proprioceptive, and efference copy signals. J Neurophysiol 116:767–775
Genzel D, Schutte M, Brimijoin WO et al (2018) Psychophysical evidence for auditory motion parallax. Proc Natl Acad Sci 115:4264–4269
Goldstein B (2013) Sensation and perception, 9th edn. Wadsworth CENAGE Learning, Belmont
Goupell MJ, Stakhovskaya OA (2018) Across-channel interaural-level-difference processing demonstrates frequency dependence. J Acoust Soc Am 143:645–658
Hartmann WM, Rakerd B, Crawford ZD, Zhang PX (2016) Transaural experiments and a revised duplex theory for the localization of low-frequency tones. J Acoust Soc Am 139:968
Helmholtz H (1863) Die lehre von den tonempfindungen als physiologische grundlage fur de theorie der musik. Longmans, Green, and Co., London
Hendrickx E, Stitt P, Messonnier JC et al (2017) Influence of head tracking on the externalization of speech stimuli for non-individualized binaural synthesis. J Acoust Soc Am 141:2011–2023
Klemm O (1920) Über den Einfluss des binauralen zeitunterschiedes aud de localisation. Arch Ges Psychol 40:123–129
Klier EM, Angelaki DE (2008) Spatial updating and the maintenance of visual constancy. Neuroscience 156:801–818
Kohlraush A, Altosaar T (2011) Early research on spatial hearing by Alvar Wilska (1911-1987). Proceedings of forum acusticum, European acoustics association Alborg, Denmark: 1103–1108
Kolarik AJ, Moore BCJ, Zahorik P, Cirstea S, Pardhan S (2016) Auditory distance perception in humans: a review of cues, development, neuronal bases, and effects of sensory loss. Atten Percept Psychophys 78:373–395
Kuhn GF (1987) Physical acoustics and measurements pertaining to directional hearing. In: Yost WA, Gourevitch G (eds) Directional hearing. Proceedings in life sciences. Springer, New York, pp 3–25
LaChance PA, Todd TA, Taube JS (2019) A sense of space in postrihnal cortex. Science 365:141
Lackner JR, DiZio P (2004) Vestibular, proprioceptive, and haptic contributions to spatial orientation. Annu Rev Psychol 56:115–147
Leutgeb S, Leutgeb JK, Moser MB, Moser EI (2005) Place cells, spatial maps and the population code for memory. Curr Opin Neurobiol 15:738–746
Litovsky RY (2012) Spatial release from masking. Acoust Today 8(2):18–25
Litovsky RY, Colburn HS, Yost WA, Guzman SJ (1999) The precedence effect. J Acoust Soc Am 106:1633–1654
Macaulay EJ, Hartmann WM, Rakerd B (2010) The acoustical bright spot and mislocalization of tones by human listeners. J Acoust Soc Am 127:1440–1449
Macaulay EJ, Rakerd B, Andrews TJ, Hartmann WM (2017) On the localization of high-frequency, sinusoidally amplitude-modulated tones in free field. J Acoust Soc Am 141:847–863
Mack A, Herman E (1973) Position constancy during pursuit eye movement: an investigation for the Filehene illlusion. Quart J Exp Psych 25:71–84
MacPherson EA (2011) Head motion, spectral cues, and Wallach's 'principle of least displacement' in sound localization. In: Suzuki Y, Brungart D, Kato H (eds) Principles and applications of spatial hearing. World Scientific, Singapore, pp 103–120
Majdak P, Goupell MJ, Laback B (2011) Two-dimensional localization of virtual sound sources in cochlear-implant listeners. Ear Hear 32(2):198–208

McAnally KI, Martin RL (2014) Sound localization with head movement: implications for 3-d audio displays. Front Neurosci 8:1–6

McCaslin DL, Dundas JA, Jacobson GP (2008) The bedside assessment of the vestibular system, 2nd edn. Plural Publishing, San Diego

Medendorp WP (2011) Spatial constancy mechanisms in motor control. Philos Trans R Soc B Biol Sci 366:476–491

Middlebrooks JC (1992) Narrow-band sound localization related to external ear acoustics. J Acoust Soc Am 92:2607–2624

Middlebrooks JC, Green DM (1991) Sound localization by human listeners. Annu Rev Psychol 42:135–159

Middlebrooks JC, Simon JZ, Popper AN, Fay RR (eds) (2017) The auditory system at the cocktail party. ASA Press/Springer-Verlag, New York

Mills AW (1972) Auditory localization. In: Tobias J (ed) Foundations of modern auditory theory (II). Academic Press, New York

Montagne C, Zhou Y (2018) Audiovisual interactions in front and rear space. Front Psychol 9:1–15

Mustari MJ, Ono S (2010) Optokinetic eye movements. Encycl Neurosci 10:285–293

Pastore MT, Natale SJ, Yost WA, Dorman MF (2018) Head movements allow listeners bilaterally implanted with cochlear implants to resolve front-back confusions. Ear Hear 39:1224–1231

Pastore MT, Yost WA, Zhou Y (2020) Cross-modal and cognitive processes in sound localization. In: Blauert J (ed) The technology of binaural understanding. Springer, Berlin

Perrett S, Noble W (1997) The contribution of head motion cues to localization of low-pass noise. Percept Psychophys 59:1018–1026

Pettorossi VE, Brosch M, Panichi R, Botti F, Grassi S, Troiani D (2005) Contribution of self-motion perception to acoustic target localization. Acta Otolaryngol 125:524–528

Pierce AH (1901) Studies in auditory and visual space perception. Longmans, Green, and Co, New York

Popper AN, Fay RR (eds) (2005) Sound source localization. Springer-Verlag, New York

Rayleigh L (1876) On our perception of the direction of a source of sound. Proced Music Assoc 2:75–84

Rayleigh L (1907) On our perception of sound direction. Philos Mag Ser 6(13):214–232

Stein BE, Meredith MA (1993) The merging of the senses. MIT press, Cambridge, MA

Steinhauser A (1879) The theory of binaural audition. A contribution to the theory of sound. Phil Mag 7:42–56

Steinman SB, Garzia RP (2000) Foundations of binocular vision: a clinical perspective. McGraw-Hill Professional, New York/London

Stevens SS, Newman EB (1936) The localization of actual sources of sound. Am J Psychol 48:297–306

Straka H, Dieringer N (2004) Basic organization principles of the VOR: lessons from frogs. Prog Neurobiol 73:259–309

Taube JS (2007) The head direction signal: origins and sensory-motor integration. Annu Rev Neurosci 30:181–207

Teuber HL (1960) Handbook of physiology. American Physiological Society, Washington D.C

Thompson SP (1878) On binaural audition. Phil Mag 2:383–391

Van Opstal AJ (2016) The auditory system and human sound-localization behavior. Academic, Amsterdam

Vliegen J (2004) Dynamic sound localization during rapid eye-head gaze shifts. J Neurosci 24:9291–9302

von Hornbostel EM, Wertheimer M (1920) Über die wahrnehmung der schallrichtung. In: Akademie der wissenschaften, pp 388–396

Wallach H (1938) Über die wahrnehmung der wehallrichtung. Psychol Forsch 22:238–266

Wallach H (1939) On sound localization. J Acoust Soc Am 10:270–274

Wallach H (1940) The role of head movements and vestibular and visual cues in sound localization. J Exp Psychol 27:339–368

Warren DH (1970) Intermodality interactions in spatial localization. Cogn Psychol 1:114–133
Wenzel EM, Arruda M, Kistler DJ, Wightman FL (1993) Localization using non-individualized head-related transfer functions. J Acoust Soc Am 94:111–123
Wigderson E, Nelken I, Yarom Y (2016) Early multisensory intergation of self and source motion in the auditory system. PNAS 113:8308–8313
Wightman FL, Kistler DJ (1997) Monaural sound localization revisited. J Acoust Soc Am 101:1050–1063
Wightman FL, Kistler DJ (1999) Resolution of front-back ambiguity in spatial hearing by listener and source movement. J Acoust Soc Am 105:2841–2853
Woodworth RS (1938) Experimental psychology. Henry Holt and Company, New York
Yost WA (2017a) History of sound source localization: 1850-1950. POMA 30:1–15
Yost WA (2017b) Sound source localization identification accuracy: envelope dependencies. J Acoust Soc Am 142:173–185
Yost WA, Popper AN, Fay RR (eds) (1993) Human psychophysics. Springer-Verlag, New York
Yost WA, Popper AN, Fay RR (eds) (2008) Auditory perception of sound sources. Springer-Verlag, New York
Yost WA, Zhong X, Najam A (2015) Judging sound rotation when listeners and sounds rotate: sound source localization is a multisystem process. J Acoust Soc Am 138:3293–3310
Yost WA, Pastore MT, Pulling KR (2019) Sound-source localization as a multisystem process: the Wallach azimuth illusion. J Acoust Soc Am 146:382–398
Zhong X, Yost WA (2012) Relationship between postural stability and spatial hearing. J Am Acad Audiol 24:782–788
Zhong X, Sun L, Yost WA (2016) Active binaural localization of multiple sound sources. Rob Auton Syst 85:83–92

Warren DH (1970) Intermodality interactions in spatial localization. Cogn Psychol 1:114–133

[illegible]

Chapter 4
Anatomy and Physiology of the Avian Binaural System

Terry Takeshi Takahashi, Lutz Kettler, Clifford Henry Keller II, and Avinash Deep Singh Bala

4.1 Introduction

Binaural hearing is the process by which the signals at the two ears are compared to extract spatial information about the acoustical environment. This is a daunting task because sound waves from multiple sound sources as well as their reflections from nearby surfaces impinge on the two ears. For those with normal hearing, sound localization is seemingly automatic, but it becomes difficult when hearing is compromised in one ear.

The extraction of information from the comparison of two sensory surfaces, such as the basilar membranes, has parallels to other sensory modalities. The most obvious is stereopsis, where the distance relative to the fixation point is computed from positional disparities on the retinae (Julesz 1971). Likewise, weakly electric fish compare the outputs of the electroreceptors on two body patches to localize conspecifics and to adjust the frequencies of their electric organ discharge to minimize interference between its discharge and that of neighboring fish (Heiligenberg 1991). Finally, recent studies suggest that olfaction-guided localization and navigation can be accomplished by comparing the activity at the left and right olfactory epithelia (Catania 2013).

This chapter reviews the structure and function of pathways involved in binaural hearing, with a focus on avian systems and the way in which the neural processes contribute to spatial hearing in both quiet and cluttered environments. Topics such as avian cochlear mechanics, plasticity, and regeneration have been amply addressed elsewhere (e.g., Köppl et al. 2000) and are not covered here. Much of the work

T. T. Takahashi (✉) · C. H. Keller II · A. D. S. Bala
Institute of Neuroscience, University of Oregon, Eugene, OR, USA
e-mail: terry@uoregon.edu; keller@uoneuro.uoregon.edu; avinash@uoregon.edu

L. Kettler
Technical University of Munich (TUM), Freising-Weihenstephan, Germany
e-mail: Lutz.kettler@tum.de

© Springer Nature Switzerland AG 2021

R. Y. Litovsky et al. (eds.), *Binaural Hearing*, Springer Handbook of Auditory Research 73, https://doi.org/10.1007/978-3-030-57100-9_4

described in this chapter are from the common barn owl (*Tyto alba*) and the American barn owl (*Tyto furcata*), which are established model systems in the study of binaural hearing, but studies in mammals and other avian species are also addressed.

4.2 Acoustics of Binaural Hearing

Acoustical waveforms, typically represented as the fluctuation of sound pressure over time, can also be represented in the frequency domain as a spectrum of sound amplitude and phase. The pure tone, an auditory physiologist's favorite tool, has a sinusoidal temporal waveform and, in the frequency domain, has but a single spectral component corresponding to the frequency of the sinusoid in its amplitude and phase spectra. At the opposite extreme is broadband noise. Broadband noise comprises an infinite number of spectral components with roughly equal amplitudes and random phase angles.

Between a listener's eardrum and a sound source are numerous objects that can alter a spectrum of the sound. For purposes of this chapter, the important objects are the subject's head, pinna, and other parts of the face. The way in which a spectrum of a sound is altered or "filtered" by these anatomical features depends on the direction of the source relative to the head. In humans, the shape of the spectrum can be a sound localization cue provided that the listener has an idea of the spectrum of the sound at the source (Wightman and Kistler 1997; Andeol et al. 2013).

As schematized in Fig. 4.1a, the direction-dependent filtering properties of the head are measured using microphones placed near each eardrum while a broadband noise is presented from a speaker from various directions in space. The difference between the signal going into the speakers and that coming out of the microphone represents the filtering properties or the head-related transfer function (HRTF; see Table 4.1 for a list of abbreviations; Wightman and Kistler 1989; Keller et al. 1998b).

An HRTF is obtained for each ear and each location in space at which the speaker was placed. By computing the differences in the amplitude and phase spectra of the left and right ears, one can derive the interaural differences in level (ILD) and phase (IPD) corresponding to a particular location. IPD can be converted into interaural time difference (ITD) by dividing the IPD by frequency.

Figure 4.1b, c, shows the distribution of ITDs and ILDs in a barn owl and Fig. 4.1d, e, show the same for a great horned owl (*Bubo virginianus*) for several frequency bands (3.5, 5.0, 6.5, and 8.0 kHz). For both species, the ITD varies along the azimuth at all frequencies (Fig. 4.1b, d). The other binaural cue, the ILD, is species specific. In the great horned owl, the ILD, like the ITD, varies with the azimuth (Fig. 4.1e). In the barn owl, however, the spatial distribution of the ILD has a strong elevational component (Fig. 4.1c). Loci above eye level have ILDs that favor the right ear (Fig. 4.1b–e, *red*), whereas loci below favor the left ear. The manner in which the ILD varies with elevation in the barn owl is frequency specific and is most marked for frequencies above 3 kHz. The ILD-elevation relationship reflects the

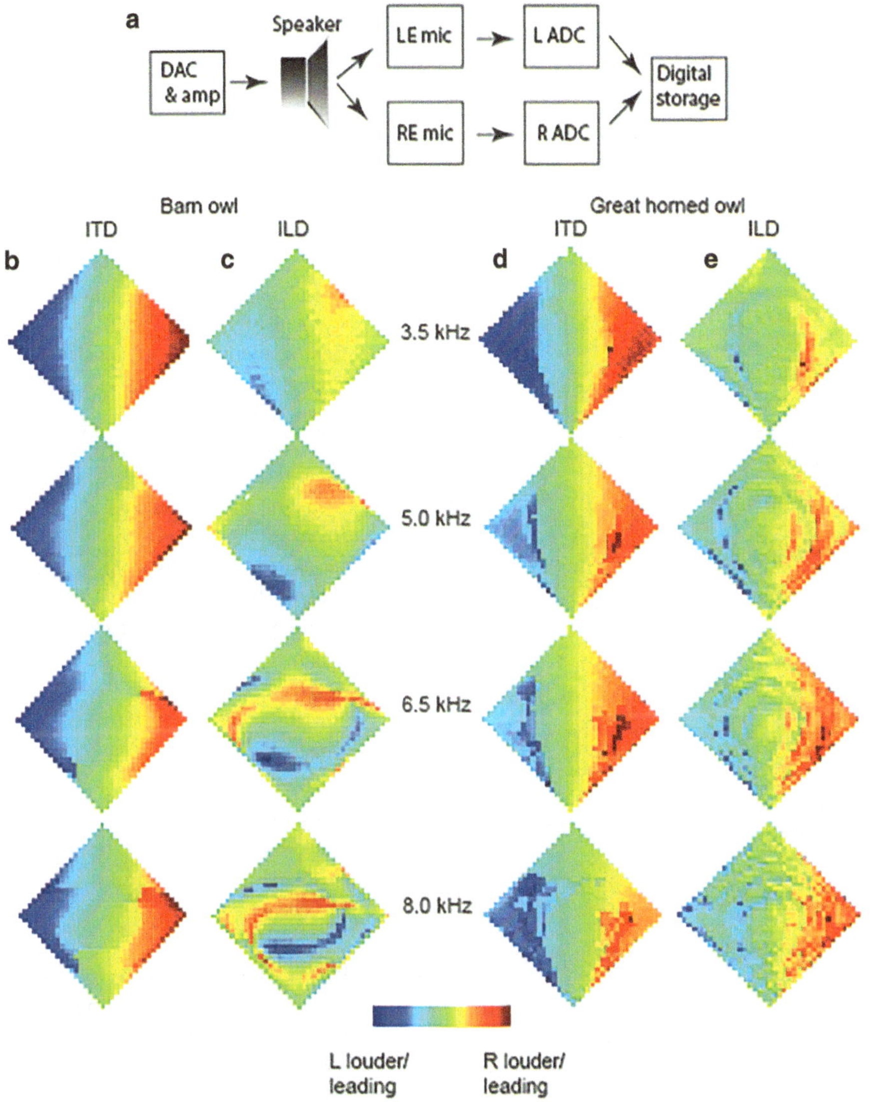

Fig. 4.1 Acoustical filtering by head structures. (**a**) System for measuring head-related transfer functions (HRTFs). A broadband signal is amplified and fed (DAC & amp) to a movable speaker. The sound is received by tiny microphones in placed near the eardrums (LE mic; RE mic), the output of which is digitized (L ADC; R ADC) and stored in computer memory (Digital storage). The difference between the broadband noise presented and the recording of that noise in the ear canals shows the manner in which the spectrum of the input sound is altered by the structures of the face and head. This difference is the HRTF. (**b** and **c**) Spatial distribution of the interaural time difference (ITD; **b**) and interaural level difference (ILD; **c**) in the barn owl and in the great horned owl (**d** and **e**) at four frequency bands (3.5, 5.0, 6.5, and 8.0 kHz). The *left* of the diamond represents the frontal space that is to the left of the owl's midline and *up* is above the owl's eye level. In both species (**b** and **d**), the ITD varies in a horizontal pattern such that sounds from the left and right lead in the left and right ears, respectively (*color bar at bottom*). The ILD, on the other hand, differs in the two species (**c** and **e**). In the barn owl, which has asymmetrical ears, the ILD varies along the vertical dimension, especially at the higher frequencies. In the great horned owl, whose ears are symmetrical, the ILD varies along the horizontal axis

Table 4.1 List of abbreviations

Abbreviation	Definition
AAr	Auditory arcopallium
AM	Amplitude modulation
AMPA	α-Amino-3-hydroxy-5-methyl-4-isoxazolepropionic acid
D	Standard separation (*d′*-like measure)
HRTF	Head-related transfer function
IACUC	Institutional Animal Care and Use Committee
ICc-core	Central inferior collicular nucleus
ICc-ls	Central inferior collicular lateral shell
ILD	Interaural level difference
ILD(*f*)	Interaural level difference for a spectral channel (*f*)
IPD	Interaural phase difference
ITD	Interaural time difference
ITD(*f*)	Interaural time difference for a spectral channel (*f*)
LLDp	Dorsal lateral lemniscal nucleus
MAA	Minimal audible angle
mTeg	Tegmental motor nuclei
NA	Nucleus angularis
NL	Nucleus laminaris
NM	Nucleus magnocellularis
OT	Optic tectum
nOv	Nucleus ovoidalis
PDR	Pupillary dilation response
PE	Precedence effect
RF	Receptive field
RLF	Rate-level function
RP	Relative probability of evoking discharge
SPL	Sound pressure level
VLVp	Nucleus ventralis lemnisci lateralis, pars posterior

vertical asymmetry in the shapes of the left and right ear canals and the position of the preaural flaps (Fig. 4.1a; Payne 1971; Keller et al. 1998a). Thus, in the barn owl, the ITD and ILD serve as cues for different spatial dimensions. In the great horned owl, the two binaural cues encode the same spatial dimension, the azimuth. The latter is reminiscent of human subjects in which Rayleigh's duplex theory (Strutt 1907) holds that the ITD and ILD are azimuthal cues for low and high frequencies, respectively.

The acoustic system of birds presents another special situation. The middle ears of avian and nonavian archosaurs (dinosaurs and crocodilians) are connected by a series of trabeculated, air-filled cavities, allowing the eardrums to act as pressure-gradient receivers (Witmer 1990; Witmer and Ridgely 2008). Theoretically, binaural cues such as the ITD and ILD can be amplified by pressure gradient receivers (Vedurmudi et al. 2016). However, data on the relationship between the internally coupled ears and the coding of binaural cues are lacking so far because, to this day, the barn owl is the most extensively investigated bird, with many model concepts in sound localization based on the owl. However, the barn owl is a remarkable auditory specialist, and its ears are functionally coupled only for low frequencies (up to 3 kHz), which are of less relevance for the behavior of the barn owl (Kettler et al. 2016). Due to the focus on the "owl model," the question of the influence of coupled ears on the coding of the ITD and ILD has so far been neglected. In the less-studied auditory generalists, the frequency span in which the coupled ears amplify the naturally occurring cues covers a large part of their auditory range, thus pressure-gradient receivers may become behaviorally relevant in these species. The neural computation of the ILD and ITD within and between spectral channels is addressed next.

4.3 Neural Coding of Monaural Phase and Amplitude

Before binaural cues can be computed, the frequency-specific amplitudes and phase angles of the signals at each ear must be encoded. Figure 4.2 shows the structures involved in binaural hearing, taken predominantly from work in the barn owl.

In the auditory nerve and in one of the two cochlear nuclei, nucleus angularis (NA), the amplitude is coded by the spike rate of frequency-selective neurons. The rate-level function (RLF), which plots spike rate against sound pressure level (SPL), is typically sigmoidal and helps to define four important parameters: the threshold of the unit, which is the lowest SPL at which firing rates are statistically greater than the rate observed when no sound is applied; saturation, the SPL above which the spike rate does not change; the dynamic range, the range of SPLs over which the spike rate changes with the SPL; and the sensitivity, the slope of the RLF over the dynamic range. Moreover, the position of the sigmoidal function can shift to higher or lower levels depending on the ambient-noise levels, thus preserving sensitivity over a wide range of SPLs (Dean et al. 2008; Keller and Takahashi 2015).

In the auditory nerve and in the other cochlear nucleus, the nucleus magnocellularis (NM), the phase angles of the spectral components are preserved by the phenomenon of phase locking, whereby spikes are evoked within a restricted range of phase angles of the stimulus sinusoid. Phase locking is quantified by the period histogram, which is the distribution of spike number relative to one cycle of the stimulus sinusoid. It is quantified by the vector strength, the degree to which spikes are confined to a restricted range of phase angles and by the mean phase angle at which the spikes are locked on average.

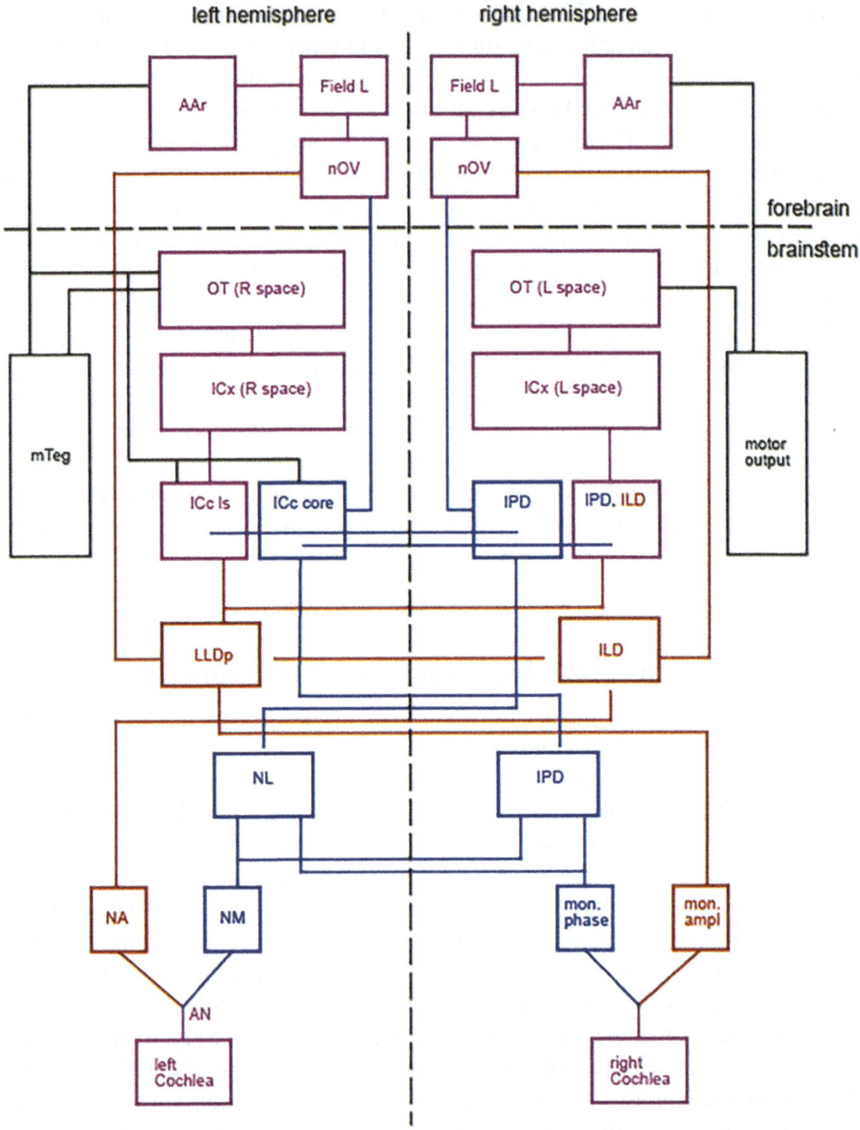

Fig. 4.2 Brainstem and forebrain auditory pathways in the barn owl. *Red and blue*, parts of the amplitude and phase pathways, respectively; *purple,* structures that carry information about both amplitude and phase. *Dashed vertical line*, the midline; *dashed horizontal line*, boundary between the forebrain and brainstem. Structures are on the *left* and their putative functions are specified on the *right*. The tectotegmental pathways culminate in the auditory space maps in the OT where, predominantly, contralateral space is represented. Additional connections within the arcopallial gaze field are not shown (e.g., from the auditory arcopallium to the lateral striatum [formerly paleostriatum augmentatum]) and from Field L to the caudolateral nidopallium (formerly neostriatum) to the auditory arcopallium (see Cohen et al. 1998). Both pathways converge onto brainstem tegmental motor nuclei (mTeg)

Not surprisingly, the ability of auditory nerve axons and NM cells to lock their spikes to a part of a stimulus cycle is limited by the frequency; the vector strength declines at higher frequencies. In the cat and the emu (*Dromaius novaehollandiae*), phase locking is observed up to about 4 kHz (Rose et al. 1974; Manley et al. 1997). By contrast, the vector strength remains above chance levels up to about 10 kHz in the barn owl (Sullivan and Konishi 1984; Köppl 1997). Alligators, like their closest extant avian relatives, are quite vocal and rely on sounds for communication and predation (Beach 1944), but their audible frequency range is lower than in most birds. Alligators can hear frequencies up to about 2.5 kHz (Higgs et al. 2002). Auditory nerve fibers and NM cells lock their responses to the stimulus frequency up to 2 kHz (Carr et al. 2009), allowing for coding the temporal structure of the signal over almost their entire hearing range. The strong phase locking in auditory nerve fibers is conserved or even enhanced in NM neurons by the big end bulb terminals onto the NM somata. Synaptic adaptations such as the large somatic synapses, fast transmitter release, AMPA receptors with rapid kinetics, large postsynaptic cells, and adapted potassium currents lead to the transmission of temporal information with little jitter (for more details, see Carr and Soares 2002).

Thus, within a frequency channel, the stimulus amplitude is encoded by the spike rate, and its phase angle is preserved by the temporal firing pattern. Interestingly, individual auditory nerve fibers, which bifurcate to innervate the two cochlear nuclei, carry both amplitude and phase information, whereas the postsynaptic targets in the cochlear nuclei appear to be specialized to carry either amplitude or phase information (Sullivan and Konishi 1984). The cochlear nuclei are the origin of separate, parallel pathways that compute the binaural differences in phase and amplitude. Once computed, the phase- and amplitude-difference channels converge in the inferior colliculus (see Fig. 4.2).

4.4 Computation of Interaural Time Difference and Interaural Level Difference

The interaural differences in ITD/IPD and ILD are computed separately in, respectively, the nucleus laminaris (NL) and the posterior segment of the dorsal lateral lemniscal nucleus (Fig. 4.2; LLDp; previously, the nucleus ventralis lemnisci lateralis, pars posterior or VLVp after Karten and Hodos 1967).

ITD is computed by a frequency-specific, cross-correlation-like process similar to that originally proposed by Jeffress (1948). This process is implemented in the NL, which, in the chicken (*Gallus gallus*), is a monolayer of neurons that are innervated by axons from the NM of both sides (Parks and Rubel 1975; Overholt et al. 1992). In the barn owl, a predator that relies heavily on sound localization, the NL is expanded in a manner suggesting that the layer of neurons in the chicken is a "module" that has been replicated repeatedly. Studies by Carr and Konishi (1990) in the owl and by Yin and Chan (1987) in the cat have provided evidence consistent with the idea of cross-correlation within each module.

Figure 4.3a gives an overview of the avian auditory hindbrain underlying ITD detection. In general, the NL encodes ITDs generated by contralateral sound sources, with some overlap between the two NLs in the frontal space. A model of the chicken NL and its replicated units are shown in Fig. 4.3b. Axons from the ipsilateral and contralateral NMs course in opposite directions, making synapses en passant. As a result of this innervation pattern, action potentials from the two sides proceed in opposite directions along the modules. They coincide at a neuron that is located at a point where the difference in axon lengths from the two sides compensates for the difference in the time of arrival of the sound waves at the two ears. ITDs, therefore, are topographically arranged. For instance, a sound source at the left will arrive first at the left ear, say, by 20 μs, and the spikes from the right NM will travel further along the module than those from the left NM. The sound wave will reach the left ear 20 μs later, and the spikes from the left NM will proceed in the

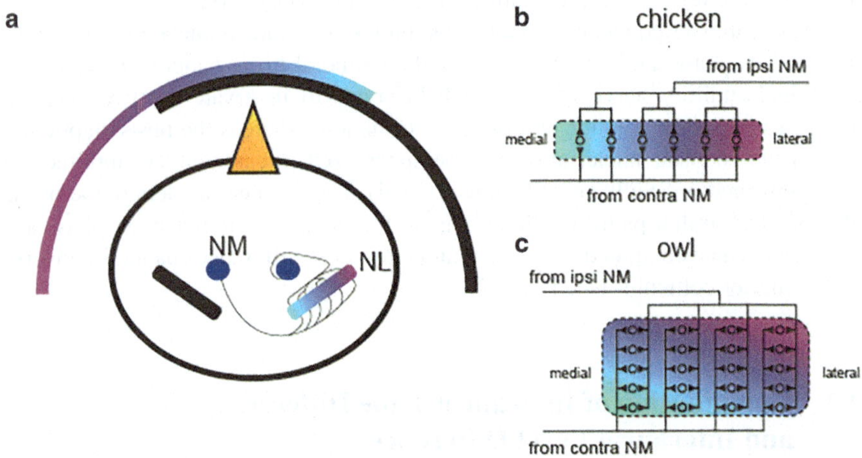

Fig. 4.3 Neural mechanism of the ITD computation in the nucleus laminaris (NL). (**a**) Cartoon of a bird's head viewed from above, including the brainstem structures involved in ITD detection. Each NL encodes ITDs that occur for contralateral sound sources with some overlap in the frontal space (*magenta*, contralateral peripheral space; *cyan*, central and ipsilateral space). The NL receives inputs from the NM of the ipsi- and contralateral sides. The left hemisphere is shown with a color gradient ranging indicating maximum negative ITDs (*magenta*) to slightly positive ITDs (*cyan*). To reduce clutter, only one hemisphere is shown with the gradient. Naturally, the NL of the other side also represents its contralateral hemifield. (**b** and **c**) NL circuitry in the chicken and owl, respectively. Cells (somata) of the left NL (*circles*) receive inputs from the NM of the left and right sides, the axons of which form en passant synapses on NL neurons. Neurons of the NL fire maximally when the spikes from the left and right arrive at the same time. The delay lines in the owl and chicken are typically generated by the axons of the contralateral NM neurons. These delay lines form maps of ITDs (*color gradient* corresponds to the *colors* in **a**). The chicken NL (**b**) is a less complex structure than the owl NL (**c**), where columns of loosely packed neurons with profusely branched dendrites form multiple maps of ITD within a frequency channel. After Ashida and Carr (2011), adjusted for findings by Carr et al. (2015b)

opposite direction. The spikes will arrive simultaneously at the NL cell body in which the ipsilateral and contralateral inputs differ by 20 μs.

In the owl, which relies heavily on sound localization (Fig. 4.3c), the NL can be thought of as multiple copies of the chicken NL arranged side by side (Carr and Konishi 1990). Such an arrangement might allow for increased precision in sound localization, but this has not been demonstrated empirically. Frontal space is over-represented in the owl because 0 μs is encoded in each module, whereas ITDs emerging from peripheral sound sources may only be driving neurons in the more lateral regions of the NL because the 0-μs locus shifts with the mediolateral position of the module in the NL (Carr et al. 2015a).

The precision of ITD computation by NL neurons and, in turn, spatial resolution will be affected by the temporal window in which coincidence occurs. In vitro studies of coincidence detection in the developing NL of chickens have shown that GABAergic inputs to the NL serve to shorten the postsynaptic potentials evoked by the inputs from the left and right NMs, which, in turn, could shorten the time window during which coincidence may occur (Funabiki et al. 1998; Yang et al. 1999). More recent studies in the chick have demonstrated other specializations, such as the localization of the sodium channels at various distances on an NL axon depending on frequency, that make the NL neurons more sensitive to small changes in the ITD (Kuba et al. 2006). The smallest detectable change in sound source location that an owl can detect is about 3° (Bala et al. 2003), which is equivalent to roughly 8 μs (Keller et al. 1998a). Whether or not this brief, permissible time window is due to the same mechanism described in chickens or involves further processing at subsequent synaptic stations is unknown.

As a result of the cross-correlation performed by the coincidence detectors within narrow frequency channels, the ITD tuning curves exhibit peaks at multiple ITDs. The distance of the response peaks corresponds to the period of the stimulus or characteristic frequency of the coincidence detector. As a side effect, the width of the ITD tuning curves increases at low frequencies, and the precision with which ITD can be encoded by the peak of the tuning curves therefore decreases. In this case, encoding the ITD in the slope of the tuning curves may be more optimal (Harper and McAlpine 2004; Fischer and Seidl 2014). In the gerbil (*Meriones unguiculatus*), a small-headed mammal, neuronal firing peaks at particular ITDs (best ITD) as in its avian analog, the NL. However, in this mammal, the peak of the ITD functions falls outside the gerbil's maximum, naturally occurring ITD, and the neurons are therefore unsuitable to place-code ITD. Within the physiological ITD range of the gerbil, the ITD tuning curves are sigmoidal functions facing in opposite directions on the left and right sides. Given this configuration, the ITD can be encoded by the difference in the firing of the neurons in the two brain hemispheres. Is the neural representation of ITD in the gerbil a general feature of the mammalian binaural system or is this representation a characteristic of small-headed mammals?

Interestingly, in the NL of barn owls (Palanca-Castan and Köppl 2015b) and especially in the chicken, a bird with a head size similar to that of the gerbil (Palanca-Castan and Köppl 2015a), the peaks of the ITD tuning curves of some NL cells responding to low frequencies are outside the physiological ITD range. However, contrary to what is found in the gerbil, the peaks or best ITDs are broadly and almost uniformly distributed in birds, arguing against a mammal-like rate code. In the alligator (*Alligator mississippiensis*), the birds' closest extant relative, ITDs are in fact topographically mapped almost over its entire hearing range down to frequencies as low as 300 Hz (Kettler and Carr 2019), quite similar to those of the chicken (Köppl and Carr 2008) and the owl at higher frequencies (Carr et al. 2015a). It is likely that the same holds for low-frequency regions in the chicken and owl NLs. Recording from low-frequency regions, especially in owls, turned out to be quite difficult (Carr et al. 2015a; Palanca-Castan and Köppl 2015c), hence the existence or nonexistence of a low-frequency ITD map has yet to be confirmed. But how can the large best ITDs be explained? The answer may be found in the internally coupled ears of birds and crocodilians. Pressure-gradient-receiving eardrums can increase the ITDs that are conveyed by the eardrums to the inner ears. This mechanism is especially effective at low frequencies as shown in a variety of birds (e.g., Calford and Piddington 1988). The ears in alligators are also highly coupled (Bierman et al. 2014), suggesting that the representation of the ITD in the early auditory pathway reflects a synapomorphic trait derived from the common archosaur ancestor of birds and crocodilians. Interestingly, this ITD code might be even older and common across all sauropsids because turtles also show the characteristic of a Jeffress-like ITD code, with best ITDs distributed within their physiological range. Additionally, their connections in the auditory midbrain are similar to those of birds and crocodilians (Willis and Carr 2017).

The ITD is further processed and its coding partially remodeled in the ascending auditory pathway. The NL projects directly to the core of the contralateral central inferior collicular nucleus (ICc-core), where the neurons are organized as tonotopic columns (see Fig. 4.4a–c). Neurons within a column are selective for different frequencies but for roughly the same ITD (Wagner et al. 1987). The neurons of the columns in the ICc-core project to the lateral shell of the contralateral central inferior collicular nucleus (ICc-ls; Takahashi et al. 1989). The cells in a column of the ICc-ls, in turn, project convergently onto a cluster of space-specific neurons in the external nucleus of the inferior colliculus (ICx), endowing the latter with a selectivity for the ITD of the input column and a more complex frequency selectivity (Takahashi and Konishi 1986; Wagner et al. 1987). Thus, a space-specific neuron is tuned to the location-specific ITD across all frequencies to which it is responsive (Takahashi and Konishi 1986; Gold and Knudsen 2000). This architecture is highly reminiscent of models of human binaural processing based on cross-correlation in frequency-specific channels and a subsequent, across-frequency integration (Shackleton et al. 1992; Gaik 1993b).

The ITD of a sound source is encoded in the firing rates of ICx neurons and their preferred IPD. From its convergent input, it follows, however, that if a limited number of ICc-core frequency channels are activated in response to pure tones or narrowband sounds, multiple response peaks in the ITD tuning of ICx and optic tectum (OT) neurons emerge (Mazer 1998; Saberi et al. 1999). Behaviorally, the owl indeed localizes illusory sound sources at locations that can be predicted by the ITDs that evoke additional response peaks in the neurons of the midbrain maps. These ITDs

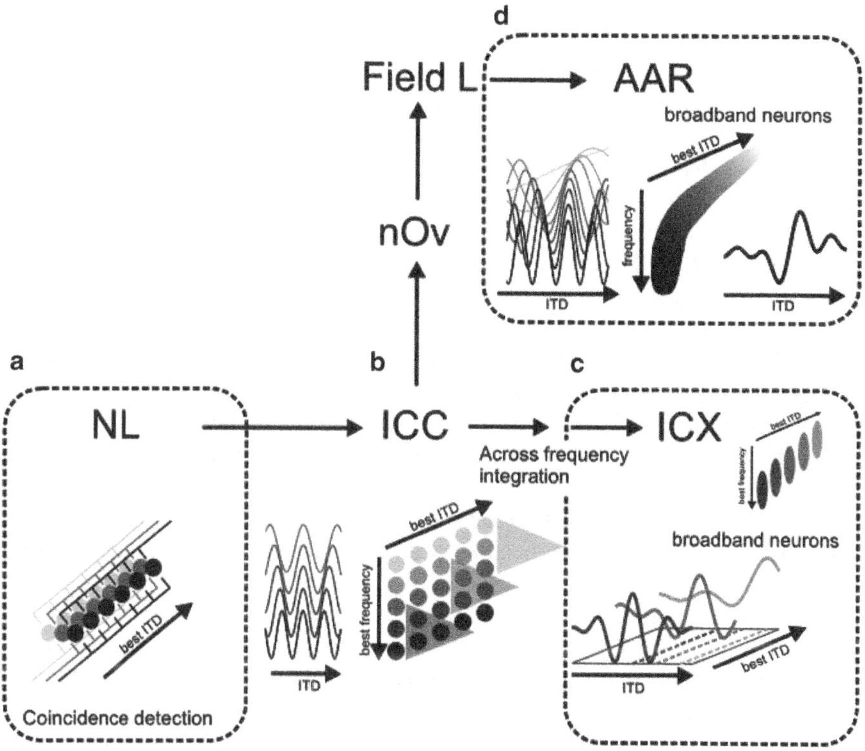

Fig. 4.4 Processing of the ITD in the brainstem and forebrain. (**a**) The ITD is detected in the NL by an array of coincidence detector neurons. NL neurons encode specific ITDs that are compensated for by internal delay lines generated by axons coming in from the ipsi- and contralateral nuclei magnocellularis (NM). NL neurons are sensitive to a narrow frequency band where the response varies cyclically with the ITD. Response peaks occur at multiples of the stimulus period. (**b**) Neurons in the central nucleus of the inferior colliculus (ICc) respond in a similar way. A column in the core of the central nucleus tonotopically represents frequencies, whereas all cells in a column respond maximally to the same ITD (Wagner et al. 1987). (**c**) Neurons in the external nucleus of the inferior colliculus (ICx) receive convergent input from a number of ICc neurons with different best frequencies but with the same best ITD. Hence, ICx neurons respond maximally to a single ITD. whereas side peaks are suppressed (Takahashi and Konishi 1986). Neurons with higher best frequencies are generally tuned to central (small) ITDs, whereas low-frequency neurons are more responsive to peripheral (large) ITDs (Cazettes et al. 2014). ICx neurons form a topographic map of ITDs. The ICc projects, ultimately, to the forebrain auditory arcopallium (AAr) via the nucleus ovoidalis (nOv) and Field L. (**d**) ITD tuning differs in the AAr. ITD tuning in the AAr is typically characterized by a peak at contralateral ITDs and a rise in response from ipsi- to contralateral ITDs. This characteristic shape of ITD tuning curves is generated by varying best ITDs at different frequencies. Although AAr neurons have maximum responses, with ITDs close to the midline if stimulated with high frequencies, the best ITD is larger with lower frequencies. Based on Vonderschen and Wagner (2012)

correspond to multiples of the stimulus period. For example, the period of a 5-kHz pure tone is 200 μs. When such a tone is presented with an ITD of 100 μs, which roughly corresponds to a true sound source 35° to the right of the owl (von Campenhausen and Wagner 2006), the owl responds with head turns not only to the true source but also to an illusory source perceived at about 35° to the left (Kettler et al. 2017). It has been shown more recently that the frequency tuning of ICx neurons varies with their best ITD (Cazettes et al. 2014). This frequency-dependent representation of space reflects the reliability of the frequency-dependent equivalent of the ITD, the IPD. The reliability, that is, the inverse of the variability, of the IPD given by the filter function of the head varies systematically with frequency. The IPD is most reliable at a low frequency for peripheral locations and at high frequencies (2–9 kHz) for frontal locations. ICx (and OT) neurons tuned to lower frequencies (<2 kHz) are also tuned to peripheral ITDs, whereas high-frequency neurons prefer frontal locations (Fig. 4.4d). These tuning properties show that the tonotopy in the ICx is shaped by the variability of spatial cues generated by the head-filtered sounds.

The representation of the ITD and its role in the creation of maps of auditory space in the owl midbrain has been known for decades. As in earlier stages of the auditory pathway, the ITD is represented by the peaks of the tuning curves and the neurons are topographically arranged in the ICx and OT. Barn owls are specialists with their rather large ITD range of ±250 μs and their ability to lock auditory nerve responses to the stimulus phase at frequencies up to about 10 kHz. These features allow for generating efficient auditory maps. On the other hand, little is known about how the ITD and auditory space are coded in the midbrain of other birds. There is evidence that in the midbrain of auditory generalists, such as chickens, the ITD is coded by the dynamic changes of the firing rate within the physiological range (Aralla et al. 2018). However, by filtering or integration over frequency or ITD, the ITD code may be refined or even reshaped along the auditory pathway (for a review, see Vonderschen and Wagner 2014). Thus, a place code found in the NL of birds and other archosaurs may still be transformed to a mammal-like rate code. Indeed, ITD coding in the owl forebrain differs from the midbrain place code. The forebrain or thalamocortical pathway branches off from the midbrain ICc-core and ends in the auditory arcopallium (AAr; formerly auditory archistriatum; Fig. 4.2). The AAr projects back to the ICx and the OT and also innervates brainstem motor nuclei (Knudsen et al. 1993). ITDs are not organized into maps as in the ICx or in the OT. In the entire population of AAr neurons, time differences are encoded by the dynamic changes of the firing rate around the midline (Vonderschen and Wagner 2009, 2012; Tellers et al. 2017), which results in a high correlation of responses of widely distributed neurons in the AAr and thus supports a rate code (Beckert et al. 2017) compared with clusters of preferred angles in Field L (Cohen and Knudsen 1996) and a spatial topography in the OT. In the barn owl, the midbrain map provides a fast and precise code of sound localization, whereas the forebrain may generate top-down control, direct auditory attention, or compensate for midbrain deficits.

Less is known about the processing of the ILD, the other binaural cue in the owl. Barn owl HRTF measures show that the ILD is closely related to the vertical coordinate of a target at high frequencies (Fig. 4.1c). The first site of binaural convergence in the ILD stream is the pontine nucleus, LLDp. The neurons of the LLDp are excited by stimulation of the contralateral ear and inhibited by stimulation of the ipsilateral ear (Takahashi and Keller 1992; Takahashi et al. 1995b). The LLDp neurons are thus "EI" cells, the discharge rates of which are sigmoidal functions of ILD (Boudreau and Tsuchitani 1968; Mogdans and Knudsen 1994). The LLDp is therefore similar to the mammalian lateral superior olivary nucleus, except that the excitatory and inhibitory ears are reversed (Tsuchitani 1977; Moiseff and Konishi 1983). Studies incorporating neurophysiological and tract-tracing methods have shown evidence that the ILD is systematically represented in the LLDp (Manley et al. 1988; Takahashi et al. 1995a).

Hodological studies suggest that the LLDp projects bilaterally to the ICc-ls, endowing the neurons in each column with sensitivity to the ILD as well as to the ITD (Adolphs 1993). A clear topographical representation of the ILD has never been observed in the ICc-ls (Mazer 1995), whereas maps of the auditory space clearly emerge from the convergence of the ITD- and ILD- processing paths in the ICx and OT. Partial lesions in the ICx indeed generates deficits in sound localization for the specific sound source direction for which the lesioned region was coding, whereas the owl still reliably localizes all other locations (Wagner 1993). The deficits disappear after some time, indicating reorganization of the topography in the midbrain representation of the auditory space or supplementing functions of the auditory midbrain by the thalamocortical pathway. The knowledge of HRTFs has made it possible to assess the contribution of the ILD to the space-specific spatial receptive field (RF) of the neuron. For any given location, the ILD varies as a function of frequency in a complex manner (see Hartman, Chap. 2). To assess ILD tuning, HRTFs were altered so that the ILD spectrum at each location was preserved, but the ITD was held constant at a preferred value of the cell. This generates the spatial RF the cell would have had were it sensitive only to the ILD (Euston and Takahashi 2002). These altered HRTFs were used to filter broadband noises that served as the stimulus. These ILD-alone RFs were generally horizontal swaths of activity at the elevation of the normal spatial RF of the cell (shown in Fig. 4.5a for three space-specific neurons). Analogously, ITD-alone RFs formed a vertical swath at the azimuth of the normal RF of the cell (Fig. 4.5b). The normal RF thus lies at the intersection of the ITD and ILD-alone RFs (Euston and Takahashi 2002). At this intersection, the optimal ITD and ILD spectra of the cell are present and are combined by a multiplication-like process (Fischer et al. 2009), the product of which generates receptive fields (Fig. 4.5c) that are very similar to those obtained neurophysiologically (Fig. 4.5d).

The shape of the ILD-alone RF is ultimately determined by the frequency bands to which the space-specific neuron has access and by the tuning of the cell for the ILD at each of these frequencies (Delgutte et al. 1999; Tollin and Yin 2002). Accordingly, a space-specific neuron should be broadly tuned to the frequency, and for each frequency band, it should be sharply tuned to the ILD value that occurs at

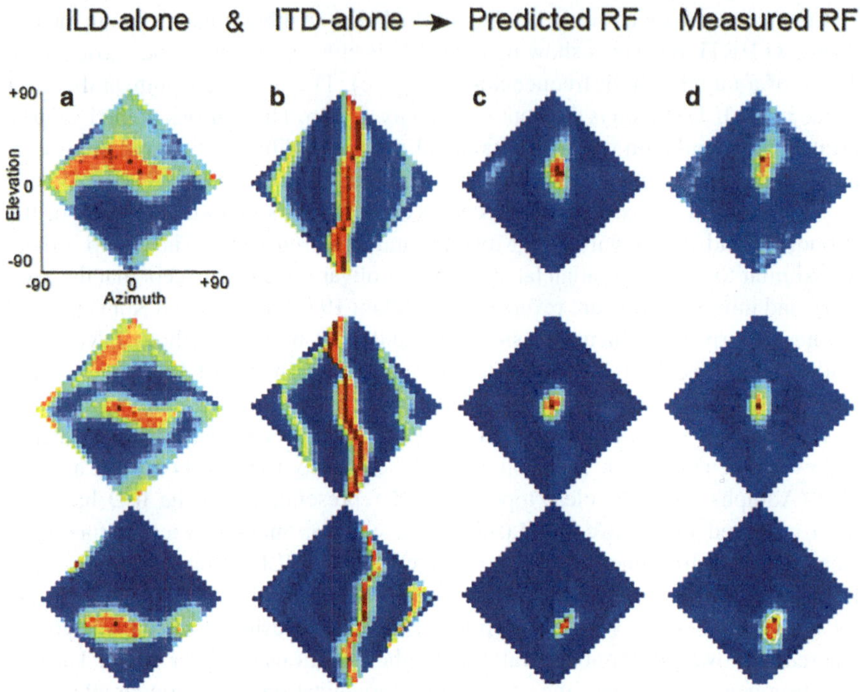

Fig. 4.5 Construction of a spatial receptive field (RF) from ILD- and ITD-alone receptive fields. Each *row* represents a different cell. *Red hues*, higher firing rates. The regions of space that evoke activity in a space-map neuron when the only spatial cues are based only on the ILD (**a**) or ITD (**b**). (**c**) The product obtained when **a** and **b** are multiplied together and adjusted for the average binaural level (ABL), which is lower in the peripheral regions of space (Keller et al. 1998). (**d**) The spatial RF measured empirically is similar to the RF computed from the ILD- and ITD-alone RFs. Based on Takahashi et al. (2003)

its spatial RF as it was for the ITD (Gaik 1993a; Knudsen 1999). Can ILD tuning, measured with tones of different frequencies, account for the ILD-alone RF? Euston and Takahashi (2002) transformed the ILD tuning curves at each frequency into ILD-alone RFs using the individual bird's HRTFs and a linear model that summed the contributions of each frequency band independently. When the frequency-specific ILDs (see Fig. 4.6a for an example) is transformed into the firing rate at various spatial loci, a broad horizontal swath of activity is presented (Fig. 4.6b) that can be compared with the measured ILD-alone RF in Fig. 4.6c.

More recently, frequency-specific ILD tuning was estimated in reverse, going from the ILD-alone spatial RF obtained with broadband noise to a frequency-specific ILD tuning curve (Spezio and Takahashi 2003). The ILD frequency plots inferred from the ILD-alone RF using noise (Fig. 4.6d) and that derived with tone bursts (Fig. 4.6e) were similar, although the ILD/frequency plot was broader when obtained using noise (Fig. 4.6d; Euston and Takahashi 2002). The plots shown in Fig. 4.6d, e, were transformed into spatial coordinates (Fig. 4.6f, g) that can be

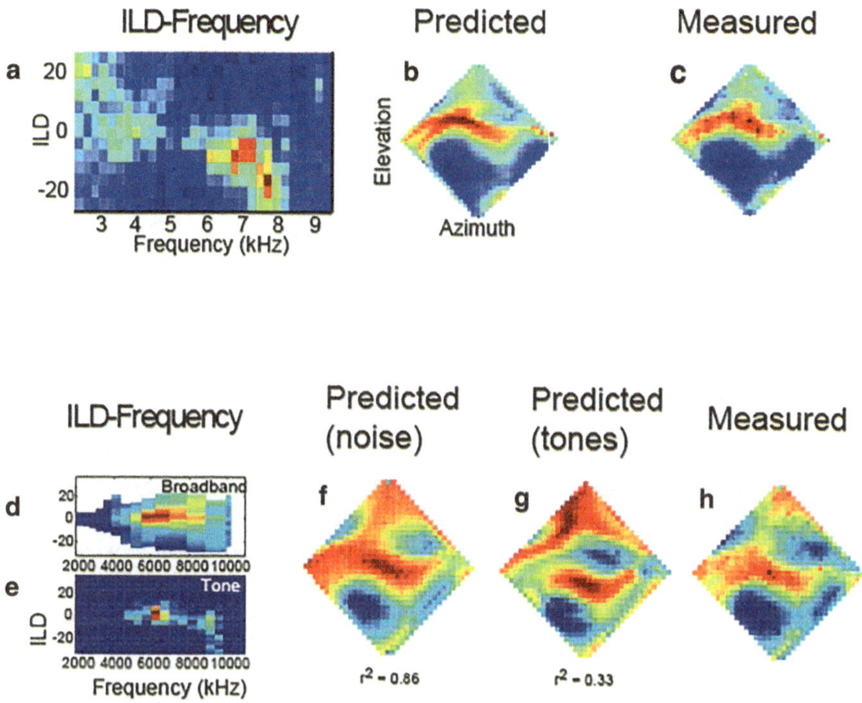

Fig. 4.6 ILD tuning of space-map neurons at different frequencies. (**a**) Plot of firing rate of a space-map cell at different combinations of ILD (*rows*) and frequency (*columns*) inferred from responses to tones. *Red hues*, higher firing rates. (**b**) Responses of the cell at different ILD and frequency combinations translated into space. (**c**) ILD-alone RF measured as in Fig. 4.5. (**d**) Responses to combinations of ILD and frequency can also be inferred from the response of a neuron to broadband noise bursts (Spezio and Takahashi 2003). (**e**) Responses of the same cell as in **d** to tones of different frequencies and ILDs. Note that the noise-evoked responses in **d** show responses to a wider range of frequencies and ILDs. (**f** and **g**) ILD-frequency maps shown in **d** and **e**, respectively, transformed into space. The r^2 values below the spatial plots (**f** and **g**) compare the predicted spatial plots to the measured one (**h**) and suggest that the noise-based method typically yields a better match to the measured ILD-alone RF than that based on tone bursts. Based on Spezio and Takahashi (2003)

compared with the measured ILD-alone RF (Fig. 4.6h). The noise-based estimate of ILD tuning predicts, on average, more of the variance of the experimentally measured ILD-alone RF than does the tone-based ILD frequency response. It also has the practical advantage that the frequency tuning of a cell, which is ILD dependent, can be deduced from the ILD-alone RF (Spezio and Takahashi 2003).

These two approaches revealed that space-specific neurons are *not* necessarily tuned sharply to the ILD over a wide, continuous frequency range. Instead, their frequency tuning seems to be patchy (e.g., Fig. 4.6a). At some frequencies, the ILD tuning functions retained the sigmoidal shape seen at lower stages, specifically in the LLDp (Euston and Takahashi 2002; Spezio and Takahashi 2003). Whether this seemingly patternless tuning has advantages over broadly continuous tuning is not known.

For humans, who have symmetrical ears, ILDs have a large azimuthal component with increasing frequency and relatively a small elevational component. For owls, in contrast, who have asymmetrical ears, there a substantial elevational component (Fig. 4.5a). A combination of ITD and ILD leads to robust azimuthal localization in cases where the stimulus ITD may be ambiguous, for example, with narrowband sounds. In this case, the systematic variations of the ILD in the equatorial plane help to determine the true azimuth of the sound source. The effect of the ILD becomes clear when comparing behavioral responses to narrowband signals presented either by free-field speakers or by headphones (Kettler et al. 2017). Free-field stimuli contain all natural cues, whereas the ILD of headphone stimuli that are HRTF filtered can be manipulated. If the ILD is constant at 0 dB and only the ITDs are changed between trials, the owl locates more phantom sources than when responding to free-field narrowband sounds. Furthermore, one can set the ILD to signal a sound source that corresponds to a phantom source for a given ITD and stimulus frequency. In this case, the animal makes even more localization errors than in the case with 0 dB ILD. This observation is strikingly similar to what can be found in humans, where experimental manipulation of the miniscule ILDs that naturally occur at low frequencies can lead to localization of ITDs that are larger or smaller by one stimulus cycle than the true ITD (Hartmann et al. 2016).

4.5 Neural Basis of Spatial Acuity

Given the fine-grained nature of the spatial RFs in the owl's ICx, it is natural to ask how the owl's perceptual spatial acuity compares with its neuronal spatial acuity. Earlier studies of sound localization precision suggest that the aim of an owl's head toward an auditory source has an error radius of about 6° for broadband stimuli (Knudsen and Konishi 1979; Whitchurch and Takahashi 2006). By comparison, visual localization in the owl is accurate to about 2° (Whitchurch and Takahashi 2006). In fact, the precision of head turns toward sound sources on the equatorial plane decreases for peripheral locations and is highest if the sound source is in front of the owl (Hausmann et al. 2009). This behavior is consistent with the fact that frontal locations are overrepresented by neurons in the OT space map (Knudsen 1982; Fischer and Pena 2011). Observations of freely hunting barn owls have shown that these birds, while tracking prey such as small rodents like voles and mice, turn their head to the prey and stay focused on it (Edut and Eilam 2004). This head tracking is likely to keep their prey in the region of their sharpest hearing and to align both the visual and auditory senses onto the target.

The relationship between spatial acuity and the representation of space in the space map was studied using the pupillary dilation response (PDR). In owls and humans, the pupil dilates on detecting a sound or on detecting a change in the features of a sound, such as the location of a sound source (Bala and Takahashi 2000; Kahneman 2003). An advantage of this method is that the PDR behavior does not involve movements of the head, so motor errors for these large movements do not occur.

To implement the PDR method, the pupils of an unanesthetized owl are monitored under infrared illumination. A sound presented from one location causes the pupils to dilate. If the sound is repeated from the same location, the PDR habituates. It recovers, however, if the location of the source is changed. The smallest change of source position that reelicits the PDR is considered the minimal audible angle (MAA). Assessed in this manner, the barn owl's MAA is about 3° horizontally and 6° vertically (Bala et al. 2003, 2007).

The behavioral MAA was explained by Bala et al. (2003, 2007) as a simple shift of the locus of activity or "neural image" on the ICx space map. Figure 4.7a shows a cartoon of a neural image. The stimulus maximally excites neuron 6 that is tuned to the speaker's location x, causing it to fire maximally (Fig. 4.7a, *red*). Cells with spatial RFs farther away fire less, forming a localized patch of activity on the ICx or, in other words, a neural image of the sound location. Note that the span of the neural image will depend on the breadth of the spatial RFs of the constituent cells. The repeated presentation of the sound from location x leads to habituation somewhere between the space map and the neurons that control the pupil size.

If the speaker is moved by an amount ($\Delta x°$), the neural image shifts, causing some cells to fire more (e.g., neuron 8; Fig. 4.7a, *bottom row, green to red*) and others to fire less (e.g., neuron 6; Fig. 4.7a, *bottom row, red to green*). It was assumed that the absolute value of the difference in firing for speaker locations x (Fig. 4.7a, *top row*) and $x + \Delta x$ (Fig. 4.7a, *bottom row*) drives the recovery from habituation. The hypothesis predicts that the smallest Δx that leads to a significant shift in the neural image will recover the habituated PDR.

By recording from many ICx neurons, firing rates at different stimulus locations can be converted to a d′-like quantity the standard separation (*D*; Sakitt 1973). The size of the PDR, which scales with the difference between the habituating and recovering stimuli (Bala and Takahashi 2000), can similarly be expressed as *D*. Figure 4.7b plots the neural and behavioral *D* values for Δx values of 1° to 6° of azimuth. The behavioral *D* for each of three birds Fig. 4.7b, *black line and circles* and the neuronal *D* Fig. 4.7b, *pink line and circles* overlie one another well, suggesting that the difference in the position of the neural image can be the basis for recovery from PDR habituation. Using an arbitrary threshold of $D = 0.8$ as the criterion, the neural and behavioral MAAs are about 3° in azimuth (Bala et al. 2003). By comparison, the azimuthal MAAs in the rhesus monkey and the human are 4° and 1°, respectively (Brown et al. 1982). It should be noted that the head widths of these primates are considerably larger, yielding maximal ITD values of 470 μs in the monkey (Spezio et al. 2000) and 700 μs in humans (Wightman and Kistler 1989) compared with roughly 250 μs in the barn owl (Keller et al. 1998a).

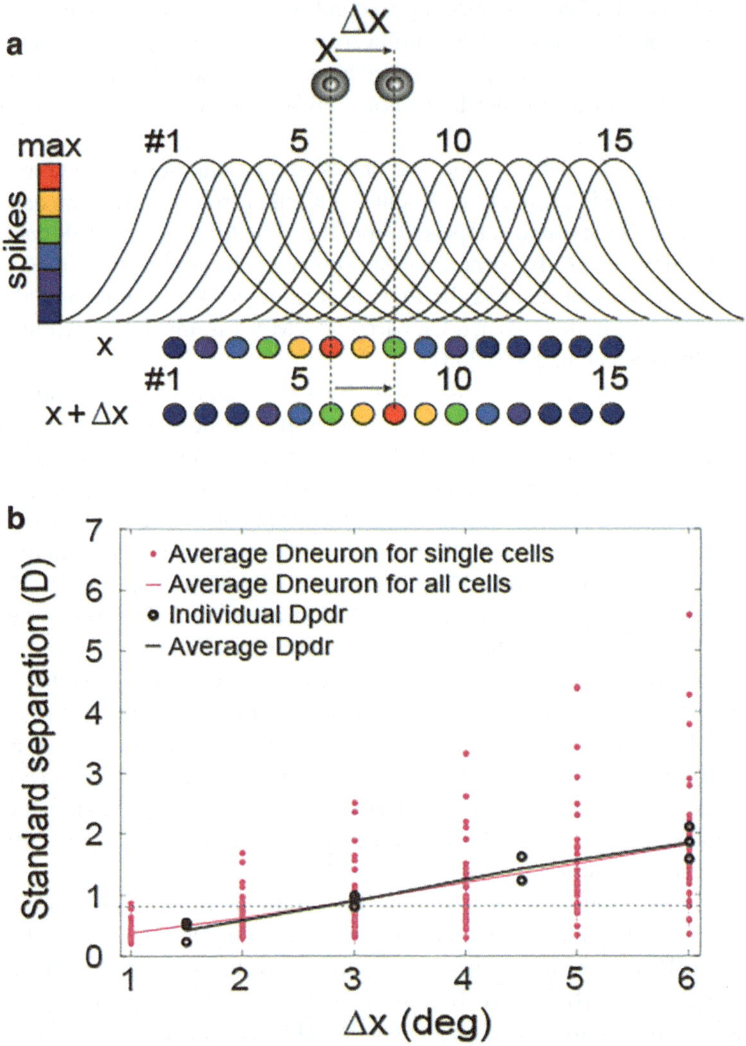

Fig. 4.7 Prediction of the MAA from space-map responses. (**a**) Cartoon of the activity on the space map when a speaker is moved from location x to location $x + \Delta x$. *Colored circles*, RFs of the neurons (*curved lines*); *colors on scale bar*, firing rates of the neurons, with *red* corresponding to the highest activity. The *curves* represent cross sections of the spatial RFs of a group of ICx neurons tuned to different spatial locations. *Top row*, cells showing the activity pattern on the map for the source at location x; *bottom row*, activity pattern when the source is shifted to location $x + \Delta x$. The hypothesis was that the behavioral MAA is a result of a significant shift of activity on the map. (**b**) Comparison of neuronal and behavioral spatial discrimination performance. Discrimination performance was quantified by D that, like d', incorporates the trial-to-trial variance in neuronal activity (D_{neuron}) or PDR (D_{pdr}; Sakitt 1973). *Pink circles*, D value for individual neurons ($n = 100$ cells); *pink line* connects the means of the D values of the cells for changes in speaker position ($\Delta x°$) ranging from 1° to 6° in azimuth; *black circles*, behavioral MAA measured by the PDR in three owls; *black line* connects the means of the individual bird's D values. If the MAA is defined as the smallest Δx at which D exceeds 0.8 (*dashed horizontal line*), both the birds and their space-map neurons have an MAA of about 3°. Based on Bala et al. (2003)

A more recent study, using the same methodology based on the PDR, showed that the MAA in elevation is 7.5°, which is consistent with the observation that the vertical dimension of the spatial RF in the typical space map cell is about twice that of its horizontal dimension (Bala et al. 2007).

4.6 Sound Localization in Complex Acoustical Environments

The putative neural mechanisms of localization of a single sound source in a quiet environment were described. In nature, however, a sound from an actively emitting source, such as a conversation partner, may arrive at the ears accompanied by reflections from nearby surfaces. Moreover, there may be other sound sources, such as a conversation in the background. In more natural environments, these reflections may be altered in an even more complex way by plants or geological structures. Here, the principles of localization gleaned from the study of single-source localization are applied to localization in these more complex acoustical environments, starting first with echoes. (See Zahorik, Chap. 9, on the effects of acoustical reflections.)

As early as 1856, Henry noted that acoustical reflections or echoes lack the perceptual saliency of the sound reaching the ears directly from the emitting source (cited in Gardener 1968). Asked to localize a source when a pair of sounds is presented from different directions with a 1- to 10-ms delay, listeners preferentially localize the earlier or leading source, a phenomenon now known as localization dominance. At the same time, the ability to discriminate changes in the location of the lagging source is compromised, leading to the phenomenon called lag-discrimination suppression. Both of these phenomena are components of the precedence effect (Wallach et al. 1949), which has been thoroughly reviewed elsewhere (e.g., Litovsky et al. 1999).

One potential mechanism for the precedence-effect phenomenology is that neurons responding to the leading source temporarily suppress the responses to sounds from locations other than those of the leading source. Indeed, a pioneering study by Yin (1994) showed that in the inferior colliculus of cats, if the lagging sound source is placed in the spatial receptive field of a cell, the response of the cell is weaker. The response of the cell recovers if the lead-lag delay is increased to about 8 ms. A similar effect was later reported in the space map of the owl (Keller and Takahashi 1996b).

The precedence effect is typically studied with brief transients, such as clicks, that avoid the overlap of the direct and reflected clicks, each of which are presented from different loci (Fig. 4.8a, b). However, natural sounds are often longer than the delay between the two sounds, so their two waveforms overlap, complicating the analysis. Nonetheless, studies have demonstrated localization dominance with overlapping leading and lagging stimuli in humans (Zurek 1980; Dizon and Colburn 2006) and in owls (Nelson and Takahashi 2008b, 2010). The increased complexity has offered novel insights into the mechanism.

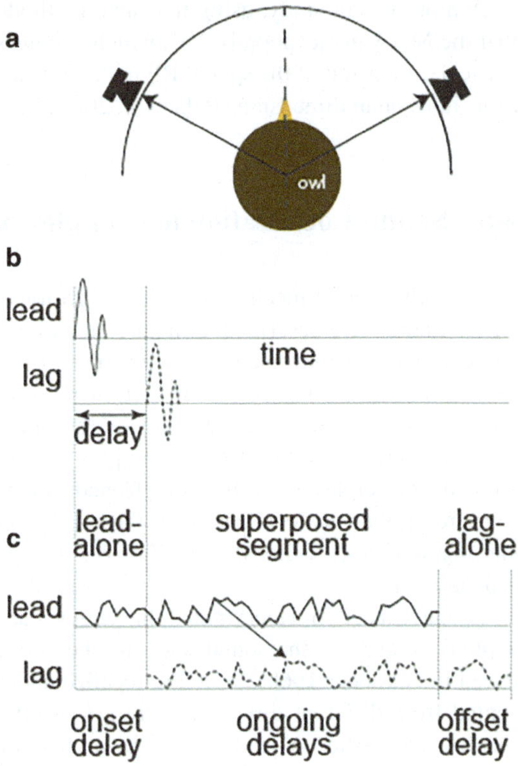

Fig. 4.8 Defining lead/lag delays in precedence effect paradigm. (**a**) Cartoon of an owl head from above shown with speakers to its left and right, one of which emits a sound a few milliseconds before the other. (**b**) Pair of short-duration clicks show the temporal relationship that can be defined by a single "delay." (**c**) When the speakers emit sounds with a longer duration, three delays can be defined. During the lead-alone and lag-alone segments, only the leading and lagging sounds, respectively, are present for an amount of time equal to the lead/lag delay. During the superposed segment, the delay can be defined from corresponding envelope features (*arrow*)

4.6.1 Does the Delay of the Echo Really Determine Echo Threshold?

By analogy to click-like stimuli, the echo delay for overlapping sounds is the interval between the onsets of the leading and lagging sounds. A longer sound and its delayed copy are shown Fig. 4.8c, *dashed line*. Because the two sounds are equally long, however, Nelson and Takahashi (2008b) argued that the echo delay can be defined by the lead-alone segment or by the lag-alone segment during which, respectively, the leading and lagging sounds are present alone. Furthermore, the superposed segment, during which both sounds are present, has ongoing delays that are defined by corresponding waveform features (Fig. 4.8c, *arrow*). Each of these segments reflects the echo delay, so which determines whether or not the echo is heard?

To answer this question, the lengths of the lead- and lag-alone segments were independently manipulated and the owl's natural head-turning behavior was observed (Nelson and Takahashi 2008a). The ongoing delays in the superposed segments were always equal to the lead-alone segment. If the delay between the onsets of the two sounds is crucial, lengthening the lead-alone segment independently of the lag-alone segment should cause an echo threshold to be reached. However, if the echo threshold is reached when the lag-alone segment is long enough, extending the lag-alone segment while holding the lead-alone segment constant should cause the echo threshold to be reached.

Experimental evidence suggests the importance of the lag-alone segment (Nelson and Takahashi 2008a). This was demonstrated by holding the lead-alone segment constant while varying the length of the lag-alone segment (see Fig. 4.9a, *right*).

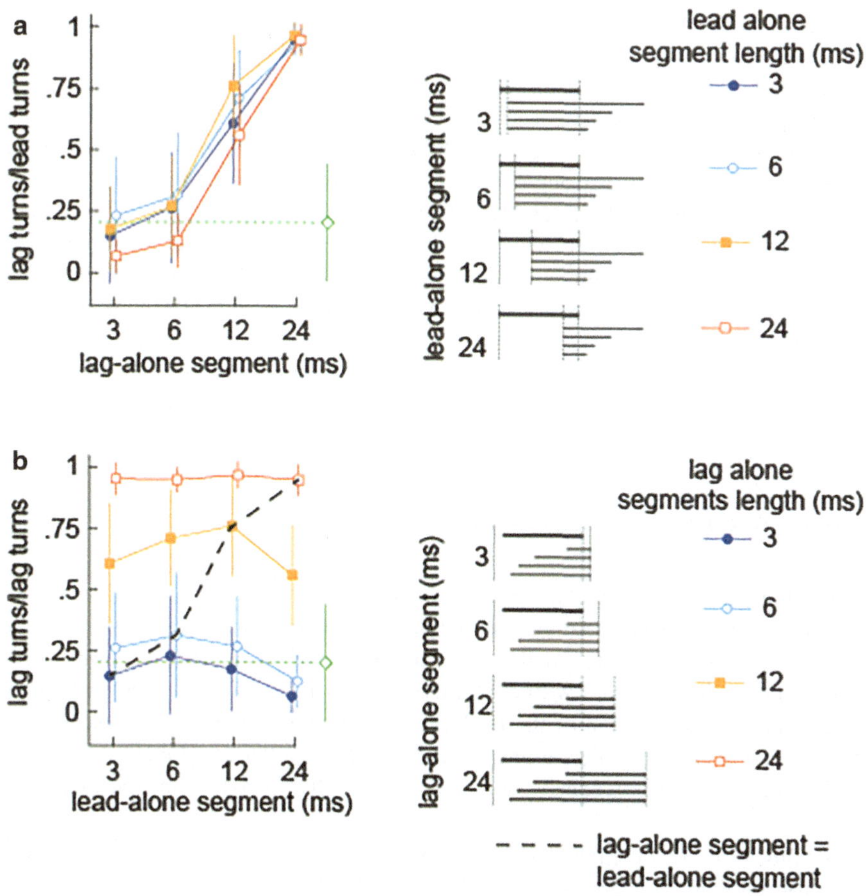

Fig. 4.9 An alternative explanation for the PE? (**a**) Plot of the ratio of lag-directed head turns to lead-directed turns (*left*) as the lag-alone segment is lengthened (*right*) while holding the lead-alone segment at 3 (*dark blue circles*), 6 (*light blue circles*), 12 (*orange squares*), and 24 (*red squares*) ms. The ratio of head turns approaches unity (number of lead turns = number of lag turns) as the lag-alone segment is lengthened. (**b**) Ratio of lag and lead turns as the lead-alone segment is lengthened (*left*) while holding the lag-alone segment constant (*right*). The ratio of lead- and lag-directed head turns for each lead-alone segment length remains fairly constant. *Dashed line* connects the points having equal lead- and lag-alone segment lengths. Based on Nelson and Takahashi (2010)

Under this condition, the lag source was localized more frequently as the lag-alone segment was lengthened while holding the lead-alone segment at a constant duration. In this paradigm, the ratio of head turns to the lagging and leading sources increased with duration of the lag-alone segment, approaching unity (lead turns = lag turns) regardless of the length of the lead-alone segment.

Nelson and Takahashi (2008a) replotted the ratio of lag-to-lead head turns against the length of the lead-alone segment (Fig. 4.9b, *left*), in which case the proportion of lag-to-lead directed turns is largely flat, suggesting that changing the duration of the lead-alone segment has little effect. These findings cast doubt on the traditional idea (e.g., Wallach et al. 1949) that the leading sound prevents responses to the lagging sound and, instead, suggest that the lagging sound is localized when the lagging segment is long enough.

The performance when the lead- and lag-alone segments are equal because they are in more standard precedence effect (PE) paradigms, is shown in Fig. 4.9b, *black dashed line*. The line rises from no turns toward the lagging sound at short delays to a nearly equal number of lag- and lead-ward turns at longer delays.

In the standard paradigm, why did the owls localize the leading sound even though the lead- and lag-alone segments were equally long? One possible reason is that neurons from the owl's space map fire a burst of spikes at the onset of the sound in its RF. This burst adapts in about 10 ms, leaving only a lower rate, sustained component that remains after the leading sound stops. Thus, the onset response evoked by the leading sound makes its representation stronger.

4.6.2 Are Echoes Suppressed?

Physiological studies of localization dominance, which typically use clicks, generally report that neuronal responses to the lagging click are weaker than those to the leading click at short delays and recover as the interclick interval increases. Such observations are consistent with a lateral inhibition-like mechanism in which neurons, tuned to the location of the leading click, preempt the responses of neurons tuned to the location of the lagging click (Yin 1994; Keller and Takahashi 1996b). The echo threshold is reached when the lagging click arrives after the inhibition subsides.

Application of this hypothesis to longer sounds, however, is not straightforward. Because the auditory system cannot "know" the length of the sound, one would have to assume that the aforementioned inhibition lasts for a finite duration, perhaps equivalent to the echo threshold. If such a mechanism operated for longer sounds, the echo threshold should be determined by the lead-alone segment, but this is not the case, at least for owls (see Sect. 4.6.1). Moreover, human listeners experience localization dominance even when the lead-alone segment, which would presumably trigger the inhibition, is removed (Zurek 1980; Dizon and Colburn 2006). Complex sounds in nature, such as vocalizations, typically have amplitude modulations (AMs). One might therefore propose that each peak in the envelope of the leading sound suppresses responses that would otherwise be evoked by corresponding peaks in the lag. But is such an "ongoing" inhibition necessary?

When the loudspeakers simultaneously emit noise bursts that are statistically uncorrelated, that is, they have the same amplitude spectra but different phase spectra, the quantities ITD(*f*) and ILD(*f*) fluctuate differently in each frequency band and do not correspond to a single, discrete locus; in other words, they are spatially incoherent. Space-map neurons, which are sensitive to the cues that correspond to their spatial RFs, respond poorly even without inhibition to spatial incoherency (Yin et al. 1987; Keller and Takahashi 1996a). Thus, when the sound pair consisted of a sound and a delayed copy, segments in the delayed copy aligned with earlier parts of the undelayed copy and the sound pair is effectively uncorrelated. (Fig. 4.8c; Keller and Takahashi 1996a). Therefore, it has been argued that the low response of cells tuned to the lagging source may be due to spatial incoherency, not to renewed inhibition by cells responding to the location of the lead source (Nelson and Takahashi 2008a).

4.6.3 A Simpler Explanation of Localization Dominance?

Nelson and Takahashi (2010) proposed an "envelope hypothesis" to explain localization dominance for long-duration sounds. The authors offered that the lead-lag relationship of direct sound and echo is determined from corresponding peaks in the ongoing envelope (Fig. 4.10a). Without peaks and troughs in the envelope, Nelson and Takahashi argued, there should be no localization dominance. Their behavioral evidence (Fig. 4.10b, c, *dot rasters*) plot the frequency of saccades for a 3-ms delay when the modulation was deep (100%; Fig. 4.10b) and when it was unmodulated (0%; Fig. 4.10c).

When the AMs were maximal (100% modulations) and the delay short (3 ms), the birds predominantly localized the leading source (Fig. 4.10b; Nelson and Takahashi 2010). Analogously, these authors showed that the neural response in the space map is considerably stronger than that to the lagging sound (Fig. 4.10b, *dot rasters*). By contrast, with minimal modulations (0%; Fig. 4.10c; 3-ms delay), the owls localized the lead and lag sounds with equal probability. Accordingly, neural responses (Fig. 4.10c, *right*) are symmetric. Thus, behavioral and neural responses are consistent with these aspects of the envelope hypothesis.

Nelson and Takahashi (2010) noted that these observations can be explained by the *conjunction* of two simple factors that trigger spiking in space-map neurons. First, they suggested, that amplitude of the target must be higher than that of the masker. Because ITD(*f*) and ILD(*f*) are vector averages of the binaural cues of each sound weighted by their amplitudes, the cues will approach those corresponding to the louder sound source (Snow 1954; Keller and Takahashi 2015). Thus, when the amplitude of the target is higher momentarily, ITD(*f*) and ILD(*f*) approximate those generated at the spatial RF of the cell. Second, although the difference of amplitudes favors the target, the amplitude of the target must be rising. Like cells in a

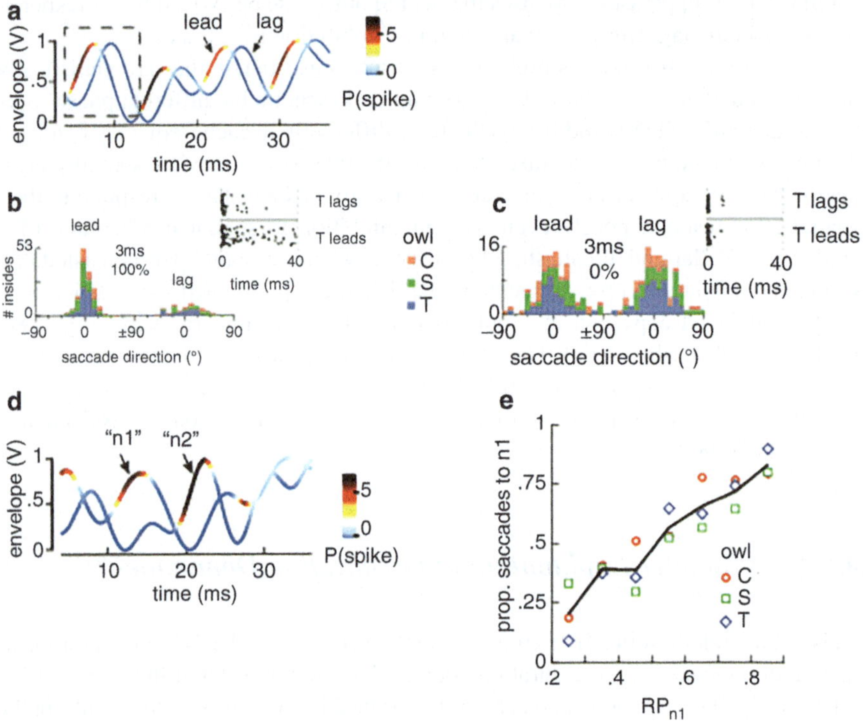

Fig. 4.10 Effect of envelope modulation depth and shape. (**a**) Envelopes of a pair of identical broadband noises (50-ms duration), one of which is delayed by 3 ms. The lead- and lag-alone segments were excised so that the two sounds start and end simultaneously. Thus, the only cue for determining the lead/lag delay is the time difference between homologous envelope features of the lead and lag sounds. *Colored lines*, envelope traces indicating the probability (*color bar*) that an envelope segment will elicit a spike in the space map of the ICx. This "envelope hypothesis" predicts that without envelope modulations, the owl will be unable to determine which of the noises led the other. (**b**) Distribution of the number of head saccades toward the leading speaker (0°; "lead") and lagging speaker (0°; "lag") for a 3-ms delay and deep (100%) modulations. Results are from 3 owls (*colored histograms*). The responses of a space-map neuron to leading (T leads) and lagging (T lags) sounds are shown as *dot-raster displays* (*right*). Note that neural response to the leading source is stronger than its response to the lagging source. Thus, when envelope modulations are present, the leading source is localized more often than is the lagging source, and, correspondingly, the region of the space map representing the leading source responds more strongly than that representing the lagging source. (**c**) Distribution of the number of head saccades toward the leading speaker (0°; lead) and lagging speaker (0°; lag) for a 3-ms delay and no (0%) modulations. When there are no modulations, the head turns to the leading and lagging sources are nearly equal, as are the responses of space-map neurons to the leading and lagging sources. (**d**) Pair of nonidentical envelopes (n1 and n2). Although the pair of envelopes is completely different, it is still possible to compute the relative probability that each envelope will evoke a spike in a space-map neuron. This relative probability is displayed in pseudocolor (*color bar*). (**e**) Proportion of head saccades to noise n1 plotted against the relative probability of spiking evoked by n1 (RP_{n1}; Eq. 4.1). *Colored circles*, data from 3 different owls; *black line* connects the means of the three birds' performances. The owls' responses are well explained by RP_{n1} ($r^2 = 0.97$). Based on Baxter et al. (2013)

variety of auditory structures of many species, cells in the owl's space map tend to respond to rising AMs (e.g., Keller and Takahashi 2015).

The probability that various envelope segments will evoke activity in space-map neurons can be seen by returning to Fig. 4.10a, *dashed-line box*. Within this time period, the leading sound is more intense than the lagging sound and its amplitude is rising. Therefore, Nelson and Takahashi (2010) argued that neurons tuned to the location of the leading source are likely to discharge (Fig. 4.10a). Of course, the lagging envelope has a corresponding peak, but its rising edge is masked by the leading peak. Moreover, the binaural cues are spatially incoherent during this overlap, so cells representing the location of the lagging sound fire less vigorously. As the delay between the two waveforms increases, the rising edge of the peak of the lagging sound emerges, strengthening the representation of the lagging sound, presumably causing the owl to also localize the lagging sound.

4.6.4 Generalizing the Envelope Hypothesis

What happens if the envelopes are different? This situation corresponds more closely to the "cocktail party" effect (Sayers and Cherry 1957), which describes an environment in which independent sounds are simultaneously active. Is it possible to generalize the envelope hypothesis to account for an owl's behavior in a cocktail-party environment?

This question was addressed by Baxter et al. (2013) for two sources of nonidentical broadband noises (n1 and n2) presented simultaneously without lead- or lag-alone segments. They computed the relative likelihood that n1 or n2 would evoke spikes in a given space-map neuron from the conjunction of a rising envelope waveform and the relative amplitudes of n1 and n2 at any given time

$$\mathrm{RP}_{\mathrm{n1}} = p(\mathrm{n}_1)/\left[p(\mathrm{n}_1)+p(\mathrm{n}_2)\right] \tag{4.1}$$

The quantities $p(\mathrm{n}_1)$ and $p(\mathrm{n}_2)$ refer to the probability that n1 and n2, respectively, would evoke a response based on the conjunction of rising envelopes and the relative amplitudes of n1 and n2. Thus, $\mathrm{RP}_{\mathrm{n1}}$ is the relative probability (RP) that n1 would evoke a spike. Figure 4.10d applies Eq. 4.1 to a pair of nonidentical envelopes used in the experiments of Baxter et al. (2013). If the envelope hypothesis is valid, they argued, the proportion of head turns toward the source of n1 or n2 should be predicted by $\mathrm{RP}_{\mathrm{n1}}$, that is, if $\mathrm{RP}_{\mathrm{n1}}$ is large, the equation predicts a higher proportion of turns toward the source of n1 than toward the source of n2. If $\mathrm{RP}_{\mathrm{n1}}$ is small, there should be more turns toward n2 than toward n1.

Baxter et al. (2013) demonstrated that this is indeed the case. Figure 4.10e plots the proportion of head turns in three owls toward n1 against $\mathrm{RP}_{\mathrm{n1}}$. As shown, they demonstrated a strong linear relationship ($r^2 = 0.97$) between the proportion of turns toward n1 and $\mathrm{RP}_{\mathrm{n1}}$. Thus, the envelope hypothesis can be generalized from an explanation of localization dominance to a way of determining which of two simultaneously active sources dominates activity on the space map.

4.7 Concluding Remarks

The barn owl has fascinated auditory researchers and served as a model organism for decades now. Some theories, such as the Jeffress model for coding ITD and maps in the midbrain for the representation of the acoustic environment, have been confirmed and studied in depth in the barn owl. However, the barn owl may not represent a "typical" animal model for sound localization because it is equipped with a number of auditory specializations that allow it to hunt at night. More recently, data from other birds and related archosaurs have become available, providing more insight into the evolution of auditory systems (e.g., MacLeod et al. 2006; Kettler and Carr 2019). Some principles, such as encoding ITD in neuronal arrays of coincidence detectors and delay lines, as proposed by Jeffress (1948) over 70 years ago, seem to be a stable feature over millions of years, whereas others, such as the topographical representations in the midbrain, may be an apomorphic trait newly evolved in barn owls. The almost crystalline beauty and predictability of neuronal responses in the midbrain of barn owls and their tractable behavioral responses, be it swift head turns toward the sound source or pupillary dilation responses, opened research in avian sound localization to a completely new field, namely, population coding theory (Pouget et al. 2000). The brain infers the stimulus probability and uncertainty from its sensory inputs and encodes these parameters in the responses of neural populations. The sound localization behavior of the barn owl can be predicted by Bayesian decoding (Fischer and Pena 2011) of the populations of neurons in the optic tectum (Rich et al. 2015) and midbrain tegmentum (Cazettes et al. 2018). This new perspective on how the brain encodes and decodes stimulus statistics will lead to insights relevant to research in hearing and sensory processing in a variety of species, including humans.

Acknowledgments We thank Rachael C. Allison for her comments on the text and graphics. Experiments conducted by the authors have been approved by their Institutional Animal Care and Use Committees (IACUC). All authors have been approved by their IACUC.

Compliance with Ethics Requirements

Terry T. Takahashi declares that he has no conflict of interest.

Lutz Kettler declares that he has no conflict of interest.

Clifford Henry Keller III declares that he has no conflict of interest.

Avinash Deep Singh Bala declares that he has no conflict of interest.

References

Adolphs R (1993) Bilateral inhibition generates neuronal responses tuned to interaural level differences in the auditory brainstem of the barn owl. J Neurosci 13(9):3647–3668

Andeol G, Macpherson EA, Sabin AT (2013) Sound localization in noise and sensitivity to spectral shape. Hear Res 304:20–27. https://doi.org/10.1016/j.heares.2013.06.001

Aralla R, Ashida G, Köppl C (2018) Binaural responses in the auditory midbrain of chicken (Gallus gallus). Eur J Neurosci. https://doi.org/10.1111/ejn.13891

Ashida G, Carr CE (2011) Sound localization: Jeffress and beyond. Curr Opin Neurobiol 21(5):745–751. https://doi.org/10.1016/j.conb.2011.05.008

Bala AD, Takahashi TT (2000) Pupillary dilation response as an indicator of auditory discrimination in the barn owl. J Comp Physiol A 186(5):425–434

Bala AD, Spitzer MW, Takahashi TT (2003) Prediction of auditory spatial acuity from neural images on the owl's auditory space map. Nature 424(6950):771–774. https://doi.org/10.1038/nature01835

Bala AD, Spitzer MW, Takahashi TT (2007) Auditory spatial acuity approximates the resolving power of space-specific neurons. PLoS One 2(7):e675. https://doi.org/10.1371/journal.pone.0000675

Baxter CS, Nelson BS, Takahashi TT (2013) The role of envelope shape in the localization of multiple sound sources and echoes in the barn owl. J Neurophysiol 109(4):924–931. https://doi.org/10.1152/jn.00755.2012

Beach FA (1944) Responses of captive alligators to auditory stimulation. Am Naturalist 78:481–505

Beckert MV, Pavao R, Pena JL (2017) Distinct correlation structure supporting a rate-code for sound localization in the owl's auditory forebrain. eNeuro 4(3). https://doi.org/10.1523/ENEURO.0144-17.2017

Bierman HS, Thornton JL, Jones HG, Koka K, Young BA, Brandt C, Christensen-Dalsgaard J, Carr CE, Tollin DJ (2014) Biophysics of directional hearing in the American alligator (*Alligator mississippiensis*). J Exp Biol 217(7):1094–1107. https://doi.org/10.1242/jeb.092866

Boudreau JC, Tsuchitani C (1968) Binaural interaction in the cat superior olive S segment. J Neurophysiol 31(3):442–454

Brown CH, Schessler T, Moody D, Stebbins W (1982) Vertical and horizontal sound localization in primates. J Acoust Soc Am 72(6):1804–1811

Calford MB, Piddington RW (1988) Avian interaural canal enhances interaural delay. J Comp Physiol A 162(4):503–510. https://doi.org/10.1007/Bf00612515

Carr CE, Konishi M (1990) A circuit for detection of interaural time differences in the brain stem of the barn owl. J Neurosci 10(10):3227–3246

Carr CE, Soares D (2002) Evolutionary convergence and shared computational principles in the auditory system. Brain Behav Evol 59(5–6):294–311. https://doi.org/10.1159/000063565

Carr CE, Soares D, Smolders J, Simon JZ (2009) Detection of interaural time differences in the alligator. J Neurosci 29(25):7978–7990. https://doi.org/10.1523/JNEUROSCI.6154-08.2009

Carr CE, Shah S, McColgan T, Ashida G, Kuokkanen PT, Brill S, Kempter R, Wagner H (2015a) Maps of interaural delay in the owl's nucleus laminaris. J Neurophysiol 114(3):1862–1873. https://doi.org/10.1152/jn.00644.2015

Carr CE, Shah S, McColgan T, Ashida G, Kuokkanen PT, Brill S, Kempter R, Wagner H (2015b) Maps of interaural delay in the owl's nucleus laminaris. J Neurophysiol 114(3):1862–1873. https://doi.org/10.1152/jn.00644.2015

Catania KC (2013) Stereo and serial sniffing guide navigation to an odour source in a mammal. Nat Commun 4:1441. https://doi.org/10.1038/ncomms2444

Cazettes F, Fischer BJ, Pena JL (2014) Spatial cue reliability drives frequency tuning in the barn Owl's midbrain. elife 3:e04854. https://doi.org/10.7554/eLife.04854

Cazettes F, Fischer BJ, Beckert MV, Pena JL (2018) Emergence of an adaptive command for orienting behavior in premotor brainstem neurons of barn owls. J Neurosci 38(33):7270–7279. https://doi.org/10.1523/JNEUROSCI.0947-18.2018

Cohen YE, Knudsen EI (1996) Representation of frequency in the primary auditory field of the barn owl forebrain. J Neurophysiol 76(6):3682–3692

Cohen YE, Miller GL, Knudsen EI (1998) Forebrain pathway for auditory space processing in the barn owl. J Neurophysiol 79(2):891–902

Dean I, Robinson BL, Harper NS, McAlpine D (2008) Rapid neural adaptation to sound level statistics. J Neurosci 28(25):6430–6438. https://doi.org/10.1523/JNEUROSCI.0470-08.2008

Delgutte B, Joris PX, Litovsky RY, Yin TC (1999) Receptive fields and binaural interactions for virtual-space stimuli in the cat inferior colliculus. J Neurophysiol 81(6):2833–2851
Dizon RM, Colburn HS (2006) The influence of spectral, temporal, and interaural stimulus variations on the precedence effect. J Acoust Soc Am 119(5 Pt 1):2947–2964
Edut S, Eilam D (2004) Protean behavior under barn-owl attack: voles alternate between freezing and fleeing and spiny mice flee in alternating patterns. Behav Brain Res 155(2):207–216. https://doi.org/10.1016/j.bbr.2004.04.018
Euston DR, Takahashi TT (2002) From spectrum to space: the contribution of level difference cues to spatial receptive fields in the barn owl inferior colliculus. J Neurosci 22(1):264–293
Fischer BJ, Pena JL (2011) Owl's behavior and neural representation predicted by Bayesian inference. Nat Neurosci 14(8):1061–1066. https://doi.org/10.1038/nn.2872
Fischer BJ, Seidl AH (2014) Resolution of interaural time differences in the avian sound localization circuit – a modeling study. Front Comput Neurosci 8:1–10
Fischer BJ, Anderson CH, Pena JL (2009) Multiplicative auditory spatial receptive fields created by a hierarchy of population codes. PLoS One 4(11):e8015. https://doi.org/10.1371/journal.pone.0008015
Funabiki K, Koyano K, Ohmori H (1998) The role of GABAergic inputs for coincidence detection in the neurones of nucleus laminaris of the chick. J Physiol 508(Pt 3):851–869
Gaik W (1993a) Combined evaluation of interaural time and intensity differences: psychoacoustic results and computer modeling. J Acoust Soc Am 94(1):98–110
Gaik W (1993b) Combined evaluation of interaural time and intensity differences: psychoacoustic results and computer modeling. J Acoust Soc Am 94:98–110
Gardener MB (1968) Historical background of the Haas and/or precedence effect. J Acoust Soc Am 43:1243–1248
Gold JI, Knudsen EI (2000) A site of auditory experience-dependent plasticity in the neural representation of auditory space in the barn owl's inferior colliculus. J Neurosci 20(9):3469–3486
Harper NS, McAlpine D (2004) Optimal neural population coding of an auditory spatial cue. Nature 430(7000):682–686. https://doi.org/10.1038/nature02768
Hartmann WM, Rakerd B, Crawford ZD, Zhang PX (2016) Transaural experiments and a revised duplex theory for the localization of low-frequency tones. J Acoust Soc Am 139(2):968–985. https://doi.org/10.1121/1.4941915
Hausmann L, von Campenhausen M, Endler F, Singheiser M, Wagner H (2009) Improvements of sound localization abilities by the facial ruff of the barn owl (*Tyto alba*) as demonstrated by virtual ruff removal. PLoS One 4(11):e7721. https://doi.org/10.1371/journal.pone.0007721
Heiligenberg W (1991) Neural nets and electric fish. MIT, Cambridge
Higgs DM, Brittan-Powell EF, Soares D, Souza MJ, Carr CE, Dooling RJ, Popper AN (2002) Amphibious auditory responses of the American alligator (*Alligator mississipiensis*). J Comp Physiol A 188(3):217–223. https://doi.org/10.1007/s00359-002-0296-8
Jeffress L (1948) A place theory of sound localization. J Comp Physiol Psychol 41:35–39
Julesz B (1971) Foundations of cyclopean perception. University of Chicago Press, Chicago
Kahneman D (2003) A perspective on judgment and choice: mapping bounded rationality. Am Psychol 58(9):697–720. https://doi.org/10.1037/0003-066X.58.9.697
Karten H, Hodos W (1967) A stereotaxic atlas of the brain of the pigeon (*Columba livia*). The Johns Hopkins Press, Baltimore
Keller CH, Takahashi TT (1996a) Binaural cross-correlation predicts the responses of neurons in the owl's auditory space map under conditions simulating summing localization. J Neurosci 16(13):4300–4309
Keller CH, Takahashi TT (1996b) Responses to simulated echoes by neurons in the barn owl's auditory space map. J Comp Physiol A 178(4):499–512
Keller CH, Takahashi TT (2015) Spike timing precision changes with spike rate adaptation in the owl's auditory space map. J Neurophysiol 114(4):2204–2219. https://doi.org/10.1152/jn.00442.2015
Keller CH, Hartung K, Takahashi TT (1998) Head-related transfer functions of the barn owl: measurement and neural responses. Hear Res 118:13–34.

Keller CH, Hartung K, Takahashi TT (1998a) Head-related transfer functions of the barn owl: measurement and neural responses. Hear Res 118(1–2):13–34. https://doi.org/10.1016/S0378-5955(98)00014-8

Keller CH, Hartung K, Takahashi TT (1998b) Head-related transfer functions of the barn owl: measurement and neural responses. Hear Res 118(1–2):13–34

Kettler L, Carr C (2019) Neural maps of interaural time difference in the American alligator: a stable feature in modern archosaurs. J Neurosci. https://doi.org/10.1523/JNEUROSCI.2989-18.2019

Kettler L, Christensen-Dalsgaard J, Larsen ON, Wagner H (2016) Low frequency eardrum directionality in the barn owl induced by sound transmission through the interaural canal. Biol Cybern 110(4–5):333–343. https://doi.org/10.1007/s00422-016-0689-3

Kettler L, Griebel H, Ferger R, Wagner H (2017) Combination of interaural level and time difference in azimuthal sound localization in owls. eNeuro 4(6). https://doi.org/10.1523/ENEURO.0238-17.2017

Knudsen EI (1982) Auditory and visual maps of space in the optic tectum of the owl. J Neurosci 2(9):1177–1194

Knudsen EI (1999) Mechanisms of experience-dependent plasticity in the auditory localization pathway of the barn owl. J Comp Physiol A 185(4):305–321

Knudsen EI, Konishi M (1979) Sound localization by the barn owl (*Tyto alba*) measured with the search coil technique. J Comp Physiol A 133:1–11

Knudsen EI, Knudsen PF, Masino T (1993) Parallel pathways mediating both sound localization and gaze control in the forebrain and midbrain of the barn owl. J Neurosci 13(7):2837–2852

Köppl C (1997) Phase locking to high frequencies in the auditory nerve and cochlear nucleus magnocellularis of the barn owl, *Tyto alba*. J Neurosci 17(9):3312–3321

Köppl C, Carr CE (2008) Maps of interaural time difference in the chicken's brainstem nucleus laminaris. Biol Cybern 98(6):541–559. https://doi.org/10.1007/s00422-008-0220-6

Köppl C, Manley GA, Konishi M (2000) Auditory processing in birds. Curr Op Neurobiol 10(4):474–481. https://doi.org/10.1016/S0959-4388(00)00110-0

Kuba H, Ishii TM, Ohmori H (2006) Axonal site of spike initiation enhances auditory coincidence detection. Nature 444(7122):1069–1072. https://doi.org/10.1038/nature05347

Litovsky RY, Colburn HS, Yost WA, Guzman SJ (1999) The precedence effect. J Acoust Soc Am 106(4 Pt 1):1633–1654

MacLeod KM, Soares D, Carr CE (2006) Interaural timing difference circuits in the auditory brainstem of the emu (*Dromaius novaehollandiae*). J Comp Neurol 495:185–201

Manley GA, Köppl C, Konishi M (1988) A neural map of interaural intensity differences in the brain-stem of the barn owl. J Neurosci 8(8):2665–2676

Manley GA, Köppl C, Yates GK (1997) Activity of primary auditory neurons in the cochlear ganglion of the emu *Dromaius novaehollandiae*: spontaneous discharge, frequency tuning, and phase locking. J Acoust Soc Am 101(3):1560–1573. https://doi.org/10.1121/1.418273

Mazer J (1995) Integration of parallel processing streams in the inferior colliculus of the owl. Doctoral thesis, California Institute of Technology, Pasadena

Mazer JA (1998) How the owl resolves auditory coding ambiguity. Proc Natl Acad Sci U S A 95(18):10932–10937

Mogdans J, Knudsen EI (1994) Representation of interaural level difference in the VLVp, the first site of binaural comparison in the barn owl's auditory system. Hear Res 74(1–2):148–164

Moiseff A, Konishi M (1983) Binaural characteristics of units in the owl's brainstem auditory pathway: precursors of restricted spatial receptive fields. J Neurosci 3:2553–2562

Nelson BS, Takahashi TT (2008a) Independence of echo-threshold and echo-delay in the barn owl. Plos One 3(10):e3598. https://doi.org/10.1371/journal.pone.0003598

Nelson BS, Takahashi TT (2008b) Independence of echo-threshold and echo-delay in the barn owl. PLoS One 3(10):e3598. https://doi.org/10.1371/journal.pone.0003598

Nelson BS, Takahashi TT (2010) Spatial hearing in echoic environments: the role of the envelope in owls. Neuron 67(4):643–655. https://doi.org/10.1016/j.neuron.2010.07.014

Overholt EM, Rubel EW, Hyson RL (1992) A circuit for coding interaural time differences in the chick brainstem. J Neurosci 12(5):1698–1708

Palanca-Castan N, Köppl C (2015a) Change in the coding of interaural time difference along the tonotopic axis of the chicken nucleus laminaris. Front Neural Circuit 9:43. https://doi.org/10.3389/fncir.2015.00043
Palanca-Castan N, Köppl C (2015b) In vivo recordings from low-frequency nucleus laminaris in the barn owl. Brain Behav Evol 85(4):271–286. https://doi.org/10.1159/000433513
Palanca-Castan N, Köppl C (2015c) In vivo recordings from low-frequency nucleus laminaris in the barn owl. Brain Behav Evol 85(4):271–286. https://doi.org/10.1159/000433513
Parks TN, Rubel EW (1975) Organization and development of brain stem auditory nuclei of the chicken: organization of projections from n. magnocellularis to n. laminaris. J Comp Neurol 164(4):435–448. https://doi.org/10.1002/cne.901640404
Payne RS (1971) Acoustic location of prey by barn owls (*Tyto alba*). J Exp Biol 54(3):535–573
Pouget A, Dayan P, Zemel R (2000) Information processing with population codes. Nat Rev Neurosci 1(2):125–132. https://doi.org/10.1038/35039062
Rich D, Cazettes F, Wang Y, Pena JL, Fischer BJ (2015) Neural representation of probabilities for Bayesian inference. J Comput Neurosci 38(2):315–323. https://doi.org/10.1007/s10827-014-0545-1
Rose JE, Kitzes LM, Gibson MM, Hind JE (1974) Observations on phase-sensitive neurons of anteroventral cochlear nucleus of the cat: nonlinearity of cochlear output. J Neurophysiol 37(1):218–253. https://doi.org/10.1152/jn.1974.37.1.218
Saberi K, Takahashi Y, Farahbod H, Konishi M (1999) Neural bases of an auditory illusion and its elimination in owls. Nat Neurosci 2(7):656–659
Sakitt B (1973) Indices of discriminability. Nature 241(5385):133–134
Sayers B, Cherry E (1957) Mechanism of binaural fusion in the hearing of speech. J Acoust Soc Am 29:973–987
Shackleton T, Meddis R, Hewitt M (1992) Across frequency integration in a model of lateralization. J Acoust Soc Am 91(4):2276–2279
Snow W (1954) The effects of arrival time on stereophonic localization. J Acoust Soc Am 26:1071–1074
Spezio ML, Takahashi TT (2003) Frequency-specific interaural level difference tuning predicts spatial response patterns of space-specific neurons in the barn owl inferior colliculus. J Neurosci 23(11):4677–4688. https://doi.org/10.1523/JNEUROSCI.23-11-04677.2003
Spezio ML, Keller CH, Marrocco RT, Takahashi TT (2000) Head-related transfer functions of the Rhesus monkey. Hear Res 144(1–2):73–88. https://doi.org/10.1016/s0378-5955(00)00050-2
Strutt JW (1907) On our perception of sound direction. Philos Mag 13:214–232. https://doi.org/10.1080/14786440709463595
Sullivan WE, Konishi M (1984) Segregation of stimulus phase and intensity coding in the cochlear nucleus of the barn owl. J Neurosci 4(7):1787–1799
Takahashi TT, Keller CH (1992) Commissural connections mediate inhibition for the computation of interaural level difference in the barn owl. J Comp Physiol A 170(2):161–169
Takahashi T, Konishi M (1986) Selectivity for interaural time difference in the owl's midbrain. J Neurosci 6(12):3413–3422
Takahashi TT, Wagner H, Konishi M (1989) Role of commissural projections in the representation of bilateral auditory space in the barn owl's inferior colliculus. J Comp Neurol 281(4):545–554. https://doi.org/10.1002/cne.902810405
Takahashi TT, Barberini CL, Keller CH (1995a) An anatomical substrate for the inhibitory gradient in the VLVp of the owl. J Comp Neurol 358(2):294–304. https://doi.org/10.1002/cne.903580210
Takahashi TT, Barberini CL, Keller CH (1995b) An anatomical substrate for the inhibitory gradient in the VLVp of the owl. J Comp Neurol 358(2):294–304
Takahashi TT, Bala AD, Spitzer MW, Euston DR, Spezio ML, Keller CH (2003) The synthesis and use of the owl's auditory space map. Biol Cybern 89:378–387
Tellers P, Lehmann J, Fuhr H, Wagner H (2017) Envelope contributions to the representation of interaural time difference in the forebrain of barn owls. J Neurophysiol 118(3):1871–1887. https://doi.org/10.1152/jn.01166.2015

Tollin DJ, Yin TCT (2002) The coding of spatial location by single units in the lateral superior olive of the cat. II. The determinants of spatial receptive fields in azimuth. J Neurosci 22:1468–1479
Tsuchitani C (1977) Functional organization of lateral cell groups of cat superior olivary complex. J Neurophysiol 40(2):296–318
Vedurmudi AP, Goulet J, Christensen-Dalsgaard J, Young BA, Williams R, van Hemmen JL (2016) How internally coupled ears generate temporal and amplitude cues for sound localization. Phys Rev Lett 116(2):028101. https://doi.org/10.1103/PhysRevLett.116.028101
von Campenhausen M, Wagner H (2006) Influence of the facial ruff on the sound-receiving characteristics of the barn owl's ears. J Comp Physiol A192(10):1073–1082. https://doi.org/10.1007/s00359-006-0139-0
Vonderschen K, Wagner H (2009) Tuning to interaural time difference and frequency differs between the auditory arcopallium and the external nucleus of the inferior colliculus. J Neurophysiol 101(5):2348–2361. https://doi.org/10.1152/jn.91196.2008
Vonderschen K, Wagner H (2012) Transformation from a pure time delay to a mixed time and phase delay representation in the auditory forebrain pathway. J Neurosci 32(17):5911–5923. https://doi.org/10.1523/JNEUROSCI.5429-11.2012
Vonderschen K, Wagner H (2014) Detecting interaural time differences and remodeling their representation. Trends Neurosci 37(5):289–300. https://doi.org/10.1016/j.tins.2014.03.002
Wagner H (1993) Sound-localization deficits induced by lesions in the barn owl's auditory space map. J Neurosci 13(1):371–386
Wagner H, Takahashi T, Konishi M (1987) Representation of interaural time difference in the central nucleus of the barn owl's inferior colliculus. J Neurosci 7(10):3105–3116
Wallach H, Newman E, Rosenzweig M (1949) The precedence effect in sound localization. Am J Psycho 62:315–336
Whitchurch EA, Takahashi TT (2006) Combined auditory and visual stimuli facilitate head saccades in the barn owl (*Tyto alba*). J Neurophysiol 96(2):730–745. https://doi.org/10.1152/jn.00072.2006
Wightman FL, Kistler DJ (1989) Headphone simulation of free-field listening. I: stimulus synthesis. J Acoust Soc Am 85(2):858–867
Wightman FL, Kistler DJ (1997) Monaural sound localization revisited. J Acoust Soc Am 101(2):1050–1063
Willis KL, Carr CE (2017) A circuit for detection of interaural time differences in the nucleus laminaris of turtles. J Exp Biol 220(Pt 22):4270–4281. https://doi.org/10.1242/jeb.164145
Witmer LM (1990) The craniofacial air sac system of Mesozoic birds (Aves). Zool J Linnean Soc 100:327–378
Witmer LM, Ridgely RC (2008) The paranasal air sinuses of predatory and armored dinosaurs (Archosauria: Theropoda and Ankylosauria) and their contribution to cephalic structure. Anat Rec 291:1362–1388. https://doi.org/10.1002/ar.20794
Yang L, Monsivais P, Rubel EW (1999) The superior olivary nucleus and its influence on nucleus laminaris: a source of inhibitory feedback for coincidence detection in the avian auditory brainstem. J Neurosci 19(6):2313–2325
Yin TC (1994) Physiological correlates of the precedence effect and summing localization in the inferior colliculus of the cat. J Neurosci 14(9):5170–5186
Yin TCT, Chan JCK (1987) Neural mechanisms underlying interaural time sensitivity to tones and noise. Auditory Function John Wiley and Sons NY
Yin TC, Chan JC, Carney LH (1987) Effects of interaural time delays of noise stimuli on low-frequency cells in the cat's inferior colliculus. III. Evidence for cross-correlation. J Neurophysiol 58(3):562–583
Zurek PM (1980) The precedence effect and its possible role in the avoidance of interaural ambiguities. J Acoust Soc Am 67(3):953–964

Chapter 5
Binaural Hearing by the Mammalian Auditory Brainstem: Joint Coding of Interaural Level and Time Differences by the Lateral Superior Olive

Zoe L. Owrutsky, Victor Benichoux, and Daniel J. Tollin

5.1 Introduction

Auditory-guided behavior depends on the fidelity by which the auditory nervous system constructs a representation of sources of sound in the environment. Thus, the auditory system must represent both *what* produced a sound and *where* the sound originated relative to the listener. Knowledge of the source of a sound and what produced it facilitates the segregation of competing sounds and also the initiation of appropriate behavioral responses (Bregman 1990). For example, a predator like a cat might move toward the sound produced by a mouse rustling in the leaves while the mouse might freeze or move away from the sounds of the approaching cat (Table 5.1).

This review considers the mechanisms by which sound location is encoded by neurons in the earliest parts of the auditory system, the brainstem, based on the two binaural acoustical cues to horizontal sound location, the interaural level (ILD) and time (ITD) differences (Figs. 5.1 and 5.2). According to the duplex theory of sound localization (Rayleigh 1907), ILDs and ITDs are used to localize the source of different kinds of sound: low-frequency sounds are located based on ITDs and high-frequency sounds are based on ILDs. For over a century, this has served as the main framework to study how sound sources are localized. In addition, decades of anatomical and physiological studies have examined how the duplex theory could be wired in the brain. These experiments focused on the part of the auditory brainstem

Z. L. Owrutsky (✉) · D. J. Tollin
Department of Physiology and Biophysics, University of Colorado Anschutz Medical Campus, Aurora, CO, USA
e-mail: zoe.owrutsky@ucdenver.edu; daniel.tollin@ucdenver.edu

V. Benichoux
Unit of Genetics and Physiology of Hearing, Department of Neuroscience, Institut Pasteur, Paris, France
e-mail: benichoux@crans.org

© Springer Nature Switzerland AG 2021

R. Y. Litovsky et al. (eds.), *Binaural Hearing*, Springer Handbook of Auditory Research 73, https://doi.org/10.1007/978-3-030-57100-9_5

Table 5.1 List of abbreviations

Abbreviation	Definition
ABR	Auditory brainstem response
ANF	Auditory nerve fiber
AM	Amplitude modulated
AVCN	Anteroventral cochlear nucleus
BIC	Binaural interaction component
BMLD	Binaural masking level difference
CN	Cochlear nucleus
DNLL	Dorsal nucleus of the lateral lemniscus
EPSP	Excitatory postsynaptic potential
GBC	Globular bushy cell
IC	Inferior colliculus
ILD	Interaural level difference
IPSP	Inhibitory postsynaptic potential
ITD	Interaural time difference
LSO	Lateral superior olive
MNTB	Medial nucleus of the trapezoid body
MSO	Medial superior olive
SBC	Spherical bushy cell
SPL	Sound pressure level

where the inputs from each ear first converge: the superior olivary complex (Fig. 5.3). This area is divided into several structures that include the lateral superior olive (LSO) and medial superior olive (MSO) on each side. Particular focus is paid here to how ILDs and ITDs are encoded in parallel but with anatomically separate neural pathways (Fig. 5.3). There have been numerous reviews in recent years (e.g., Joris and Yin 2007; Grothe et al. 2010) on encoding of ITDs in the temporal fine structure (Fig. 5.2) of low-frequency sounds by neurons comprising the brainstem pathways through the MSO (Fig. 5.3b), so ITD coding by the MSO is covered only briefly here. Instead, the focus is on the encoding of ILDs and ITDs by neurons comprising the brainstem pathways through the LSO (Fig. 5.3a).

The motivation for the narrow focus of this review is the relatively recent revelation that the intricate specializations in the neural circuits responsible for the initial encoding of ILDs by the LSO allows these neurons to also be exquisitely sensitive to various types of ITDs, including temporal fine structure, envelope, and transient stimuli (Fig. 5.2). This new knowledge is important because it provides possible neuroanatomical bases for the perceptual sensitivity to these ITDs, which represents a departure from the classical duplex theory of sound localization, as well as an enhanced sensitivity to ILDs for stimuli with temporal fluctuations (see Stecker, Bernstein, and Brown, Chap. 6). For example, Zhang and Wright (2009) reported lower ILD thresholds for sinusoidally amplitude-modulated tones than for pure tones. Laback et al. (2017) and Brown and Tollin (2016) recently examined temporal constraints on the behavioral sensitivity to the ILD. Via computational models of LSO function, these latter studies attributed their results to the mechanisms by

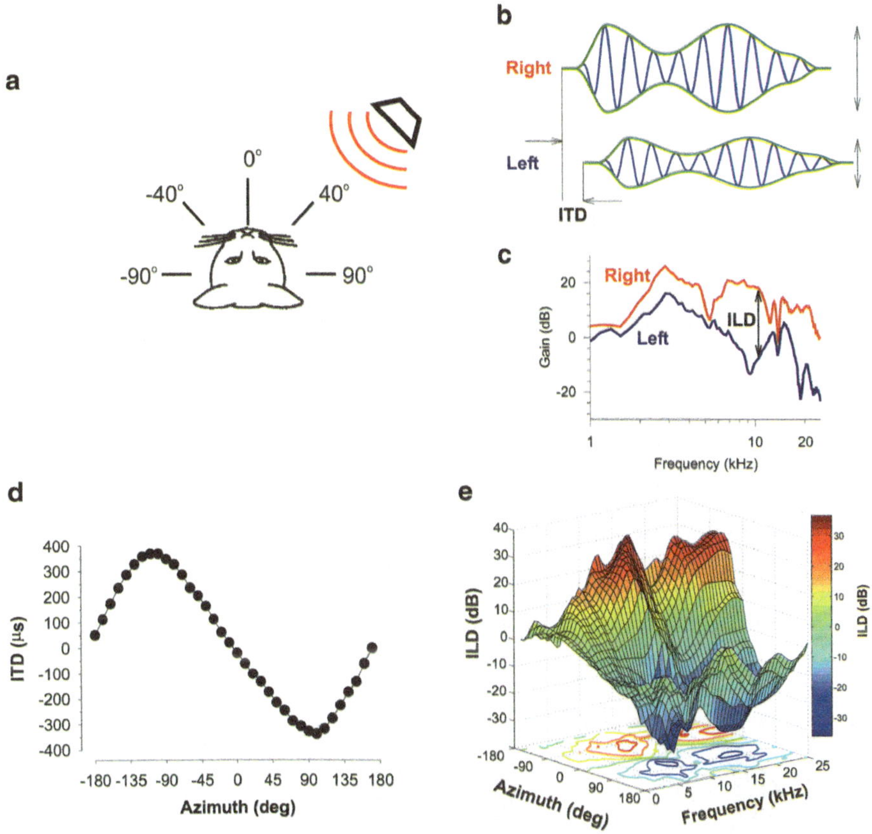

Fig. 5.1 The acoustical cues to sound location. (**a**) Broadband sounds are presented repeatedly from a loudspeaker at 40°. (**b**) Acoustical responses recorded by microphones near the eardrums in the left and right ears to sound presented at 40° illustrate two cues to location. The interaural time difference (ITD) is represented by the relative differences in the onset times of the impulse responses at each ear (*right and left gray arrows*). The interaural level difference (ILD) is represented by the difference in amplitude of the sound at each ear. (**c**) Head-related transfer function (HRTF) represents the spectrum of the acoustical responses in terms of the acoustical gain introduced by the presence of the head and pinnae relative to the sound level in response to the broadband sound recorded in the absence of the subject. The monaural spectral cues are captured by the changes in the shapes of the spectra as a function of source location. *Red*, right HRTF; *blue*, left HRTF. (**d**) ITD as a function of sound source azimuth. (**e**) Joint spatial and frequency dependence of ILDs. The ILD is indicated by the relative differences in the gains of the HRTFs as a function of frequency (e.g., as shown in **c** for a source at 40°). ILDs are a complicated function of azimuth and frequency for high-frequency stimuli. For illustrative purposes, the sources were restricted to the horizontal plane and only in the frontal hemisphere for frequencies between 3 and 30 kHz. The HRTFs, ITDs, and ILDs were derived directly from the acoustical measurements of Tollin and Koka (2009)

which the neural circuits that encode ILDs can also encode particular kinds of ITDs (see Dietz and Ashida, Chap. 10), following recent biophysically inspired models of LSO that have focused on their abilities to encode envelope ITDs (Ashida et al. 2016, 2017).

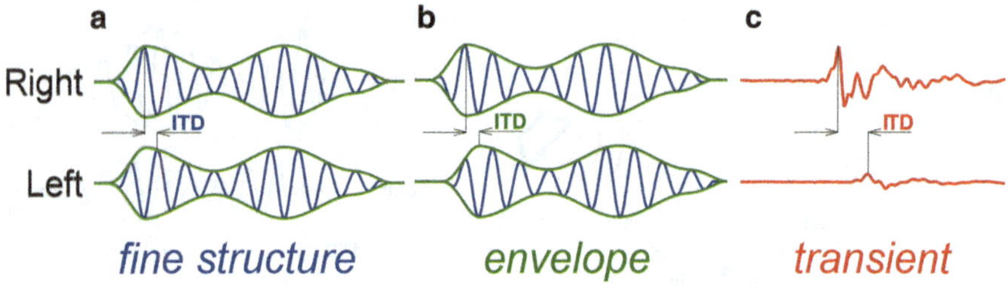

Fig. 5.2 ITDs represent the difference between when a sound reaches the left ear and the right ear. ITDs can occur between the fine structure of the sound (*blue*), the amplitude envelope of the sound (*green*), or for differences in the onset times of abrupt transient sounds (*red*)

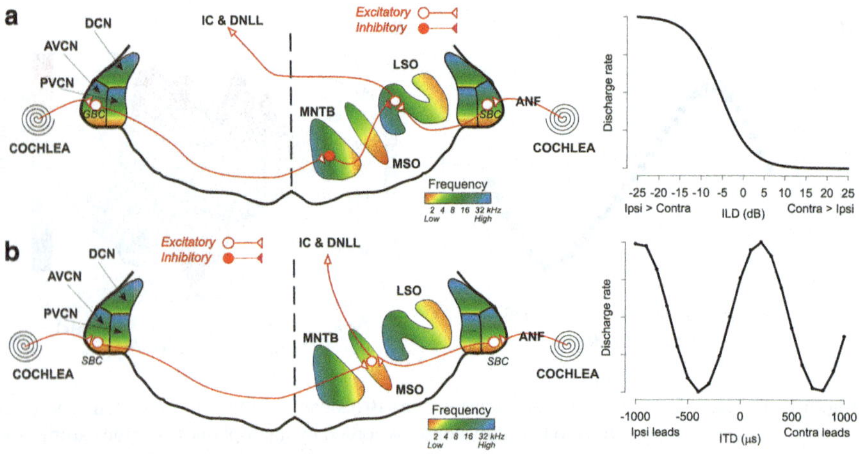

Fig. 5.3 Illustration of a frontal section through the auditory brainstem showing the ascending pathways through the lateral superior olive (LSO; **a**) and medial superior olive (MSO; **b**). (**a**) LSO neurons receive bilateral inputs from the left and right ears. The input from the ipsilateral ear (right ear in the figure) via the spherical bushy cells (SBCs) is excitatory (*open circles*). The input from the contralateral ear (left ear in the figure) comes via the globular bushy cells (GBCs) of the contralateral anteroventral cochlear nucleus (AVCN) and is ultimately inhibitory (*solid circles*) due to the additional synapse with the ipsilateral medial nucleus of the trapezoid body (MNTB). The interplay of ipsilateral (Ipsi) excitation and contralateral (Contra) inhibition confers ILD sensitivity by LSO neurons (*right*). LSO neurons send excitatory projections across the midline to the contralateral IC and dorsal nucleus of the lateral lemniscus (DNLL) and inhibitory projections (not shown) to the ipsilateral IC and DNLL. *Colors*, tonotopic organization (i.e., frequency sensitivity) that shows that the neurons comprising the MNTB and LSO are sensitive to predominantly high-frequency sounds. The SBCs and GBCs receive excitatory input from the auditory nerve fibers (ANFs). (**b**) MSO neurons receive excitatory inputs from the two ears via the SBCs. MSO neurons respond when they receive coincident arrival of excitatory inputs from the SBCs and are thus exquisitely sensitive to ITDs. The SBCs receive excitatory input from the ANFs. DCN dorsal cochlear nucleus, PVCN posteroventral cochlear nucleus

A clearer understanding of the neural circuits that encode fine structure, envelope, and transient ITDs may also help to understand the neuroanatomical bases for binaural hearing deficits and optimize binaural hearing aids and prostheses. For example, ILDs and envelope ITDs appear to be the only binaural cues to location that users of bilateral cochlear implants can utilize (see Ricketts and Kan, Chap. 13). Additionally, early temporary conductive hearing loss, hearing impairment, aging, and disease can severely impair the utilization of certain kinds of ITDs (see Gallun, Srinivasan, and Diedesch, Chap. 11). Finally, some of these binaural hearing impairments have been shown to be predicted by a noninvasive biomarker to binaural hearing, the binaural interaction component (BIC) of the auditory brainstem response (ABR; Laumen et al. 2016). The BIC of the ABR has its origin in the circuit comprising the LSO (Benichoux et al. 2018), thus shedding new light on the importance of this traditionally understudied and certainly underappreciated circuit.

5.2 The Acoustical Cues to Sound Source Location

There are three primary acoustical characteristics of sounds, or cues, which are used to determine the spatial position of a sound source in mammals (see also Hartmann, Chap. 2). The cues are generated by the interaction of propagating sound waves with the anatomical structures supporting the receptors of the ear, namely, the head and external ear (or pinnae), and not by the topographic organization of the receptors themselves in the cochlea. Examples of the three cues are shown in Fig. 5.1. Two cues, ITD (Fig. 5.1b, d) and ILD (Fig. 5.1c, e), rely on the comparison of acoustic input to the two ears and thus result directly from the fact that the two ears themselves are sampling the sound waves at different spatial locations due to their positions on opposite sides of the head. These two *binaural* cues specify primarily sound location in the horizontal dimension or azimuth. The third cue, spectral shape (Fig. 5.1c), results from the way that sound waves are distorted or filtered by the complicated convolutions and folds of the pinna. Spectral shape cues, which specify primarily sound location in elevation and whether a sound is in the front of or behind the observer, are not discussed in this review (see Young and Davis 2002).

5.2.1 *Interaural Level Differences*

The two ears are separated by the head, which is a solid obstacle. Consequently, for sounds with short wavelengths (i.e., high frequency), the head essentially creates an acoustic shadow for the far ear because sounds with wavelengths on the order of the diameter of the head and smaller are reflected off the near side of the head. Thus, sound arriving at the ear farthest from the source is attenuated, as evidenced by the acoustical gain, the sound level measured at the eardrum relative to the sound level measured in the absence of a subject, being less than 0 dB for most frequencies

(Fig. 5.1c). Additionally, the pinna nearest the source amplifies sound for some frequencies, evidenced by acoustic gains being greater than 0 dB for frequencies greater than ~2 kHz. Attenuation of sound at the far ear and an increase in gain at the near ear create direction-dependent *differences* in the amplitudes or levels of the sounds that reach the two ears, the ILD. Because of the joint effects of sound level at the two ears, ILD magnitudes are a complex function of both source azimuth and sound frequency (Fig. 5.1e). ILDs are small in magnitude (<6 dB) for sources near the midline and for low-frequency sounds at any location and increase in magnitude (>30 dB) for high-frequency sounds and more lateral locations (Fig. 5.1e). Therefore, ILDs are believed to be primarily useful for localization of high-frequency sounds. ILDs that are large enough to be useful for sound localization are available even in small mammals with small heads provided that the sound source contains high enough frequencies with wavelengths shorter than the head diameter.

5.2.2 *Interaural Time Differences: Fine Structure, Envelope, and Transient*

ITDs occur because the ears are physically separated in space by the head so that direction-dependent differences in path lengths to each ear from a source will generate different times of arrival of the sound at the two ears (Fig. 5.1b). For example, a sound from a source on the right side (Fig. 5.1a) will arrive at the right ear a short time before it arrives at the left ear, whereas a sound from directly ahead will arrive at the ears at the same time. The range of ITD magnitudes is proportional to the head diameter, so species with larger heads, hence larger distances between the two ears, exhibit a larger range of ITDs. For example, in cats, which have head diameters of 6–7 cm (Tollin and Koka 2009), ITDs increase monotonically, with the azimuth along the horizontal plane up to a maximum of 350–400 μs (Fig. 5.1d). The diameter of the human head is nearly twice that of the cat, on the order of 15 cm. Consequently, the maximum ITD in humans is ~800 μs (Benichoux et al. 2016). Yet smaller mammals, such as some species of bats and small rodents like mice and rats (Koka et al. 2008), have head diameters of just a few centimeters or less, resulting in maximum ITDs that can be as small as a few tens of microseconds.

Natural sounds are composed of complex mixtures of abrupt onsets (i.e., transients) and amplitude and frequency modulations. These complex temporal modulations in sound create different kinds of ITD cues. The first type of ITD, *fine-structure ITD*, refers to differences in timing of the fine structure of the sounds arriving at the two ears (Fig. 5.2a). Fine-structure ITDs depend on both sound azimuth and frequency (Kuhn and Burnett 1977; Benichoux et al. 2016), and for a given source location, can be nearly 50% larger in magnitude for low frequencies than for higher frequencies. In small mammals such as mice and rats, fine-structure ITDs are not useful for sound localization (Wesolek et al. 2010). A second type of ITD, *envelope ITD,* occurs in the low-frequency amplitude envelopes of typically higher frequency sounds (Fig. 5.2b). Finally, a third type of ITD, *transient ITD*, occurs in transient sounds (Fig. 5.2c) such as clicks that are difficult to characterize in terms of fine structure or envelope. Recent evidence suggests that envelope and transient ITDs can be used by small mammals such as rats that cannot use fine-structure ITDs (Li et al. 2019).

5.3 Neural Processing of Sound Localization Cues in the Auditory Brainstem

Given that there are two physically distinct binaural cues to the horizontal position of sound, ILD and ITD, naturally suggests that two different kinds of neural mechanisms are needed to encode them: one group of neurons to extract differences in the timing of sounds at the ears (Fig. 5.3b) and another to extract differences in intensity (Fig. 5.3a). Stimulus timing and intensity are two fundamentally different physical quantities that are encoded by the auditory periphery in different ways. Stimulus intensity is encoded by systematic increases in the number of action potentials fired or discharge rate (Fig. 5.4). Stimulus timing is encoded via neurons locking the timing of action potentials to amplitude fluctuations, a process called phase locking. Motivated by the duplex theory and also because of the different requirements for the initial encoding of intensity and timing, it has long been thought that ILDs and ITDs must logically be encoded in anatomically separate but parallel pathways in the auditory system (Fig. 5.3; see Grothe et al. 2010 for a review). The first places in the auditory pathway to receive large-scale and systematic converging inputs from both ears are two of the primary nuclei comprising the superior olivary complex, the MSO (Fig. 5.3b) and the LSO (Fig. 5.3a). Historically, low-frequency fine-structure ITDs were thought to be coded in the MSO and high-frequency ILDs in the LSO, providing the anatomical and physiological bases for the well-known duplex theory of sound localization (Tollin 2003; Grothe et al. 2010). However, new evidence has emerged to suggest that all three types of ITDs are also encoded by the LSO.

5.3.1 Coding of Interaural Level Differences in the Lateral Superior Olive

LSO neurons are sensitive to ILDs because they are inhibited by sounds presented to the contralateral ear and excited by sounds presented to the ipsilateral ear (Fig. 5.3a; Boudreau and Tsuchitani 1968). Many neurons in higher auditory areas, including the dorsal nucleus of the lateral lemniscus (DNLL), inferior colliculus (IC), medial geniculate body, and auditory cortical areas, are sensitive to changes in sound location, typically for sounds in the contralateral hemifield (Irvine 1986).

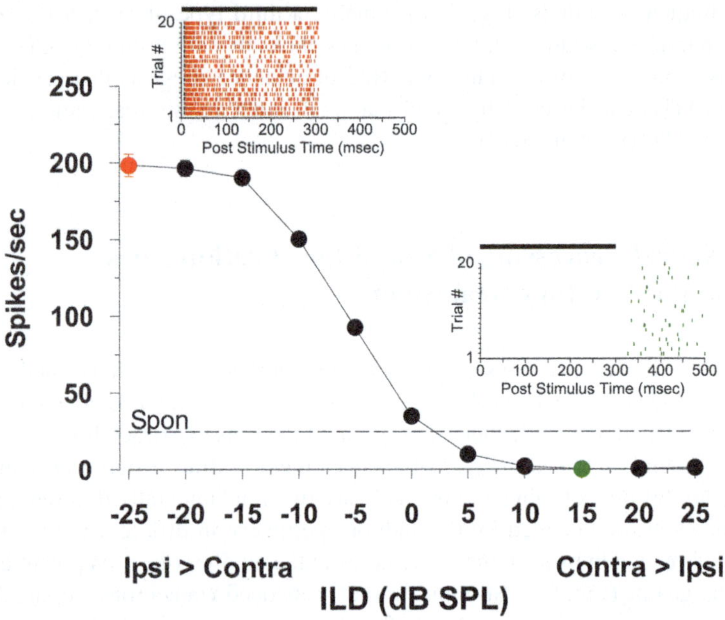

Fig. 5.4 ILD sensitivity of an LSO neuron. The response to 300-ms-duration tones at the best frequency of the neuron as a function of the ILD is shown as average spikes per second. The ILD was varied by holding constant the sound level to the ipsilateral excitatory ear and varying the sound level of the stimulus presented to the contralateral inhibitory ear. The ILD is defined as the difference between the sound level at the contralateral ear and that at the ipsilateral ear, so negative ILDs indicate greater sound levels at the excitatory ear. LSO neurons respond robustly when ILDs favor the ipsilateral ear (*red circle*) and are often completely inhibited even below spontaneous activity levels when ILDs favor the contralateral ear (*green circle*). *Horizontal line*, spontaneous firing rate (Spon). SPL, sound pressure level. *Insets:* dot rasters of responses to 20 presentations of the stimuli at an ILD of −25 dB (*red*) and + 15 dB (*green*), the latter demonstrating that spontaneous spiking can be inhibited. Data replotted from Tollin and Yin (2002a)

These neurons likely derive their spatial sensitivity in large part from the initial processing of ILD in LSO neurons (Tollin and Yin 2002a,b), which, as indicated in Fig. 5.3a, project excitatory afferents to the contralateral DNLL and IC and inhibitory afferents (not shown) to the ipsilateral IC and DNLL (Glendenning and Masterton 1983; Loftus et al. 2004). The initial processing of ILDs by the LSO may place limitations on the spatial selectivity exhibited in higher order nuclei and, thus, ultimately limit behavioral binaural and spatial hearing performance (e.g., Jones et al. 2015; Brown and Tollin 2016).

Figure 5.3a summarizes the major afferent inputs to LSO neurons (see Yin 2002; Tollin 2003). Principal LSO neurons receive excitatory inputs from the ipsilateral ear via the spherical bushy cells (SBCs) of the anteroventral cochlear nucleus (AVCN) and inhibitory inputs from the contralateral ear via neurons in the ipsilateral medial nucleus of the trapezoid body (MNTB). MNTB neurons receive excitatory input from the globular bushy cells (GBCs) of the AVCN. Finally, the SBCs and GBCs receive excitatory input from the auditory nerve.

The hypothesis that LSO neurons might be the earliest to encode ILDs was first inferred from the patterns of neural activity elicited when ILDs were systematically varied by presenting sounds to experimental animals over headphones (Boudreau and Tsuchitani 1968). Neural responses evoked by the sound presented to the ipsilateral ear could be systematically reduced by increasing the level of sound presented to the contralateral ear (Figs. 5.3a and 5.4). Although this hypothesis was attractive, a change in the discharge rate of an LSO neuron with changes in the ILD does not by itself directly implicate the LSO as the actual site where the ILD is first computed; the computation could have occurred more peripherally and simply been inherited by the LSO. Moore and Caspary (1983) provided conclusive evidence that ILD sensitivity results from the direct interaction of excitatory and inhibitory synaptic inputs on single LSO neurons. These experiments demonstrated that input from the ipsilateral MNTB to the LSO was glycinergic and inhibitory by showing that inhibition evoked by sound presented to the contralateral ear was blocked by application of the glycine receptor antagonist strychnine. Additionally, an inhibitory effect comparable to that evoked by presentation of sound to the contralateral ear could be elicited by the application of glycine to LSO neurons. Immunocytochemical studies confirm that MNTB principal neurons are glycinergic and that glycine receptors are present in large quantities on principal neurons of the LSO (Bledsoe et al. 1990).

As further evidence of the nature of binaural interaction at LSO neurons, in vivo intracellular recordings revealed that sound presented to the ipsilateral ear generates excitatory postsynaptic potentials (EPSPs) and that sound presented to the contralateral ear generates inhibitory potentials (IPSPs), demonstrating that individual LSO neurons are directly innervated by excitatory and inhibitory afferents and do not inherit their ILD sensitivity from some earlier source (Finlayson and Caspary 1989). Finally, in vitro recordings in brainstem slices, which allow the afferent pathways to the LSO from each ear to be isolated and electrically stimulated directly and independently, also showed that ipsilateral inputs generated glutamatergic EPSPs, whereas contralateral inputs generated glycinergic IPSPs (Caspary and Faingold 1989; Sanes 1990). However, exceptions were noted; some LSO neurons could be inhibited by ipsilateral electrical stimulation from glycinergic inputs directly to LSO neurons themselves (Wu and Kelly 1991, 1994). Such inputs were also inferred from in vivo studies where some LSO neurons could be inhibited by the ipsilateral presentation of tones with frequencies away from the best frequency, consistent with side-band inhibition (Brownell et al. 1979; Caird and Klinke 1983). Sideband inhibition could play a role in sharpening the frequency tuning in this pathway (Koka and Tollin 2014). The source of the ipsilateral inhibitory input remains unclear and its potential role in behavior remains unknown (Greene and Davis 2012; Ashida et al. 2016).

The firing rate responses of LSO neurons to changes in the ILD can be described by a sigmoidal curve (Fig. 5.4). LSO neurons fire action potentials when sound at the ipsilateral ear is greater than the sound at the contralateral excitatory ear and are often completely inhibited, even suppressing spontaneous activity, when sound at the contralateral inhibitory ear exceeds that at the ipsilateral (Tollin et al. 2008; Tsai et al. 2010). Virtually all principal LSO neurons exhibit this stereotypical sensitivity to ILD, so it is unsurprising that the LSO has been typically described as the site where ILD is first extracted and represented for over 60 years. However, this is an over simplification.

Neural responses to identical sounds (i.e., sounds containing the same ILD) in LSO neurons are quite variable from trial to trial, which causes uncertainty in the encoding of ILD magnitude via spikes. This initial ILD coding uncertainty likely sets limits on psychophysical performance. Yet psychophysical studies indicate that ILDs must still be represented with remarkable precision since the just-noticeable differences for changes in ILD (i.e., ILD acuity) range from just 0.5 to 4 dB across most mammalian species tested (reviewed by Greene et al. 2018). Consistent with psychophysical performance, Tollin et al. (2008) and Tsai et al. (2010) used detection theoretic methods to reveal that *neural* thresholds for ILD discrimination, determined from distributions of discharge rates and associated response variabilities of single LSO neurons in response to tones, were comparable with or better than behavior over a wide range of frequencies (0.3–35 kHz) and baseline or pedestal ILDs (±25 dB) tested. With a pedestal ILD of 0 dB, ILD increments of just 1 dB could be discriminated by some neurons. For pedestal ILDs away from 0 dB, the best threshold ILDs were as low as 0.5 dB. Psychophysically, ILD acuity is robust to the pedestal ILD about which it is measured, at least for small to moderately sized ILDs (Brown et al. 2018). Moreover, behavioral ILD acuity is relatively frequency invariant, consistent with neural ILD acuities found in the LSO (Tollin et al. 2008) and IC (Jones et al. 2015). These findings support the hypothesis that the LSO plays an essential role in the initial extraction of the ILD across the audible range for frequencies and pedestal ILDs (a proxy for different spatial positions) and that the representation of the ILD by the LSO may set a lower bound on the behavioral sensitivity to ILDs.

Although individual LSO neurons appear to have the capacity to collectively discriminate ILDs on par with behavioral abilities under controlled and fixed stimulus conditions (e.g., fixed stimulus level and ILD), other observations indicate that single LSO neurons cannot encode the ILD in an absolute sense. For example, the rate versus ILD functions (e.g., Fig. 5.4) of LSO neurons shift considerably with even slight variations in the overall level of sounds presented to the two ears (Tsai et al. 2010). That is, a LSO neuron response to, for example, a 5 dB ILD with a 60 dB average sound pressure level (SPL) at the ears can be quite different from the response to a 5 dB ILD with an 80 dB SPL. Thus, responses of single LSO neurons cannot assign a fixed discharge rate to a given ILD magnitude (i.e., 5 dB in this example) independent of sound levels (see also Benichoux et al. 2017). These results contrast with those observed in psychophysical studies of ILD discrimination thresholds under experimental conditions where the overall stimulus level is roved from trial to trial are virtually invariant to overall changes in stimulus level (Grantham 1984; Hartmann and Constan 2002).

In contrast to the LSO, sound level invariant ILD processing has been shown to be present in many neurons of the IC, which is the major site of ascending projections from the LSO (Park et al. 2004). Tsai et al. (2010) demonstrated that a simple computational model, which incorporated the known antagonistic inputs of bilateral LSO nuclei as well as those of the DNLL to the IC, produced a more robust and level invariant encoding of the ILD even in the setting of a roving stimulus level, consistent with empirical results in the coding of the ILD by neurons in the IC. It is thus conceivable that the ILD is processed in a more complicated network, including the neurons from the LSO and IC (see Brown and Tollin 2016).

5.3.2 Coding of Interaural Time Differences in the Lateral Superior Olive

Natural sounds, like speech and animal vocalizations, are modulated over time in both frequency and amplitude. Other sounds are brief and transient, like rustling leaves and snapping twigs. To compute the ILD present in these complex signals at the LSO requires special care because the inputs from the two ears must encode the temporal variations in the spectra with high fidelity. Surprisingly, neurons comprising the ascending pathway through the LSO (Fig. 5.3a) contain remarkable anatomical and biophysiological specializations that appear suited to accurate encoding of temporal information. This is particularly interesting because it is contrary to the duplex theory-motivated hypothesis that the ILD pathway is concerned only with the encoding of sound spectra (reviewed by Tollin 2003; see also Koka and Tollin 2014).

As illustrated in the circuits in Fig. 5.3, the auditory nerve fiber (ANF) synapses onto the SBCs and GBCs of the CN along with the intrinsic biophysical properties of the bushy cells themselves are specialized to permit the accurate preservation of temporal information (Trussell 2002). Most notable of these are low-voltage-activated K^+ currents, which minimize action potential duration, prevent multiple spike generation and enable rapid firing rates (Brew and Forsythe 2005; Song et al. 2005). Consistent with these biophysical specializations, the encoding of temporal variations in the fine structure of low-frequency sounds is enhanced in the CN neurons over that of their ANF input (Joris et al. 1994). Yet at high frequencies where ILDs are acoustically useful, ANFs do not encode temporal fine structure (Johnson 1980), but they can encode the amplitude modulations (Joris and Yin 1992) and frequency modulations of the stimuli (Britt and Starr 1976). Amplitude-modulated and frequency-modulated stimuli give rise to temporal changes in the envelopes of sounds (e.g., Fig. 5.2), and representations of these envelopes are preserved in the bushy cells and MNTB neurons (Joris and Yin 1998). MNTB neurons have similar membrane properties as bushy cells, with short membrane time constants and non-linear current-voltage relationships (Wu and Kelly 1991; Banks and Smith 1992), also making them well suited to preserve temporal information (Joris and Trussell 2018). Together, these specializations ensure that rapid time-varying changes in the amplitude and frequency content of stimuli at each ear are accurately represented through the LSO pathway. As a result of these remarkable specializations in *temporal* processing, LSO neurons are sensitive to ITDs as well as ILDs. In Sect. 5.3.2, we review the evidence that LSO neurons are sensitive to each of the three different types of ITD cues: fine structure, envelope, and transient (Fig. 5.2).

5.3.2.1 Coding of Fine-Structure Interaural Time Differences

Sounds comprised of predominantly low frequencies or the low-frequency components of broadband sounds contain fine-structure ITDs. LSO neurons are sensitive to ILDs and fine-structure ITDs of low-frequency sounds. The sensitivity to ILDs in low-frequency sounds arises due to the mechanisms discussed in Sect. 5.3.1. Sensitivity to fine-structure ITDs requires at least three additional properties: (1) that the afferents to the LSO phase lock their responses to the ongoing sounds at the two ears; (2) that the excitation and inhibition arrive at the LSO from the two ears in an approximate temporal register; and (3) that the temporal integration of excitation and inhibition at LSO neurons is sufficiently brief. All of these conditions are satisfied, as shown by Tollin and Yin (2005). Figures 5.5 and 5.6 summarize these findings.

Satisfying the first requirement for ITD sensitivity, Fig. 5.5a shows the responses of a low-frequency-sensitive MNTB neuron (center frequency = 317 Hz), demonstrating precise phase locking by the inhibitory afferents to the LSO. Afferent excitatory inputs to the ipsilateral LSO from the SBCs (see Fig. 5.3a) and the LSO neurons themselves phase lock similarly well (Fig. 5.5e). SBCs and GBCs exhibit enhanced phase locking that exceeds that in the auditory nerve (see Joris and van der Heijden 2019). Figure 5.5b summarizes the degree of phase locking, expressed as the synchronization index, as a function of the SPL of tones presented at the best frequency of the neuron (Fig. 5.5a). The synchronization index quantifies how precisely neurons fire action potentials on each cycle of the ongoing pure tone; one cycle of the stimulus is illustrated in Fig. 5.5d, and the histogram indicates the phase of the cycle at which the MNTB neuron fired action potentials over all of the cycles and all of the stimulus repetitions in Fig. 5.5c. Moderate (Fig. 5.5c, d, *top*) and exceptional (Fig. 5.5c, d, *bottom*) synchrony by an MNTB neuron is indicated by

Fig. 5.5 (continued) increase with increasing level of a 50-ms-duration best-frequency tone. (**c**) responses of the same MNTB neuron as in **a** and **b** plotted as dot rasters to 200 presentations of a 50-ms-duration best-frequency tone. (**d**) Associated period histograms plotting the timing of action potentials during each cycle of the pure tone at two different stimulus levels. Responses are phase locked to the stimulus frequency as evidenced by the tendency of spike times to line up vertically in the rasters (**c**) and for the responses to occur at particular phase angles resulting in peaked period histograms (**d**) *Black lines*, one cycle of the ongoing pure tone. r, Synchronization coefficients of the responses in **c** and **d**. (**e**) Low-frequency-sensitive MNTB and LSO neurons exhibit phase locking to best-frequency tones that is enhanced over those seen in ANFs. Maximum (Max) synchronization coefficients are plotted as a function of the characteristic frequencies for MNTB (*solid circles*) and Ipsi stimulated LSO (*solid triangles*) neurons (Tollin and Yin 2005) are plotted along with a population of ANFs (*open circles*; Johnson 1980). (Data replotted from Tollin and Yin 2005)

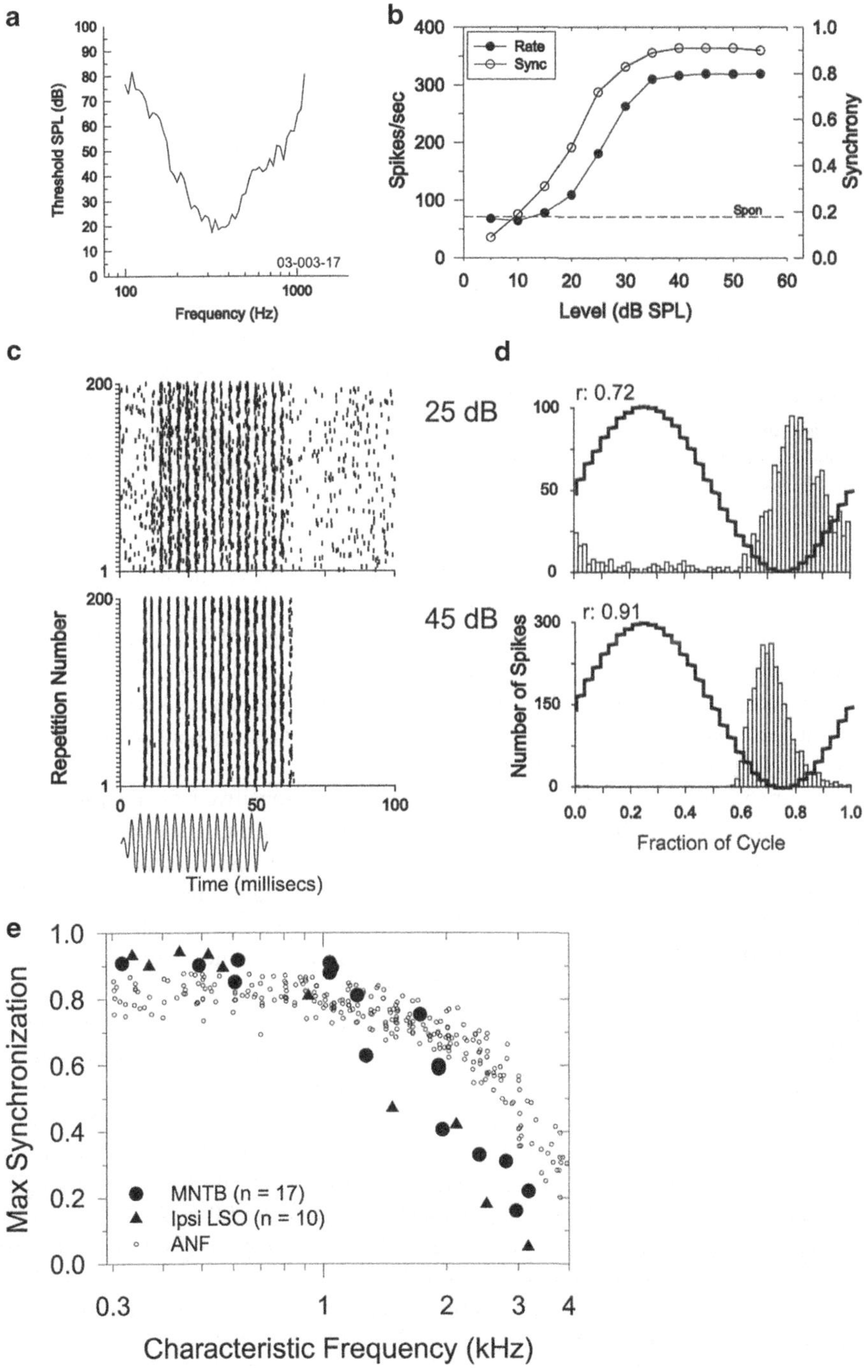

Fig. 5.5 Illustration of exquisite phase-locked responses of low-frequency-sensitive MNTB and LSO neurons. (**a**) Frequency tuning curve (best frequency, 317 Hz) of an MNTB neuron. (**b**) The neuron discharge rate (*left axis; solid circles*) and response synchrony (Sync; *right axis; open circles*)

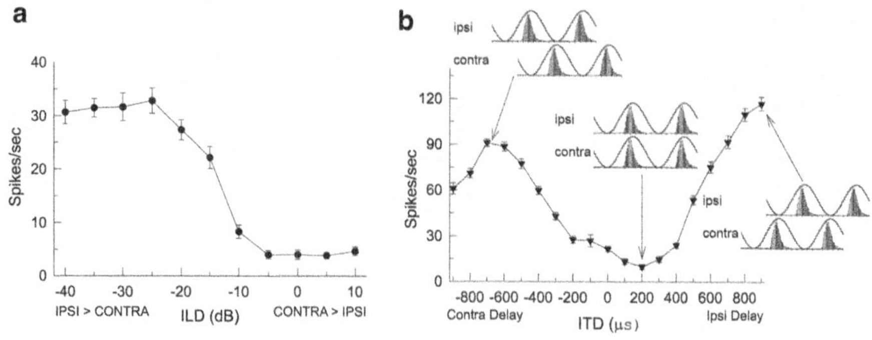

Fig. 5.6 Example of a low-frequency-sensitive (best frequency, 566 Hz) LSO neuron that is sensitive to both ILDs (**a**) and fine-structure ITDs (**b**). Responses to the ITD are minimal when excitation and inhibition from the two ears arrive in the temporal register, here at 200 μs. Values are means ± SE. (**b**) *Insets*, overlap of ipsilateral excitation and contralateral inhibition at particular ITDs. (Data replotted from Tollin and Yin 2005)

wider and narrower distributions of spiking, respectively. Figure 5.5e summarizes the phase locking capabilities of neurons in the LSO pathway, demonstrating that for frequencies less than about 1.2 kHz, neurons in the MNTB and LSO, when stimulated by sounds presented to the ipsilateral ear, exhibit phase locking that exceeds that in comparable ANFs. For frequencies greater than 1.2 kHz, phase locking in these pathways is not as precise as in ANFs, consistent with prior results in bushy cells (Joris and van der Heijden 2019).

Figure 5.6 illustrates the tuning of a low-frequency-sensitive LSO neuron to ILD (A) and fine structure ITDs (B). In addition to the exquisite phase locking of the afferents to the LSO, inhibition and excitation arrive at LSO neurons within just 200 μs of each other, with the input from the contralateral inhibitory ear typically being delayed (Joris and Yin 1995; Tollin and Yin 2005; reviewed by Joris and Trussell 2018). LSO neurons integrate excitatory and inhibitory inputs with very short time constants, less than 1 ms (Tollin 2003; Joris and Trussell 2018). In Fig. 5.6b, two cycles of the stimulus at the two ears as well as the phase locking of afferent responses are shown corresponding to the excitatory ipsilateral and inhibitory contralateral ear for three different ITDs. When fine-structure ITD favors the contralateral (contra) inhibitory ear relative to the ipsilateral (ipsi) excitatory ear by 200 μs, ipsilateral excitation and contralateral inhibition arrive at the LSO at exactly the same time (Fig. 5.6b). As a consequence, spiking by the LSO is maximally inhibited. Because ITDs have values greater or less than 200 μs in this example, spiking increases because there is progressively less overlap between the ipsilateral excitation and contralateral inhibition, and thus the excitation alone causes the LSO neuron to fire action potentials. Thus, low-frequency LSO neurons are sensitive to fine-structure ITDs in low-frequency sounds in addition to ILDs.

5.3.2.1.1 Coding of Low-Frequency Fine-Structure Interaural Time Differences in the Medial Superior Olive

Although the focus of this chapter is on the relatively new discovery of ITD sensitivity and its mechanisms by LSO neurons, this chapter would be remiss if it did not include a review of the mechanisms of ITD sensitivity by MSO neurons. MSO neurons are the earliest in the auditory system to extract primarily low-frequency fine-structure ITDs (e.g., Fig. 5.3b). MSO neurons receive excitatory inputs from both ears via the bushy cells of the AVCN on both sides (Fig. 5.3b). The bushy cells of both sides receive excitatory inputs from the cochlea via the ANFs via specialized synaptic terminals called the endbulbs of Held (Young and Oertel 2004). The AVCN is organized tonotopically, with most neurons sensitive to lower sound frequencies (Fig. 5.3b; Goldberg and Brown 1969; Yin and Chan 1990). Most low-frequency-sensitive MSO neurons are sensitive to ITDs in the fine structure of sounds (Fig. 5.2a) presented to the two ears, and there is some evidence that higher frequency MSO neurons can be sensitive to envelope ITDs (Fig. 5.2b; Yin and Chan 1990).

ITD sensitivity in MSO results from a three-stage process as first hypothesized by a model proposed by Jeffress (Jeffress 1948; see Joris et al. 1998 for a review). First, the afferent inputs to MSO neurons from the ANFs and from the bushy cells of the AVCN (see Fig. 5.3b) carry timing information in the form of phase-locked neural responses to the fine structure of sound at the two ears (see Fig. 5.5c, d, for a phase-locking example). Recall that from Sect. 5.3.2.1 phase locking refers to the tendency of a neuron to fire action potentials at particular phases of an ongoing periodic sound waveform, such as the sinusoidal waveforms that are typically used in physiological studies of the auditory system. Phase locking is the mechanism by which the peripheral auditory system keeps track of the times of occurrence of the ongoing amplitude fluctuations in sounds. Second, the MSO neurons behave like coincidence detectors, responding maximally only when action potentials from the SBC inputs from the left ear and the right ear arrive nearly simultaneously (e.g., within a few hundreds of microseconds; Joris et al. 1998) at the MSO neuron. Third, some mechanism along the afferent input pathway to the MSO from the bushy cells essentially forms temporal "delay lines" along one axis of the nucleus. As illustrated in Fig. 5.7a, the mechanism producing these so-called *internal* delays is thought to offset the physical acoustical delay between the two ears due to the ITD.

Jeffress (1948) originally postulated that differences in the physical lengths of the neural pathways from the two ears to the MSO would result in differences in neural conduction times to the MSO from the two ears, which would then offset the acoustic ITD cue, as illustrated in Fig. 5.7b. The way that internal delays work in practice is illustrated in Fig. 5.7a. The stimulus in the example contains an ITD favoring the right ear such that the phase-locked action potentials, or spike trains, encode the timing of the acoustic inputs at the two ears. Next, a mechanism producing an internal delay in the right ear neural pathway to the MSO (e.g., longer neural path length; Fig. 5.7b) compensates for the acoustic ITDs in the stimulus such that

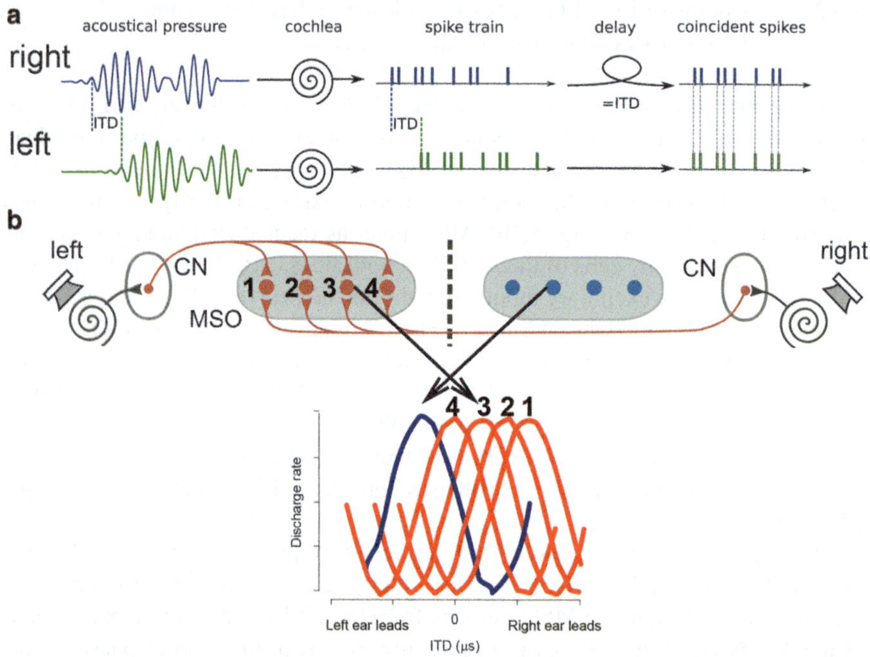

Fig. 5.7 Illustration of mechanisms the produce sensitivity to low-frequency fine structure ITDs by MSO neurons. (**a**) Sound waveforms at the left and right ears are separated by an ITD, with the sound to the right ear leading the sound to the left ear. These stimuli are processed in the cochlea and peripheral parts of the neural circuits leading to the MSO (shown in Fig. 5.3b). Neurons in the SBCs of the cochlear nucleus (CN) phase lock their action potentials to the ongoing sound waveforms, thereby keeping track of the relative timing of the sound inputs to the right and left ears. Note that the spike trains in the neural pathways to the MSO from the two ears are essentially separated by the ITD, with the spikes in the right ear pathway leading those in the left. ITD coding in MSO posits an "internal delay" mechanism that effectively delays the incoming spikes to the MSO from the right ear (in this example) to compensate for the ITD. After the internal delay mechanism, spikes from the right and left ears arrive at the MSO neuron at the same time, or in coincidence. (**b**) Jeffress (1948) hypothesized that the internal delays were produced by longer axonal pathlengths. In this example, the pathlengths to the MSO on the left side from the right ear are longer than those from the left ear. As such, an ITD favoring the right ear as in **a** would be compensated for by the circuit in **b** because it would take longer for the spikes to propagate over the longer axon length from the right ear than the spikes traveling along the axons from the left ear. These axonal delay lines compensate for the physical ITD so that the action potentials from the two ears arrive at a MSO neuron at the same time, causing the MSO neuron to respond maximally (*arrow* and *red curve*). An example ITD tuning curve is also shown for a corresponding MSO neuron (*right*), which fires maximally for an ITD favoring the left ear. *Numbers*, individual MSO neurons*; blue circles*, MSO neurons from the right side; *blue curve*, an ITD tuning curve from a right MSO neuron

the action potentials from the two ears arrive at the MSO neuron at virtually the same time, causing the MSO neuron to fire maximally.

The ITD at which an MSO neuron fires the most action potentials (e.g., the peak response in the rate vs. ITD function in Fig. 5.7b) is called the best delay or best ITD. As acoustic ITDs are changed away from the best ITD, there would be less temporal overlap between the action potentials arriving at MSO from the two ears, resulting in the MSO neuron becoming less responsive. Jeffress (1948) speculated that MSO neurons receive systematically delayed inputs from the two ears via these axonal delay lines, with the net result being that different MSO neurons (e.g., neurons labeled 1–4 in Fig. 5.7b) would be maximally sensitive to different ITDs. Neurons in the opposite MSO would be sensitive to ITDs favoring the other ear (e.g., Fig. 5.7b, *blue curve*, favoring ITDs leading to the left ear). Thus, the population of MSO neurons would be able to encode the physiological range of ITDs that the animal would be expected to experience because the distribution of best ITDs of the neurons would collectively span the range of physical ITDs experienced.

Although there is reasonable evidence for Jeffress-like axonal delay lines in some species like birds (Takahashi, Kettler, Keller, and Bala, Chap. 4), there is little evidence for this in mammals (Smith et al. 1993; Karino et al. 2011). Instead, several other mechanisms recently reviewed by Joris and van der Heijden (2019) have been proposed to produce internal delays necessary for ITD coding, such as delays due to cochlear processing (Joris et al. 2006), delays resulting from the timing of synaptic interactions between excitatory and inhibitory inputs to MSO (Grothe 2003), or delays resulting from differences in the rise times of contralateral and ipsilateral synaptic events in MSO neurons (Jercog et al. 2010). Regardless of mechanism, empirical measurements of ITD sensitivity in neurons comprising the MSO and its primary target nucleus, the IC, reveal that the best ITDs of the neurons are generally positive (i.e., the neurons are most responsive for ITDs favoring the ear contralateral to the neuron being recorded) and that across neurons the distributions of best ITDs generally covers the range of physical acoustical ITDs that the animals experience given the size of their heads.

More recent analysis has revealed that the range of best ITDs is actually dependent on the frequency selectivity of the neurons (i.e., the characteristic frequency; McAlpine et al. 2001; Grothe et al. 2010). For some small-headed mammals such as the gerbil, the dependence of the best ITD with the center frequency can cause the best delays to have values of ITD that exceed the physical size of the ITD that the animal would experience. How could neurons that are responsive to physiologically implausible ITDs be used for sound localization? This observation led to the development of a model whereby the encoded ITD is read out via the relative differences of neural activity between neurons that are tuned to opposite ITDs (e.g., the two ITD tuning curves shown in Fig. 5.7b), the so-called two-channel or opponent hemifield model (McAlpine et al. 2001; Harper and McAlpine 2004). Here, regardless of whether the best delays are within or exceed the physiological range of ITDs, the *difference* between two such neurons will have the maximum sensitivity for ITDs near 0 μs, which is consistent with behavioral observations that the best ITD acuity occurs at 0 μs (Hartmann, Chap. 2; Dietz and Ashida, Chap. 10). Although much has been learned about low-frequency fine-structure ITD coding by MSO neurons and its output targets such as the IC, the precise mechanism by which ITDs are encoded and used for behavior still eludes researchers.

5.3.2.2 Coding of Envelope Interaural Time Differences

Sounds that have rapid amplitude and/or frequency modulations give rise to ITDs in their low-frequency envelopes (Fig. 5.2b). LSO neurons are sensitive to envelope ITDs because of the same mechanisms for low-frequency fine-structure ITD sensitivity described in Sect. 5.3.2.1. Figure 5.8 shows responses to an amplitude-modulated (AM) tone by the afferent pathways to the LSO (Fig. 5.3a). In Fig. 5.8, *bottom right*, three periods of an AM waveform are shown as arriving at the ears with zero ITD, which approximates a sound source at the midline. The temporal fluctuations in the amplitude envelopes of the sound stimulus are conveyed in the timing of the discharges of the ANFs of the two ears, the SBCs and GBCs of the ipsilateral and contralateral ears, respectively, and the ipsilateral MNTB neurons because these afferents to the LSO phase lock their responses to the AM fluctuations with extreme precision (see Joris et al. 2004 for a review). LSO neurons integrate excitatory and inhibitory inputs within a very short amount of time, less than 1 ms

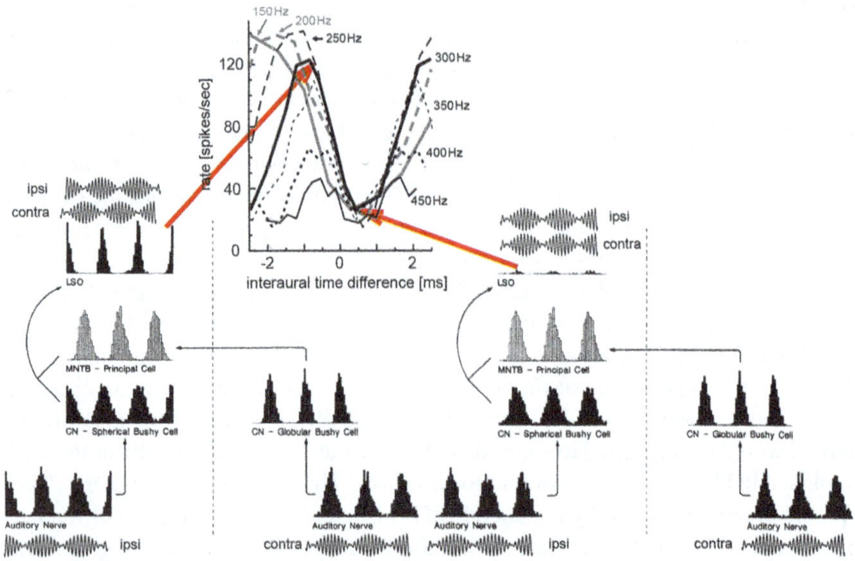

Fig. 5.8 Example of a high-frequency LSO neuron that is sensitive to envelope ITDs. *Bottom right*: hypothetical responses of an LSO neuron and its afferents, the auditory nerve, cochlear nucleus bushy cells, and the MNTB (see Fig. 5.3) to amplitude-modulated stimuli presented to the ipsilateral and contralateral ears with an ITD of 0 µs in the envelopes. In this case, the excitatory and inhibitory inputs arrive at the LSO neuron in a temporal register, resulting in a minimal response (indicated via *red arrow at the bottom*). *Bottom left*: hypothetical responses of the same LSO neuron and its afferents to the AM stimuli but where the envelope to the contralateral ear has been delayed in time relative to the envelope at the ipsilateral ear. The time delays are accurately preserved by the responses of the contralateral afferents. Consequently, the excitation and inhibition do not arrive in temporal register at the LSO, and there is thus a release from inhibition, resulting in an increased response (indicated via *red arrow at the peak*). (Data replotted from Ashida et al. 2016)

(Tollin 2003; Joris and Trussell 2018). Thus, for envelope ITDs, the inhibition of the discharge rate of the LSO should be nearly maximal when these stimuli are presented to the two ears, with approximately zero delay between their onsets (i.e., a source at the midline), as demonstrated in Fig. 5.8, *bottom right*. When the ITDs are such that the envelopes are nearly out of phase at the two ears, such as the example shown in Fig. 5.8, *left,* the inhibitory input from the MNTB to the LSO no longer aligns temporally with the excitatory input from the SBCs. Thus, the LSO neuron responds robustly to the input to the ipsilateral ear. In summary, changing the ITD in the low-frequency *envelopes* of sounds presented to the two ears systematically changes the temporal overlap of excitation and inhibition arriving at the LSO within a small integration window of about 1 ms or less (Ashida et al. 2016, 2017), which causes the discharge rates of LSO neurons to be systematically modulated by envelope ITDs. The unique and remarkable anatomical and biophysical specializations observed in the long contralateral pathway through the GBCs and MNTB to the LSO likely serve to minimize both the relative timing delays and the jitter in the synaptic delays incurred along the ILD pathway so that ILDs in the stimulus, particularly for rapidly varying envelopes, can be accurately represented (Joris and Trussell 2018).

5.3.2.3 Coding of Transient Interaural Time Differences

Transient sounds in the environment, such as rustling leaves from animal locomotion or snapping twigs, generate ITDs but are too brief to contain meaningful temporal information in the envelope or ongoing fine structure. Localization of such transients has been proposed to occur by ITD processing in the LSO (Joris and Trussell 2018). The mechanisms enabling sensitivity to transient ITDs in the LSO are the same as those enabling sensitivity to fine-structure and envelope ITDs (reviewed in Sects. 5.3.2.1 and 5.3.2.2). These include the precise encoding of stimulus onset by the afferents to the LSO, the nearly coincident arrival of excitation and inhibition to the LSO from the two ears, and the integration of excitation and inhibition with a short time frame that shows modulation by changes in stimulus ITD (see Joris and Trussell 2018).

Although there had been some hints from earlier studies that LSO neurons can encode transient ITDs (e.g., Joris and Yin 1995; Park 1998), this was demonstrated conclusively in a recent study by Beiderbeck et al. (2018). By varying the ITDs in transients presented to the two ears, the authors showed that the precisely timed inhibitory input from the MNTB to the LSO suppressed action potential generation only during a period of a few hundred microseconds, corresponding to the point of functional coincidence of excitation and inhibition (Fig. 5.9). The ITD required to maximally inhibit LSO neurons was reported to vary from neuron to neuron for fine-structure (Tollin and Yin 2005) and envelope (Joris and Yin 1995) ITDs but averaged about 200 μs, favoring the contralateral inhibitory ear. Beiderbeck et al. (2018) reported a similar range of ITDs necessary to produce maximum inhibition for transient ITDs. This ITD corresponds to the functional coincidence between

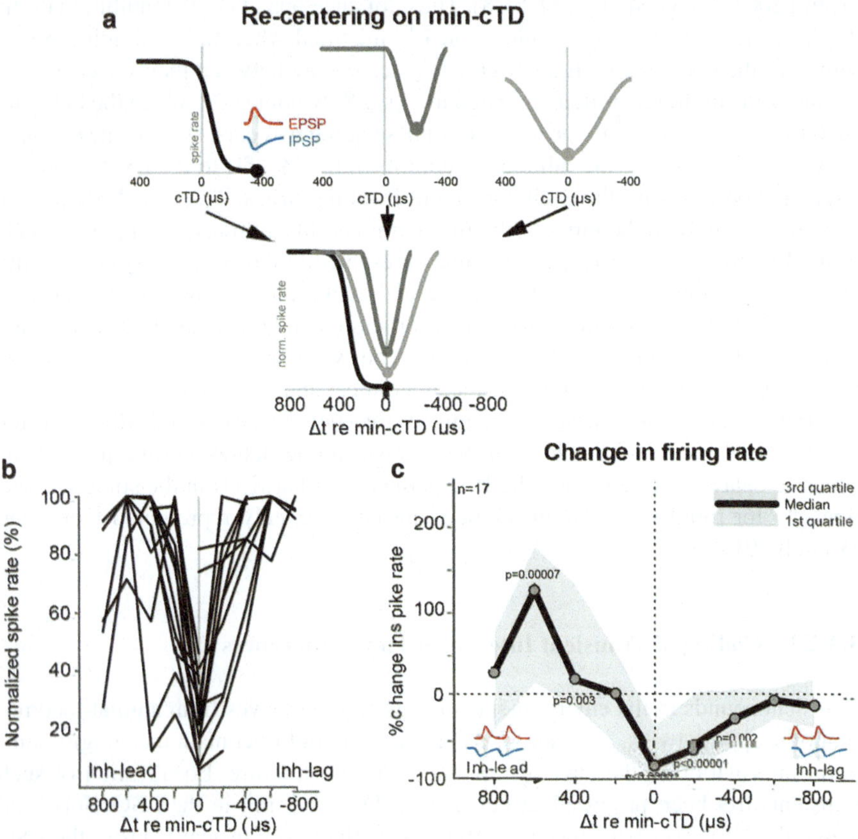

Fig. 5.9 Discharge rates of LSO neurons are modulated by microsecond differences in transient ITDs. (**a**) Composite timing delays (cTD) represent the total change in timing of inputs resulting from the ITD and latency (as a function of sound intensity). Re-centering the cTD-discharge rate function relative to the minimum cTD of each neuron normalizes the function to the point of functional coincidence between excitation and inhibition. (**b**) Normalized spike rates of a population of LSO neurons reveal that they are maximally inhibited at 0-μs cTD. As the time difference (Δt) values deviate from 0 μs, the overlap of excitation and inhibition (inh) decreases and normalized spike rate increases. (**c**) Median changes in spike rate relative to the respective response rate of each neuron during excitation-only stimulation for re-centered Δt-discharge rate functions of all LSO neurons tested (n = 17). LSO neurons show minimum discharge rates at 0 μs cTD and an increase in relative discharge rate for Δt values of 600 μs. (Reprinted from Beiderbeck et al. (2018), with permission)

excitation and inhibition (Fig. 5.9). Overall, these studies reveal that the discharge rates of LSO neurons are strongly modulated by microsecond changes in the ITDs of transients (Fig. 5.9; Beiderbeck et al. 2018). Figure 5.10 summarizes graphically how the response rate functions in LSO and MSO neurons are modulated by the temporal overlap of IPSPs and EPSPs.

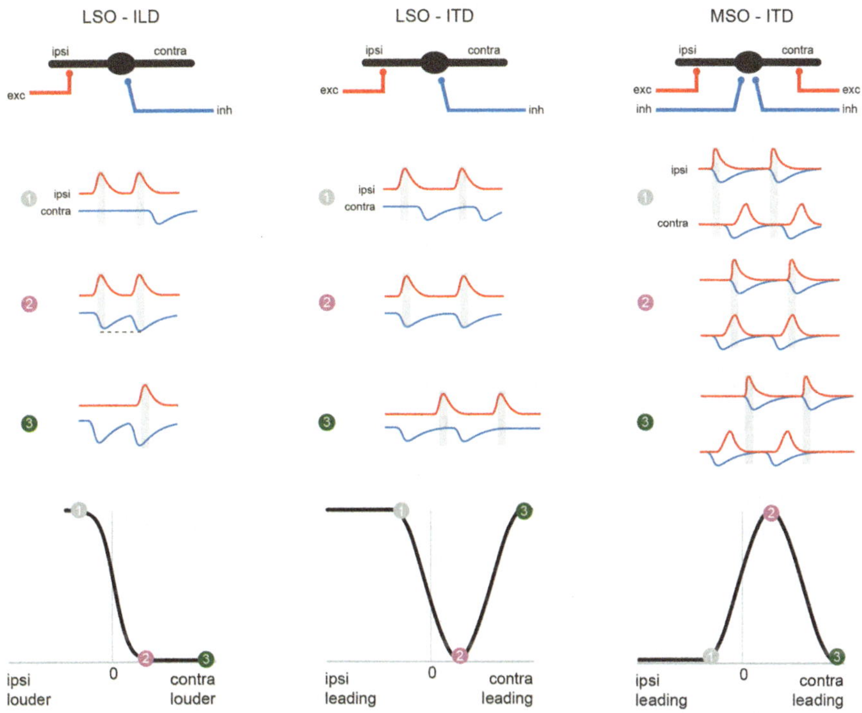

Fig. 5.10 Common mechanisms of coincidence detection are shaped by spatial distributions of excitatory and inhibitory inputs onto MSO and LSO neurons. Temporal overlap of inhibitory postsynaptic potentials (IPSPs; inh) and excitatory postsynaptic potentials (EPSPs; exc) gives rise to the spatial tuning functions underlying ILD (*left*) and ITD (*middle*) sensitivity in LSO neurons. The coincidence detection mechanism for detecting the ITD in MSO neurons is similar to that in LSO neurons, except with additional contralateral excitatory and ipsilateral inhibitory inputs (*right*). Numbers, relative overlap of excitation and inhibition mapped onto ILD and ITD tuning curves for an LSO and MSO neuron. (Reprinted from Grothe and Pecka (2014), with permission)

The view that the LSO can be sensitive to ITDs even in transients is further supported by in vivo recordings by Franken et al. (2018), which revealed that the LSO principal neurons have fast membrane kinetics and can thus resolve individual synaptic events, such as the excitatory and inhibitory inputs resulting from transient sounds, on a rapid timescale. The authors also found that all principal LSO neurons fired a single spike at the onset of an ipsilateral tone stimulus. This onset-response type contradicts previous reports in principal neurons, which described a "chopper" or tonic response to ipsilateral stimulation (Boudreau and Tsuchitani 1968). Rather, Franken et al. (2018) found that this sustained response type is reserved for nonprincipal neurons in the LSO. Taken together, the Franken et al. (2018) and Beiderbeck et al. (2018) results strongly imply that the principal neurons of the LSO, which comprise over 75% of cells within the nucleus (Helfert and Schwartz 1986; Franken et al. 2018), preserve the timing of the stimulus onset at the two ears, relayed by their excitatory and inhibitory inputs. This sensitivity to the stimulus onset shown by LSO neurons resembles that of the principal neurons in the MSO, which fire maximally when bilateral excitatory inputs are received within an approximate temporal register (e.g., Fig. 5.7), thus acting as "coincidence detectors" (Roberts et al. 2013; Franken et al. 2015). It appears that the principal neurons in the LSO are performing a similar coincidence computation (see Benichoux and Tollin 2018), except rather than firing maximally when two excitatory inputs are received in temporal register as MSO neurons do, LSO neurons are maximally inhibited when the excitatory and inhibitory inputs arrive in the temporal register. As such, LSO neurons are performing an anticoincidence detection computation.

5.4 Features of the Lateral Superior Olive Circuitry Favor Detection of Interaural Time Differences

Principal neurons in the LSO fire action potentials with smaller amplitudes than nonprincipal neurons, making them difficult to detect using standard electrophysiological methods (Franken et al. 2018). As a result, early electrophysiological studies mischaracterized the principal LSO neurons as having a more temporally integrative tonic or chopper response pattern to auditory stimuli. This led many to assume that LSO neurons were integrating information over a few milliseconds or more, reinforcing the hypothesized role of the LSO as posited by the duplex theory, to localize high frequencies by ILD detection (see Benichoux and Tollin 2018 for a review). Given that neurons in the LSO are biased to higher frequencies (e.g., Fig. 5.3a), one plausible hypothesis was that these neurons integrate information over multiple milliseconds to compare sound level differences conveyed via the discharge rates of binaural inputs. However, assigning the LSO to the role of a level-difference detector fails to account for the striking temporal specializations observed in its afferent circuitry.

One alternative hypothesis is that the highly specialized MNTB-LSO synapse evolved to detect ITDs of transient, high-frequency stimuli under evolutionary pressure to lateralize predator or other animal movement (Joris and Trussell 2018). This view is supported by morphological and biophysical features of the MNTB-LSO synapse that as well as the fact that neurons in the LSO are biased toward high frequencies and display sensitivity to multiple types of ITD (discussed in Sect. 5.3.2). Moreover, the traditional role of the LSO as posited by the duplex theory is refuted by recent evidence from Beiderbeck et al. (2018) revealing that LSO neurons have much shorter integration times than previously thought, on the order of just a few hundred microseconds.

Accounting for human psychophysical ILD acuity appears to require integration of information about sound level at the two ears over a significantly longer period time. Hartmann and Constan (2002) suggested integration times on the order of hundreds of milliseconds for broadband ILD computation. More recently, Brown and Tollin (2016) modeled an ideal level detector based on empirical recordings from the LSO and also the IC and reported that to account for behavioral ILD sensitivity, a temporal integration window of approximately 3–5 ms was required, which is much longer than that observed in the LSO, about 1 ms. Brown and Tollin showed that neurons in the IC effectively integrate excitatory and inhibitory inputs over temporal windows spanning 3–5 ms, consistent with the modeling results. Moreover, it has also been shown that human ILD sensitivity is preserved despite a near complete decorrelation of temporal information between binaural inputs (Hartmann and Constan 2002; Brown and Tollin 2016). This suggests that the neural correlate of ILD may not depend on precise temporal information relayed by the LSO and may actually be computed later in the ascending auditory pathway by integrating inputs from the two LSO over longer time periods. A similar explanation was put forth by Tsai et al. (2010) to account for the finding that behavioral sensitivity to the ILD is invariant to overall level, whereas LSO neuron coding of ILD is not (see Sect. 5.3.1).

5.4.1 Precisely Timed Glycinergic Inhibition Shapes Processing of Interaural Time Differences

To detect fine-structure ITDs, neurons in the MSO (Fig. 5.3b) act as "coincidence detectors," integrating excitatory inputs from the two ears and firing an output spike when the inputs are received within a brief window of about 200 μs (Fig. 5.10c). Similarly, the response rates of LSO neurons (Fig. 5.3a) are maximally inhibited when excitatory and inhibitory inputs are received within a comparable window of coincidence, thereby acting as "anticoincidence detectors" (Fig. 5.10b). Therefore, it appears that the LSO and MSO are performing equal and opposite tasks: LSO neurons fire unless binaural coincidence occurs, whereas MSO neurons fire only if binaural coincidence occurs. In both cases, however, precisely timed glycinergic inputs from the MNTB play a fundamental role in shaping submillisecond sensitivity to ITD.

As reviewed in Sect. 5.3.2.1.1 in the MSO, individual MSO neurons are tuned to a "best ITD" at which maximal firing occurs (Fig. 5.7b). It was postulated that a difference in the physical length of axons from the two ears accounted for this tuning (Jeffress 1948). Although there is evidence for such anatomical delay lines in birds (see Takahashi, Kettler, Keller, and Bala, Chap. 4), this mechanism does not appear to play a role in mammalian sound localization. Rather, well-timed feedforward inhibition from the MNTB could potentially determine the window for coincident detection of bilateral excitatory inputs (Fig. 5.10c). In vivo studies in gerbil MSO principal neurons reveal that preceding glycinergic inhibition lowers spike probability, effectively narrowing the window for coincidence detection (Roberts et al. 2013; van der Heijden et al. 2013). Pharmacological blockade of glycine-mediated transmission with the application of strychnine, a glycine receptor antagonist, broadened the window for coincidence detection and shifted it toward 0 μs

(Brand et al. 2002; Pecka et al. 2008). Together, these results implicate a deterministic role for glycinergic signaling in setting the short window of coincidence detection seen in the MSO.

Similarly, Beiderbeck et al. (2018) demonstrate that the firing rates of LSO neurons are most reduced when contralateral inhibition preceded ipsilateral excitation by approximately 200 μs (Fig. 5.9). This reveals a mechanism by which glycinergic inhibition modulates spiking in LSO neurons to enable ITD sensitivity with submillisecond precision. Beiderbeck et al. (2018) also noted that the well-timed and short-acting glycinergic inhibition to the LSO produced "postinhibitory facilitation" (Fig. 5.9c) that likely occurs by lowering the threshold for action potential firing. Together, these findings point to a common mechanism facilitated by temporally precise, feedforward glycinergic inhibition in tuning sensitivity to the ITD in both the MSO and LSO (Fig. 5.10). Yet, in order for the arrival of excitatory and inhibitory inputs to coincide in temporal register at the LSO, there must be a mechanism to counteract the longer axonal path traveled from the contralateral AVCN as well as the delay accrued by an additional synapse via the MNTB. This is accomplished by highly specialized features of the MNTB (Joris and Trussell 2018), which include high conduction velocities enabled by large axon diameters and heavy myelination (Ford et al. 2015). Moreover, inhibitory contacts to the LSO from the MNTB are restricted almost entirely to the soma (Fig. 5.10b; Grothe and Pecka 2014), reducing the time required for inhibitory postsynaptic potentials (IPSPs) to summate at the cell body where action potentials are generated. This allows for rapid control over the resting membrane potential of the neuron and, consequently, the likelihood that the neuron will fire an action potential, thus providing a basis for the remarkable submillisecond sensitivity to ITD. Together, these features aid in overcoming the longer path traveled by the contralateral inhibitory input to allow coincident arrival of the binaural inputs at LSO neurons.

5.5 The Binaural Interaction Component of the Auditory Brainstem Response: A Noninvasive Window into Binaural Brainstem Function

Knowledge of the circuitry comprising the mammalian LSO and its function has not only scientific value but also potential clinical value. The sensitivity to transient ITDs by LSO neurons provides the basis for a well-known and studied noninvasive measure or biomarker of binaural brainstem function, the binaural interaction component (BIC) of the auditory brainstem response (ABR) in mammals (see Laumen et al. 2016 for a review). ABRs are measured by presenting broadband transient or click stimuli to the ears, triggering synchronized activity in large of groups of neurons in the brainstem that can be recorded from surface electrodes placed on the scalp (Fig. 5.11a, b; Jewett et al. 1970). Binaural evoked ABRs can be measured by presenting stimuli to both ears (Fig. 5.11c). Subtracting binaurally evoked ABRs

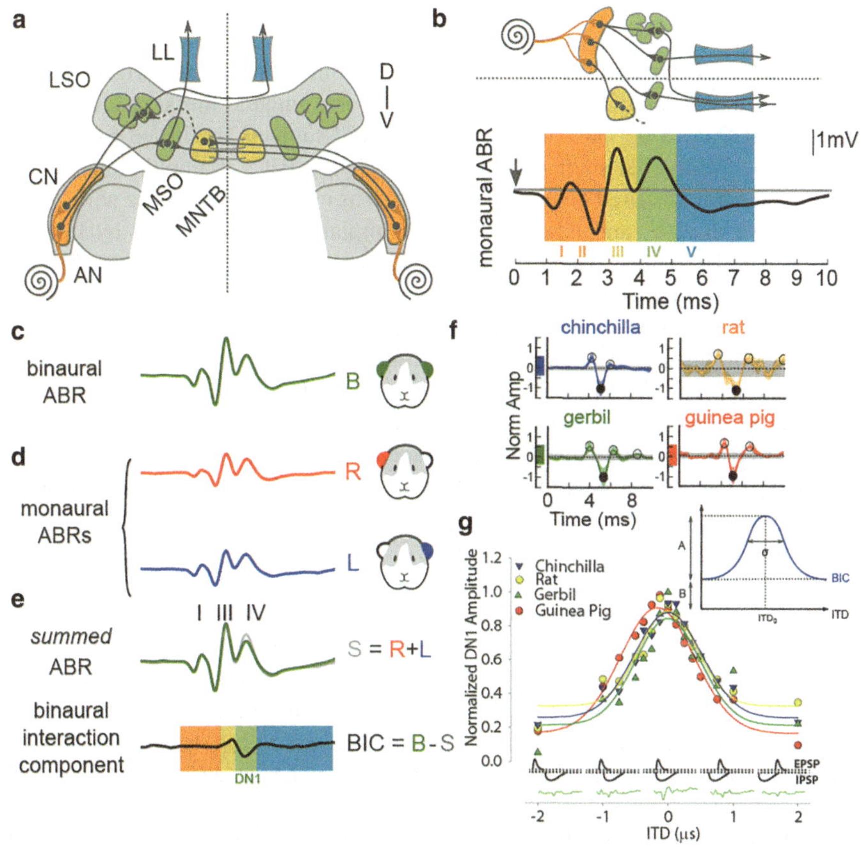

Fig. 5.11 (**a**) Brainstem binaural pathways. LL lateral lemniscus, AN auditory nerve, D/V dorsal/ventral. (**b**) *Top*, schematics of a monaural pathway; *bottom*, monaural auditory brainstem response (ABR) from the guinea pig. Background colors indicate rough correspondence between different ABR waves and brainstem nuclei. (**c**) Binaural ABRs obtained by presenting a click to both ears with 0-μs ITD. (**d**) Monaural ABRs. (**e**) Binaural interaction component (BIC) is computed by taking the difference between the binaural and the sum (S) of the monaural ABRs (*top, gray curve*). The DN1 wave is the first negative peak in the BIC. (**f**) BIC waveforms measured at 0-μs ITD in 4 species. BIC waveform morphology is similar across species. *Solid circles*, DN1. Norm Amp, normalized amplitude. (**g**) Normalized DN1 amplitude as a function of ITD measured in four species. Note the similar morphology of the fitted Gaussian functions across species ($r^2 > 0.9$ for all fits). (Figure adapted from Benichoux et al. 2018)

from the sum of monaurally evoked ABRs (Fig. 5.11c–e) results in a residual trace that is not fully explained by activation of the monaural pathways (Fig. 5.11e). This residual component, referred to as the BIC, has generated intense interest as a prospective clinical biomarker of binaural hearing (Dobie and Berlin 1979; Laumen et al. 2016).

The most prominent feature of the BIC waveform is a negative peak, often referred to as DN1 (Fig. 5.11e), emerging from ABR wave IV in small mammals (e.g., Fig. 5.11e) and wave V in humans (not shown), indicating a *reduced* binaural response relative to the sum of the monaural responses. The latencies and amplitudes of DN1 have been shown to be predictive of binaural hearing performance for a variety of tasks, lateralization, discrimination, and binaural masking level differences (BMLDs), in normal- and hearing-impaired subjects (reviewed by Laumen et al. 2016). Additional studies have revealed altered BICs in populations of subjects that exhibit impaired binaural hearing such as in the aged, in those with neurodegenerative disease, and bilateral cochlear implant users (reviewed by Laumen et al. 2016). Thus, the neural circuits and mechanisms that produce the BIC are likely to be essential for binaural hearing performance for a wide range of tasks.

There are several clues regarding the neuroanatomical source of the BIC. First, the latency of the DN1 peak of the BIC corresponds to the latencies of the LSO and MSO neurons (Fig. 5.11a, b, *green shading*). Second, consistent with an MSO or LSO source, a substantial body of literature (e.g., Dobie and Berlin 1979; Laumen et al. 2016) has shown that the DN1 amplitude is modulated by ITDs and, to a lesser extent, by ILDs. DN1 is maximal for ITDs (and ILDs) of zero and can no longer be detected for large ILDs or ITDs (Fig. 5.11g). Modulation of the BIC DN1 by the ITD cannot conclusively rule in or out the LSO or MSO because while MSO has been known for decades to be sensitive to ITDs, the evidence that the LSO is also sensitive to ITDs has only been recently discovered and appreciated, as reviewed in this chapter. Third, the amplitude of DN1 is negative (Fig. 5.11e, f), indicating a smaller response to binaural than for monaural stimulation, consistent with LSO-like inhibitory mechanisms (e.g., Fig. 5.10a, b). Based on this evidence and other empirical experimental as well as modeling studies, it has been long debated whether the source of the BIC was the MSO and/or the LSO (see Laumen et al. 2016 for the history).

To more comprehensively examine the source of the BIC, Benichoux et al. (2018) studied the characteristics of the BIC across multiple species of mammals. The rationale for the study was that the relative size and numbers of neurons comprising the MSO and LSO vary significantly across mammalian species (Glendenning and Masterton 1998), which allowed correlation of the morphology of the BIC waveform with the relative sizes of the MSO and LSO. Masterton et al. (1975) dubbed this type of experimental approach "natural ablation," reasoning that behaviors and physiological responses that rely on a particular nucleus, MSO or LSO, would scale with its size across species. For example, mice and rats have large and well-developed LSOs but small or nonexistent MSOs. Consistently, although both species can use high-frequency ILDs and likely envelope and transient ITDs (Li et al. 2019), neither can use low-frequency fine-structure ITDs for sound localization because of the apparent lack of the MSO in those species (Wesolek et al. 2010). In low-frequency hearing mammals including cats, chinchillas, guinea pigs, gerbils, and humans, both the LSO and MSO nuclei are well developed. Consequently, these species can use high-frequency ILDs, envelope ITDs, and transient ITDs as well as low-frequency fine-structure ITDs for sound localization. Benichoux et al. (2018) showed that not only was the morphology of the BIC waveform similar across species (e.g., Fig. 5.11f) but also the function relating the BIC amplitude (DN1) and

ITD was statistically the same in seven species studied, from mouse and rat to cat and human (four species shown in Fig. 5.11g). Although all species tested by Benichoux et al. have a well-developed LSO, mice and rats do not have a binaurally functional MSO (Masterton et al. 1975; Glendenning and Masterton 1998), indicating that the origin of the BIC is the LSO. Benichoux et al. (2018) confirmed this by constructing a model of LSO responses similar to those of Ashida et al. (2016, 2017) and demonstrated that the BIC amplitude and latency versus ITD data in all mammalian species studied could be accounted for via the same set of LSO neuron biophysical parameters.

Finally, the dependence and sharpness of the BIC amplitude with changes in the ITD (e.g., curve width in Fig. 5.11g, *inset*) virtually mimic the functions relating LSO spiking and ITD with transient stimuli (Fig. 5.9). This fact provides a functional explanation for the BIC. For example, for a 0-μs ITD, excitation and inhibition arrive at the LSO at virtually the same time (Fig. 5.11g, *bottom,* excitatory and inhibitory inputs [EPSPs and IPSPs, respectively]), resulting in maximum suppression of LSO responses (Fig. 5.9). In terms of the ABRs in response to transient stimuli, there would thus be substantially reduced LSO neuron responses for binaural stimulation relative to monaural stimulation and thus the amplitude of the ABR corresponding to wave IV would be smaller for binaural than monaural. For larger ITDs away from zero, there would be less overlap between excitation and inhibition at the LSO (Fig. 5.11g, *bottom*), thus freeing the LSO neurons on both sides to respond similarly to monaural stimulation. In terms of the ABR, the binaural response amplitude at wave IV approaches the sum of the monaural as ITDs are increased from zero. For ITDs that exceed the integration window for excitation and inhibition at the LSO, about 1 ms (Fig. 5.9), there would no longer be any binaural interaction at the LSO and thus the amplitude of the BIC DN1 approaches zero. These data together strongly support the hypothesis that the LSO pathway generates the BIC of the ABR.

The BIC of the ABR could provide a noninvasive assay of binaural hearing functions that seems to rely on the processing of sound through the circuit comprising the LSO. In addition, the detailed knowledge of the circuitry of the LSO and its ultimate function reviewed in this chapter is required to place the BIC of the ABR in its context. Why is this important? Many models of binaural psychophysical phenomena require inhibitory-excitatory binaural mechanisms consistent with LSO function rather than excitatory-excitatory mechanisms consistent with MSO function. For example, models of binaural unmasking assume a subtractive process, including the equalization-cancellation model (Culling and Lavandier, Chap. 8). Models of binaural speech segregation and illusory binaural pitches (Huggin's, binaural edge, Fourcin) require subtractive mechanisms (see Tollin and Yin 2005 for a review). These models require fast inhibitory interaction of temporally structured or phase-locked inputs to stimulus fine structure, envelope, or transients. This is precisely the type of interaction that appears to be created in the circuitry and synaptic specializations of the LSO. A major goal of this review is to shed new light on the importance of this traditionally understudied and certainly underappreciated neural circuit.

5.6 Summary

The two binaural acoustical cues to the horizontal location of sound sources, ILD and ITD, have been traditionally thought to be initially encoded in two anatomically separate but parallel circuits in the auditory brainstem, the LSO and MSO, respectively. This view has been recently challenged by observations that the LSO circuit contains similar biophysical specializations as the MSO circuit and thus allows LSO neurons to be sensitive not only to ILDs but also to ITDs. Here we reviewed the specializations that allow LSO neurons to be sensitive to ITDs in the fine-structure and low-frequency envelopes of sound as well as a sensitivity to ITDs in brief transient sounds. Explanations for many binaural and spatial hearing phenomena require subtraction-like mechanisms consistent with the LSO, but not necessarily the MSO, circuitry. Thus, the circuitry comprising the LSO could provide the bases for these binaural and spatial hearing capabilities.

Acknowledgments The work during the preparation of this chapter was supported by Grants R01-DC-011555 and R01-DC-017924 from the National Institute of Deafness and Other Communication Disorders (NIDCD), National Institutes of Health. We thank John Peacock and Monica Benson for comments.

Compliance with Ethics Requirements

Zoe L. Owrutsky declares that she has no conflict of interest.

Victor Benichoux declares that he has no conflict of interest.

Daniel J. Tollin declares that he has no conflict of interest.

References

Ashida G, Kretzberg J, Tollin DJ (2016) Roles for coincidence detection in coding amplitude-modulated sounds. PLoS Comput Biol 12(6):e1004997

Ashida G, Tollin DJ, Kretzberg J (2017) Physiological models of the lateral superior olive. PLoS Comput Biol 13(12):e1005903. https://doi.org/10.1371/journal.pcbi.1005903

Banks MI, Smith PH (1992) Intracellular recordings from neurobiotin-labeled cells in brain slices of the rat medial nucleus of the trapezoid body. J Neurosci 12(7):2819–2837

Beiderbeck B, Myoga MH, Muller NIC, Callan AR, Friauf E, Grothe B, Pecka M (2018) Precisely timed inhibition facilitates action potential firing for spatial coding in the auditory brainstem. Nat Commun 9(1):1771. https://doi.org/10.1038/s41467-018-04210-y

Benichoux V, Tollin DJ (2018) These are not the neurons you are looking for. *eLife* 7:e39244

Benichoux V, Rebillat M, Brette R (2016) On the variation of interaural time differences with frequency. J Acoust Soc Am 139(4):1810. https://doi.org/10.1121/1.4944638

Benichoux V, Brown AD, Anbuhl KL, Tollin DJ (2017) Representation of multidimensional stimuli: quantifying the most informative stimulus dimension from neural responses. J Neurosci 37(31):7332–7346

Benichoux V, Ferber AT, Hunt SD, Hughes EG, Tollin DJ (2018) Across species 'natural ablation' reveals the brainstem source of a non-invasive biomarker of binaural hearing. J Neurosci 38:1211–1218

Bledsoe SC Jr, Snead CR, Helfert RH, Prasad V, Wenthold RJ, Altschuler RA (1990) Immunocytochemical and lesion studies support the hypothesis that the projection from the medial nucleus of the trapezoid body to the lateral superior olive is glycinergic. Brain Res 517(1–2):189–194

Boudreau JC, Tsuchitani C (1968) Binaural interaction in the cat superior olive S segment. J Neurophysiol 31(3):442–454

Brand A, Behrend O, Marquardt T, McAlpine D, Grothe B (2002) Precise inhibition is essential for microsecond interaural time difference coding. Nature 417:543–547

Bregman AS (1990) Auditory scene analysis: the perceptual organization of sound. Bradford Books, MIT Press, Cambridge, MA

Brew HM, Forsythe ID (2005) Systematic variation of potassium current amplitudes across the tonotopic axis of the rat medial nucleus of the trapezoid body. Hear Res 206(1–2):116–132

Britt R, Starr A (1976) Synaptic events and discharge patterns of cochlear nucleus cells. I. Steady-frequency tone bursts. *J Neurophysiol* 39(1):162–178

Brown AD, Tollin DJ (2016) Slow temporal integration enables robust neural coding and perception of a cue to sound source location. J Neurosci 36(38):9908–9921

Brown AD, Benichoux V, Jones HG, Anbuhl KL, Tollin DJ (2018) Spatial variation in signal and sensory precision both constrain auditory acuity at high frequencies. Hear Res 370:65–73

Brownell WE, Manis PB, Ritz LA (1979) Ipsilateral inhibitory responses in the cat lateral superior olive. Brain Res 177(1):189–193

Caird D, Klinke R (1983) Processing of binaural stimuli by cat superior olivary complex neurons. Exp Brain Res 52(3):385–399

Caspary DM, Faingold CL (1989) Non-N-methyl-D-aspartate receptors may mediate ipsilateral excitation at lateral superior olivary synapses. Brain Res 503(1):83–90

Dobie RA, Berlin CI (1979) Binaural interaction in brainstem-evoked responses. Arch Otolaryngol 105(7):391–398

Finlayson PG, Caspary DM (1989) Synaptic potentials of chinchilla lateral superior olivary neurons. Hear Res 38(3):221–228

Ford MC, Alexandrova O, Cossell L, Stange-Marten A, Sinclair J, Kopp-Scheinpflug C, Pecka M, Attwell D, Grothe B (2015) Tuning of Ranvier node and internode properties in myelinated axons to adjust action potential timing. Nat Commun 6:8073. https://doi.org/10.1038/ncomms9073

Franken TP, Roberts MT, Wei L, Golding NL, Joris PX (2015) In vivo coincidence detection in mammalian sound localization generates phase delays. Nat Neurosci 18:444–452

Franken TP, Joris PX, Smith PH (2018) Principal cells of the brainstem's interaural sound level detector are temporal differentiators rather than integrators. eLife 7. https://doi.org/10.7554/eLife.33854

Glendenning KK, Masterton RB (1983) Acoustic chiasm: efferent projections of the lateral superior olive. J Neurosci 3(8):1521–1537

Glendenning KK, Masterton RB (1998) Comparative morphometry of mammalian central auditory systems: variation in nuclei and form of the ascending system. Brain Behav Evol 51(2):59–89

Goldberg JM, Brown PB (1969) Response of binaural neurons of dog superior olivary complex to dichotic tonal stimuli: some physiological mechanisms of sound localization. J Neurophysiol 22:613–636

Grantham DW (1984) Discrimination of dynamic interaural intensity differences. J Acoust Soc Am 76(1):71–76

Greene NT, Davis KA (2012) Discharge patterns in the lateral superior olive of decerebrate cats. J Neurophysiol 108(7):1942–1953

Greene NT, Anbuhl KL, Ferber AT, DeGuzman M, Allen PD, Tollin DJ (2018) Spatial hearing ability of the pigmented guinea pig (*Cavia porcellus*): minimum audible angle and spatial release from masking in azimuth. Hear Res 365:62–76

Grothe B (2003) New roles for synaptic inhibition in sound localization. Nat Rev Neurosci 4:540–550

Grothe B, Pecka M (2014) The natural history of sound localization in mammals – a story of neural inhibition. Front Neural Circuits 8:116. https://doi.org/10.3389/fncir.2014.00116
Grothe B, Pecka M, McAlpine D (2010) Mechanisms of sound localization in mammals. Physiol Rev 90(3):983–1012
Harper NS, McAlpine D (2004) Optimal neural population coding of an auditory spatial cue. Nature 430:682–686
Hartmann WM, Constan ZA (2002) Interaural level differences and the level-meter model. J Acoust Soc Am 112:1037–1045
Helfert RH, Schwartz IR (1986) Morphological evidence for the existence of multiple neuronal classes in the cat lateral superior olivary nucleus. J Comp Neurol 244:533–549
Irvine DR (1986) The auditory brainstem: processing of spectral and spatial information. Springer, Berlin
Jeffress LA (1948) A place theory of sound localization. J Comp Physiol Psychol 41(1):35–39
Jercog PE, Svirskis G, Kotak VC, Sanes DH, Rinzel J (2010) Asymetric excitatory synaptic dynamics underlie interaural time difference processing in the auditory system. PLoS Biol 8(6):e1000406. https://doi.org/10.1371/journal.pbio.1000406
Jewett DL, Romano MN, Williston JS (1970) Human auditory evoked potentials: possible brain stem components detected on the scalp. Science 167:1517–1518
Johnson DH (1980) The relationship between spike rate and synchrony in responses of auditory-nerve fibers to single tones. J Acoust Soc Am 68(4):1115–1122
Jones HG, Brown AD, Koka K, Thornton JL, Tollin DJ (2015) Sound frequency-invariant neural coding of a frequency-dependent cue to sound source location. J Neurophysiol 114(1):531–539
Joris PX, Trussell LO (2018) The Calyx of held: a hypothesis on the need for reliable timing in an intensity-difference encoder. Neuron 100(3):534–549
Joris PX, van der Heijden M (2019) Early binaural hearing: the comparison of temporal differences at the two ears. Annu Rev Neurosci 42:433–457
Joris PX, Yin TCT (1992) Responses to amplitude-modulated tones in the auditory nerve of the cat. J Acoust Soc Am 91:215–232
Joris PX, Yin TC (1995) Envelope coding in the lateral superior olive. I. Sensitivity to interaural time differences. J Neurophysiol 73(3):1043–1062
Joris PX, Yin TCT (1998) Envelope coding in the lateral superior olive. III. Comparison with afferent pathways. *J Neurophysiol* 79(1):253–269
Joris P, Yin TCT (2007) A matter of time: internal delays in binaural processing. Trends Neurosci 30(2):70–78
Joris PX, Carney LH, Smith PH, Yin TCT (1994) Enhancement of neural synchronization in the anteroventral cochlear nucleus. I. Responses to tones at the characteristic frequency. *J Neurophysiol* 71(3):1022–1036
Joris PX, Smith PH, Yin TCT (1998) Coincidence detection in the auditory system: 50 years after Jeffress. Neuron 21:1235–1238
Joris PX, Schreiner CE, Rees A (2004) Neural processing of amplitude-modulated sounds. Physiol Rev 84:541–577
Joris PX, Van de Sande B, Louage DH, van der Heijden M (2006) Binaural and cochlear disparities. Proc Natl Acad Sci 103:12917–12922
Karino S, Smith PH, Yin TCT, Joris PX (2011) Axonal branching patterns as sources of delay in the mammalian auditory brainstem: a re-examination. J Neurosci 31:3016–3031
Koka K, Tollin DJ (2014) Linear coding of complex sound spectra by discharge rate in neurons of the medial nucleus of the trapezoid body (MNTB) and its inputs. Front Neural Circuits 8:144. https://doi.org/10.3389/fncir.2014.00144
Koka K, Read HL, Tollin DJ (2008) The acoustical cues to sound location in the rat: measurements of directional transfer functions. J Acoust Soc Am 123(6):4297–4309
Kuhn GF, Burnett ED (1977) Acoustic pressure field alongside a manikin's head with a view towards in situ hearing-aid tests. J Acoust Soc Am 62(2):416–423

Laback B, Dietz M, Joris PX (2017) Temporal effects in interaural and sequential level difference perception. *J Acoust Soc Am* 142(5):3267–3283

Laumen G, Ferber AT, Klump GM, Tollin DJ (2016) The physiological basis and clinical use of the binaural interaction component of the auditory brainstem response. Ear Hear 37(5):e276–e290. https://doi.org/10.1097/AUD.0000000000000301

Li K, Chan CHK, Rajendran VG, Meng Q, Rosskothen-Kuhl N, Schnupp JWH (2019) Microsecond sensitivity to envelope interaural time differences in rats. J Acoust Soc Am 145(5):EL341. https://doi.org/10.1121/1.5099164

Loftus WC, Bishop DC, Saint Marie RL, Oliver DL (2004) Organization of binaural excitatory and inhibitory inputs to the inferior colliculus from the superior olive. J Comp Neurol 472(3):330–344

Masterton B, Thompson GC, Bechtold JK, RoBards MJ (1975) Neuroanatomical basis of binaural phase-difference analysis for sound localization: a comparative study. J Comp Physiol Psychol 89(5):379–386

McAlpine D, Jiang D, Palmer AR (2001) A neural code for low-frequency sound localization in mammals. Nat Neurosci 4:396–401

Moore MJ, Caspary DM (1983) Strychnine blocks binaural inhibition in lateral superior olivary neurons. J Neurosci 3(1):237–242

Park TJ, Klug A, Oswald JP, Grothe B (1998) A novel circuit in the bat's midbrain recruits neurons into sound localization processing. Naturwissenschaften 85(4): 176–179

Park TJ, Klug A, Holinstat M, Grothe B (2004) Interaural level difference processing in the lateral superior olive and the inferior colliculus. J Neurophysiol 92(1):289–301

Pecka M, Brand A, Behrend O, Grothe B (2008) Interaural time difference processing in the mammalian medial superior olive: the role of glycinergic inhibition. J Neurosci 28(27):6914–6925

Roberts MT, Seeman SC, Golding NL (2013) A mechanistic understanding of the role of feedforward inhibition in the mammalian sound localization circuitry. Neuron 78(5):923–935

Sanes DH (1990) An in vitro analysis of sound localization mechanisms in the gerbil lateral superior olive. J Neurosci 10(11):3494–3506

Smith PH, Joris PX, Yin TCT (1993) Projections of physiologically characterized spherical bushy cell axons from the cochlear nucleus of the cat: evidence for delay lines to the medial superior olive. J Comp Neurol 331:245–260

Song P, Yang Y, Barnes-Davies M, Bhattacharjee A, Hamann M, Forsythe ID, Oliver DL, Kaczmarek LK (2005) Acoustic environment determines phosphorylation state of the Kv3.1 potassium channel in auditory neurons. Nat Neurosci 8(10):1335–1342

Rayleigh L (1907) On our perception of sound direction. Philos Mag 6:13–74

Tollin DJ (2003) The lateral superior olive: a functional role in sound source localization. Neuroscientist 9(2):127–143

Tollin DJ, Koka K (2009) Postnatal development of sound pressure transformations by the head and pinnae of the cat: binaural characteristics. J Acoust Soc Am 126(6):3125–3136

Tollin DJ, Yin TCT (2002a) The coding of spatial location by single units in the lateral superior olive of the cat. I. Spatial receptive fields in azimuth. *J Neurosci* 22(4):1454–1467

Tollin DJ, Yin TCT (2002b) The coding of spatial location by single units in the lateral superior olive of the cat. II. The determinants of spatial receptive fields in azimuth. *J Neurosci* 22(4):468–1479

Tollin DJ, Yin TCT (2005) Interaural phase and level difference sensitivity in low-frequency neurons in the lateral superior olive. J Neurosci 25:0648–10657

Tollin DJ, Koka K, Tsai JJ (2008) Interaural level difference discrimination thresholds for single neurons in the lateral superior olive. J Neurosci 28(19):4848–4860

Trussell LO (2002) Modulation of transmitter release at giant synapses of the auditory system. Curr Opin Neurobiol 12(4):400–404

Tsai JJ, Koka K, Tollin DJ (2010) Varying overall sound intensity to the two ears impacts interaural level difference discrimination thresholds by single neurons in the lateral superior olive. J Neurophysiol 103(2):875–886 van der Heijden M, Lorteije JA, Plauska A, Roberts MT, Golding NL, Borst JG (2013) Directional hearing by linear summation of binaural inputs at the medial superior olive. Neuron 78(5):936–948

Wesolek CM, Koay G, Heffner RS, Heffner HE (2010) Laboratory rats (Rattus norvegicus) do not use binaural phase differences to localize sound. Hear Res 265(1–2):54–62

Wu SH, Kelly JB (1991) Physiological properties of neurons in the mouse superior olive: membrane characteristics and postsynaptic responses studied in vitro. J Neurophysiol 65(2):230–246

Wu SH, Kelly JB (1994) Physiological evidence for ipsilateral inhibition in the lateral superior olive: synaptic responses in mouse brain slice. Hear Res 73(1):57–64

Yin TCT (2002) Neural mechanisms of encoding binaural localization cues. In: Oertel D, Fay RR, Popper AN (eds) Integrative functions in the mammalian auditory pathway. Springer, New York, pp 99–159

Yin TCT, Chan JC (1990) Interaural time sensitivity in medial superior olive of cat. J Neurophysiol 64:465–488

Young ED, Davis KA (2002) Circuitry and function of the dorsal Cochlear nucleus. In: Oertel D, Fay RR, Popper AN (eds) Integrative functions in the mammalian auditory pathway. Springer New York, New York, pp 160–206

Young ED, Oertel D (2004) Cochlear nucleus. In: Shepherd GM (ed) The synaptic organization of the brain. Oxford University Press, New York, pp 125–163

Zhang Y, Wright BA (2009) An influence of amplitude modulation on interaural level difference processing suggested by learning patterns of human adults. J Acoust Soc Am 126(3):1349–1358

Chapter 6
Binaural Hearing with Temporally Complex Signals

G. Christopher Stecker, Leslie R. Bernstein, and Andrew D. Brown

6.1 Introduction

Even with eyes closed, one can understand the scene: a woman, standing at arm's length directly ahead, relates a clever story to a man standing off to the left. His deeper voice occasionally interjects: "yes," "I see," "oh wow." Soft cello music emanates from a small loudspeaker on the kitchen counter behind the woman; its sound seems to fill the space, highlighting the reverberation introduced by the tile walls and floor. One imagines that echoes of the voices are also present, although for some reason they are barely noticeable; instead, the location of each voice seems stable and clear.

What are the acoustic features of this scene that allow one to locate, segregate, and understand its contents? How does the brain determine which features are most reliable, and combine them to perceive accurately the surrounding acoustic world? Particularly, what role is played by the dynamic changes–temporal complexity–in each sound? Sounds and voices change continuously in amplitude and frequency over several relevant time scales. The changes occur slowly as the cello glides from note to note or as a voice rises and falls in emphasis to convey the

G. C. Stecker (✉)
Center for Hearing Research, Boys Town National Research Hospital, Omaha, NE, USA

Department of Hearing and Speech Sciences, Vanderbilt University School of Medicine, Nashville, TN, USA
e-mail: cstecker@spatialhearing.org

L. R. Bernstein
Department of Neuroscience and Department of Surgery (Otolaryngology), University of Connecticut Health Center, Farmington, CT, USA
e-mail: LBernstein@uchc.edu

A. D. Brown
Department of Speech and Hearing Sciences, University of Washington, Seattle, WA, USA
e-mail: andrewdb@uw.edu

© Springer Nature Switzerland AG 2021

R. Y. Litovsky et al. (eds.), *Binaural Hearing*, Springer Handbook of Auditory Research 73, https://doi.org/10.1007/978-3-030-57100-9_6

punchline of a joke, more rapidly as bursts of laughter add to the sound wave in segments a few hundred milliseconds long, and even on shorter time scales such as the fundamental period of the voice pitch, some 5–10 ms. Echoes and reverberation produce other forms of temporal complexity, adding copies of each sound that repeat every few milliseconds and die away over long fractions of a second.

This chapter reviews some of the research regarding how temporal complexity affects binaural hearing. In some cases, such as those of echoes and reverberation, temporal complexity adds challenge to the brain's task. But for the most part, this is a story about how temporal complexity supports access to binaural information, and of the brain mechanisms that exploit temporal complexity to achieve robust and stable perception of auditory space.

Beginning with the observations of Lord Rayleigh (1907), psychoacousticians have considered the nature of acoustical cues for the localization of sounds. Rayleigh's "Duplex Theory," in particular, considered the nature of cues available at different frequencies. Rayleigh's argument emphasized, based on the physics of sound, that pure tones are localized on the basis of two types of interaural cue: ITD (interaural time differences or interaural temporal disparities) (see Table 6.1 for list of abbreviations) cues are responsible for localization at low frequencies and ILD (interaural level differences or interaural intensitive differences [IID]) cues are responsible for localizing high-frequency pure tones. Over time, this view expanded to recognize the listener's ability to discriminate ITD cues within *complex*

Table 6.1 List of abbreviations

AM	Amplitude modulation
AMBB	Amplitude-modulated binaural beat
CF	Center frequency
GCT	Gabor click train
IAC	Interaural coherence
IC	Inferior colliculus
ICI	Interclick interval
IID	Interaural intensitive difference
ILD	Interaural level difference
ITD	Interaural time difference
ITD*env*	Envelope ITD
ITD*fs*	Fine-structure ITD
LSO	Lateral superior olive
RS	Raised sine
RESTART	Reliable envelope-slope-triggered auditory representation theory
SAM	Sinusoidal amplitude modulation
S/N	Signal-to-noise ratio
TT	Transposed tone
TWF	Temporal weighting function

high-frequency waveforms such as sinusoidally amplitude-modulated (SAM) tones, filtered transients, and bands of noise (e.g., Klumpp and Eady 1956; David et al. 1959). The results of those studies were bolstered by a later round of investigations (e.g., Henning 1974; McFadden and Pasanen 1976). Taken together, those studies established firmly that sensitivity to ITDs conveyed by high-frequency complex waveforms is mediated by interaural delays within the time-varying amplitude envelopes of the waveforms and not by the cycle-by-cycle or fine-structure disparities within those waveforms. Those studies spawned many others detailing the sometimes complex interactions between binaural performance and specific temporal aspects of complex waveforms.

The foregoing example demonstrates why an in-depth understanding of binaural processing requires the use of stimuli that are temporally (and, concomitantly, spectrally) complex. Such stimuli share important features with "real-world" sounds, such as speech, which contain a variety of temporal complexities. For example, they are characterized by amplitude modulation (AM), the rate and depth of which may vary among different sounds, and within sounds across time. In addition, the temporal characteristics of sounds emitted by naturally occurring acoustic sources are typically altered by interactions with the human head or with the surfaces defining sound-fields, such as in rooms.

Beyond consideration of the temporal properties of individual ("monaural") waveforms that reach the ears, yet another level of temporal complexity concerns the binaural disparities themselves. Specifically, the effective ITDs and ILDs conveyed by sounds may vary over time, for example when a sound source moves in space. Clearly discriminable auditory motion (see Yost, Pastore, and Zhou, Chap. 3), however, requires that such changes occur slowly–over at least a few hundred milliseconds (Grantham and Wightman 1978; Grantham 1986). This relative inability to follow rapidly fluctuating binaural cues, termed "binaural sluggishness," provides an upper limit to changes that are perceived as object motion (i.e., changes that can be explicitly tracked by a listener). More rapid fluctuations in ITD and ILD are less likely to be produced by source motion, but commonly arise as a result of room acoustics (reflections and reverberation). The acoustic effects of rapid fluctuations may be more clearly understood as reducing the similarity of sound at the two ears (i.e., the interaural coherence [IAC], or maximum interaural correlation coefficient) and thereby increasing perceived source width or diffuseness, which are the primary perceptual correlates of IAC. In what follows, we describe how various types of "diagnostic" stimuli embodying the types of temporal complexities described here affect binaural performance (e.g., discrimination of ITD, ILD, and IAC values) and, perhaps more importantly, how such stimuli have been and can be used to provide a greater understanding of binaural processing in general. In some instances, we extend that understanding to consider consequences of such processing in "real-world" scenarios.

6.1.1 Taxonomy and Terminology of Binaural Cues

A consideration of binaural hearing with temporally complex sounds concerns two types of temporal complexity. The first is complexity of the waveform common to both ears, particularly fluctuations in the temporal envelope resulting from AM that occurs within or across frequency bands. The second is temporal variation in the binaural cues themselves, resulting from interaural decorrelation or dynamic cues (e.g., motion). Note that both types of temporal complexity are naturally accompanied by spectral complexity (e.g., in spectral shape and/or variation in binaural cues across frequency). In this chapter, the relationships between temporal and spectral complexities are addressed briefly where relevant, but the focus remains largely on the role of temporal complexity.

For all signals, binaural hearing depends on the nature and variety of binaural cues available. Three main cue types may be identified: ITD, ILD, and IAC (for reviews, see Durlach and Colburn 1978; Hafter and Trahiotis 1997). As illustrated in Fig. 6.1 and described in Sect. 6.1, two different types of ITD cues can be distinguished: those carried by the temporal fine structure of the waveform (ITD*fs*) and those carried by the envelope (ITD*env*). In the free-field, the ITD affects the whole waveform equally and the two types of ITD agree. Yet in reverberant listening and in headphone-based experiments ITD*fs* and ITD*env* can, or can be made to, disagree

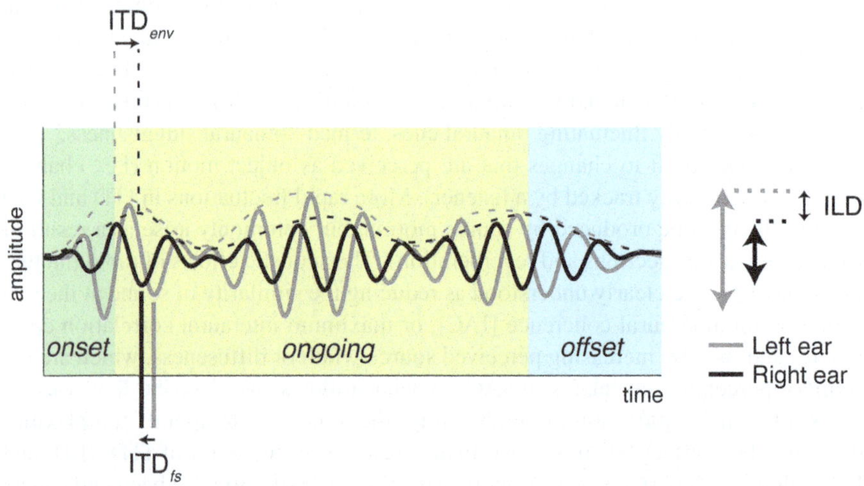

Fig. 6.1 Illustration of interaural time difference (ITD) and interaural level difference (ILD) cues. *Black solid and dashed lines*, right ear waveform of a brief complex sound and the corresponding amplitude envelope. *Gray solid and dashed lines*, left ear waveform and the corresponding amplitude envelope. At sound onset, the waveform carries initial ILD and envelope ITDenv cues that favor the left ear. The initial fine-structure ITD*fs* favors the right ear by a smaller amount. Each of the cues evolves over time, such that onset (*green box*), ongoing (*yellow box*), and offset (*blue area*) portions of the sound each carry different ITD and ILD values. In this example, ILD diminishes toward zero and ITD*fs* takes on increasingly large values later in the sound

substantially. Several studies have specifically investigated the relative influences of ITD*fs* and ITD*env* cues by applying independent ITD cues to the ongoing envelope, the envelope onset and/or offset ("gating delays"), or the fine structure (e.g., Henning 1980; Buell et al. 1991). It is clear that sensitivity to the ITD*env* depends on the temporal features of the envelope, which is addressed in some detail here. In contrast, there is no *a priori* reason to expect ITD*fs* or ILD sensitivity to depend on the envelope features. Nevertheless, numerous data suggest a similarly important role of the envelope in accessing ITD*fs* and ILD cues (see Sect. 6.4; e.g., Stecker and Brown 2012; Stecker and Bibee 2014). Thus, a major focus of this chapter is on the effects of envelope fluctuations on the processing of ITD*env*, ITD*fs*, and ILD cues.

Another focus of the chapter is the role of the interaural correlation (IAC), the meaning of which bears some explanation. As discussed in detail by Trahiotis et al. (2005), the relevant metric or "index" of IAC is, technically speaking, the *coefficient of the normalized interaural correlation*. It is computed as the simple correlation of left and right-ear signals (acoustical waveforms or as processed by the auditory system). The IAC ranges from 1.0 (for identical signals; L = R) to −1.0 (if one signal is inverted; L = −R). The IAC is zero if the two waveforms are statistically independent of each other.

In order to account for the localization and lateralization of sounds, however, it is important to consider not only the single-valued IAC, but the interaural cross-correlation *function* obtained across a range of relative time delays, called "lags," between the left- and right-ear waveforms. ITDs shift the peak correlation away from zero lag without reducing the peak correlation value, i.e., although the correlation at zero lag (the IAC) decreases with nonzero ITDs, the peak value of the cross-correlation function is maintained, occurring at a location that reflects the imposed delay.

Figure 6.2 illustrates the interaural cross-correlation function for a 900-Hz wide band of Gaussian noise centered at 500 Hz and presented identically at the two ears (i.e., diotically; blue) or with an ITD of 500 μs leading in the right ear (dotted red line). The IAC (computed at zero lag) is 1.0 and 0.0, respectively, in the two cases. The peak correlation, however, is the same in both cases, and occurs at the lag matching the ITD (0 or 500 μs). The maximum interaural correlation along the lag-axis defines the *interaural coherence*, which is 1.0 in both cases. In contrast, when waveforms are statistically independent across the ears (black line), the interaural coherence (and the IAC, by definition) is zero because the waveforms are uncorrelated at all lags.

A major emphasis of the binaural literature has been the distribution and weighting (i.e., relative importance) of the various cues across the frequency spectrum. In particular, it has been shown that ITD*fs* cues dominates the localization and lateralization of low-frequency sounds. "Lateralization" refers to the perceived displacement of intracranial (within the head) images for sounds presented via earphones. In contrast to the dominance of ITD*fs* at low frequencies, ITD*env* and ILD cues mediate the localization and lateralization of high- frequency sounds for which ITD*fs* cues are not resolved. For broadband complex sounds, within which ITD*fs*, ITD*env*, and ILD cues are all potentially available to the listener, studies suggest a dominant

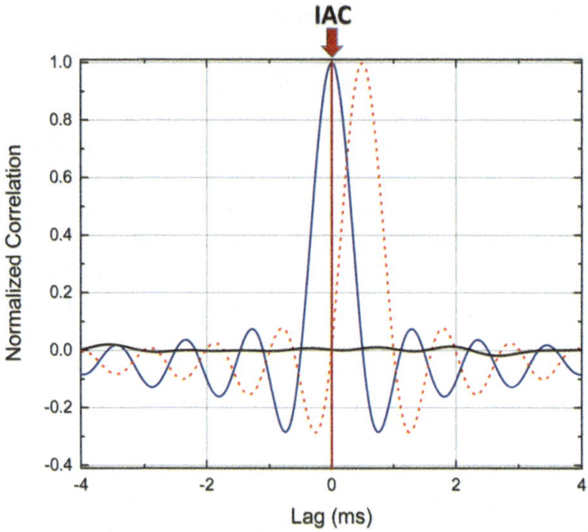

Fig. 6.2 Example of cross-correlation functions. *Red dashed line*, a binaurally coherent signal presented with an ITD of 0.5 ms. The peak of the cross-correlation function occurs at the lag corresponding to the ITD. The correlation coefficient computed at that lag (the *normalized interaural correlation*) defines the interaural coherence (IAC). In this case, IAC = 1.0, just as for identical (diotic) stimulation of the two ears (*blue line*). *Black line*, cross-correlation function for interaurally uncorrelated noise (IAC = 0); the dissimilar signals are uncorrelated at all possible lags

role of low-frequency ITD*fs* cues (e.g., Wightman and Kistler 1992; Macpherson and Middlebrooks 2002). That dominance appears consistent with classic data demonstrating relatively poor sensitivity to high-frequency ITD*env* cues. Compared to low-frequency ITD*fs* cues, ITD*env* cues typically yield higher discrimination thresholds (e.g., Zwislocki and Feldman 1956; McFadden and Pasanen 1976), reduced binaural masking release (see Durlach and Colburn 1978; Zurek and Durlach 1987), and limited extents of lateralization (i.e., intracranial images appear closer to the midline; see Blauert 1983; Bernstein and Trahiotis 1985). Yet, interpreting such differences requires a careful consideration of the stimuli *as processed*; i.e., as transformed by the auditory periphery and pre-binaural neural mechanisms. As noted by Colburn and Equissaud (1976), differences in binaural processing at high versus low frequencies may emerge from differences in the effective input to binaural mechanisms in the two frequency regions. For example, neural impulses synchronize to both fine-structure and envelope features of low-frequency waveforms, but only to the envelopes of high-frequency waveforms. Careful consideration of these effects has helped to demonstrate similar precision and salience in the binaural processing of low- and high-frequency sound (see Sect. 6.5 and Fig. 6.3; e.g., van de Par and Kohlrausch 1997; Bernstein and Trahiotis 2002).

Just as the various binaural cues are distributed in frequency across the stimulus spectrum, they are also distributed in time over the stimulus duration. Thus, one can contrast "early" cues (i.e., those available in close proximity to sound onset) versus

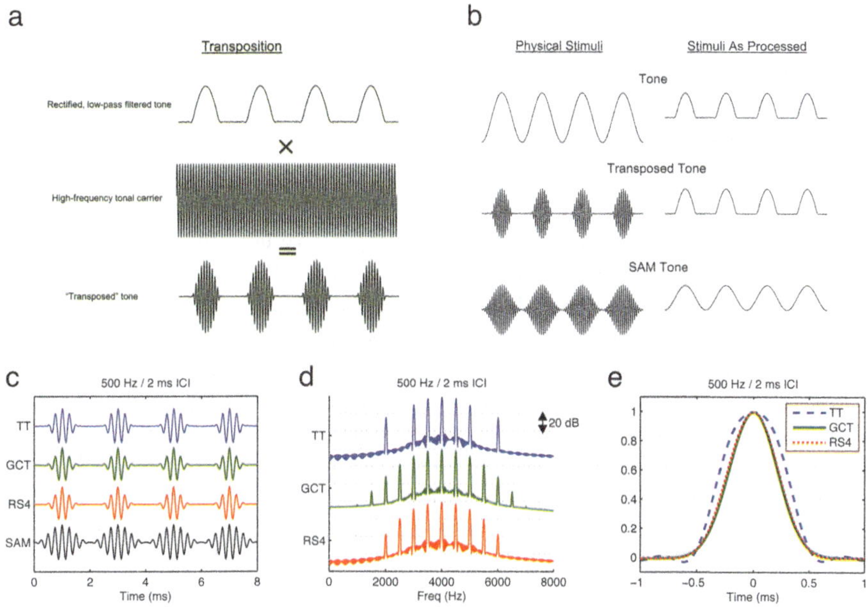

Fig. 6.3 Stimuli used to explore the role of envelope shape. (**a**) Transposed tones are synthesized by modulating a high-frequency tonal carrier with a rectified, low-pass-filtered, low-frequency signal. (**b**) Waveforms of low-frequency tones, transposed tones, and SAM tones before and after transformation in the auditory periphery. A variety of pulsatile stimuli are compared in terms of their waveforms (**c**), spectra (**d**), and details of the envelope shape (**e**). Transposed tones (TT), Gabor click trains (GCT), raised-sine stimuli with exponent 4 (RS4), and SAM tones are depicted with center/carrier frequency of 4 kHz and AM frequency of 500 Hz / 2-ms interclick interval (ICI). (**a** and **b,** Adapted from Bernstein and Trahiotis (2002), with the permission of the Acoustical Society of America; **c–e**, adapted from Bibee and Stecker (2014), with the permission of the Acoustical Society of America)

"late" cues, as well as "onset" cues versus "ongoing" and "offset" cues, etc. For non-moving sounds presented in the free field, the ITD and ILD do not change over time. In that case, the onset, ongoing, and offset cues do not differ in direction or magnitude. Reverberant listening environments, simultaneous or moving sound sources, and a variety of other scenarios occurring naturally or simulated in the laboratory, however, produce dynamic cues that evolve over time. Any given cue may thus be more reliable during different temporal epochs (e.g., at sound onset) than others. Stimulus transformations in the peripheral and central auditory system, such as basilar-membrane compression or neuronal adaptation, also contribute to the temporal distribution of effective cues available for binaural processing, reinforcing the importance of considering the representations of stimuli as processed by the auditory periphery. Numerous studies of the temporal weighting of binaural information have quantified listeners' use of these cues to reveal phenomena such as rate-dependent dominance of onset cues (e.g., Saberi 1996; Stellmack et al. 1999) and integration of ongoing cues for sounds with slow or temporally irregular envelopes.

A brief comment should be made about terminology in order to avoid confusion. Many studies have used the terms "onset," "ongoing," and "offset" to refer to early, middle, and late cues, respectively. Hence, "onset dominance" (Houtgast and Aoki 1994, Freyman et al. 1997) typically refers to the relative importance of cues carried by the first-arriving sound, an aspect of the precedence effect (Wallach et al. 1949; see Brown et al. 2015 for a recent review). Others have used the term "onset cue" specifically for the ITD of the envelope onset itself (Buell et al. 1991). This chapter follows the convention of McFadden and Pasanen (1976), using the terms "onset" to refer to the initial segment as it reaches the two ears, "offset" to refer to the final segment, and "ongoing" to refer to the intervening quasi-steady-state segment. These are illustrated in Fig. 6.1. Importantly, binaural cues of any type can be associated with each segment: "onset ITD*env*" identifies the delay carried by the initial envelope attack, "onset ITD*fs*" refers to the simultaneous fine-structure delay, and "offset ILD" identifies the ILD at termination of sound, etc. Similarly, ongoing cues can reflect ILD ("ongoing ILD"), interaural delays of the steady waveform ("ongoing ITD*fs*"), or interaural delays of an ongoing periodic or aperiodic envelope ("ongoing ITD*env*"). Even this taxonomy leaves something to be desired: when ongoing envelope fluctuations occur at slow rates, their neural representations and psychophysical characteristics do not differ, functionally, from onsets. Thus, the temporal weighting of binaural cues is intimately tied to the rate of envelope fluctuations (see Sects. 6.2 and 6.3).

6.2 Binaural Sensitivity Declines at High Rates of Envelope Fluctuation

As discussed in Sect. 6.1, sensitivity to ITDs conveyed by high-frequency, temporally complex waveforms is limited to interaural delays within the time-varying amplitude envelopes (ITD*env*) of such stimuli. The results of a number of studies suggested that the efficacy of such envelope-based ITD processing depends on the rate of fluctuation of the envelope (see McFadden and Pasanen 1976; Nuetzel and Hafter 1981). Those studies demonstrated that sensitivity to the ongoing ITD*env* cue declines rapidly when the rate of fluctuation of the envelope increases beyond some "rate-limitation" or envelope low-pass "cutoff" (between about 250 and 500 Hz). Subsequent studies demonstrated that above that rate, threshold ITD*env* and ILD values improve minimally with increasing stimulus duration (Hafter and Dye 1983; Hafter et al. 1983). This result suggests that rate limitations affect both ongoing ITD and ILD cues, although the nature of that limitation might differ across cues (see Sect. 6.3.2).

Sensitivity to ITD*env* cues logically depends on the presence of fluctuations in the amplitude envelope. Thus, it is reasonable to expect peripheral processing to degrade the ITD*env* sensitivity at high rates. Increasing the modulation rate of an AM signal beyond a certain value increases the frequency separation between spectral sideband components beyond the width of a peripheral auditory filter. This attenuates the sidebands and reduces the degree to which they interact, such that the effective modulation depth is reduced as the modulation rate increases.

Auditory filter widths increase with center frequency (CF). For example, the auditory filter centered at 8 kHz is roughly twice as wide as the one centered at 4 kHz. Thus, on the basis of peripheral auditory filtering alone, one would expect the envelope low-pass cutoff to be proportional to the center (carrier) frequency of high-frequency modulated stimuli. Numerous studies, however, have demonstrated the opposite relation. Bernstein and Trahiotis (1994, 2002) measured ITD*env*-discrimination thresholds as a function of AM rate for high-frequency complexes ranging from 4 to 10 kHz. In both studies, the rate of modulation beyond which the envelope information was no longer useful (i.e., the envelope low-pass cutoff) was inversely related to CF. Although opposite to expectations based on auditory filtering, this inverse relationship was also found in Bernstein and Trahiotis' (2014) measurement of threshold-ITDs for filtered transients centered at 4600, 6500, and 9200 Hz. Figure 6.4 is taken from Fig. 1 of that study. The symbols represent mean ITD*env* thresholds normalized to remove inter-individual differences among four

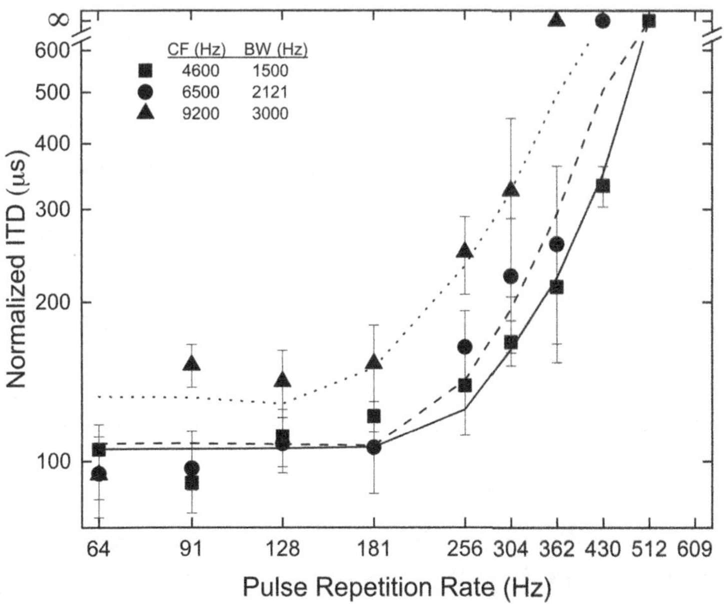

Fig. 6.4 Threshold (normalized) ITD values plotted versus pulse repetition rate for transients bandpass filtered at three combinations of center frequency (CF) and bandwidth (BW). The upper limit of "good" performance declines from about 250 Hz at 4600 Hz CF to about 180 Hz at 9200 Hz. *Symbols*, mean normalized ITD*env* thresholds across four normal-hearing listeners; *lines*, predictions obtained via a cross-correlation-based model. (Figure from Bernstein and Trahiotis (2014), with permission of the Acoustical Society of America)

normal-hearing listeners. The lines represent predictions obtained via a cross-correlation-based model. The predictions account for over 90% of the variance in the data, with best-fitting envelope low-pass cutoffs of 300, 195, and 125 Hz for the center frequencies of 4600, 6500, and 9200 Hz, respectively. Interestingly, attempting to substitute the derived low-pass cutoff for one of those center frequencies with that derived for one of the other center frequencies, resulted in the variance dropping from about 90% to 0%. The inverse relationship between spectral frequency and the envelope low-pass cutoff is inconsistent with the effects of peripheral auditory filtering. Rather, it suggests a central origin for the rate limitation in binaural processing.

Neurophysiological data appear consistent with the behavioral results of Bernstein and Trahiotis (2014). Geis and Borst (2009) measured the intracellular responses of units in the cat inferior colliculus (IC) to SAM tones as a function of the rate of modulation. They observed a shared gradient of increasing center frequency (CF) and increasing membrane time-constants consistent with a decreased ability to follow rapid envelope fluctuations at higher CF. Rodríguez et al. (2010) and Middlebrooks and Snyder (2010) measured the responses of neural units within the IC of cats to acoustic and electrical stimulation, respectively. Both sets of results demonstrated envelope low-pass cutoffs that decreased systematically with increases in the frequency to which the unit was best tuned. In fact, the cutoff values of the neurophysiological data were remarkably similar to those derived from behavioral data in human listeners. Remme et al. (2014) used in vitro techniques to measure the intrinsic filtering properties of high-frequency neurons of the guinea pig lateral superior olive (LSO), units that are responsive to the temporal properties of the envelope of stimuli. Their findings were consistent with a "gradient in temporal processing that runs counter to the tonotopic gradient;" in other words, an inverse relationship between the CF and the envelope low-pass cutoff.

Thus, the relationship between carrier frequency and rate limitation suggests that these effects do not reflect lost sensitivity to ongoing ITD*env* resulting from a reduced modulation depth in the auditory periphery. As shown in Sects. 6.3.2 and 6.3.3, other types of studies suggest that such rate limitations also apply to ILD cues and that the same mechanisms may limit access to ongoing ITD*fs* cues as well. Together, the psychophysical and neurophysiological data point to a central rather than a purely peripheral origin of envelope-rate limitations on binaural processing.

6.3 Temporal Weighting Reveals Binaural Dominance of Onset-Like Events

Intimately tied to the issue of rate limitation in binaural processing is the temporal dynamics of sensitivity to binaural cues. Do listeners integrate binaural information effectively over the duration of a sound? If not, are some temporal epochs, such as the sound onset, more salient? Three approaches have been used to study this question, which we refer to as the temporal-integration approach, the dynamic-cues approach, and the direct temporal-weighting approach. The following sections review evidence for rate-dependent onset dominance in the weighting of three types of binaural cues: high-frequency ITD*env* (Sect. 6.3.1), ILD (Sect. 6.3.2), and low-frequency ITD*fs* (Sect. 6.3.3), along with evidence against such dominance in aperiodic sounds (Sect. 6.3.4).

*6.3.1 Evidence of Rate-Dependent Onset Dominance in the Temporal Weighting of ITD*env *Cues at High Frequency*

In the temporal-integration approach, binaural sensitivity (e.g., threshold ITD) is measured as a function of sound duration (see Fig. 6.5a, b). From a signal-detection perspective, "threshold" is tied to the variance inherent in the processing, which becomes less as more samples are combined (i.e., as the sound gets longer). Under some basic assumptions (stationarity of the cue, temporally independent internal noise, and total duration shorter than the maximum integration time), optimal thresholds are expected to decrease in proportion to the square root of duration. A plot of log-threshold versus log-duration would, thus, be expected to be linear with a slope of −0.5 (Hafter and Dye 1983). In contrast to that expectation, numerous studies have demonstrated shallower slopes (indicating suboptimal threshold improvement): Houtgast and Plomp (1968) used this approach to study ITD discrimination in narrow bands of noise centered at 500 Hz. For noise bands presented in isolation, they found shallow threshold-duration slopes on the order of −0.2, suggesting very little improvement with the duration. Houtgast and Plomp (1968) concluded, in part, that "...the onset of the signal contributes much more to the lateral position perceived than the ongoing part does (onset effect)."

Hafter and Dye (1983) and Buell and Hafter (1988) applied the temporal-integration approach to ITD*env* discrimination in trains of narrowband-filtered clicks (approximately ½-octave centered at 4 kHz). They observed theoretically optimal threshold-duration slopes when the click rate was lower than ~100 Hz (interclick interval [ICI] >10 ms). At rates of 500 Hz or above (ICI = 1–2 ms), slopes were similarly shallow in comparison to those of Houtgast and Plomp (1968), and thresholds for trains of 16 or 32 clicks were only marginally lower than for single clicks. Hafter and Dye (1983) concluded that listeners lateralize high-rate (short ICI) stimuli on the basis of the initial clicks, but at lower rates, the ongoing information carried by later clicks also contributes. That finding appears to be entirely consistent with rate-limited processing of ongoing ITD*env* cues as described in Sect. 6.2: as listeners lose access to the ongoing cue, the overall onset cue contributes more.

In the dynamic-cues approach, stimuli are designed so that the binaural cues change over time, to probe the listener's access to binaural information within

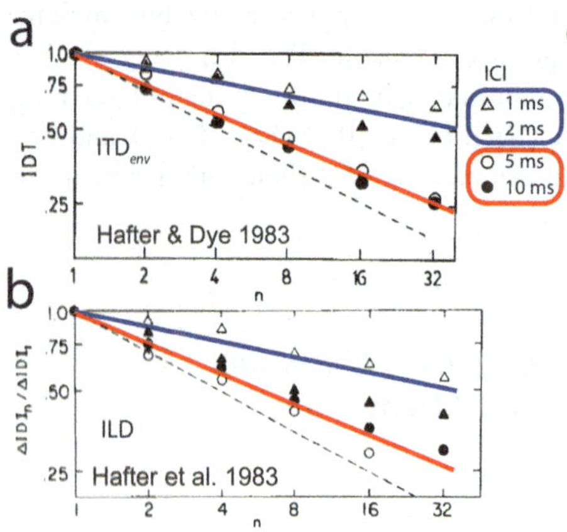

c

	Study	Threshold-Duration Slope (Log-Log)		
		Min	Max	Mean
ITD	Houtgast and Plomp (1968) 500 Hz narrow-band noise	-0.23	-0.25	-0.24
ITD	Hafter and Dye (1983) 4000 Hz click trains, 2 ms ICI	-0.08	-0.33	-0.22
ITD	Stecker and Bibee (2014) 500 Hz pure tones	-0.09	-0.27	-0.18
ILD	Hafter et al (1983) 4000 Hz click trains, 2 ms ICI	-0.20	-0.37	-0.29

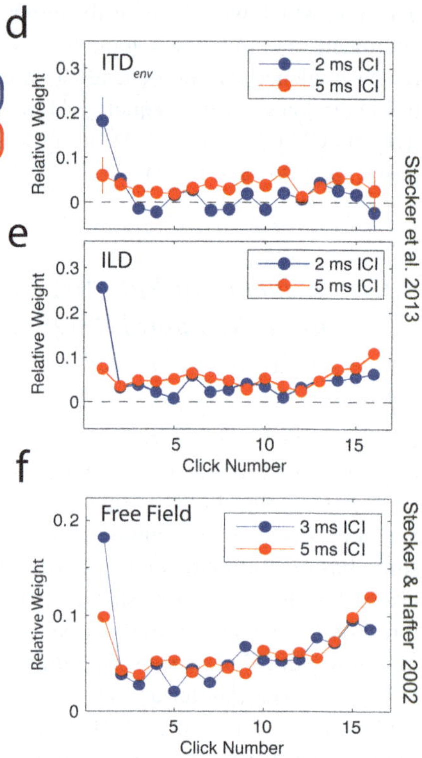

Fig. 6.5 Temporal weighting of ITD cues. (**a**) Hafter and Dye (1983) used the temporal-integration approach to measure threshold ITD versus duration in trains of 4000-Hz narrowband clicks. Shallow threshold-duration slopes at short ICI (*blue line*) reflect dominance of onset clicks. The slope was closer to optimal (*dashed line*) for long ICI values (*red line*). (**b**) Hafter et al. (1983) adapted the same approach using the ILD as the cue. Threshold-duration slopes were steeper in some conditions but were similarly dependent on ICI. (**c**) Table of threshold-duration slopes for narrow bands of low-frequency noise (Houtgast and Plomp 1968), high-frequency click trains (Hafter and Dye 1983; Hafter et al. 1983), and 500-Hz pure tones (Stecker and Bibee 2014). Similar values suggest similar behavior across cue type and frequency. (**d** and **e**) Example temporal weighting functions for ITD- and ILD-based lateralization. (**f**) Temporal weighting functions for sound localization in the free field. (**a**, Adapted from Hafter and Dye (1983), with the permission of the Acoustical Society of America; **b,** adapted from Hafter et al. (1983), with the permission of the Acoustical Society of America; **d** and **e**, adapted from Stecker et al. (2013), with the permission of the Acoustical Society of America; **f**, adapted from Stecker and Hafter (2002), with the permission of the Acoustical Society of America)

particular segments of the sound duration. For example, measurements of threshold-ITD and threshold-ILD conveyed by brief target "probes" flanked by variable-length leading and/or trailing temporal "fringes" have consistently revealed a short period of relative binaural insensitivity starting 1–5 ms after onset and recovering over the duration of the trailing fringe, for example 50 ms (Zurek 1980; Stellmack et al. 2005). Stecker and Brown (2010) used a different approach, presenting trains of

4-kHz-centered Gabor clicks identical to Buell and Hafter (1988) but with the ITD*env* (or ILD; see Sec. 6.3.2) sweeping linearly from or toward zero, so as to position the peak value at the sound onset or offset in different conditions. Lending strong support to the rate-dependent onset effect described by Hafter and Dye (1983), Stecker and Brown (2010) observed a profound rate-dependent asymmetry in ITD*env* thresholds: at high click rates, listeners exhibited a robust ITD*env* sensitivity at the sound onset, and a relative inability to access ITD*env* cues late in the ongoing sound.

Finally, in the temporal-weighting approach, random variation is applied to binaural information across temporal segments of a sound. Statistical regression is used to associate trial-to-trial changes in a listener's responses with trial-to-trial changes in the binaural cues presented in each segment. Regression weights obtained for each segment illustrate the time-course of cue effectiveness (the temporal weighting function [TWF]; Fig. 6.5d–f). For example, Saberi (1996) and Brown and Stecker (2010) measured ITD*env* discrimination in 4-kHz-centered Gabor click trains in which ITD*env* values varied randomly from click to click. The results of both studies were, again, consistent with the rate-dependent onset dominance reported by Hafter and Dye (1983): at high click rates ($\geq$500/s), only the initial (onset) click received substantial regression weight, as if later ongoing clicks did not contribute to the listener's judgments. At click rates $\leq$100/s, the weights were approximately equal across clicks–as expected if listeners integrate binaural cue values over the full sound duration. Importantly, these and other temporal-weighting studies (e.g., Stecker et al. 2013; Stecker 2014) showed that onset dominance enhances the weight of only the very first click—not of early clicks in general. That finding strongly suggests that the onset *per se*—perhaps the initial rise of the amplitude envelope—mediates the salience of binaural cues in the initial portion of a sound.

6.3.2 Evidence of Rate-Dependent Emphasis of Interaural Level Difference Cues Near Onset and Offset

Hafter et al. (1983) investigated whether rate-dependent onset dominance was specific to ITD*env* processing by replicating the study of Hafter and Dye (1983) with ILD as the cue to be discriminated. Other aspects of the experimental stimuli and procedure were identical. The slopes of the functions relating ILD thresholds to duration were similar to, albeit slightly steeper than, the slopes for ITD*env* stimuli (see Fig. 6.5a–c). This led the authors to suggest that similar onset dominance occurs for both ITD*env* and ILD, perhaps mediated by a common central mechanism. Additional studies, however, have suggested a more complex relationship between sensitivity and duration for ILD than for ITD*env* cues.

As was shown to be the case for ITD, ILD cues conveyed at the onset of rapidly fluctuating envelopes appear to be particularly salient (Brown and Stecker 2010; Stecker et al. 2013). In contrast to what was observed for ITD, sensitivity appears to be also influenced by post-onset segments of the envelope, particularly by those near the stimulus offset. Performance in tasks where ILD cues are available–including free-field localization (Stecker and Hafter 2009), ILD-based lateralization (Stecker et al. 2013), and dynamic-ILD discrimination (Stecker and Brown 2012)–all reveal the enhanced weighting of cues at the sound onset and offset (see Fig. 6.5e, f). ILD discrimination is impaired, but not eliminated, during "middle" segments (Stecker and Brown 2012), and weights obtained for those segments are typically small but not zero (Fig. 6.5e, f). The long-term average ILD is also predictive of performance (Brown and Stecker 2010). In contrast, the onset appears to be truly dominant in ITD*env*–based lateralization (Fig. 6.5d), which does not reveal significant weights for post-onset stimulus segments (see Sect. 6.8.3).

Additional studies are required to untangle the roles of modulation rate, total stimulus duration, and/or the number of informative events conveyed by any particular stimulus (see Hafter et al. 1983). Consistent with the positive impact of envelope fluctuations on ongoing-cue processing (see Sect. 6.4), threshold ILDs are lower for modulated tones than for steady-state tones (Dietz et al. 2013a; Laback et al. 2017). ILD discrimination performance changes minimally with increasing modulation rate when the stimulus duration is held constant (Laback et al. 2017; Stecker and Brown 2010), even with modulation rates as high as 800 Hz (Laback et al. 2017). Although the modulation rate does seem to affect the relative weighting of onset and ongoing ILD cues (see Fig. 6.5), such results suggest that rate limitation exhibits a much shallower low-pass characteristic for ILD than for ITD cues (cf. McFadden and Pasanen 1976; Bernstein and Trahiotis 2002).

6.3.3 Evidence of Onset Dominance in the Temporal Weighting of Low-Frequency Interaural Time Difference Fine Structure Cues

The "textbook" account of binaural sensitivity to ITD is that of a low-frequency (e.g., 500 Hz) pure tone, delayed to one ear by a small amount (e.g., Hafter and Trahiotis 1997). Each cycle of the waveform elicits a response in the auditory nerve, and the relative timing of responses in the two ears encodes the ITD. Because ITD*fs* is evident in the cycle-by-cycle phase difference of the tone, it seems reasonable to expect each successive cycle to be equally efficacious, and there is little reason to expect onsets or other envelope fluctuations to have any substantive effect on low-frequency ITD sensitivity. Empirical studies, however, reveal that the temporal weighting of ITD*fs* cues does not differ markedly from the weighting of high-frequency envelope ITD and ILD.

Stecker and Bibee (2014) used the temporal-integration approach to study ITD*fs* discrimination in 500-Hz tones that were gated on and off simultaneously in the two ears so that the envelope itself did not provide an ITD cue. Threshold-duration slopes obtained in that study were nearly identical to those reported by Hafter and

Dye (1983) for high-rate high-frequency click trains. That study and a follow-up (Diedesch and Stecker 2015), also used the dynamic-cues approach to confirm higher ITD*fs* sensitivity at the sound onset than at the sound offset. That is, onset dominance was observed for pure tones, even when the envelope onset itself was diotic. A similar observation was made by Dietz et al. (2013b), who used a periodic variant of the dynamic-cues approach to demonstrate the primacy of ITD*fs* cues conveyed during positive-going envelope fluctuations (i.e., the local "onsets") of periodically modulated low-frequency tones (see Sect. 6.4.1). Thus, it appears that low-frequency ITD*fs* processing, like high-frequency ITD*env* and ILD processing, is enhanced during onset-like events in the overall envelope and in slow periodic modulations.

6.3.4 Evidence Against Onset Dominance for "Noise"

In contrast to the periodic signals described in Sects. 6.3.1, 6.3.2 and 6.3.3 (tones and periodic click trains), aperiodic stimuli have in many studies revealed greater salience of *ongoing* than *onset* ITD cues. Tobias and Schubert (1959), for example, asked listeners to lateralize broadband noise with the onset ITD*env* and ongoing ITD leading to opposite ears. For durations longer than roughly 100 ms, judgments consistently followed the ongoing ITD and were unaffected by the onset ITD. Similarly, Freyman et al. (1997) presented trains of 1-ms noise bursts, repeating every 2 ms, in which the initial (onset) burst carried an ITD opposite to the remaining (ongoing) bursts. When the waveforms of the bursts were identical (so that the sound was in fact periodic), the listeners' judgments were ambiguous: sometimes consistent with the onset, and sometimes with the ongoing ITD. However, when each burst was a new sample of noise, and the sound was thus truly aperiodic, the listeners consistently lateralized the sound in the direction of the ongoing ITD. Consistent with that result, Stecker (2018a) reported that TWFs measured for such stimuli exhibited strong onset dominance for trains of repeating noise bursts but much weaker onset dominance for trains of non-repeating noise bursts.

Once again, interpreting the differences between periodic and aperiodic stimuli requires a careful consideration of the stimuli *as processed* by the auditory periphery: although the broadband "envelopes" of such sounds appear flat, the effective envelope within any single region along the basilar membrane is temporally irregular for aperiodic sounds. These narrowband envelopes randomly include a range of envelope features, e.g., intense or rapid fluctuations, quiet segments, etc., which are occasionally similar to those responsible for effective binaural processing of sound onsets or slow AM. As discussed further in the next section, it may be argued that such fluctuations explain the difference in the temporal weighting of binaural information for periodic and aperiodic sounds (Stecker 2018a). According to that view, sensitivity to ongoing cues is driven by aperiodic features of the narrowband envelope–including onset-like fluctuations–not by the broadband spectrum of the noise itself.

6.4 Envelope Fluctuations Improve Sensitivity to Ongoing Cues

Section 6.3 considered the relative importance of binaural information conveyed in different segments of the duration of a sound. A common observation has been the apparent dominance of binaural cues carried by, or temporally proximal to, the overall onset of a high-rate periodic sound. But what exactly counts as an onset with respect to these phenomena? Are onset and ongoing cues fundamentally different sorts of things, or does the rate limitation (failure to encode ongoing cues at high rates) itself play a role in defining the boundary between onset and ongoing? Particularly in more natural sounds that include modulations across a wide range of rates, which envelope features count as onsets and which do not?

Consider a train of 16 clicks presented at an ICI of 100 ms (10-Hz rate). Although perceptual grouping may encourage the clicks to form an overall gestalt, or perceptual "whole," each individual click can be reliably identified as a separate perceptual event. Similarly, recording in the auditory nerve would reveal substantially equivalent responses to each click. In this case, the first (onset) click has no special meaning; each subsequent ongoing click acts as its own independent onset. If we measured a TWF for such a stimulus, we would expect a flat TWF with all clicks weighted equally. Consider, in contrast, a train of 16 clicks presented at 1-ms ICI (1000-Hz rate). The clicks are no longer individually perceived but are instead fused into a single harmonic complex tone with a single readily identifiable onset. Auditory-nerve responses to successive clicks would be expected to overlap in time, and the overall response to diminish over time as the nerve fibers adapt to the constant stimulus. TWFs measured for such stimuli reveal clear onset dominance (Saberi 1996), and temporal-integration measurements reveal thresholds that are nearly independent of duration (Hafter and Dye 1983), as if the entire train provides a *single quantum of binaural information* rather than 16 such quanta as in the 10-Hz train. What boundary separates these two extremes and defines whether a new "onset" occurs? At least in the context of binaural processing, one possibility is that the envelope low-pass cutoff defines this boundary. Above the cutoff, successive events fail to evoke independent responses. Below it, each ongoing event acts as an independent "onset". In this view, ongoing-cue sensitivity may be mediated entirely by onset-like responses to slow (<100 Hz) amplitude fluctuations. Sections 6.4.1 to 6.4.4 consider whether access to ongoing binaural cues benefits from (or requires) the presence of such fluctuations.

6.4.1 *Binaural "Readout" Window*

A particularly revealing demonstration of the importance of envelope fluctuations in sensitivity to ongoing cues was made by Dietz et al. (2013b; see Sect. 6.3.3), using a periodic variant of the dynamic-cue approach for low-frequency ITD*fs* cues. Pure tones of 484 and 516 Hz were presented to the two ears, resulting in a "binaural beat" whose ITD*fs* sweeps repeatedly from −1 to +1 ms, 32 times per second. SAM was applied at 100% depth and a modulation frequency of 32 Hz (matching the ITD*fs* sweep rate), resulting in a spatially compact intracranial percept unlike the dynamic blur of a rapid unmodulated binaural beat. Dietz and colleagues asked listeners to judge the lateral position as they varied the relative phase of the SAM and the binaural beat from trial to trial. Judgments consistently reflected the ITD*fs* that coincided with the rising portion of the sinusoidal envelope rather than the peak or the falling portion. That is, the envelope fluctuations imposed by SAM enhanced sensitivity to ITD*fs* within a brief "readout window" triggered by the local onset of each modulation cycle, similar to the effects of overall onset in single tone bursts (Stecker and Bibee 2014) and high-rate click trains (Stecker and Brown 2010).

Stecker (2018a) used the temporal-weighting approach to quantify the time course of cue weighting with SAM imposed on trains of broadband clicks presented at 2-ms ICI. Across envelope modulation rates of 31.25–125 Hz, the highest weight was observed consistently for the initial click of the SAM envelope despite its minimal amplitude. Clicks aligned with the amplitude maximum or with the falling phase of the SAM envelope were consistently low. The result supports the suggestion of Dietz et al. (2013b) that the readout window is brief and occurs early within the rising phase of the AM envelope.

The results of Dietz et al. (2013b) suggest that the timing of the readout window varies across listeners and with aspects of the stimulus such as modulation rate. Hu et al. (2017) extended that work by investigating the effects of carrier rate on the binaural readout in electric hearing and acoustical simulations using filtered pulse trains. They found that reducing the carrier rate from 600 to 200 pulses per second increased listeners' sensitivity to the ITD applied at the envelope maximum. The authors concluded that the readout window shifts later in time at lower carrier rates. A different explanation, however, is suggested by consideration of rate effects on the temporal weighting of ITD (see Sects. 6.2 and 6.3). Flatter TWFs and steeper threshold-duration slopes at slow rates suggest that successive pulses are treated independently–in other words, that each pulse triggers an independent readout of binaural information. Stecker (2018b) tested that hypothesis by measuring TWFs for SAM pulse trains similar to those used by Hu et al. (2017). The results, shown in Fig. 6.6e, illustrate the dominance of early clicks at 2-ms ICI. At 5-ms ICI, all clicks received significant and approximately equal weights. Thus, the readout window does not appear to shift later in time at low rates; rather, ongoing ITD cues are effective across more of the stimulus duration when the carrier pulse rate falls below the applicable limit (see Fig. 6.4).

6.4.2 Binaural "Restarting"

Hafter and Buell (1990) investigated the nature of what might constitute an "onset" in high-rate click trains by using a simple modification of the approach used by Hafter and Dye (1983). They inserted brief silent gaps between clicks with the effect of transiently increasing the ICI from 2.5 ms to 7.5 or 10 ms. That resulted in the

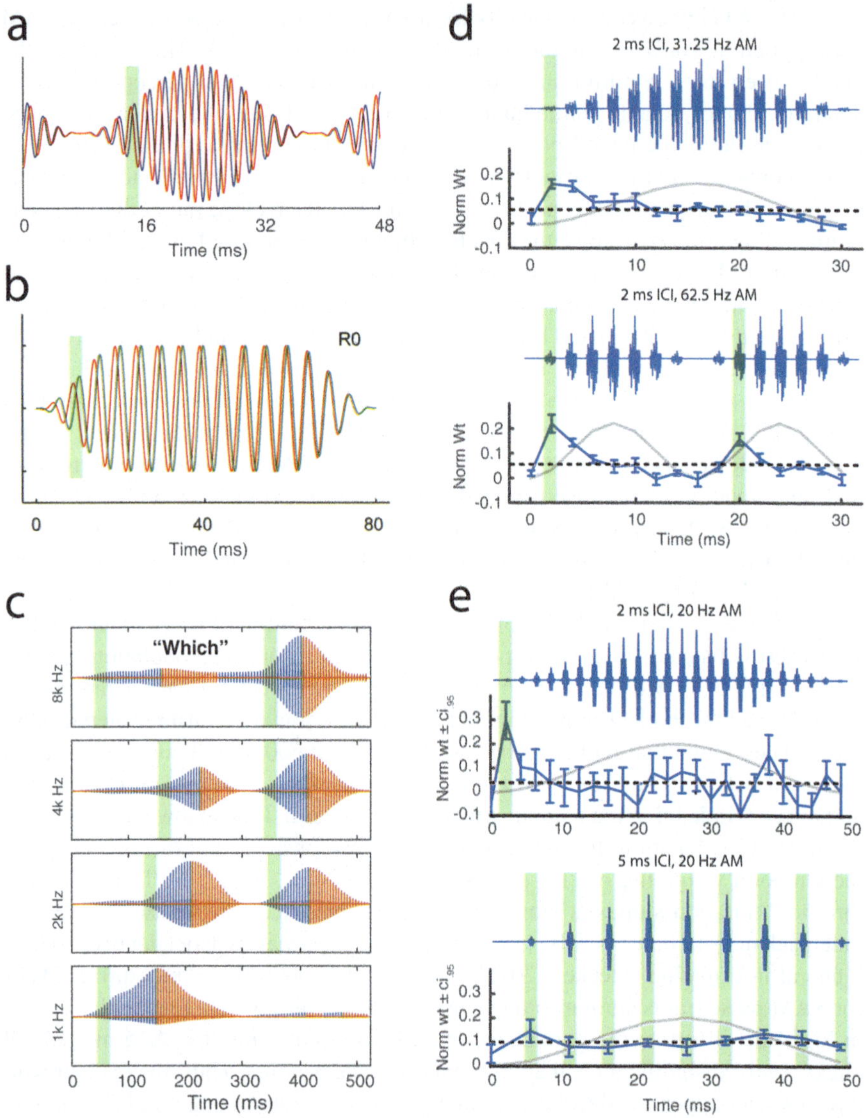

Fig. 6.6 Stimuli used to investigate binaural readout. (**a**) Amplitude-modulated binaural-beat (AMBB) stimuli used by Dietz et al. (2013b). A 32-Hz binaural beat, centered at 500 Hz, is paired with a 32-Hz SAM. Listeners tend to lateralize the AMBB stimulus according the ITD*fs* occurring early in the rising phase of each SAM period (*green area*). (**b**) Dynamic-ITD stimuli used by Stecker and Bibee (2014) demonstrate greater sensitivity to ITD*fs* occurring early versus late in a 500-Hz pure tone. Results were consistent with readout during the early rising phase. (**c**) Multi-band click-vocoded speech sounds used by Stecker (2016) to demonstrate localization dominance of spatial cues presented in the rising phase (*blue area*) in each vocoder band. Results were, again, consistent with a readout window during the early rising phases. (**d**) Temporal weighting functions (TWF) for SAM noise-burst trains (Stecker 2018a) plot normalized weight (*norm wt*) over time,

click train being separated into 2 or 4 "clusters" of clicks. Hafter and Buell found that each new cluster provided a nearly optimal improvement in the ITD*env* threshold (i.e., thresholds improved by the square root of the cluster count), as if each cluster added an independent quantum of binaural information even though the number of clicks remained unchanged. Termed "binaural restarting" by Hafter and Buell, the phenomenon is highly consistent with a readout of binaural information during the envelope fluctuation resulting from each gap. Evidence from some temporal-weighting studies, in the form of increased weight following brief gaps, supports this view (Stecker and Hafter 2002), although other studies reported no effect of such gaps (Saberi 1996). Altogether, the data suggest that sensitivity to ongoing binaural cues could be mediated by transient fluctuations of the amplitude envelope, each of which potentially contributes a quantum of binaural information.

6.4.3 Temporal Jitter

An interesting variation is the effect of introducing temporal irregularity to an ongoing stimulus, for example by randomizing the timing of clicks in an otherwise-periodic train. Such binaurally synchronized temporal "jitter" of the ICI produces an irregular temporal envelope, improves ITD thresholds (Goupell et al. 2009), and reduces onset dominance (Brown and Stecker 2011). The phenomenon has been explored, in clinical populations, as a means to enhance ongoing ITD sensitivity for users of cochlear implants (Laback and Majdak 2008; Srinivasan et al. 2018). Although a number of explanations of the phenomenon have been proposed, one parsimonious explanation for such effects is that temporal jitter results in aperiodic amplitude modulation in each band (cochlear place or auditory nerve fiber). The resulting envelope includes fluctuations over a wide range of rates, which in turn enhance ongoing ITD sensitivity in auditory neurons (Hancock et al. 2012).

6.4.4 Implications for Access to Ongoing Cues in Noise

The similar effects of adding temporal irregularity to a narrowband (or even electrical) stimulus suggest a new perspective on the relative accessibility of ongoing binaural cues in noise: in each case, the narrowband envelope features randomly timed fluctuations that may be sufficient for triggering readout of ongoing binaural cues.

Fig. 6.6 (continued) revealing dominance of binaural cues carried by initial clicks of each SAM period, although readout window might shift to later clicks across successive AM cycles. (**e**) TWFs for 4000-Hz click trains with SAM applied at 20 Hz (Stecker 2018b). Lengthening the ICI from 2 (*top*) to 5 ms (*bottom*) increases sensitivity to ITD of post-onset clicks and thus flattens the TWF. ci_{95}, 95% Confidence interval. (**d**, Adapted from Stecker (2018a), with the permission of the Acoustical Society of America; **e**, data from Stecker (2018b))

Reduced onset dominance and better overall sensitivity result from integration across multiple quanta of binaural information that are not available from rapid periodic sounds. That view is supported by Houtgast and Plomp's (1968) observation of better relative sensitivity to ongoing ITD under noise-masking conditions, and by several studies that specifically manipulated the temporal regularity of trains of noise bursts presented at high rate (500 / s) (Freyman et al. 1997, 2010; Stecker 2018a). In those studies, strong onset dominance was observed when successive noise bursts repeated identical noise tokens so that the stimulus was in fact periodic. Greater ongoing sensitivity was observed when noise tokens changed from burst to burst to create a truly aperiodic noise stimulus with a temporally irregular envelope at each frequency. Changing noise tokens infrequently during the stimulus resulted in "clusters" of periodic sound (similar to Hafter and Buell 1990) that were each dominated by the ITD of the initial token. That result, termed the "ongoing precedence effect" (Freyman et al. 2010), can also be observed in studies with long-duration overlapping stimuli (e.g., Dizon and Colburn 2006) and appears consistent with mechanisms that weight binaural processing by the dynamic envelope (Wolf 1991; Nelson and Takahashi 2010) or interaural coherence (Faller and Merimaa 2004; Dietz et al. 2011) of ongoing sound.

6.5 Evidence That Envelope Shape Influences Sensitivity to Binaural Cues

A critical issue raised in Sect. 6.1 is the importance of considering the stimuli *as processed* by the auditory system. That consideration has been important for understanding how stimulus features such as envelope depth, fluctuation rate, and interaural correlation are affected by transformations in the auditory periphery. Of particular importance are differences in the representation of high- and low-frequency sounds. Processes beginning with hair cell transduction first effectively rectify the input signal (i.e., by removing the negative parts of the waveform) and then smooth the result in time. In high-frequency channels, the result is a continuous representation of the temporal envelope. In low-frequency channels, smoothing has a smaller effect and the representation is, essentially, that of the half-wave rectified waveform itself. Colburn and Equissaud (1976) hypothesized that apparent differences in binaural processing at low and high frequencies might be explained via a single binaural mechanism, if such differences are taken into account.

The most direct evaluations of Colburn and Equissaud's (1976) hypothesis were made using high-frequency "transposed stimuli" (van de Par and Kohlrausch 1997). These were designed such that, subsequent to peripheral auditory processing, the neural temporal information serving as an input to high-frequency binaural channels would effectively mimic that conveyed to low-frequency channels by conventional low-frequency stimuli. Figure 6.3a illustrates the generation of high-frequency transposed stimuli. Importantly, the temporal envelope of the transposed tone is the rectified and low-pass filtered low-frequency tone. Figure 6.3b depicts the effects of peripheral processing on a low-frequency tone, its transposed counterpart, and a high-frequency SAM tone. Generalizing from comparisons among the waveforms in Fig. 6.3b, *right*, it can be understood how, subsequent to peripheral processing (1) high-frequency, transposed stimuli would result in neural inputs to the binaural processor that mimic or are quite similar to those conveyed by their low-frequency counterparts; and (2) neural responses conveyed by "conventional" stimuli, such as high-frequency SAM tones, would not be characterized by the sharp "peaks" and distinct "off times" that characterize the other types of stimuli.

Consistent with Colburn and Equissaud's (1976) hypothesis, several studies have demonstrated comparable binaural performance for low-frequency stimuli and their transposed counterparts, along with poorer performance for high-frequency SAM stimuli. van de Par and Kohlrausch (1997) found that binaural releases from masking using high-frequency transposed noises and transposed tonal signals stimuli were similar in magnitude to those obtained with their low-frequency counterparts. The binaural masking-level differences they measured were substantially larger than those obtained using conventional high-frequency noise maskers and tonal signals. Bernstein and Trahiotis (2002) obtained similar threshold-ITDs with low-frequency pure tones and with transpositions of them to 4 kHz. Bernstein and Trahiotis (2003) demonstrated that, for a given ITD, perceived extents of laterality of intracranial images were essentially equivalent for low-frequency bands of noise and their counterparts transposed to 4 kHz. In contrast, intracranial images were perceived to be substantially closer to the midline when measured with conventional bands of noise centered at 4 kHz. Thus, it appears that the relatively poor binaural performance commonly found at high frequencies does not stem from any "deficit" in the central binaural comparator at high vs. low frequencies. Rather, it appears to stem, primarily, from differences in the nature of the peripherally processed neural inputs to an essentially frequency-independent binaural processor, just as Colburn and Equissaud (1976) hypothesized.

In an effort to better understand which aspects of the envelopes of high-frequency complex stimuli are most salient for binaural processing, Bernstein and Trahiotis (2009, 2012) employed high-frequency "raised sine" stimuli (John et al. 2002). Briefly, generating a high-frequency raised-sine stimulus involves raising a DC-shifted sine wave to an exponent, n, and using the result to modulate a high-frequency carrier ("RS4" in Fig. 6.3c–e illustrates a raised-sine stimulus with $n = 4$ exponent). Increasing the exponent from 1.0 (equivalent to a SAM tone) to greater values increases the "peakedness" of the raised-sine envelope. In that manner, one can produce high-frequency stimuli with envelope features falling in between those of SAM and transposed tones. Bernstein and Trahiotis (2009, 2011) demonstrated that threshold ITD*env* decreased, and extents of the ITD*env*-based laterality increased, systematically as n increased from 1 to 8.

What are the key features of transposed and raised-sine stimuli that give rise to potent binaural cues? Already noted are their sharp peaks and distinct off times. The slope of the rising and falling envelope edges is also steeper for transposed and raised-sine ($n > 1$) tones than for SAM tones. Other stimuli that also support enhanced sensitivity to high-frequency ITD*env* cues include trains of Gabor clicks (Hafter and Buell 1990) or noise bursts (Freyman et al. 1997), and the envelopes of those stimuli are also characterized by steep slopes and distinct off times. Figure 6.3c–e directly compares the envelopes, waveforms, and spectra of a family of such "pulsatile" stimuli: transposed tones, Gabor click trains, raised-sine stimuli, and SAM tones. Analysis of these stimuli by Bibee and Stecker (2016) demonstrated, for example, that a Gabor click train at 2-ms ICI (i.e., as used by Buell and Hafter 1988; Stecker and Hafter 2002; and others) is nearly identical to a 500-Hz raised-sine stimulus with an $n = 4$ exponent. Unlike raised-sine stimuli, however, changing the click-train ICI does not alter the shape of each AM cycle, providing independent control of off time (via ICI) and peakedness/slope (via click duration). The importance of off time was considered in Sect. 6.3: studies using Gabor click trains to measure ITD*env* sensitivity have consistently found thresholds to improve (Buell and Hafter 1988; Stecker and Brown 2010), and onset dominance to diminish, as ICI is increased from 1–2 ms [0 ms off time] to 10–12 ms [8–10 ms off time] (Saberi 1996; Stecker 2014). That is, manipulating off time provides a clear illustration of the underlying rate limitation in sensitivity to ongoing ITD*env* cues.

Klein-Hennig et al. (2011) took a systematic approach to measuring the contributions of periodic envelope slopes (attack and decay), on-time (hold duration), and off- time (pause duration) to ongoing ITD*env* sensitivity. They found that threshold-ITDs improved with increasing off time from 0 to 8.8 ms, consistent with studies using Gabor click trains. Effects of envelope slope were also consistent with measurements of raised-sine stimuli (Bernstein and Trahiotis 2011): threshold-ITDs improved as attack time decreased from 10 to 1.3 ms. Klein-Hennig et al. (2011) also observed a marked (~three-fold) advantage for envelopes with steep attack (1.3 ms) and slow decay (18.8 ms) than vice-versa, suggesting the specific importance of positive-going envelope fluctuations, further supported by Dietz et al. (2013b) and Stecker (2018a).

Finally, while signals with fluctuating envelopes appear to elicit lower ILD thresholds than steady-state tones (Dietz et al. 2013a; Laback et al. 2017), clear effects of envelope shape on ILD sensitivity like those observed for ITD sensitivity have not been demonstrated. In one study (Dietz et al. 2013a), SAM tones and transposed tones were found to yield similar ILD thresholds across a variety of presentation levels.

6.6 Evidence for Sensitivity to Binaural Cues Conveyed by Interaurally Decorrelated Signals

Most data described to this point were obtained using stimuli that carried static ITD or ILD, but were otherwise perfectly interaurally correlated. That is, they had an interaural coherence of 1.0. Such stimuli are encountered only in artificial settings such as anechoic environments and headphone listening. Real sounds are never perfectly coherent across the ears. Reverberation and competing sounds (see Zahorik, Chap. 9) can reduce interaural coherence substantially. With decreasing interaural coherence, the phase relation of the signals at the two ears becomes increasingly random, and the ITD*fs cues* become correspondingly dynamic. The listener's ability to detect and utilize ITD*fs* cues is reduced or eliminated under such conditions (e.g., Jefress et al. 1962; Trahiotis et al. 2001). Rakerd and Hartmann (2010) measured threshold-ITDs for noise bands varying in interaural coherence. Figure 6.7a replots data from that study for a band centered at 715 Hz. Threshold-ITDs increased steadily as interaural coherence decreased from 1.0; the increase was particularly rapid below an interaural coherence of ~0.5 and thresholds could not be determined reliably for values of interaural coherence <0.2. Temporal variation in ITD*fs* reduces the detectability of ITD cues, although ITD variability may provide an "image width" cue (Blauert and Lindemann 1986) based on differences in IAC (see Sect. 6.7). In the extreme case of interaurally uncorrelated stimuli (IAC and interaural coherence = 0), the ITD*fs* is undefined or indeterminate (see Fig. 6.2).

In contrast, the ILD can be computed even when imposed on a signal having an interaural coherence of zero because the physical ILD represents a relatively long-term amplitude difference at the two ears. That is, if the intensity of the signal at one ear is greater than the intensity at the other, the long-term ILD by definition favors that ear whether the two signals are perfectly interaurally coherent, independent, or antiphasic. Although the short-term ILD fluctuates with the brief,

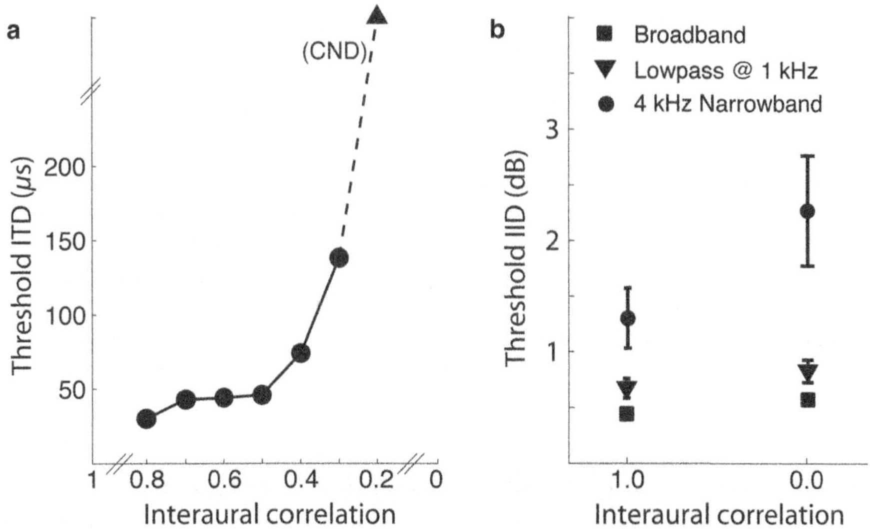

Fig. 6.7 Effects of interaural correlation on binaural discrimination. (**a**) Threshold ITD increases precipitously as IAC decreases below 0.5; a threshold could not be determined (CND) for IAC below 0.2. Data replotted from Rakerd and Hartmann (2010). (**b**) Threshold ILD, in contrast, decreases only slightly (less than 1 dB) as IAC drops from 1.0 to 0.0. (Broadband and lowpass data replotted from Hartmann and Constan (2002), with the permission of the Acoustical Society of America; narrowband data replotted from Brown and Tollin (2016), with permission)

moment-to-moment changes in level at each ear, an underlying "average" ILD could remain discernable given a sufficiently long "averaging window". Thus, the effects of interaural decorrelation of the signals on ILD processing should depend on the duration and bandwidth of the signal, and the duration of the temporal window over which ILD is computed. Such reasoning motivates an empirical question: over what time scale does the binaural system extract ILD and do rapid fluctuations in ILD (e.g., stemming from decoherence) negatively impact ILD sensitivity?

Early forays into the topic by Nuetzel (1982) and Grantham and Ahlstrom (1982) suggested very slight effects of decoherence on ILD thresholds. These results, in stark contrast to those obtained for ITD imposed on incoherent stimuli, led Nuetzel (1982) to surmise that "data are consistent with binaural models in which [ILDs] are processed independently of interaural time differences." Hartmann and Constan (2002) later measured ILD thresholds for broadband or lowpass noise, again finding a relatively small (<0.5 dB) elevation of thresholds for independent (interaurally uncorrelated) versus perfectly interaurally correlated stimuli (Fig. 6.7b). Brown and Tollin (2016) later evaluated the influence of interaural coherence on both ILD discrimination and lateralization of suprathreshold ILDs using narrow bands of noise and Gabor click trains. When signals at the two ears were interaurally incoherent, the extent of ILD-based lateralization was slightly reduced, and ILD discrimination thresholds were slightly increased (by roughly 1 dB, see Fig. 6.7b), relative to those for interaurally coherent signals (IAC = 1). Thus, although reduced interaural correlation produced a measurable effect, listeners remained acutely sensitive to ILD even for temporally independent signals at the two ears. Brown and Tollin (2016) were able to model the data using a simple level comparison between left- and right-ear inputs after temporal averaging within a running ~3 ms temporal window. This window duration was found to be further consistent with empirical measurements of temporal integration windows in auditory midbrain neurons (IC neurons of the chinchilla).

In summary, a reduction in interaural coherence degrades ITD*fs* cues, but has relatively little effect on discrimination or lateralization of ILD. These results suggest that ILD may provide a usable cue for discrimination or localization even when ITD*fs* cues are not available. Data suggest that a few milliseconds of temporal averaging, like those observed in the auditory midbrain, may be sufficient to extract usable ILD cues from even minimally correlated inputs such as those observed in highly reverberant environments (e.g., Devore and Delgutte 2010).

6.7 Sensitivity to Interaural Correlation in Binaural Detection Tasks

This chapter has focused primarily on the processing of binaural information in terms of the major binaural cues: ITD and ILD. Yet another impact of temporal complexity on binaural performance involves the IAC cue itself; in other words, the overall similarity (or dissimilarity) of sound at the two ears. Unlike ITD and ILD, which convey stimulus lateralization or azimuthal localization, changing the interaural coherence affects the apparent width or diffuseness of the binaural image (Blauert and Lindemann 1986). The results of many binaural detection studies, conducted over decades using *low-frequency stimuli*, have shown that listeners' performance can be explained quite well in terms of their sensitivity to changes in the IAC of the waveform (e.g., Pollack and Trittipoe 1959; Robinson and Jeffress 1963).

As an example, consider the binaural detection of a tonal signal masked by a noise in the classic NoSπ configuration. In that configuration, the noise masker is presented identically at each ear (diotically) and the tonal signal is presented antiphasically across the ears (i.e., interaurally inverted). Because the masking noise is identical at the two ears, when it is presented in the absence of the signal, the IAC is +1.0. The interaurally phase-inverted signal (Sπ) has an IAC of −1.0. Introducing the signal to the noise therefore reduces the IAC of the combined stimulus by an amount proportional to the power of the signal. The studies cited above suggest that listeners use this change in the IAC to perform such tasks, and with remarkable sensitivity: a change in the IAC from 1.0 to about 0.995 is typically sufficient for accurate detection.

Recall from Sect. 6.1 that when high-frequency stimuli are employed, interaural information is conveyed via the time-varying *envelopes* of the waveforms. To the degree that the normalized IAC also could account for data obtained using high-frequency stimuli, it would have to be computed on the basis of the *envelopes* of the waveforms at the left and right ears and not the waveforms themselves. Bernstein and Trahiotis (1992, 1996a) demonstrated that the normalized IAC, when computed on the envelopes of high-frequency stimuli, could, indeed, account for binaural detection at high frequencies.

In a later study, Bernstein and Trahiotis (1996b) measured listeners' ability to discriminate the NoSπ configuration from the NoSo (diotic noise and signal) configuration, using a tonal signal added at the center frequency of a 100-Hz-wide band of noise. Data were obtained as a function of the signal-to-noise (S/N) ratio at center frequencies ranging from 500 Hz to 2000 Hz. Figure 6.8 displays data from that study. For each CF, percent correct detection is plotted as a function of signal-to-noise ratio. Squares represent the obtained data; solid lines represent predictions based on the assumption of an underlying, frequency-independent, psychometric function relating detection to change in interaural correlation. The key to producing successful predictions was the computation of the normalized interaural correlation on representations of the waveforms *as they would be expected to be processed within the auditory periphery* and not on the waveforms themselves. Bernstein and Trahiotis (2017) further expanded this approach to include in the model (1) stages emulating peripheral auditory processing, (2) internal noise, and (3) a "decision variable" that explicitly takes into account the sampling variance of the normalized correlation for noise-alone and signal-plus-noise waveforms. The expanded model was able to account for binaural detection data even in experiments employing narrowband, partially interaurally correlated maskers, for which IAC changes had previously provided poor predictions.

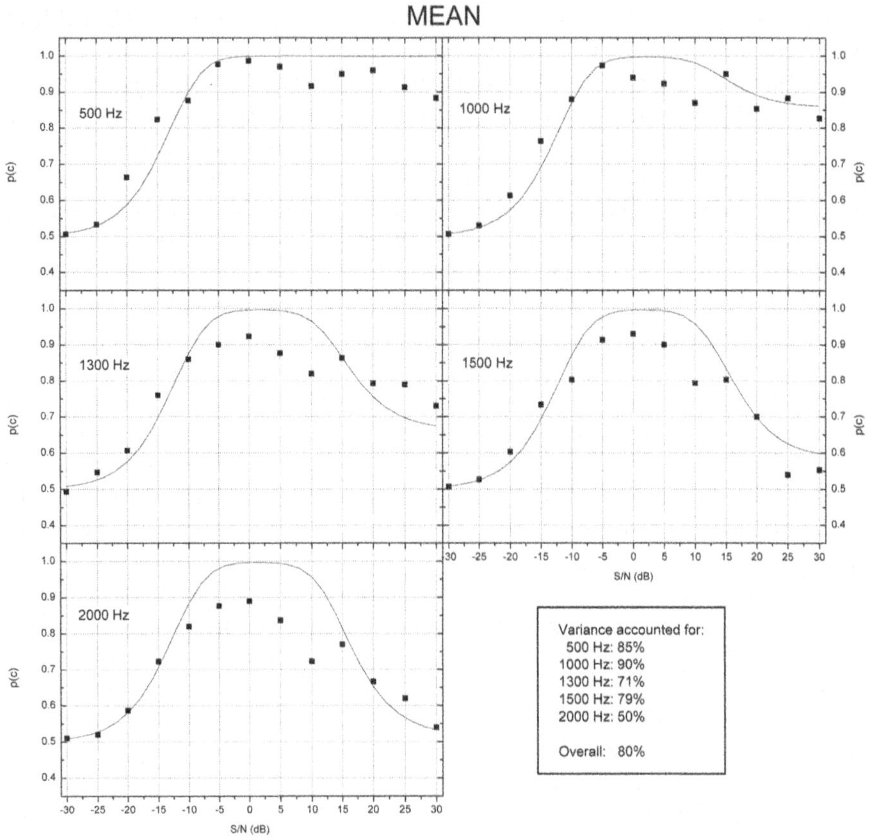

Fig. 6.8 NoSπ / NoSo discrimination from Bernstein and Trahiotis 1996b. Plots of percent correct [p(c)] as a function of signal-to-noise (S/N) ratio in dB. *Squares*, obtained data; *solid lines*, predicted values of p(c). Each panel displays data and predictions for a different center frequency. *Bottom right*, table displays the amount of variance accounted for by the predictions. (Figure adapted from Bernstein and Trahiotis (1996b), with the permission of the Acoustical Society of America)

The foregoing, along with the discussion in Sect. 6.2 regarding rates of envelope fluctuation, demonstrates how the normalized interaural correlation of stimuli *as processed by the auditory periphery* successfully accounts for how changes in the temporal characteristics of complex stimuli affect binaural performance. It should be noted, however, that for certain high-frequency stimuli characterized by asymmetric envelopes, the normalized correlation computed over the entire stimulating waveform cannot account for binaural discrimination performance (e.g., Klein-Hennig et al. 2011). For such stimuli, time-reversing their presentation can yield substantially different binaural discrimination performance, whereas the normalized correlation computed over the duration of the stimulus would not be affected. Apparently consistent with the role of positive-going envelope fluctuations in lateralization and binaural discrimination (Dietz et al. 2013b; Stecker and Bibee 2014), performance improves when binaural information is presented at or near the sound onset compared with when it is presented near the offset. In other words, binaural processing appears to be dominated by the interaural correlation cue proximal to positive-going envelope fluctuations, such as onsets, similarly to the weighting of ITD and ILD (Hafter and Dye 1983; Brown and Stecker 2010). It remains to be determined whether incorporating appropriate "temporal weighting" within a correlation-based approach would allow for the prediction of data obtained with the types of stimuli used by Klein-Hennig et al. (2011).

6.8 Accounting for the Effects of Envelopes on Binaural Hearing: RESTART Theory

As conveyed in Sects. 6.2, 6.3, 6.4, and 6.5, numerous observations suggest that envelope fluctuations exert a profound influence on binaural processing across frequency and across cue type. Many of these observations can be synthesized into a unified theoretical framework. Reliable Envelope-Slope-Triggered Auditory Representation Theory (RESTART; Stecker and Diedesch 2014; Stecker 2016) posits a single mechanism, which emphasizes binaural processing during infrequent rising-envelope events in each auditory band, to account for many of the phenomena described in Sects. 6.2, 6.3, 6.4, 6.5, 6.6 and 6.7. The key parameters of RESTART theory encompass those observations.

6.8.1 Sampling of Binaural Cues During Rising-Envelope Events

First, the RESTART mechanism samples binaural information primarily at moments of rising-envelope events such as sound onsets, as suggested by work on binaural restarting (see Sect. 6.4.2; Hafter and Buell 1990) and readout (Sect. 6.4.1), Because such events dominate binaural-cue processing regardless of cue type (ITD*fs*, ITD*env*, ILD, and likely IAC) or spectral frequency, RESTART theory posits a single mechanism, which could be an active central process that triggers a binaural readout (Hafter and Buell 1990; see also Patterson et al. 1995) or an intrinsic property of binaural-cue processing. For example, many auditory neurons exhibit transient-adapting responses that selectively emphasize rising-envelope events and respond weakly to steady-state sound. These include bushy cells of the ventral cochlear nucleus, which provide the bulk of input to binaural neurons in the superior olivary complex and respond to sound onsets with single spikes at short, stable latencies (Schwarz et al. 1998; Ashida et al. 2019).

6.8.2 Rate Limitation of Binaural Sampling

To account for rate limitations in binaural processing (Sect. 6.2), RESTART theory posits a refractory period of 1–3 ms following each rising-envelope event. Successive events that are separated by longer intervals (e.g. 5 ms; Fig. 6.6e *bottom*) are represented independently in the binaural system, with each event contributing an independent quantum of binaural information that can be integrated across events. Successive events that occur within the refractory period are not represented independently. Rather, a rapid sequence of events is treated as unitary and localized on the basis of the overall onset, which contributes a single quantum of binaural information (Fig. 6.6e, *top*). Slower modulations imposed on such sequences can provide additional quanta (e.g., if the AM click train of Fig 6.6e was repeated over multiple 50-ms cycles). Integration *across*, but not *within*, these events can be optimal in a statistical sense (Hafter and Buell 1990).

The rate limitation of this process places significant constraints on how binaural information may be accessed from ongoing sounds. In particular, RESTART theory predicts greater sensitivity to ongoing cues when the ongoing envelope carries slow or temporally irregular fluctuations–such as in noise (Stecker 2018a) or aperiodic click trains (Brown and Stecker 2011)–than for steady-state envelopes (see also Pastore and Braasch 2019).

6.8.3 Reduced Sensitivity to Ongoing Cues in Steady-State Sounds

Compared with onset-like events, steady-state cues appear to contribute relatively little to binaural detection, lateralization, and localization. Nevertheless, some evidence suggests that ongoing cues may contribute to a degree even in purely steady-state sounds. Temporal-weighting studies generally reveal small but positive post-onset weights (Fig. 6.5), and a few studies have even demonstrated binaural sensitivity in sounds designed carefully to eliminate onsets (Hafter et al. 1979; Macauley et al. 2010). Such evidence suggests that the absence of envelope fluctuations does not preclude binaural sensitivity. Yet caution in that interpretation is urged by the consideration of cues as processed by the auditory system. Given the exquisite temporal fidelity of auditory responses to even relatively slow onsets (Heil 2001; also see Fig. 6.6e), along with the possibility that envelope fluctuations may be introduced to an ongoing sound via head movements, physiological variation (e.g., swallowing), and modulation of spontaneous activity (e.g., Luczak et al. 2009; Zhao and Dhar 2011), it is difficult to rule out the possibility that induced rising-slope events contribute during the nominally steady-state portions of sounds with flat or rapidly modulated envelopes. According to RESTART theory, ongoing cues contribute primarily through fluctuations that are magnified by the transient sampling process described in Sect. 6.8.1.

An important aspect of ongoing-cue sensitivity is the cue-dependence described in Sect. 6.3.2. The salience of ILD cues in ongoing and, particularly, offset segments suggests that multiple cue-specific mechanisms may be at work. RESTART theory posits a common mechanism that emphasizes onsets and low-rate ongoing envelope fluctuations prior to binaural cue extraction. The data, however, suggest that additional processing effectively integrates ongoing ILD cues but not ITD cues. One possibility is suggested by differences in the temporal precision of ITD and ILD mechanisms: whereas ITD processing requires precisely synchronized responses in the two ears, ILD processing appears tolerant of mismatches of at least a few milliseconds (Remme et al. 2014; Brown and Tollin 2016). By integrating over that range, the ILD mechanism maintains sensitivity to ongoing cues even in flat or high-rate envelopes that elicit binaurally incoherent responses from which the ITD cannot be computed.

Another possibility is that the processing of ongoing ILD cues is supplemented by entirely different mechanisms. That is, ILD cues–but not ITD cues–can potentially be extracted at later stages in the auditory pathway that remain sensitive to the ear of stimulation but lack the temporal fidelity to resolve sub-millisecond ITD cues. Such mechanisms, unaffected by temporal limitations of the RESTART mechanism, could help explain "recency" effects that emphasize late-arriving cues (Stecker and Hafter 2009).

6.8.4 Independent Binaural Sampling Across Frequency Bands

Data suggest that the mechanism(s) pertinent to RESTART theory act independently within frequency bands. For example, Hafter and Wenzel (1983) showed that temporally interleaving two click trains (i.e., reducing the overall ICI by half) does not affect the temporal integration of the ITD if the alternating clicks occupy different spectral regions (e.g., 4 and 6 kHz center frequency). In that case, the two click trains are processed independently by the binaural system. A further demonstration by Stecker (2016; see Fig. 6.6c) used a four-band vocoder to reveal the dominance of rising-envelope clicks even when modulations occurred at different times in different frequency bands. As argued by Hafter et al. (1988), independent operation across frequency bands suggests a peripheral or early central mechanism, such as neuronal adaptation prior to across-frequency integration of binaural information.

6.9 Summary and Conclusions

The results reviewed in this chapter reveal five key stimulus features that significantly affect the binaural hearing of temporally complex sounds. First, envelope fluctuations play a dominant role in binaural processing. Temporal weighting studies reveal greater utilization of binaural information at moments of positive-going envelope slope such as sound onsets (Hafter and Buell 1990, Stecker 2018a), along with enhanced sensitivity to ongoing cues accompanied by temporally irregular envelopes (Goupell et al. 2009, Brown and Stecker 2011) or slow periodic envelopes (Dietz et al. 2013b). The importance of envelope fluctuations is not limited to the processing of ITD*env* cues–which logically requires them–but extends to ITD*fs*, ILD, and possibly even IAC cues. Studies reveal that a wide range of binaural information is accessed most readily at moments of positive-going envelope fluctuations in high- (Klein-Hennig et al. 2011; Stecker and Brown 2012) and low-frequency sounds (Dietz et al. 2013b; Stecker and Bibee 2014).

Second, the shape of the envelope fluctuation matters. ITD*env* sensitivity in particular appears to be enhanced during moments of positive-going envelope slope relative to moments of flat or negative slopes. Correspondingly, greater binaural sensitivity is observed for classes of stimuli (described Sect. 6.5) designed to contain steep rising envelope slopes preceded by sufficient "dead time" (Bernstein and Trahiotis 2011; Klein-Hennig et al. 2011).

Third, and closely related to the previous point, the efficient processing of binaural information at high frequencies appears to require stimulus envelope fluctuations with rates below 150 Hz or so (Bernstein and Trahiotis 2002). This "rate limitation" is not explained by reduced modulation depth due to cochlear filtering (Nuetzel and Hafter 1981). Instead, it appears to reflect a fundamental property of binaural processing independent of cue type or frequency range.

Fourth, the auditory system exhibits greater access to ongoing ILD cues than to ongoing ITD*env* or ITD*fs* cues (Brown and Stecker 2010; Stecker and Brown 2010). Temporal-weighting measures reveal "recency" effects that suggest temporal integration of the ongoing ILD, which is not observed for the ongoing ITD*env* (Stecker and Hafter 2009; Stecker et al. 2013). Robust ILD discrimination for temporally interaurally uncorrelated signals (Hartmann and Constan 2002; Brown and Tollin 2016) further suggests that ILD, unlike ITD, is computed via an averaging window at least a few milliseconds long.

Fifth, the similarity of sound waveforms at the two ears provides a unique cue apart from the ITD and ILD. Unlike those cues, which convey stimulus lateralization or azimuthal localization, changing the IAC affects the apparent width or diffuseness of the binaural image. Binaural detection studies have repeatedly demonstrated listeners' exquisite sensitivity to changes in IAC, for example resulting from the addition of an antiphasic target to a diotic masker (Bernstein and Trahiotis 1996b, 2017). More importantly, computing the IAC based on the stimulus *as processed* by the auditory periphery (i.e., the effective input to binaural comparison) supports accurate predictions of performance using envelope and fine-structure cues over a wide range of frequencies.

6.9.1 Implications and Applications for Real-World Listening

The findings reviewed in this chapter can help to understand the kitchen-conversation scenario described in Sect. 6.1. Envelope fluctuations occurring at time scales ranging from seconds (pitch variation in music and speech) to milliseconds (successive echoes) and in between (syllables) briefly enhance access to binaural cues (see Fig. 6.6 and Sect. 6.4.1). Such enhancements may recur with repeating envelope fluctuations, but not beyond a rate of 200 Hz or so. This is fast enough to track successive cello notes (seconds), syllabic bursts of speech and laughter (hundreds of milliseconds), and even the fundamental period of voiced speech (5–10 ms), but not to follow successive echoes arriving <5 ms apart.

Each of the talkers in the scene is localized on the basis of binaural cues (ITD*fs*, ITD*env*, and ILD) that are processed mainly during the rising slopes of each talker's envelope. Because rising envelopes are associated with the direct sound path (Nelson and Takahashi 2010), the process reduces the effects of echoes, reverberation, and competing sound present at other moments in time. The result is a temporally sparse representation that helps stabilize the spatial image of each voice (cf. Patterson et al. 1995) and prevent confusion of locations with other voices, echoes, or extraneous sounds.

Reflections of sound from hard surfaces such as kitchen counters and tiled walls introduce rapid fluctuations in ITD and ILD that are not strongly related to sound-source locations (see Zahorik Chap. 9). Temporally sparse sampling of these cues reduces the effects of reflections on sound localization, but rapid fluctuations contribute to perception in other important ways. Echoes of the cello music reduce the IAC of that signal and enhance its apparent source width even when the echoes themselves are not separately localized or strongly perceived, providing information about the size and materials within the listening room and supporting the sense of acoustic space that defines the overall scene.

A better understanding of how the binaural system deals with temporally complex sound in real-world listening could also motivate advances in binaural and spatial hearing technology for use by both normal-hearing and hearing-impaired individuals (see Gallun, Srinivasan, and Diedesch Chap. 11 and Ricketts and Kan Chap. 13). For example, enhanced binaural sensitivity during envelope fluctuations suggests that disorders or signal-processing algorithms that obscure envelope features could produce spatial-hearing deficits even when binaural fidelity is otherwise preserved. That relationship might also help to explain the reported concordance between the results of ITD-discrimination and gap-detection tasks (Strouse et al. 1998, Ochi et al. 2014)–two "temporal" tasks that operate on drastically different time scales and would otherwise seem unrelated.

Conversely, it may not be necessary for devices to preserve binaural detail during steady-state segments of sound that are not associated with good binaural sensitivity. Optimizing the timing of binaural information could reduce storage and processing demands to allow more complex processing in low-power devices (Stecker and Diedesch 2016). Similarly, algorithms that aim to extract spatial information from binaural signals, or predict human perception, could capitalize on these data by focusing on cues *as processed* by the auditory system–including the effects of envelope-triggered readout–rather than the raw acoustical cues themselves (Stecker 2016).

Compliance with Ethics Requirements
G. Christopher Stecker declares no conflict of interest.
Leslie R. Bernstein declares no conflict of interest.
Andrew D. Brown declares no conflict of interest.

References

Ashida G, Heinermann HT, Kretzberg J (2019) Neuronal population modeling of globular bushy cells. https://github.com/pinkbox-models/GBC2019. Accessed 16 Aug 2019

Bernstein LR, Trahiotis C (1985) Lateralization of low-frequency, complex waveforms: the use of envelope-based temporal disparities. J Acoust Soc Am 77:1868–1880

Bernstein LR, Trahiotis C (1992) Discrimination of interaural envelope correlation and its relation to binaural unmasking at high frequencies. J Acoust Soc Am 91:306–316

Bernstein LR, Trahiotis C (1994) Detection of interaural delay in high-frequency sinusoidally amplitude-modulated tones, two-tone complexes, and bands of noise. J Acoust Soc Am 95:3561–3567

Bernstein LR, Trahiotis C (1996a) On the use of the normalized correlation as an index of interaural envelope correlation. J Acoust Soc Am 100:1754–1763

Bernstein LR, Trahiotis C (1996b) The normalized correlation: accounting for binaural detection across center frequency. J Acoust Soc Am 100:3774–3784

Bernstein LR, Trahiotis C (2002) Enhancing sensitivity to interaural delays at high frequencies using 'transposed stimuli'. J Acoust Soc Am 112:1026–1036

Bernstein LR, Trahiotis C (2003) Enhancing interaural-delay-based extents of laterality at high frequencies by using 'transposed stimuli'. J Acoust Soc Am 113:3335–3347

Bernstein LR, Trahiotis C (2009) How sensitivity to ongoing interaural temporal disparities is affected by manipulations of temporal features of the envelopes of high-frequency stimuli. J Acoust Soc Am 125:3234–3242

Bernstein LR, Trahiotis C (2011) Lateralization produced by envelope-based interaural temporal disparities of high-frequency, raised-sine stimuli: empirical data and modeling. J Acoust Soc Am 129:1501–1508

Bernstein LR, Trahiotis C (2012) Lateralization produced by interaural temporal and intensitive disparities of high-frequency, raised-sine stimuli: data and modeling. J Acoust Soc Am 131:409–415

Bernstein LR, Trahiotis C (2014) Sensitivity to envelope-based interaural delays at high frequencies: Center frequency affects the envelope rate-limitation. J Acoust Soc Am 135:808–816

Bernstein LR, Trahiotis C (2017) An interaural-correlation-based approach that accounts for a wide variety of binaural detection data. J Acoust Soc Am 141:1150–1160

Bibee JM, Stecker GC (2016) Spectrotemporal weighting of binaural cues: effects of a diotic interferer on discrimination of dynamic interaural differences. J Acoust Soc Am 140:2584–2592

Blauert J (1983) Spatial hearing. MIT Press, Cambridge, MA

Blauert J, Lindemann W (1986) Spatial mapping of intracranial auditory events for various degrees of interaural coherence. J Acoust Soc Am 79:806–813

Brown AD, Stecker GC (2010) Temporal weighting of interaural time and level differences in high-rate click trains. J Acoust Soc Am 128:332–341

Brown AD, Stecker GC (2011) Temporal weighting functions for interaural time and level differences. II. The effects of binaurally synchronous temporal jitter. J Acoust Soc Am 129:293–300
Brown AD, Tollin DJ (2016) Slow temporal integration enables robust neural coding and perception of a cue to sound source location. J Neurosci 36:9908–9921
Brown AD, Stecker GC, Tollin DJ (2015) The precedence effect in sound localization. [review]. J Assoc Res Otolaryngol 16:1–28. https://doi.org/10.1007/s10162-014-0496-2
Buell TN, Hafter ER (1988) Discrimination of interaural differences of time in the envelopes of high-frequency signals: integration times. J Acoust Soc Am 84:2063–2066
Buell TN, Trahiotis C, Bernstein LR (1991) Lateralization of low-frequency tones: relative potency of gating and ongoing interaural delay. J Acoust Soc Am 90:3077–3085
Colburn HS, Equissaud P (1976) An auditory-nerve model for interaural time discrimination of high-frequency complex stimuli. J Acoust Soc Am 59:S23
David E, Guttman N, Bergeijk WV (1959) Binaural interaction of high-frequency complex stimuli. J Acoust Soc Am 31:774–782
Devore S, Delgutte B (2010) Effects of reverberation on the directional sensitivity of auditory neurons across the tonotopic axis: influences of interaural time and level differences. J Neurosci 30:7826–7837
Diedesch AC, Stecker GC (2015) Temporal weighting of binaural information at low frequencies: discrimination of dynamic interaural time and level differences. J Acoust Soc Am 138:125–133
Dietz M, Ewert S, Hohmann V (2011) Auditory model based direction estimation of concurrent speakers from binaural signals. Speech Comm 53:592–605
Dietz M, Bernstein LR, Trahiotis C, Ewert SD, Hohmann V (2013a) The effect of overall level on sensitivity to interaural differences of time and level at high frequencies. J Acoust Soc Am 134:494–502
Dietz M, Marquardt T, Salminen NH, McAlpine D (2013b) Emphasis of spatial cues in the temporal fine structure during the rising segments of amplitude-modulated sounds. Proc Natl Acad Sci 110:15151–15156
Dizon RM, Colburn HS (2006) The influence of spectral, temporal, and interaural stimulus variations on the precedence effect. J Acoust Soc Am 119:2947–2964
Durlach NI, Colburn HS (1978) Binaural phenomena. In: Carterette EC, Friedman MP (eds) Handbook of perception, vol 4. Academic Press, New York pp, pp 365–466
Faller C, Merimaa J (2004) Source localization in complex listening situations: selection of binaural cues based on interaural coherence. J Acoust Soc Am 116:3075–3089
Freyman RL, Zurek PM, Balakrishnan U, Chiang YC (1997) Onset dominance in lateralization. J Acoust Soc Am 101:1649–1659
Freyman RL, Balakrishnan U, Zurek PM (2010) Lateralization of noise-burst trains based on onset and ongoing interaural delays. J Acoust Soc Am 128:320–331
Geis H-R, Borst JGG (2009) Intracellular responses of neurons in the mouse inferior colliculus to sinusoidal amplitude-modulated tones. J Neurophysiol 101:2002–2016
Goupell MJ, Laback B, Majdak P (2009) Enhancing sensitivity to interaural time differences at high modulation rates by introducing temporal jitter. J Acoust Soc Am 126:2511–2521
Grantham DW (1986) Detection and discrimination of simulated motion of auditory targets in the horizontal plane. J Acoust Soc Am 79:1939–1949
Grantham DW, Ahlstrom JB (1982) Interaural intensity discrimination of noise as a function of center frequency, duration, and interaural correlation. J Acoust Soc Am 71:S86
Grantham DW, Wightman FL (1978) Detectability of varying interaural temporal differences. J Acoust Soc Am 63:511–523
Hafter ER, Buell TN (1990) Restarting the adapted binaural system. J Acoust Soc Am 88:806–812
Hafter ER, Dye RHJ (1983) Detection of interaural differences of time in trains of high-frequency clicks as a function of interclick interval and number. J Acoust Soc Am 73:644–651
Hafter ER, Trahiotis C (1997) Functions of the binaural system. In: Crocker M (ed) Encyclopedia of acoustics, vol 3. Wiley, New York, pp 1461–1479

Hafter ER, Wenzel EM (1983) Lateralization of transients presented at high rates: site of the saturation effect. In: Klinke R, Hartmann R (eds) Hearing–physiological basis and psychophysics. Springer-Verlag, Berlin, pp 202–208
Hafter ER, Dye RHJ, Gilkey RH (1979) Lateralization of tonal signals which have neither onsets nor offsets. J Acoust Soc Am 65:471–477
Hafter ER, Dye RHJ, Wenzel E (1983) Detection of interaural differences of intensity in trains of high-frequency clicks as a function of interclick interval and number. J Acoust Soc Am 73:1708–1713
Hafter ER, Buell TN, Richards VM (1988) Onset-coding in lateralization: its form, site, and function. In: Edelman GM, Gall WE, Cowan WM (eds) Auditory function: neurobiological bases of hearing. Wiley, New York, pp 648–676
Hancock KE, Chung Y, Delgutte B (2012) Neural ITD coding with bilateral cochlear implants: effect of binaurally coherent jitter. J Neurophysiol 108:714–728
Hartmann WM, Constan ZA (2002) Interaural level differences and the level-meter model. J Acoust Soc Am 112:1037–1046
Heil P (2001) Representation of sound onsets in the auditory system. Audiol Neurootol 6:167–172
Henning GB (1974) Detectability of interaural delay in high-frequency complex waveforms. J Acoust Soc Am 55:84–90
Henning GB (1980) Some observations on the lateralization of complex waveforms. J Acoust Soc Am 68:446–453
Houtgast T, Aoki S (1994) Stimulus-onset dominance in the perception of binaural information. Hear Res 72:29–36
Houtgast T, Plomp R (1968) Lateralization threshold of a signal in noise. J Acoust Soc Am 44:807–812
Hu H, Ewert SD, McAlpine D, Dietz M (2017) Differences in the temporal course of interaural time difference sensitivity between acoustic and electric hearing in amplitude modulated stimuli. J Acoust Soc Am 141:1862–1873
Jeffress LA, Blodgett HC, Deatherage BH (1962) Effect of interaural correlation on the precision of centering a noise. J Acoust Soc Am 32:1122–1123
John MS, Dimitrijevic A, Picton TW (2002) Auditory steady-state responses to exponential modulation envelopes. Ear Hear 23:106–117
Klein-Hennig M, Dietz M, Hohmann V, Ewert SD (2011) The influence of different segments of the ongoing envelope on sensitivity to interaural time delays. J Acoust Soc Am 129:3856–3872
Klumpp RG, Eady HR (1956) Some measurements of interaural time difference thresholds. J Acoust Soc Am 28:859–860
Laback B, Majdak P (2008) Binaural jitter improves interaural time-difference sensitivity of cochlear implantees at high pulse rates. Proc Natl Acad Sci 105:814–817
Laback B, Dietz M, Joris P (2017) Temporal effects in interaural and sequential level difference perception. J Acoust Soc Am 142:3267–3283
Luczak A, Bartho P, Harris KD (2009) Spontaneous events outline the realm of possible sensory responses in neocortical populations. Neuron 62:413–425
Macauley EJ, Hartmann WM, Rakerd B (2010) The acoustical bright spot and mislocalization of tones by human listeners. J Acoust Soc Am 127:1440–1449
Macpherson EA, Middlebrooks JC (2002) Listener weighting of cues for lateral angle: the duplex theory of sound localization revisited. J Acoust Soc Am 111:2219–2236
McFadden DM, Pasanen EG (1976) Lateralization of high frequencies based on interaural time differences. J Acoust Soc Am 59:634–639
Middlebrooks JC, Snyder RL (2010) Selective electrical stimulation of the auditory nerve activates a pathway specialized for high temporal acuity. J Neurosci 30:1937–1946
Nelson BS, Takahashi TT (2010) Spatial hearing in echoic environments: the role of the envelope in owls. Neuron 67:643–655. https://doi.org/10.1016/j.neuron.2010.07.014
Nuetzel JM (1982) Sensitivity to interaural intensity differences in tones and noises as measured with a roving level procedure. J Acoust Soc Am 71:S47

Nuetzel JM, Hafter ER (1981) Discrimination of interaural delays in complex waveforms: spectral effects. J Acoust Soc Am 69:1112–1118

Ochi A, Yamasoba T, Furukawa S (2014) Factors that account for inter-individual variability of lateralization performance revealed by correlations of performance among multiple psychoacoustical tasks. Front Neurosci 8:1–10

Pastore MT, Braasch J (2019) The impact of peripheral mechanisms on the precedence effect. J Acoust Soc Am 146:425–444

Patterson RD, Allerhand MH, Giguere C (1995) Time-domain modeling of peripheral auditory processing: a modular architecture and a software platform. J Acoust Soc Am 98:1890–1894

Pollack I, Trittipoe WJ (1959) Binaural listening and interaural noise cross correlation. J Acoust Soc Am 31:1250–1252

Rakerd B, Hartmann WM (2010) Localization of sound in rooms. V. Binaural coherence and human sensitivity to interaural time differences in noise. J Acoust Soc Am 128:3052–3063

Rayleigh [Strutt JW] (1907) On our perception of sound direction. Philos Mag 13:214–232

Remme MWH, Donato R, Mikiel-Hunter J, Ballestero JA, Foster S, Rinzel J, McAlpine D (2014) Subthreshold resonance properties contribute to the efficient coding of auditory spatial cues. Proc Natl Acad Sci 111:E2339–E2348

Robinson D, Jefress LA (1963) Effect of varying the interaural noise correlation on the detectability of tonal signals. J Acoust Soc Am 35:1947–1952

Rodríguez FA, Read HL, Escabí MA (2010) Spectral and temporal modulation tradeoff in the inferior colliculus. J Neurophysiol 103:887–903

Saberi K (1996) Observer weighting of interaural delays in filtered impulses. Percept Psychphys 58:1037–1046

Schwarz DW, Tennigkeit F, Adam T, Finlayson P, Puli E (1998) Membrane properties that shape the auditory code in three nuclei of the central nervous system. J Otolaryngol 27:311–317

Srinivasan S, Laback B, Majdak P, Delgutte B (2018) Introducing short interpulse intervals in high-rate pulse trains enhances binaural timing sensitivity in electric hearing. J Assoc Res Otolaryngol 19:301–315

Stecker GC (2014) Temporal weighting functions for interaural time and level differences. IV. Effects of carrier frequency. J Acoust Soc Am 136:3221–3232. https://doi.org/10.1121/1.4900827

Stecker GC (2016) Exploiting envelope fluctuations to achieve robust extraction and intelligent integration of binaural cues. In: Proceedings of the 22nd International Congress on Acoustics. 22:445

Stecker GC (2018a) Temporal weighting functions for interaural time and level differences. V. Modulated noise carriers. J Acoust Soc Am 143:686–695

Stecker GC (2018b) Temporal weighting functions for lateralization of amplitude-modulated click trains. Assoc Res Otolaryngol Abs 41:259

Stecker GC, Bibee JM (2014) Nonuniform temporal weighting of interaural time differences in 500 Hz tones. J Acoust Soc Am 135:3541–3547. https://doi.org/10.1121/1.4876179

Stecker GC, Brown AD (2010) Temporal weighting of binaural cues revealed by detection of dynamic interaural differences in high-rate Gabor click trains. J Acoust Soc Am 127:3092–3103

Stecker GC, Brown AD (2012) Onset and offset-specific effects in interaural level difference discrimination. J Acoust Soc Am 132:1573–1580

Stecker GC, Diedesch AC (2014) The role of onsets and envelope fluctuations in binaural cue use. J Acoust Soc Am 136:2308

Stecker GC, Diedesch AC (2016) Perceptual weighting of binaural information: toward an auditory perceptual "spatial codec" for auditory augmented reality. In: Proceedings of the AES International Conference on Audio for Virtual and Augmented Reality 2016

Stecker GC, Hafter ER (2002) Temporal weighting in sound localization. J Acoust Soc Am 112:1046–1057. https://doi.org/10.1121/1.1497366

Stecker GC, Hafter ER (2009) A recency effect in sound localization? J Acoust Soc Am 125:3914–3924

Stecker GC, Ostreicher JD, Brown AD (2013) Temporal weighting functions for interaural time and level differences. III. Temporal weighting for lateral position judgments. J Acoust Soc Am 134:1242–1252

Stellmack MA, Dye RH Jr, Guzman SJ (1999) Observer weighting of interaural delays in source and echo clicks. J Acoust Soc Am 105:377–387

Stellmack MA, Viemeister NF, Byrne AJ (2005) Comparing monaural and interaural temporal windows: effects of a temporal fringe on sensitivity to intensity differences. J Acoust Soc Am 118:3218–3228

Strouse A, Ashmead DH, Ohde RN, Grantham DW (1998) Temporal processing of the aging auditory system. J Acoust Soc Am 104:2385–2399

Tobias JV, Schubert ER (1959) Effective onset duration of auditory stimuli. J Acoust Soc Am 31:1595–1605

Trahiotis C, Bernstein LR, Akeroyd MA (2001) Manipulating the "straightness" and "curvature" of patterns of interaural cross-correlation affects listeners' sensitivity to changes in interaural delay. J Acoust Soc Am 109:321–330

Trahiotis C, Bernstein LR, Stern RM, Buell TN (2005) Interaural correlation as the basis of a working model of binaural processing: an introduction. In: Popper AN, Fay RR (eds) Sound source localization, Springer handbook of auditory research series. Springer, USA, pp 238–271

van de Par S, Kohlrausch A (1997) A new approach to comparing binaural masking level differences at low and high frequencies. J Acoust Soc Am 101:1671–1680

Wallach H, Newman EB, Rosenzweig MR (1949) The precedence effect in sound localization. Am J Psych 62:315–336

Wightman FL, Kistler DJ (1992) The dominant role of low-frequency interaural time differences in sound localization. J Acoust Soc Am 91:1648–1661

Wolf S (1991) Untersuchungen zur Lokalisation von Schallquellen in geschlossenen Räumen, Dissertation, Ruhr-Universitaät Bochum

Zhao W, Dhar S (2011) Fast and slow effects of medial olivocochlear efferent activity in humans. PLoS One 6:e18725

Zurek PM (1980) The precedence effect and its possible role in the avoidance of interaural ambiguities. J Acoust Soc Am 67:952–964

Zurek PM, Durlach NI (1987) Masker-bandwidth dependence in homophasic and antiphasic tone detection. J Acoust Soc Am 81:459–464

Zwislocki J, Feldman RS (1956) Just noticeable differences in dichotic phase. J Acoust Soc Am 28:860–864

Chapter 7
Binaural Hearing and Across-Channel Processing

Virginia Best, Matthew J. Goupell, and H. Steven Colburn

7.1 Introduction

To localize sounds in the horizontal plane, listeners make use of differences in timing and level between the signals reaching the two ears. Interaural time differences (ITDs) arise due to differences in the path length that a sound must travel to reach each ear. Interaural level differences (ILDs) result from the shadowing effect of the head, which attenuates a signal at the ear furthest from the source. For pure-tone stimuli presented in the free field, it has long been known that ITDs are only effective at low frequencies, whereas ILDs are most effective at high frequencies, an observation that led Lord Rayleigh to propose the "duplex theory" for sound localization (Strutt 1907). It is now known that the effectiveness of these fine-structure ITDs is reliant on the coding of cycle-by-cycle timing information in the auditory periphery, which diminishes at higher frequencies. ILDs, on the other hand, are encoded by the auditory system at all frequencies, but they primarily occur for higher frequency sounds, whose short wavelengths are more strongly diffracted by the head (for more details on interaural differences, see Hartmann, Chap. 2). Although the duplex theory is still broadly accepted, it has been refined and

V. Best (✉)
Department of Speech, Language and Hearing Sciences, Boston University, Boston, MA, USA
e-mail: ginbest@bu.edu

M. J. Goupell
Department of Hearing and Speech Sciences, University of Maryland, College Park, MD, USA

Department of Biomedical Engineering, Boston University, Boston, MA, USA
e-mail: goupell@umd.edu

H. S. Colburn
Department of Biomedical Engineering, Boston University, Boston, MA, USA
e-mail: colburn@bu.edu

© Springer Nature Switzerland AG 2021

R. Y. Litovsky et al. (eds.), *Binaural Hearing*, Springer Handbook of Auditory Research 73, https://doi.org/10.1007/978-3-030-57100-9_7

modified based on a wealth of data. For example, Hartmann et al. (2016) demonstrated the effects of ILDs at low frequencies, which may help to resolve ambiguities in the ITD. Other studies have used stimuli with broader bandwidths and more complex temporal characteristics and have shown that ITD information is also available to listeners through sensitivity to time delays in the amplitude envelopes of high-frequency sounds (e.g., Klumpp and Eady 1956; Henning 1980). High-frequency envelope ITDs are generally less salient than low-frequency fine-structure ITDs, although this depends strongly on the characteristics of the envelope (e.g., Bernstein and Trahiotis 2002; Monaghan et al. 2015).

Although much of what is known about human sensitivity to ITD and ILD comes from careful experimentation using narrowband stimuli, most sounds in the real world contain energy at multiple frequencies, and thus spatial perception can be based on binaural information available across the spectrum. For a single sound source presented under ideal (quiet, anechoic) conditions, one might imagine that the binaural cues are independent of frequency. Although this might be an appropriate first-order approximation for ITDs, the ITDs for a fixed location change with frequency (Kuhn 1977; Benichoux et al. 2016; see also Hartmann, Chap. 2). ILDs are well-known to vary with frequency, and for some frequencies, the ILD is non-monotonic as a function of azimuth (Macaulay et al. 2010). Therefore, for a broadband sound, there is a relatively complicated pattern of binaural information that must be combined to arrive at an estimate of sound source location, even for a single source. Moreover, for sounds that arise from multiple sources, that are corrupted by noise, or that contain reflections or echoes, it would be a poor strategy to simply combine all of the available information. Rather, it would be optimal to aggregate binaural cues only over those channels containing reliable, target-related information.

This chapter considers how binaural cues are combined across frequency for signals containing energy across more than one spectral region. Included are both cases in which the binaural cues are *constant* across frequency (i.e., having the same physical value) and cases in which cues are *not constant* across frequency (as would be expected for multiple sources at different azimuths). Similarly, cases are discussed in which the binaural cues are *consistent* with those naturally produced by the head or *inconsistent*. The combination of binaural information has been studied using a variety of perceptual measures. Perhaps the most common is the binaural discrimination threshold (or just-noticeable difference), which describes the smallest change in ITD or ILD that is detectable by listeners for a given signal. Other experiments have measured the "extent of laterality" produced by different signals by asking listeners to judge the lateral position of a sound (usually perceived within the head) as the ITD or ILD is varied. Still others have measured localization in the free field for sounds containing natural combinations of ITD and ILD. In the rest of the chapter, Sect. 7.2 describes the combination of cues across frequency for single sources, Sect. 7.3 considers stimuli with conflicting cues across frequency, and Sect. 7.4 considers models of across-frequency processing of localization information. Note that extensive experimental and modeling work has explored across-channel processing as it applies to binaural detection and unmasking of tones and speech, but this is covered in Chap. 8 by Culling and Lavandier.

7.2 Integration Across Frequency for Binaural Discrimination and Lateralization

7.2.1 *Increasing the Bandwidth of a Noise*

It is useful to start with a summary of how the bandwidth affects the basic waveform characteristics of a narrowband signal. In general, a narrowband waveform can be visualized as a modulated sinusoidal signal, with both the amplitude and the phase of the sinusoid fluctuating with rates that are constrained by the bandwidth of the signal. As the bandwidth increases, the amplitude and phase fluctuations become faster. This basic conceptualization applies to both noise bands and tone complexes and to any stimulus for which the bandwidth is narrow compared with the center frequency. For wider bandwidths, these narrowband conceptualizations are still relevant in the context of a bank of filters because the outputs of the individual filters are narrowband.

Because the auditory system is thought to process acoustic signals through a bank of band-pass filters, across-channel binaural processing requires a consideration of the individual filter outputs and their combination. For narrowband inputs, specifically those narrower than the "critical-band filters" (Moore and Glasberg 1983), binaural processing can operate directly on the outputs of corresponding filters from the left and right ears. When the bandwidth of the stimulus exceeds the bandwidth of a single filter or, more precisely, when the frequency spread of a signal is greater than accommodated that by a single filter, then multiple filters are excited. Assuming that there is only one sound source present, then binaural information can be combined across the different filters. The multiple-source case is discussed in Sects. 7.3 and 7.4.6.

ITD sensitivity is extremely good for narrowband sounds at low frequencies where listeners rely primarily on fine-structure delays, with most studies reporting the lowest discrimination thresholds (on the order of 10–20 μs) for frequencies in the region of 600–1000 Hz (e.g., Brughera et al. 2013; Thavam and Dietz 2019). ITD thresholds do not improve much from increases in the bandwidth in this region (e.g., McFadden and Pasanen 1976). For narrowband noises with higher center frequencies (4 kHz and above), however, improvements in ITD sensitivity occur with increases in bandwidth (e.g., McFadden and Pasanen 1976; Bernstein and Trahiotis 1994). These improvements generally come from concomitant changes in the envelope features because there is no sensitivity to fine-structure delays in this high-frequency region (see Stecker, Bernstein, and Brown, Chap. 6).

Considering the extent of laterality, Fig. 7.1, *circles*, shows that increasing the bandwidth of a low-frequency narrowband noise yields no (or only modest) changes in the extent of laterality produced by ITDs within the human physiological range (e.g., Schiano et al. 1986; Trahiotis and Bernstein 1986). For high-frequency noises, the extents of laterality do increase with bandwidth (Fig. 7.1, *diamonds*), but again this can largely be explained by changes in the envelope fluctuations. Overall, these experiments suggest that the bandwidth effects for the ITD relate more to useful

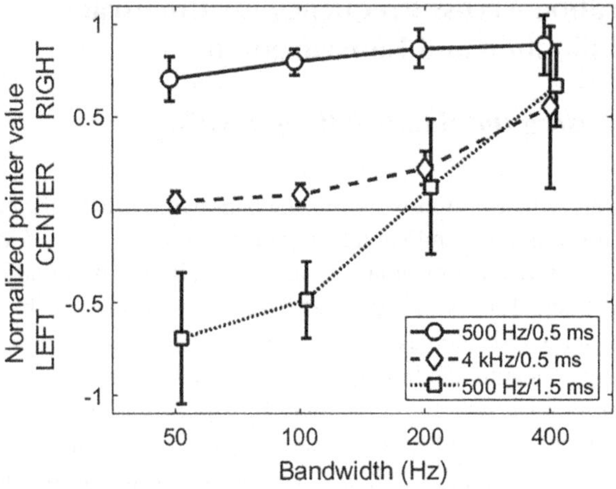

Fig. 7.1 Lateralization of narrow bands of noise by ITD as a function of bandwidth. Listeners gave responses by adjusting an ILD-based acoustic pointer to match the perceived intracranial position of the target. Data were extracted by eye from the published data for the four listeners in those studies and were normalized by dividing by the maximum pointer value for each listener. Values are the means ± SD of these normalized values for a noise centered on 500 Hz with a 0.5-ms ITD (*circles*), a noise centered on 4 kHz with a 0.5-ms ITD (*diamonds*), and noise centered on 500 Hz with a 1.5-ms ITD (*squares*). The data suggest that across-frequency processing has a modest effect for ITDs within the physiological range but a dramatic effect for ITDs beyond that range. (Data from Trahiotis and Bernstein (1986) and Trahiotis and Stern (1989))

envelope information than to a true combination of information across channels (Trahiotis and Bernstein 1986).

An influential set of studies has examined across-frequency processing of ITDs by systematically varying the bandwidth of narrowband noise stimuli with ITDs that are larger than the human physiological range (which is approximately ±750 μs; see Hartmann, Chap. 2). In one striking demonstration, the perceived lateral position of a stimulus can flip from one side of the head to the other as the bandwidth is increased (e.g., Trahiotis and Stern 1989; Yost et al. 2007). For example, a noise centered at 500 Hz with a right-leading ITD of 1.5 ms is perceived to the left when it has a very narrow bandwidth but is perceived to the right as the bandwidth is extended (Fig. 7.1, *squares*). The explanation for this phenomenon lies in the ambiguity of the interaural phase difference for an ITD of 1.5 ms, which for a pure tone at 500 Hz is equivalent to that produced by an ITD of 0.5 ms to the opposite side. Given this ambiguity, pure tones or noises with bandwidths less than a critical band are perceived at the location of the ITD closest to zero (see discussion of "centrality" in Sect. 7.4.4). With an increase in bandwidth, there is the possibility for a combination of information across frequency and a resolution of the ambiguity. It has been proposed that the perception of laterality in this case follows the ITD that is consistent across the full range of frequencies (see discussion of "straightness" in ITD-frequency patterns in Sect. 7.4.4).

For ILDs, Hartmann and Constan (2002) reported better discrimination thresholds for broadband noise (bandwidth of 10 kHz) than for low-passed noise (bandwidth of 1 kHz). This bandwidth effect was explained in terms of a reduction in the internal noise gained by summing over multiple auditory channels, with independent internal noise in each channel. The presence of summation in this study, but not for the aforementioned studies of ITD, may in part reflect the very large changes in bandwidth examined by Hartmann and Constan (2002).

In the free field, where ITDs and ILDs are both available, there are conflicting results regarding whether or not localization accuracy improves with increasing bandwidth. Yost and Zhong (2014) showed that listeners localized noises as narrow as one-twentieth octave significantly better than pure tones in the horizontal plane and that there were continued improvements as the bandwidth was extended to two octaves. It was not clear whether the improvement from the narrowest bandwidth (tones) to broader band noises was attributable to the increased bandwidth per se or to differences in the envelopes of tones and noises. The authors speculated that for high-frequency stimuli (>2 kHz) larger bandwidths may provide more robust ILD information than pure tones (because a given ILD can map to more than one azimuth; see Macaulay et al. 2010). On the other hand, Middlebrooks (1992) found very little difference in horizontal localization accuracy between very narrow bands of noise (one-sixth octave at center frequencies from 6 to 12 kHz) and a broadband noise. In general, it seems that the benefits of increasing the bandwidth are most evident in cases where there are ambiguities in the binaural cues for the narrowband case.

7.2.2 Increasing the Number of Components in a Complex

Another way to study across-frequency processing is to measure ITD and/or ILD sensitivity in a multicomponent stimulus and compare that with sensitivity for each component when it is presented in isolation. In this way, one can test various hypotheses about how information is combined across the components. For example, if binaural sensitivity for a multicomponent stimulus is better than for any component alone, it suggests that there is a summation of information across frequency. Alternatively, sensitivity may be limited to that of the best single component. It is also possible that components in regions with poorer sensitivity will limit performance for a multicomponent stimulus if there is an obligatory averaging of information. Note that the spacing of components in a multicomponent stimulus is important because multiple components within a single filter create temporal envelope modulation patterns that depend on the spacing of the components. To avoid this complicating factor, most studies of across-channel processing (see below) have used widely spaced components that are assumed to activate largely independent filters.

Studies examining this issue for the detection of ITDs in low-frequency stimuli generally suggest that there is a summation of information when two tones are added with the same ITD (e.g., Buell and Hafter 1991). Dye (1990) used combinations of three tones centered on 750 Hz, and although he did not report improved thresholds for the complexes relative to isolated 750-Hz tones, inspection of the raw data suggests that there may be an improvement for some listeners for the larger frequency separations that would have resolved bands. Buell and Hafter (1991) concluded that the detection of ITDs in two-tone complexes is well explained by optimal weighted summation across the components (with weights inversely proportional to their variances). They found this to be true regardless of whether the components were harmonically related or not. Summation has also been reported for high-frequency noise bands (Buell and Trahiotis 1993), with improved thresholds for pairs of bands at 2 and 4 kHz relative to either band alone. Saberi (1995) found that comodulated bands showed more summation than independent bands. He also reported that his listeners outperformed a simple model of optimal summation, which may reflect an additional level of information that can be gained by across-frequency processing (e.g., see discussion in Sect. 7.4.4 of straightness). Buchholz et al. (2018) collected an extensive set of data that allowed them to relate ITD thresholds for narrowband noises with center frequencies ranging from 148 to 1572 Hz to the lateralization of "broadband" combinations of these bands. ITD thresholds for the individual bands varied (depending on the center frequency, reference ITD, and interaural correlation), and lateral position estimates for the broadband noises were well predicted by an optimal variance-weighted combination across bands.

Although there are much less data available investigating across-frequency processing of ILDs, one study found no evidence for summation. Goupell and Stakhovskaya (2018a) measured ILD discrimination thresholds for single narrowband noises in different frequency ranges (centered on 750, 2000, and 4000 Hz) and for triplets of narrowband noises within those ranges carrying physically constant (common) ILDs. For low-frequency stimuli, the thresholds were equivalent for single- and multiband conditions. For mid- and high-frequency stimuli, the multiband thresholds were better on average than the single-band thresholds but no better than the threshold of the best single band. Using similar stimuli, Stakhovskaya and Goupell (2017) also measured lateralization by ILD in single and multiband conditions. Listeners were presented with single bands of noise (centered at 500, 750, or 1000 Hz) or a stimulus containing all three bands. Lateralization functions were very similar for the single- and multiband conditions, providing no evidence for change in the lateral percept based on across-frequency integration.

Bilateral cochlear-implant (CI) users are a unique population in which to study the combination of binaural cues across frequency (see Ricketts and Kan, Chap. 13). These listeners have a limited number of stimulating electrodes that contact different places on the cochlea and as such represent discrete frequency "channels." For research purposes, each electrode, or channel, can be activated selectively using research processors. This strategy has the specific advantage that ITDs can be delivered to a bilateral electrode pair with microsecond precision, whereas this timing information is generally lost when CIs are controlled by their clinical sound processors. ITD sensitivity in bilateral CI users presented with low-rate (e.g., 100-Hz) electrical pulse trains does not appear to improve by activating multiple electrode pairs compared with the single best electrode pair (e.g., Francart et al. 2015; Egger et al. 2016) and sensitivity can even decrease (e.g., Kan et al. 2015). Moreover, lateralization functions are quite similar for single- and multielectrode stimulation for both ITD (Kan et al. 2016) and ILD (Stakhovskaya and Goupell 2017). Thus, there is little evidence for a summation of binaural information across channels in these cases. Although this suggests a difference in the efficacy of across-frequency integration relative to normal-hearing (NH) listeners (at least for ITDs), the current spread between electrodes and the generally poorer sensitivity of CI listeners make direct comparisons difficult.

7.2.3 Across-Frequency Weighting of Interaural Time Difference and Interaural Level Difference

For broadband stimuli containing binaural information across the spectrum, it is possible to investigate how different frequencies are weighted when making judgments about the lateral position. Bilsen and Raatgever (1973) compared the lateralization of wideband stimuli containing ITDs to that of band-pass filtered clicks of different center frequencies and concluded that frequency components around 600 Hz are weighted more heavily than are frequency components in spectrally adjacent regions. To investigate this "dominant region" more closely, Raatgever (1980) presented wideband noises in which one narrow band was interaurally delayed and the rest of the noise was diotic (or vice versa). He then measured the increase in level required for the narrow band to dominate the lateral perception, testing center frequencies of 300, 600, 900, and 1200 Hz. This increase in level was found to vary with center frequency and had an estimated minimum at around 600 Hz for white noise and for clicks (with a slight shift downward in frequency when pink noise was used).

When considering frequency weighting for the ITD, it is useful to consider the reliability of the physiological representation of the ITD. As analyzed by several studies over the years (e.g., Colburn 1973; Brughera et al. 2013), the synchronization of neural activity to the stimulus fine structure is independent of frequency at low frequencies, which implies improved ITD resolution as frequency increases (and the periods become shorter). Above about 600 Hz, the synchrony observed physiologically decreases, reaching negligible values at high frequencies. The upper frequency limit in humans is not known physiologically and is a topic of much debate (Verschooten et al. 2019), with estimates ranging from 1500 to 10,000 Hz. Nevertheless, these characteristics of neural coding, in combination, may explain the observed variations in sensitivity to ITDs and provide a physiological basis for the dominant region. It is worth noting that this region is also thought to dominate pitch perception (Cariani and Delgutte 1996; Dai 2000). An analogous dominant region has not been reported for ILDs, which do not have the same reliance on temporal precision as ITDs and pitch.

Another study presented complex stimuli containing 11 narrowband noises with center frequencies from 442 to 5544 Hz (Ahrens et al. 2015). They varied the magnitude of the ITD or ILD in each band independently and then used laterality judgments and regression analysis to derive spectral weighting functions. They found that listeners placed the highest weight on the edge bands (the lower edge in the case of ITD; the higher edge in the case of ILD) and that this "edge effect" persisted when the bandwidth was reduced by removing the two outer bands. The authors interpreted their data as suggesting that contextual information (in this case, the information related to the frequency range of the stimulus) can influence the weighting of binaural information.

One problem with many studies of across-frequency binaural processing is that they do not consider the fact that ITDs and ILDs vary with frequency for a given source location in external space (see Hartmann, Chap. 2). For example, a source at a 76° azimuth has a decreasing ITD as a function of frequency, from about 0.8 ms at low frequencies to about 0.6 ms at high frequencies (Kuhn 1977). Similarly, the ILDs for this source would be less than a few dB at low frequencies and would be greater than 10 dB at many high frequencies but would also exhibit nonsystematic variations as a function of frequency (Macaulay et al. 2010). Thus, headphone-presented stimuli containing ITDs or ILDs of a fixed or constant value across a wide frequency range do not represent the range of binaural cues that would occur for such a stimulus presented from a single external location. In other words, physically constant ITDs and ILDs are inconsistent with the ITDs and ILDs produced by the head.

Only a few behavioral studies have addressed the issue of across-frequency variations in the binaural cues. Constan and Hartmann (2003) explored whether listeners were sensitive to natural variations in ITD across frequency and, in particular, whether these variations lead to broader or more diffuse percepts than for stimuli with a constant ITD across frequency. They found that physiologically consistent ITDs (as derived from a spherical head model; Kuhn 1977) led to shifts in the lateral position of a noise stimulus but produced no other changes in the diffuseness or other qualities compared to the constant ITD case. Moreover, they concluded that the lateral position estimates for the physiologically consistent ITDs were well accounted for by a weighted average across frequency that emphasizes the 600-Hz dominant region. Another view is that ITDs are not combined across frequency directly but rather ITDs are mapped to azimuth *within* frequency channels before information is pooled across frequency (McFadden 1981). Goupell and Stakhovskaya (2018b) measured lateralization functions for narrowband stimuli covering a wide range of center frequencies (500–5000 Hz), with natural patterns of ITD and ILD derived from head-related transfer functions. Contributions of ITD and ILD were estimated by including conditions in which the other cue was set to zero, and across-frequency processing was examined by presenting the bands in combination. The results indicated that ITD-based lateralization was dominated by the lower frequency bands, which were both closest to the dominant region and which provided the largest extents of laterality in isolation. ILD-based lateralization was consistent with averaging across frequency, with a larger contribution of higher frequency bands arising because of the larger ILDs occurring there.

The *relative* strengths of ITD and ILD cues have generally been measured using "time-intensity trading" experiments (e.g., Harris 1960). In a typical experiment, an ITD pointing to one side is combined with an ILD pointing to the opposite side and the relative magnitudes needed to arrive at a balanced or centered image are measured. Given that these experiments apply both cues to the *same* stimulus, they do not address across-frequency weighting. There appears to be a lack of studies that have put the two cues in conflict across *different* frequency bands. One reason for this may be that listeners tend to perceive multiple objects or split images when across-frequency conflicts become too large (see Sect. 7.3).

For free-field sounds of sufficient bandwidth, both ITD and ILD cues should be available, although the salience of each cue (in absolute terms and relative to each other) will vary in different frequency regions. For broadband noise stimuli, listeners appear to show a dominance of low-frequency (<1500Hz) ITD over ILD because sound localization responses will follow the ITD when the two cues conflict (Wightman and Kistler 1992; Macpherson and Middlebrooks 2002). An issue that has received surprisingly little attention is how the relative weighting of the ITD and ILD changes in the presence of background noise and/or reverberation. This seems to be a relevant issue given the ever-growing interest in understanding perception in realistic listening situations. One study examined the effect of broadband noise on time-intensity trading ratios for clicks (Gaskell and Henning 1981), and the results indicated that listeners were more strongly influenced by the ILD cue, probably as a result of shifting their perceptual weight to higher frequencies where the ITD cue is less salient. In reverberation, which acts to interaurally decorrelate the target sound, a similar shift in favor of the ILD has been reported for one-third octave-band noises centered at 715 and 2850 Hz (Rakerd and Hartmann 2010). On the other hand, Shinn-Cunningham et al. (2005) analyzed the available binaural cues for a broadband noise target under various reverberant conditions and concluded that ITDs may be more robust than ILDs if across-frequency integration is possible.

Overall, the way in which binaural information in broadband stimuli is weighted across frequency appears to depend primarily on the availability and reliability of ITD and ILD cues. Contextual effects may also play a role, especially in real-world situations, although this is an area that requires further investigation.

7.2.4 Binaural Fusion

An issue that has been given some consideration is how binaural information is encoded when the frequencies differ at the two ears. A prerequisite for binaural processing would seem to be binaural fusion, in which the signals at the two ears give rise to the percept of a single sound source (see Sayers and Cherry 1957). Early reports of the "binaural critical band," derived by measuring ITD thresholds as a function of the interaural frequency difference of a pair of tone bursts, suggested that it had a width and frequency dependence similar to the monaural critical band (Scharf et al. 1976). Later studies obtained subjective estimates of "binaural nonfusion" and physiological measures of binaural interaction that are broadly consistent with these early estimates (e.g., Zhou and Durrant 2003), although studies using a masked-tone detection method have estimated slightly wider binaural than monaural bands (e.g., Kollmeier and Holube 1992). Although the frequency tuning of binaural fusion is narrow in NH listeners, some studies suggest that it is broader in listeners with sensorineural hearing loss. It has been suggested that a broader fusion region may help these listeners cope with interaural asymmetries, but it may also have negative consequences for pitch and speech perception (e.g., see Reiss et al. 2017; Oh and Reiss 2017).

Interaural frequency mismatch is an important problem for bilateral CI users for whom the same stimulus might stimulate two quite different places in the left and right ears due to differences in electrode placement. Several studies in bilateral CI users have shown that increasing the frequency mismatch between electrodes reduces the tendency for listeners to hear a single, fused sound and decreases ITD and ILD sensitivity (e.g., Laback et al. 2004; Kan et al. 2013). Similarly, in NH listeners given band-limited pulse trains or noises, the systematic introduction of a frequency mismatch across the ears reduces subjective judgments of fusion and reduces sensitivity to ITD and ILD (e.g., Francart and Wouters 2007; Goupell et al. 2013). It is interesting to consider how the binaural system would interpret signals with interaural frequency mismatch. Assuming a nominal ITD and ILD of zero, these values would occur within any regions of spectral overlap. Moving out of that region, the ILD would vary because one ear would have a larger amplitude than the other, and in regions with no appreciable spectral overlap, there would be an effectively infinite ILD. If such sounds are heard as a signal auditory object, localization would have to be based on some across-frequency combination of these disparate values (see Sect. 7.3.2).

Overall, these results are consistent with physiological demonstrations of very narrow frequency tuning of binaural neurons (e.g., Fischer and Peña 2009) and reinforce the idea that binaural hearing relies first and foremost on within-channel processes, with across-channel processes playing a secondary role.

7.3 Effect of Conflicting Binaural Cues Across Frequency

7.3.1 Multiple Sound Sources

When multiple sound sources are present in the environment, listeners are generally able to separate the sounds arriving from each source into a combination of individual perceptions (as opposed to perceiving sources as a single fused percept, a phenomenon called grouping). Nevertheless, even when the separation into distinct images is clear and easy, there can be interference of one source on the localization of the other. Reverberant environments represent a particularly difficult case in which correlated sounds (i.e., a direct sound and its reflections) arrive at the listener from different directions and present conflicting binaural cues (see also Zahorik, Chap. 9). In Sects. 7.3.2 and 7.3.3, experiments are discussed that have used simplifications of this interaction between multiple sound sources to study across-channel processing of binaural information.

7.3.2 Studies of Across-Frequency Binaural Interference

Many studies have investigated across-frequency processing of binaural information by presenting conflicting binaural information to different spectral regions and measuring interference effects. This approach is referred to here as "across-frequency binaural interference" but is often less precisely called "binaural interference." Typically, sensitivity to the binaural properties of a target signal is measured in quiet and in the presence of one or more (usually diotic) spectrally remote interferers. Figure 7.2a, *two top boxes*, provides an illustration of these basic conditions.

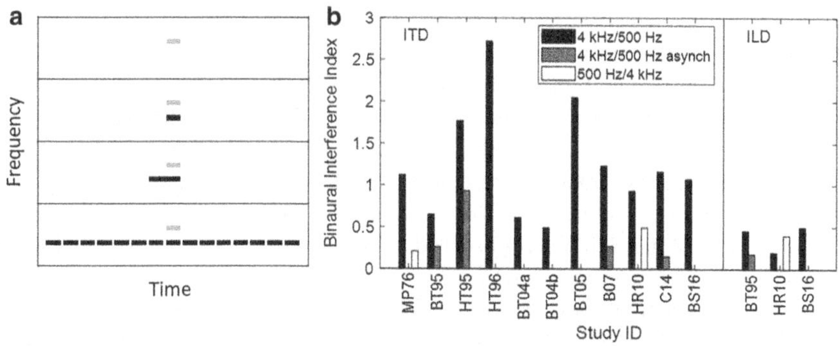

Fig. 7.2 (**a**) Illustration of stimulus arrangements that have been used to investigate across-frequency binaural interference. A high-frequency target (*gray bars*) is presented alone or with a low-frequency interferer (*black bars*) that is completely synchronous or has an onset asynchrony. In the study of Best et al. (2007), the interferer was perceptually captured into a repeating stream of identical tokens. (**b**) Binaural interference index (BII) calculated from data reported for 10 different studies. In all cases, the target and interferer were narrowband sounds centered at 500 Hz and 4 kHz. The task was either ITD discrimination/lateralization (*left*) or ILD discrimination/lateralization (*right*). *Black bars*, BII for a target at 4 kHz and a synchronous interferer at 500 Hz; *gray bars*, BII when the target and interferer were asynchronous (asynch); *white bars*, BII for a target at 500 Hz and a synchronous interferer at 4 kHz. Study IDs: MP76, McFadden and Pasanen (1976); BT95, Bernstein and Trahiotis (1995), averaged over three bandwidth conditions; HT95, Heller and Trahiotis (1995), averaged over in- and out-of-phase interferers; HT96, Heller and Trahiotis (1996); BT04a, Bernstein and Trahiotis (2004), SAM tone target; BT04b, Bernstein and Trahiotis (2004), noise target; BT05, Bernstein and Trahiotis (2005), averaged over two bandwidth conditions; B07, Best et al. (2007); HR10, Heller and Richards (2010); C14, Camalier et al. (2014), continuous condition; and BS16, Bibee and Stecker (2016)

Across-frequency binaural interference is defined as any loss of binaural sensitivity for the target in the presence of the interferer. The majority of studies on this topic have examined the phenomenon using ITD discrimination (e.g., McFadden and Pasanen 1976). There have also been studies that used a lateralization task and have shown that a diotic interferer tends to reduce the perceived extent of laterality of a target carrying a nonzero ITD (e.g., Heller and Trahiotis 1996; Bernstein and Trahiotis 2005). Across-frequency binaural interference has also been demonstrated for ILD-based discrimination (e.g., Bernstein and Trahiotis 1995; Heller and Richards 2010) and for location discrimination in the free field (Croghan and Grantham 2010).

One study observed interference in an unusual case where a low-frequency narrowband noise target (with various bandwidths but confined to the region 223–315 Hz) was presented simultaneously with a monotic narrowband noise interferer that occupied a spectral region above or below the target (Brown and Yost 2015). Target ITD thresholds increased dramatically in the presence of the interferer, particularly if it was higher in frequency (with various bandwidths but confined to the region of 397–1000 Hz). Although such an interferer contains no ITD information, it carries an infinite ILD, and thus this interference may represent a form of across-frequency and across-*cue* binaural interference.

Several characteristics of across-frequency binaural interference are worth noting. First, numerous studies have shown that interference is reduced when the interferer is gated on earlier than the target rather than synchronously with it (Fig. 7.2a, *third box*; see, e.g., Heller and Trahiotis 1995; Camalier et al. 2014). Second, interference is usually greater for harmonically related targets and interferers (Buell and Hafter 1991; Hill and Darwin 1996). Third, interference with ITDs exhibits a strong frequency asymmetry. Interference can be quite dramatic in the case of a high-frequency target (e.g., 4 kHz) and a low-frequency interferer (e.g., 500 Hz), whereas much less (or no) interference is observed for the reverse case (e.g., McFadden and Pasanen 1976; Buell and Trahiotis 1993).

To illustrate some of these points, Fig. 7.2b shows data gathered from 10 different studies that all examined across-frequency binaural interference under similar conditions. Specifically, each of these studies used narrowband targets and interferers centered at 500 Hz and 4 kHz and a task that was based on ITD or ILD discrimination or lateralization. Plotted for each study is the "binaural interference index" or BII (Bibee and Stecker 2016). For ITD discrimination, the BII represents the base-two log ratio of thresholds obtained with and without the interferer present. For ITD lateralization, the index uses the inverse of the slopes of the lateralization functions instead of the thresholds. For ILD discrimination, the BII is simply the difference in dB between the thresholds obtained with and without the interferer. Note that BII = 0 indicates no interference and BII > 0 indicates interference. BIIs for the classic case of a high-frequency target and a synchronous low-frequency diotic interferer are shown in Fig. 7.2b, *black bars*. The BII varies considerably across studies, which is not surprising given the many differences in experimental details and the use of different groups of listeners. Interference for ILD (Fig. 7.2b, *right*) tends to be smaller than for ITD (Fig. 7.2b, *left*), at least for the BII measures used here. For several of these studies, a condition was also included in which the interferer was gated on earlier than the target (Fig. 7.2b, *gray*), and in all cases, this manipulation dramatically reduced the BII (see Sect. 7.3.3).

Also shown in Fig. 7.2b, *white bars* are data from two studies of ITD interference in which the center frequencies of the target and interferer were reversed (i.e., the target was at 500 Hz and the interferer was at 4 kHz). These points demonstrate the frequency asymmetry of ITD-based interference (where the BII is much smaller for low-frequency targets). This frequency asymmetry is often assumed to be related to the much greater salience of low-frequency relative to high-frequency ITD information. Indeed, no clear asymmetry is observed when all of the components occupy a low-frequency region and do not differ greatly in their salience (Dye et al. 1996). Moreover, when the method of "transposed tones" is used to enhance the envelopes of high-frequency stimuli and thus increase the salience of ITDs in that region, the interference exerted by a low-frequency interferer is reduced (e.g., Bernstein and Trahiotis 2004). Finally, no frequency asymmetry (or a slight asymmetry in the opposite direction) has been reported for ILDs (Fig. 7.2b, *right*, Study HR10; Heller and Richards 2010) using narrowband noises centered at 500 Hz and 4 kHz. Another study that examined across-frequency binaural interference using a more extensive set of target and interferer frequencies found that the amount of interference varied in a complicated way depending on the particular combination of frequencies (Goupell and Stakhovskaya 2018a).

7.3.3 The Role of Grouping in Across-Frequency Binaural Interference

Many of the characteristics of across-frequency binaural interference described above can be explained in terms of auditory grouping (e.g., Woods and Colburn 1992; Best et al. 2007). According to this explanation, listeners are obliged to perceptually group targets and interferers at different frequencies, especially when they are gated on and off synchronously and/or are harmonically related. Once grouping occurs, it becomes difficult to judge the binaural parameters of a single component (e.g., the target) so that binaural judgments are made based on the grouped object. Logically, the lateral position of the grouped object represents some combination of the interaural differences of each of the components. Thus, for example, a lateralized target will be drawn toward the center by a diotic interferer that is grouped with the target. Best et al. (2007) explicitly tested the grouping hypothesis by "capturing" the interferer in an ongoing stream of identical tones and thus reducing the tendency of the target and interferer to group (see Fig. 7.2a, *bottom box*). This manipulation led to a reduction in interference in most listeners (these data are included in Fig. 7.2b, *left*, Study B07). The grouping explanation provides a framework for understanding the frequency asymmetries observed in binaural interference with

ITDs. In the grouping explanation, it can be assumed that more salient cues contribute more to the perceived lateral position of the combined object.

These effects of grouping suggest that across-frequency processing of binaural information cannot be understood solely in terms of peripheral interactions but is also affected by relatively central neural mechanisms. Further evidence for this idea comes from data showing that the addition of spectral uncertainty (i.e., uncertainty about at which frequency the target will occur) greatly increases across-frequency binaural interference (Buell and Trahiotis 1994). Moreover, there are hints in the literature that the susceptibility to interference varies greatly across listeners and may relate to their listening experience, again suggesting that central factors play a role (e.g., Woods and Colburn 1992; Best et al. 2007).

7.4 Models of Binaural Integration of Localization Information

7.4.1 Overview

The perception of location in response to binaural acoustic stimuli in arbitrary environments is a complex process, a process that integrates a priori knowledge as well as immediate inputs. In anechoic environments with single sources, the problem is simplified dramatically. In the discussion of models given here, this simple case of a single source with negligible reflections is discussed first. This relatively straightforward case has been the focus of most models of localization. In Sects 7.4.6 and 7.4.7, attention is given to the more general problem of localizing multiple, possibly unknown, sources in more complex acoustic spaces. Most models of localization have considered only the horizontal plane (with several exceptions; e.g., Baumgartner et al. 2014) and relate to the estimation of location with very few a priori assumptions. These models generally assume that the information from the acoustic inputs is limited to ITDs and ILDs as a function of frequency.

7.4.2 Cross-Correlation Models for Interaural Time Difference

An early discussion of localization modeling, the "duplex theory" of Lord Rayleigh (Strutt, 1907), suggested that low-frequency localization is based on ITD and that high-frequency localization is based on ILD. This general idea was noted by Jeffress (1948) when he postulated a mechanism for localizing sound on the basis of low-frequency ITD, specifically a mechanism built around a neural coincidence mechanism (see Hartmann, Chap. 2). This idea continues to be regarded as an insightful speculation, and the concept of a population of left-right coincidence detectors with a distribution of delays is a concept that is closely related to cross-correlation functions. The cross-correlation function for narrowband filtered stimuli can be thought of as a representation of ITD information in the two-dimensional space of time and center frequency. This conceptualization is an important part of many models, starting with Cherry and Sayers (1956) and Sayers and Cherry (1957). A physiologically based model supporting this approach was developed by Colburn (1973, 1977). Note that the across-frequency interactions are generally assumed to come after this processing, consistent with the integrative characteristics of neurons at the midbrain level (e.g., Peña and Konishi 2000).

7.4.3 *Intensity-Weighted Cross-Correlation Models*

Because lateralization is based on both ITD and ILD, models of binaural localization are often based on a temporal cross-correlation function that is weighted by an intensity function. The model of Cherry and Sayers (1956; see also Sayers and Cherry 1957) has this general structure, where the left-leading half of the time axis is weighted by the amplitude of the signal at the left ear and the right-leading half of the time axis is weighted by the amplitude of the signal at the right ear. A closely related model of lateralization was developed and tested by Stern and Colburn (1978) and Stern and Shear (1996). In this model, the cross-correlation time axis (τ) is weighted by an ILD-dependent function of τ (giving more weight to a positive or negative τ for appropriate ILDs and considering the laterality to depend on the center of gravity of the full, weighted function along the internal time axis). When applied to empirical lateralization data, this model provided excellent fits to data for narrowband stimuli. Another way to develop this model, which has been discussed by Harper et al. (2014), is to specify the distributions of time delays along the τ axis in a species-dependent manner so as to optimize the availability of relevant information for each species, depending on the size and physical acoustics. Finally, some authors have argued for an opponent-hemifield model in which a statistic is derived for the left and right hemispheres and compared to determine laterality (Harper and McAlpine 2004; McAlpine 2005), a model conceptually similar to that of Sayers and Cherry (1957) as briefly described above. Each statistic would involve combined timing activity from one polarity of τ and the corresponding intensity weighting. It is worth noting that none of these models have been widely tested.

7.4.4 *Across-Frequency Processing of Interaural Time Difference*

When the stimulus has a wider bandwidth or multiple-frequency components, the perceptual task must incorporate a variety of frequencies that may be analyzed as a single unit or as multiple objects. The importance of across-frequency processing for explaining the lateral perception of narrowband and broadband sounds was first considered by Jeffress (1972). Later, this idea was elaborated in the very influential study of Trahiotis and Stern (1989), which demonstrated how the bandwidth of a narrowband noise containing a larger than physiologically possible ITD can influence its perceived laterality (see Fig. 7.1). This result, which was described in Sect. 7.2.1, has provided the basis for several attempts to model across-frequency processing of ITDs.

The data motivated an extension of the weighted cross-correlation model discussed in Sect. 7.4.3 (Stern et al. 1988; Stern and Trahiotis 1996). This extension incorporated the key processes of centrality and straightness, which are depicted in Fig. 7.3. Figure 7.3a shows normalized cross-correlation functions for 5 pure tones

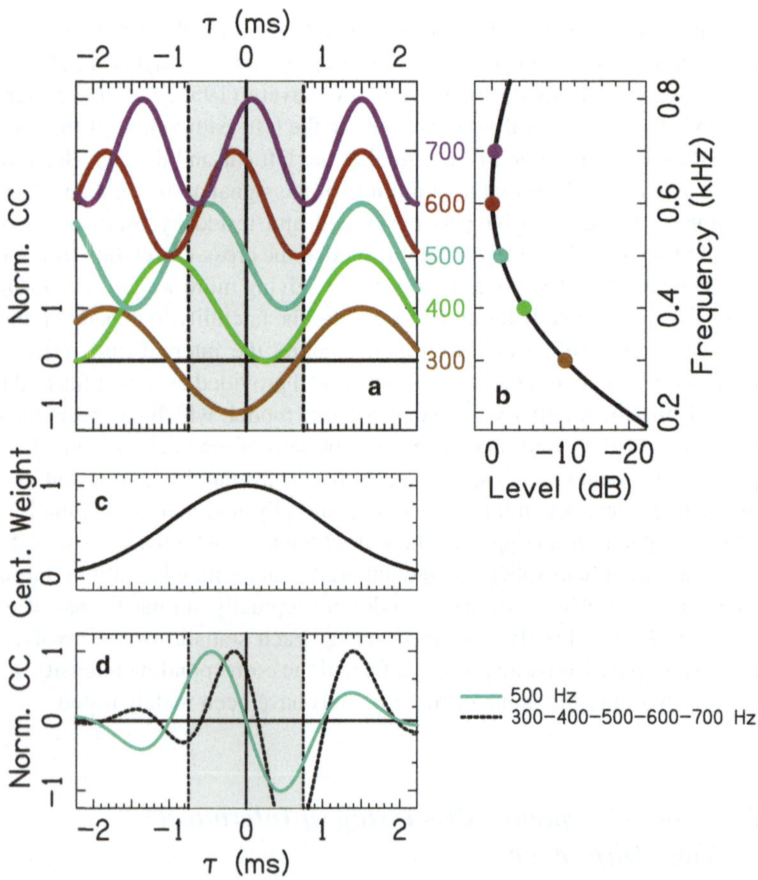

Fig. 7.3 Key components of across-frequency processing in cross-correlation models of lateralization. (**a**) Normalized cross-correlation (Norm. CC) functions for five single frequencies between 300 and 700 Hz. (**b**) An example of a frequency-weighting function. (**c**) An example of centrality weighting (Cent. Weight). (**d**) Resultant cross-correlation function after the application of frequency and centrality weightings for the 500-Hz tone (*blue*) and for the five-tone complex (*black*)

(from 300 to 700 Hz), which can be considered either in isolation or as components of a complex stimulus. The applied ITD in this example is +1.5 ms, as was used in the experiment of Trahiotis and Stern (1989). Considering first the lateralization of an individual tone (or narrowband noise), the ITD is ambiguous within multiples of a period; for example, the 500-Hz tone with an ITD of +1.5 ms (Fig. 7.3a, *blue curve*) produces the same pattern of cross-correlation peaks as the same tone with a delay of −0.5 ms because the period is 2 ms. The tendency to perceive this sound to the side with the smaller delay is attributed to a centrality weighting whereby cross-correlation peaks closer to the center contribute more to the perceived location than more lateral peaks. The exact shape of this weighting function varies in different models, but an example is given in Fig. 7.3c. The effect of applying a centrality weighting on the cross-correlation function for the 500 Hz is depicted in Fig. 7.3d, *blue curve*, which shows a dominant peak on the left side.

For a wider band of frequencies, each with an interaural delay of +1.5 ms, the cross-correlation functions have peaks at +1.5 ms for all frequencies, but the location of the other peaks varies with frequency. Thus, although the 500-Hz component also has a secondary peak at −0.5 ms, the other frequencies have secondary peaks displaced according to their periods. Thus, there is a *straight* line of peaks across frequency at +1.5 ms (Fig. 7.3a, *right side*) as well as a *curved* line of peaks passing through different ITDs (Fig. 7.3a, *left side*). The summed cross-correlation function in this case still has a dominant peak on the left side but also has a prominent peak on the right side (Fig. 7.3d, *black curve*) as a result of the alignment of the cross-correlation peaks. The tendency to perceive the wideband stimulus to the right-hand side has been attributed to an additional weighting based on the straightness of the peaks of the cross-correlation functions on the right side (e.g., Stern et al. 1988; Trahiotis and Stern 1994), which essentially competes with the centrality weighting. We note, however, that Shackleton et al. (1992) proposed that this additional weighting function is not essential and that, with the right choice of parameters, a simple summation across frequency will produce a larger peak on the right side. Note that both of these models incorporate a frequency weighting that applies more weight to the dominant region (around 600 Hz; Fig. 7.3b). Although different implementations of these models have been shown to account for several key features of the empirical data, there has been no extensive quantitative evaluation comparing predicted and perceived extents of laterality.

The basic perceptual experiment of Trahiotis and Stern (1989) has been replicated for tone complexes (Hill and Darwin 1996) and for other low-frequency regions (Yost et al. 2007). It has also been shown that ITD sensitivity is better for stimuli having straighter trajectories in the time-frequency plane (Trahiotis et al. 2001). This latter experiment used bands of noise centered on 500 Hz that contained different combinations of ITD and interaural phase disparities to systematically reduce the straightness (and increase the "curvature" or variability of ITD over frequency). Interestingly, similar effects of bandwidths for large ITDs have also been observed using high-frequency stimuli (Bernstein and Trahioti 2003), where the auditory filters are wide enough that across-channel processing may not need to be involved. Moreover, there is a scarcity of data demonstrating the utility of straightness for natural stimuli containing natural ITDs within the physiological range. Indeed, the concept of straightness and, in particular, its reliance on representations that include large ITDs, have been the source of considerable debate. One competing idea, in which only ITDs within the "π limit" (i.e., ITDs that are no greater than half a cycle for a given frequency) are represented in the auditory system, has been argued for on the grounds of its theoretical plausibility and based on neuroimaging data (Thompson et al. 2006; see also Stern et al. 2019).

7.4.5 Modeling Interaural Level Difference-Based Lateralization

There is no well-accepted model that predicts the across-frequency processing of ILDs for lateralization, perhaps due to the relative scarcity of empirical data that is available for ILDs compared with ITDs. ILDs could be averaged across frequency (Kelvasa and Dietz 2015; Stakhovskaya and Goupell 2017), but it is not clear whether across-frequency weighting occurs prior to averaging. Stakhovskaya and Goupell (2017) proposed such an averaging model to predict the lateralization of multiband stimuli from lateralization estimates measured using single bands. They tested their model on data from bilateral CI users and found that it explained 84% of the variance with no weighting applied, with an improvement to 90% if an optimal across-frequency weighting was applied. Because the latter approach added free parameters to the model, the increase in variance explained is not necessarily strong evidence for across-frequency weighting.

7.4.6 Accounting for Multiple Simultaneous Sources

In modeling of localization and discrimination, most of the work has addressed the problem of single sources, even though stimuli in natural environments often include multiple, simultaneously present sources. When the stimulus is processed as two (or more) sources, the assumption is usually that the same kind of processing that occurs for a single source is applied independently to the components of each of the sources. An example of modeling applied to this case assumes that frequency components are given a binary weighting according to whether or not they are included in the target object (Woods and Colburn 1992). The model was tested on harmonic complexes in which the segregation of one component was promoted by giving it a different onset time to the other components. Results using an ITD discrimination task showed more interference than expected in the condition that promoted segregation as well as marked individual differences. Another study by Hill and Darwin (1996) took a similar theoretical approach using a lateralization paradigm. They also reported residual interference when a target component was mistuned in frequency or when its onset was shifted from the other components. These reports of partial interference are incompatible with the idea that a binary weighting is applied and further suggest that listeners are not optimal in their use of segregation cues to parse acoustic scenes. These efforts demonstrate the difficulty in modeling complex behaviors involving auditory scene analysis.

Across-frequency binaural interference (see Sect. 7.3) represents an extreme case in which listeners fail to segregate simultaneous components. In these situations, it may be that the listener does not have enough evidence to suggest that there are multiple sound sources, and thus energy at different spectral regions is assumed to arise from a single source. Consequently, when listeners are asked to judge the lateral position of one frequency region, binaural information is obligatorily combined across frequency, even when this combination is detrimental to the task at hand. The models that have been used to predict interference are conceptually similar to those that have been used to predict lateralization based on the combination of consistent binaural information across frequency. These combination models typically involve a form of variance-weighted binaural cue integration (e.g., Buell and Hafter 1991; Heller and Richards 2010) and can account for some characteristics of interference such as the frequency asymmetry observed for ITD judgments (Sect. 7.3.2). On the other hand, when interference is measured for the task of ILD discrimination, complex patterns of frequency dependence are observed that are not easily accounted for by a simple weighted combination model (e.g., Goupell and Stakhovskaya 2018a).

7.4.7 Modeling Localization in Complex Environments

Now consider the case of a single source in a room with significant reflections off the walls and other surfaces. For a source in a given position, the waveforms reaching the ears are generated by a combination of waveforms, both direct and reflected, that reach the ears via a variety of paths and therefore differ in their levels and phases (see Zahorik, Chap. 9). For a fixed configuration of source, reflections, and receivers, the transformation from source to each ear can be considered a stable relationship and can be described as a linear time-invariant system. Therefore, the location can be deduced from the received signal when the source signal is known. The processing to deduce the location must use knowledge of the source waveform characteristics as well as knowledge of the room characteristics to accomplish this task. This a priori information about the environment can come from, for example, knowledge of the source, experience in the room, visual scanning of the room, and knowledge of the voice of a speaker in the room.

The general representation of the sound processing that is required to describe these more complicated cases is illustrated in Fig. 7.4 (Colburn and Kulkarni 2005, Fig. 8.4). As shown in Fig. 7.4, the real sources in the real environment generate acoustic input signals that are processed by the peripheral auditory system to extract relevant information about the stimulus, including estimates of ITD and ILD versus

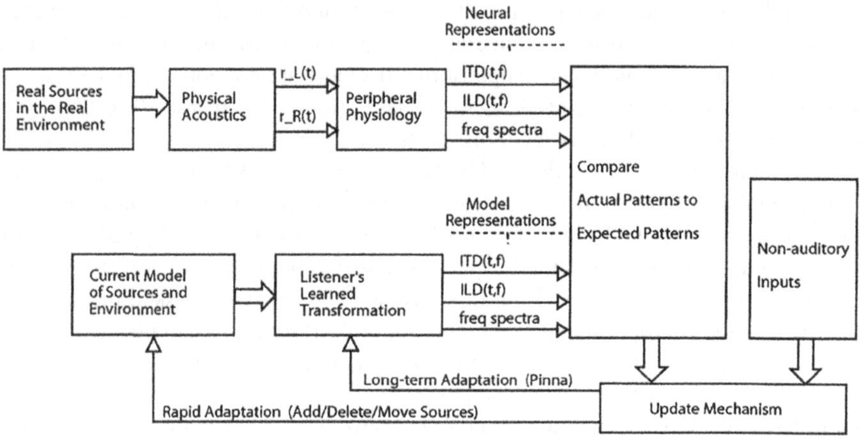

Fig. 7.4 Block diagram of a general model of sound localization. r_L(t) left ear signal, r_R(t) right ear signal, time, and frequency. (Reprinted from Colburn and Kulkarni (2005), with permission)

time and frequency, as represented in brainstem neurons. Note that spectral shape information is also available from the monaural pathways. For judgments about source location, these multiple cues are combined in the central auditory system to suggest stimulus sources with specific locations (Fig. 7.4, *top pathway*). For the simplest cases, such as a single source in an anechoic room, source location can be deduced from these parameters with little confusion. (Note that there is no specification of how across-frequency processing is handled explicitly in the model.) In more general cases, model performance can be conceptualized as the development of an internal model of the sources and the environment, which would also build on a priori information. The predicted neural representation of the model room and model environment (Fig. 7.4, *bottom pathway*) can be compared with the actual neural patterns (presumably an exercise done in the central auditory system, as in cortical levels). Then, any disagreements can be used to refine the model; this is the "update mechanism." Clearly, the process is complex and depends on previous perceptual experiences in different types of rooms and with different kinds and number of sources.

Several detailed models have been proposed that focus specifically on how ITD and ILD are combined across time and frequency in a way that supports accurate sound localization in complex environments (e.g., Faller and Merimaa 2004; Dietz et al. 2011). These models are consistent with the general outline of Fig. 7.4; however, they are also explicit and testable. In general, they include physiologically inspired front-end processing, followed by binaural cue extraction and summation across frequency bands (and time frames). The different models use slightly different approaches to select or weight different bands based on the reliability of binaural information. For example, interaural coherence might be measured in each frequency band and only those bands in which the coherence passes a threshold are used to derive ITD and ILD estimates for that time frame (e.g., Faller and Merimaa

2004). These estimates are then simply averaged over frequency and time to localize the source (or sources) of sound in the environment. It has since been suggested that using the full range of interaural coherence values to weight ITD and ILD estimates from different bands provided better predictions of the empirical data than applying a threshold (Kayser et al. 2015). In another model designed to explain localization in a multitalker mixture (Josupeit et al. 2016), frequency bands and time frames are selected using monaural templates (or "masks") based on a priori knowledge of the target signal. Note that in other related models (e.g., Roman et al. 2003; Mi et al. 2017), the process is essentially reversed, and binaural waveforms in local time-frequency regions are used to segregate mixtures of competing talkers and to estimate speech signals from sources in a given direction.

A more formal and mathematically specific approach to modeling localization is provided by Reijniers et al. (2014). Their approach is based on explicit assumptions about the availability of information from the ITD, ILD, and spectral level, all as functions of frequency. The general decision process is based on the optimum decision theory, which allows the decision to be formulated as making a judgment of the location that makes the observed data most likely. (This approach is also referred to as maximum a posteriori probability decision making.) This method of location estimation explicitly incorporates a priori knowledge about the signal spectrum, acoustic environment, and location likelihood into the decision process. For example, any information that is known or believed to be known about the source location (such as assuming that the source is always in the frontal horizontal plane) is accommodated into the decision. Reijniers et al. (2014) compared model predictions to a meta-analysis of localization data provided by Best et al. (2011) and found good general agreement. As Reijniers et al. point out, however, their computations do not specify explicit neural mechanisms or even the details of how the computations might be done. It seems that their approach could be developed further in a way that could accommodate the available neural data, specific computational details (including those related to across-frequency processing), and specific assumptions about how much a priori information is used in different circumstances. This approach has been used in much simpler situations (Colburn 1973; Daback and Johnson 1993) in ways that integrate neural modeling into optimum process modeling.

7.5 Summary

This chapter summarizes empirical data and modeling work related to the across-frequency processing of binaural information. Experiments using relatively simple stimuli have demonstrated how ITD and ILD information are combined across frequency for single sources that cover multiple critical bands, where physically constant and/or physiologically consistent binaural information is available. These data have been addressed using relatively simple models of localization and discrimination. Fewer empirical data are available for the case of multiple-source processing. In this more general case, the processing has to include an allocation of binaural information to individual source representations. The data and modeling suggest that even in simple cases this allocation is not always done optimally, and in more complex situations, performance is more dependent on a priori information.

The picture is far from complete and several areas can be identified where more research is needed.

- There appears to be a gap in the literature between the studies that have examined across-channel processing for simple, well-controlled stimuli and modeling work that attempts to predict localization of competing speech sounds in noisy and reverberant environments. For example, at the time of writing, no data were available on the across-frequency processing of binaural information for speech stimuli presented in quiet or in noise, considering both what information is available and how listeners weight that information.
- There are many interesting questions that could be asked that relate to contextual effects in across-frequency processing of binaural information. For example, how does a priori information about the number and identity of competing sounds in the environment affect the allocation of different frequency bands to different perceptual objects? In addition, how do a priori knowledge and assumptions affect the weighting of binaural information in different bands?
- The across-frequency weighting of ITD and ILD, although addressed in some studies, remains to be fully characterized. Studies that pit the two cues against each other in different frequency bands would be valuable. Interactions between ITD and ILD are also not yet satisfactorily addressed in lateralization models.
- It is generally assumed that coherent information is available at the two ears within a channel, and the critical role of binaural fusion is often overlooked. New experiments that consider binaural fusion together with lateralization or localization are needed, and modeling efforts in these two areas could be integrated. One motivation for this direction is that binaural fusion may differ in listeners with hearing loss and particularly in CI users (see Ricketts and Kan, Chap. 13).
- Although not covered in this chapter, many of the empirical studies that were reviewed here reported marked individual differences (e.g., in weighting patterns, susceptibility to across-frequency binaural interference). The sources of this variation are not at all clear, and more focused experiments on this issue are needed to further our understanding. Moreover, most of the models discussed in this chapter would need significant modifications to include listener-specific factors that could account for individual variability.

Acknowledgments We appreciate the helpful input provided by Mathieu Lavandier. This work was supported by Grants R01-DC-000100, R01-DC-014948, and R01-DC-015760 from the National Institute on Deafness and Other Communication Disorders, National Institutes of Health.

Compliance with Ethics Requirements

Virginia Best declares that she has no conflict of interest.
Matthew J. Goupell declares that he has no conflict of interest.
H. Steven Colburn declares that he has no conflict of interest.

References

Ahrens A, Joshi SN, Epp B (2015) Spectral weighting of binaural cues: effect of bandwidth and stream segregation. Poster presented at the Association for Research in Otolaryngology MidWinter Meeting, Baltimore, 21–25 Feb 2015

Baumgartner R, Majdak P, Laback B (2014) Modeling sound-source localization in sagittal planes for human listeners. J Acoust Soc Am 136(2):791–802

Benichoux V, Rébillat M, Brette R (2016) On the variation of interaural time differences with frequency. J Acoust Soc Am 139:1810–1821

Bernstein LR, Trahioti C (2003) Enhancing interaural-delay-based extents of laterality at high frequencies by using "transposed stimuli". J Acoust Soc Am 113(6):3335–3347

Bernstein LR, Trahiotis C (1994) Detection of interaural delay in high-frequency sinusoidally amplitude-modulated tones, two-tone complexes, and bands of noise. J Acoust Soc Am 95(6):3561–3567

Bernstein LR, Trahiotis C (1995) Binaural interference effects measured with masking-level difference and with ITD- and IID-discrimination paradigms. J Acoust Soc Am 98(1):155–163

Bernstein LR, Trahiotis C (2002) Enhancing sensitivity to interaural delays at high frequencies by using "transposed stimuli". J Acoust Soc Am 112(3 Pt 1):1026–1036

Bernstein LR, Trahiotis C (2004) The apparent immunity of high-frequency "transposed" stimuli to low-frequency binaural interference. J Acoust Soc Am 116(5):3062–3069

Bernstein LR, Trahiotis C (2005) Measures of extents of laterality for high-frequency "transposed" stimuli under conditions of binaural interference. J Acoust Soc Am 118(3):1626–1635

Best V, Gallun FJ, Shinn-Cunningham BG, Carlile S (2007) Binaural interference and auditory grouping. J Acoust Soc Am 121:1070–1076

Best V, Brungart D, Carlile S, Jin C, Macpherson E, Martin R, McAnally K, Sabin A, Simpson B (2011) A meta-analysis of localisation errors made in the anechoic free field. In: Suzuki Y, Brungart D, Iwaya Y, Iida K, Cabrera D, Kato H (eds) Principles and applications of spatial hearing. World Scientific, Singapore, pp 14–23

Bibee JM, Stecker GC (2016) Spectrotemporal weighting of binaural cues: effects of a diotic interferer on discrimination of dynamic interaural differences. J Acoust Soc Am 140(4):2584–2592

Bilsen FA, Raatgever J (1973) Spectral dominance in binaural lateralization. Acust Acta Acust 28:131–132

Brown CA, Yost WA (2015) Spectral overlap and interaural time difference sensitivity: possible role of binaural interference. J Acoust Soc Am 137(5):EL374–EL380

Brughera A, Dunai L, Hartmann WM (2013) Human interaural time difference thresholds for sine tones: the high-frequency limit. J Acoust Soc Am 133(5):2839–2855

Buchholz JM, Le Goff N, Dau T (2018) Localization of broadband sounds carrying interaural time differences: effects of frequency, reference location, and interaural coherence. J Acoust Soc Am 144(4):2225–2237

Buell TN, Hafter ER (1991) Combination of binaural information across frequency bands. J Acoust Soc Am 90:1894–1900

Buell TN, Trahiotis C (1993) Interaural temporal discrimination using two sinusoidally amplitude-modulated, high-frequency tones: conditions of summation and interference. J Acoust Soc Am 93(1):480–487

Buell TN, Trahiotis C (1994) Detection of interaural delay in bands of noise: effects of spectral interference combined with spectral uncertainty. J Acoust Soc Am 95:3568–3573

Camalier CR, Grantham DW, Bernstein LR (2014) Binaural interference: effects of temporal interferer fringe and interstimulus interval. J Acoust Soc Am 135:789–795

Cariani PA, Delgutte B (1996) Neural correlates of the pitch of complex tones. II. Pitch shift, pitch ambiguity, phase invariance, pitch circularity, rate pitch, and the dominance region for pitch. J Neurophysiol 76(3):1717–1734

Cherry EC, Sayers BM (1956) "Human 'Cross-Correlator'" – a technique for measuring certain parameters of speech perception. J Acoust Soc Am 28:888–895

Colburn HS (1973) Theory of binaural interaction based on auditory-nerve data. I. General strategy and preliminary results on interaural discrimination. J Acoust Soc Am 54(6):1458–1470
Colburn HS (1977) Theory of binaural interaction based on auditory-nerve data. II. Detection of tones in noise. J Acoust Soc Am 61(2):525–533
Colburn HS, Kulkarni A (2005) Models of sound localization. In: Popper A, Fay R (eds) Sound source localization, Springer handbook on auditory research. Springer, New York, pp 272–316
Constan ZA, Hartmann WM (2003) On the detection of dispersion in the head-related transfer function. J Acoust Soc Am 114(2):998–1008
Croghan NB, Grantham DW (2010) Binaural interference in the free field. J Acoust Soc Am 127(5):3085–3091
Daback AG, Johnson DH (1993) Function-based modeling of binaural processing: level and time cues. J Acoust Soc Am 94(5):2604–2616
Dai H (2000) On the relative influence of individual harmonics on pitch judgment. J Acoust Soc Am 107:953–959
Dietz M, Ewert SD, Hohmann V (2011) Auditory model based direction estimation of concurrent speakers from binaural signals. Speech Commun 53(5):592–605
Dye RH (1990) The combination of interaural information across frequencies: lateralization on the basis of interaural delay. J Acoust Soc Am 88(5):2159–2170
Dye RH, Stellmack MA, Grange AN, Yost WA (1996) The effect of distractor frequency on judgements of laterality based on interaural delays. J Acoust Soc Am 99:1096–1107
Egger K, Majdak P, Laback B (2016) Channel interaction and current level affect across-electrode integration of interaural time differences in bilateral cochlear-implant listeners. J Assoc Res Otolaryng 17(1):55–67
Faller C, Merimaa J (2004) Source localization in complex listening situations: selection of binaural cues based on interaural coherence. J Acoust Soc Am 116(5):3075–3089
Fischer BJ, Peña JL (2009) Bilateral matching of frequency tuning in neural cross-correlators of the owl. Biol Cybern 100:521–531
Francart T, Wouters J (2007) Perception of across-frequency interaural level differences. J Acoust Soc Am 122(5):2826–2831
Francart T, Lenssen A, Büchner A, Lenarz T, Wouters J (2015) Effect of channel envelope synchrony on interaural time difference sensitivity in bilateral cochlear implant listeners. Ear Hear 36(4):e199–e206
Gaskell H, Henning GB (1981) The effect of noise on time-intensity trading in lateralization. Hear Res 4(2):161–174
Goupell MJ, Stakhovskaya OA (2018a) Across-channel interaural-level-difference processing demonstrates frequency dependence. J Acoust Soc Am 143(2):645–658
Goupell MJ, Stakhovskaya OA (2018b) Across-frequency processing of interaural time and level differences in perceived lateralization. Acust Acta Acust 104:758–761
Goupell MJ, Stoelb C, Kan A, Litovsky RY (2013) Effect of mismatched place-of-stimulation on the salience of binaural cues in conditions that simulate bilateral cochlear-implant listening. J Acoust Soc Am 133(4):2272–2287
Harper NS, McAlpine D (2004) Optimal neural population coding of an auditory spatial cue. Nature 430:682–686
Harper NS, Scott BH, Semple MN, McAlpine D (2014) The neural code for auditory space depends on sound frequency and head size in an optimal manner. PLoS One 9(11):e108154
Harris GG (1960) Binaural interactions of impulsive stimuli and pure tones. J Acoust Soc Am 32(6):685–692
Hartmann WM, Constan ZA (2002) Interaural level differences and the level-meter model. J Acoust Soc Am 112(3 Pt 1):1037–1045
Hartmann WM, Rakerd B, Crawford ZD, Zhang PX (2016) Transaural experiments and a revised duplex theory for the localization of low-frequency tones. J Acoust Soc Am 139(2):968–985
Heller LM, Richards VM (2010) Binaural interference in lateralization thresholds for interaural time and level differences. J Acoust Soc Am 128(1):310–319

Heller LM, Trahiotis C (1995) Interference in detection of interaural delay in a sinusoidally amplitude-modulated tone produced by a second, spectrally remote sinusoidally amplitude-modulated tone. J Acoust Soc Am 97(3):1808–1816

Heller LM, Trahiotis C (1996) Extents of laterality and binaural interference effects. J Acoust Soc Am 99(6):3632–3637

Henning GB (1980) Some observations on the lateralization of complex waveforms. J Acoust Soc Am 68:446–454

Hill NI, Darwin CJ (1996) Lateralization of a perturbed harmonic: effects of onset asynchrony and mistuning. J Acoust Soc Am 100(4):2352–2364

Jeffress LA (1948) A place theory of sound localization. J Comp Physiol Psychol 41:35–30

Jeffress LA (1972) Binaural signal detection. In: Tobias JV (ed) Foundations of modern auditory theory, vol 2. Academic, New York, pp 349–368

Josupeit A, Kopčo N, Hohmann V (2016) Modeling of speech localization in a multi-talker mixture using periodicity and energy-based auditory features. J Acoust Soc Am 139(5):2911–2923

Kan A, Stoelb C, Litovsky RY, Goupell MJ (2013) Effect of mismatched place-of-stimulation on binaural fusion and lateralization in bilateral cochlear-implant users. J Acoust Soc Am 134(4):2923–2936

Kan A, Jones HG, Litovsky RY (2015) Effect of multi-electrode configuration on sensitivity to interaural timing differences in bilateral cochlear-implant users. J Acoust Soc Am 138(6):3826–3833

Kan A, Jones HG, Litovsky RY (2016) Lateralization of interaural timing differences with multi-electrode stimulation in bilateral cochlear-implant users. J Acoust Soc Am 140(5):EL392–EL398

Kayser H, Hohman V, Ewert SD, Kollmeier B, Anemüller J (2015) Robust auditory localization using probabilistic inference and coherence-based weighting of interaural cues. J Acoust Soc Am 138(5):2635–2648

Kelvasa D, Dietz M (2015) Auditory model-based sound direction estimation with bilateral cochlear implants. Trends Hear 19. https://doi.org/10.1177/2331216515616378

Klumpp RG, Eady HR (1956) Some measurements of interaural time difference thresholds. J Acoust Soc Am 28(5):859–860

Kollmeier B, Holube I (1992) Auditory filter bandwidths in binaural and monaural listening conditions. J Acoust Soc Am 94(4 Pt 1):1889–1901

Kuhn G (1977) Model for interaural time difference in azimuthal plane. J Acoust Soc Am 62:157–167

Laback B, Pok SM, Baumgartner WD, Deutsch WA, Schmid K (2004) Sensitivity to interaural level and envelope time differences of two bilateral cochlear implant listeners using clinical sound processors. Ear Hear 25(5):488–500

Macaulay EJ, Hartmann WM, Rakerd B (2010) The acoustical bright spot and mislocalization of tones by human listeners. J Acoust Soc Am 127(3):1440–1449

Macpherson EA, Middlebrooks JC (2002) Listener weighting of cues for lateral angle: the duplex theory of sound localization revisited. J Acoust Soc Am 111(5 Pt 1):2219–2236

McAlpine D (2005) Creating a sense of auditory space. J Physiol 566:21–28

McFadden D (1981) The problem of different interaural time differences at different frequencies. J Acoust Soc Am 69(6):1836–1837

McFadden D, Pasanen EG (1976) Lateralization at high frequencies based on interaural time differences. J Acoust Soc Am 59(3):634–639

Mi J, Groll M, Colburn HS (2017) Comparison of a target-equalization-cancellation approach and a localization approach to source separation. J Acoust Soc Am 142(5):2933–2941

Middlebrooks JC (1992) Narrow-band sound localization related to external ear acoustics. J Acoust Soc Am 92(5):2607–2624

Monaghan JJ, Bleeck S, McAlpine D (2015) Sensitivity to envelope interaural time differences at high modulation rates. Trends Hear 19. https://doi.org/10.1177/2331216515619331

Moore BCJ, Glasberg BR (1983) Suggested formulae for calculating auditory-filter bandwidths and excitation patterns. J Acoust Soc Am 74:750–753

Oh Y, Reiss LA (2017) Binaural pitch fusion: pitch averaging and dominance in hearing-impaired listeners with broad fusion. J Acoust Soc Am 142(2):780–791

Peña JL, Konishi M (2000) Cellular mechanisms for resolving phase ambiguity in the owl's inferior colliculus. Proc Natl Acad Sci U S A 97:11787–11792

Raatgever J (1980) On the binaural processing of stimuli with different interaural phase relations. Delft University of Technology, Dissertation

Rakerd B, Hartmann WM (2010) Localization of sound in rooms. V. Binaural coherence and human sensitivity to interaural time differences in noise. J Acoust Soc Am 128(5):3052–3063

Strutt JW (1907) On our perception of sound direction. Philos Mag 13:214–232

Reijniers J, Vanderelst D, Jin C, Carlile S, Peremans H (2014) An ideal-observer model of human sound localization. Biol Cybern 108:169–181

Reiss LA, Shayman CS, Walker EP, Bennett KO, Fowler JR, Hartling CL, Glickman B, Lasarev MR, Oh Y (2017) Binaural pitch fusion: comparison of normal-hearing and hearing-impaired listeners. J Acoust Soc Am 141(3):1909–1920

Roman N, Wang D, Brown GJ (2003) Speech segregation based on sound localization. J Acoust Soc Am 114(1):2236–2252

Saberi K (1995) Lateralization of comodulated complex waveforms. J Acoust Soc Am 98(6):3146–3156

Sayers BM, Cherry EC (1957) Mechanism of binaural fusion in the hearing of speech. J Acoust Soc Am 29:973–987

Scharf B, Florentine M, Meiselman CH (1976) Critical band in auditory lateralization. Sens Processes 1(2):109–126

Schiano JL, Trahiotis C, Bernstein LR (1986) Lateralization of low-frequency tones and narrow bands of noise. J Acoust Soc Am 79(5):1563–1570

Shackleton TM, Meddis R, Hewitt MJ (1992) Across frequency integration in a model of lateralization. J Acoust Soc Am 91(4):2276–2279

Shinn-Cunningham BG, Kopčo N, Martin T (2005) Localizing nearby sound sources in a classroom: binaural room impulse responses. J Acoust Soc Am 117:3100–3115

Stakhovskaya OA, Goupell MJ (2017) Lateralization of interaural level differences with multiple electrode stimulation in bilateral cochlear-implant listeners. Ear Hear 38(1):e22–e38

Stern RM, Colburn HS (1978) Theory of binaural interaction based on auditory-nerve data. IV. A model for subjective lateral position. J Acoust Soc Am 64(1):127–140

Stern RM, Shear GD (1996) Lateralization and detection of low-frequency binaural stimuli: effects of distribution of internal delay. J Acoust Soc Am 100(4 Pt 1):2278–2288

Stern RM, Trahiotis C (1996) Models of binaural perception. In: Gilkey R, Anderson TR (eds) Binaural and spatial hearing in real and virtual environments. Lawrence Erlbaum Associates, New York, pp 499–531

Stern RM, Zeiberg AS, Trahiotis C (1988) Lateralization of complex binaural stimuli: a weighted-image model. J Acoust Soc Am 84(1):156–165

Stern RM, Colburn HS, Bernstein LR, Trahiotis C (2019) The fMRI data of Thompson et al. (2006) do not constrain how the human midbrain represents interaural time delay. J Assoc Res Otolaryng 20(4):305–311

Thavam S, Dietz M (2019) Smallest perceivable interaural time differences. J Acoust Soc Am 145:458–468

Thompson SK, von Kriegstein K, Deane-Pratt A, Marquardt T, Deichmann R, Griffiths TD, McAlpine D (2006) Representation of interaural time delay in the human auditory midbrain. Nat Neurosci 9:1096–1098

Trahiotis C, Bernstein LR (1986) Lateralization of bands of noise and sinusoidally amplitude-modulated tones: effects of spectral locus and bandwidth. J Acoust Soc Am 79(6):1950–1957

Trahiotis C, Stern RM (1989) Lateralization of bands of noise: effects of bandwidth and differences of interaural time and phase. J Acoust Soc Am 86(5):1285–1293

Trahiotis C, Stern RM (1994) Across-frequency interaction in lateralization of complex binaural stimuli. J Acoust Soc Am 96:3804–3806

Trahiotis C, Bernstein LR, Akeroyd MA (2001) Manipulating the "straightness" and "curvature" of patterns of interaural cross correlation affects listeners' sensitivity to changes in interaural delay. J Acoust Soc Am 109(1):321–330
Verschooten E, Shamma S, Oxenham AJ, Moore BCJ, Joris PX, Heinz MG, Plack CJ (2019) The upper frequency limit for the use of phase locking to code temporal fine structure in humans: a compilation of viewpoints. Hear Res 377:109–121
Wightman FL, Kistler DJ (1992) The dominant role of low-frequency interaural time differences in sound localization. J Acoust Soc Am 91(3):1648–1661
Woods WS, Colburn HS (1992) Test of a model of auditory object formation using intensity and interaural time difference discrimination. J Acoust Soc Am 91:2894–2902
Yost WA, Zhong X (2014) Sound source localization identification accuracy: bandwidth dependencies. J Acoust Soc Am 136(5):2737–2746
Yost WA, Dye RH, Sheft S (2007) Interaural time difference processing of broadband and narrowband noise by inexperienced listeners. J Acoust Soc Am 121(3):EL103–EL109
Zhou J, Durrant JD (2003) Effects of interaural frequency difference on binaural fusion evidenced by electrophysiological versus psychoacoustical measures. J Acoust Soc Am 114:1508–1515

Chapter 8
Binaural Unmasking and Spatial Release from Masking

John F. Culling and Mathieu Lavandier

8.1 Binaural Unmasking

Binaural unmasking (BU) is a phenomenon in which differences in the sound arriving at the two ears, *interaural* differences, can assist in the detection or identification of sound in noise. In general, the threshold for detecting or identifying a signal in background noise is lower (better) if the interaural parameters of the signal differ from those of the noise. These parameters include those of interaural time difference (ITD), interaural phase difference (IPD) or, in certain cases, interaural level difference (ILD). They correspond with the interaural cues that may be generated by sounds that differ in source direction. Away from the midline, the ear on the side toward the sound source will receive a waveform that arrives earlier and is more advanced in phase than that in the opposite ear, due to a time delay in traveling between the ears. It will also receive a waveform that is higher in sound level, due to the acoustic obstruction of the head. In consequence, it is thought that BU contributes to the human ability to hear sounds in background noise when the target and masking sounds are spatially separated, called "spatial release from masking" (SRM; see Sect. 8.2).

The distinction between BU and SRM is that in BU, improvements in the detection/identification threshold occur without any improvement in the signal-to-noise ratio (SNR). In SRM, improvements in the SNR at one ear can also contribute to the overall effect. For instance, introducing an ITD to the tone by delaying it at one ear will produce unmasking, even though the SNR is unchanged at either ear, and this

J. F. Culling (✉)
School of Psychology, Cardiff University, Cardiff, Wales, UK
e-mail: CullingJ@cardiff.ac.uk

M. Lavandier
Laboratoire Génie Civil et Bâtiment, École Nationale des Travaux Publics de l'État, Université de Lyon, Vaulx-en-Velin Cedex, France
e-mail: mathieu.lavandier@entpe.fr

© Springer Nature Switzerland AG 2021
R. Y. Litovsky et al. (eds.), *Binaural Hearing*, Springer Handbook of Auditory Research 73, https://doi.org/10.1007/978-3-030-57100-9_8

will be pure BU. In contrast, moving either the speech or the noise source away from the other, such that it comes from a different direction, will improve the SNR at one ear, mainly because the head will block sound from the masking noise and will produce SRM, which may also include an element of BU in addition to the improved SNR.

In the earliest known demonstration of BU, Hirsh (1948) found that detection thresholds for pure tones in noise were lower when the tones had an IPD of 180 ° (π radians; see Table 8.1), whereas the noise had zero IPD. To create an IPD of π radians, the waveform was simply inverted at one ear. In spatial terms, a tonal signal with a π radian IPD is equivalent to one that has an ITD equal to half the period of the tone, whereas a noise with zero IPD is like a noise coming from somewhere on the midline.

However, in many cases, the stimuli used in these headphone experiments are not "ecologically valid," meaning that they cannot be produced by any spatial arrangement of sound sources around the listener. If the speed of sound is 340 m/s and the distance between the ears is 0.2 m, the largest possible ITD should be comparable with the time it takes for sound to travel that distance in free space, or about 590 μs. Direct measurements with a head between the ears give values extending up to 800 μs (Kuhn 1977; see also Hartmann, Chap. 2). In Hirsh's (1948) experiment, this means that only tones of 625 Hz and higher would have produced ecologically valid ITDs for which the π radian IPD is equivalent to an ITD below 800 μs (half the period of 625 Hz is 800 μs). At lower frequencies, ITDs exceeded 800 μs and so were too large for the separation of human ears to have created them. The fact that these experiments can produce large effects, sometimes larger than those that can be produced by ecologically valid stimuli, suggests that the experiments are tapping an underlying mechanism that does not depend on the interpretation of interaural differences in terms of specific spatial percepts. The key factor is that tone and noise have different interaural parameters, so if both have the same IPD, ITD, or ILD, the thresholds are high; otherwise, the threshold is lower. This improvement in threshold is called the binaural masking level difference (BMLD). The largest BMLDs for tones in broadband noise are around 15 dB (Hirsh 1948).

Table 8.1 List of symbols

Symbols	Meaning	Interaural correlation, ρ
0	No interaural difference	$\rho = 1$
π	π-Radian phase difference	$\rho = -1$
τ	Interaural time delay	
ρ	Controlled interaural noise correlation	$-1 \leq \rho \leq 1$
u	Uncorrelated noise	$\rho = 0$
m	Monaural (i.e., an infinite ILD)	
φ	Interaural phase difference	

These are commonly used and their meanings for use in representing features of binaural conditions. Conditions are labeled using the form N_xS_y, where *x* and *y* can be one of the listed symbols. The interaural correlation, particularly that of the noise is often also of interest, so these values are given where appropriate

8.1.1 Basic Phenomena and Terminology

A standard nomenclature has been adopted by most authors to represent different BU conditions. Using this nomenclature, the case described in Sect. 8.1 is called N_0S_π. Because identical interaural parameters generally result in no unmasking, Hirsh (1948) found that N_0S_0, $N_\pi S_\pi$, and N_mS_m all had about the same threshold, whereas N_0S_π, $N_\pi S_0$, and N_0S_m all had substantially lower thresholds. The only clear exception to this pattern is that N_0S_0 gives lower thresholds than N_mS_m in the specific case of a low-frequency tone, such as 150 Hz, in noise at a low sound level, such as 20 dB/Hz (Dolan 1968). Much early work on BU explored parametric variations in the interaural parameters for broadband noise maskers as well as in the signal frequency (Fig. 8.1).

Hirsh and Burgeat (1958) examined the effect of signal frequency on the BMLDs for N_0S_π, and $N_\pi S_0$ (Fig. 8.1a). They found that the BMLD for N_0S_π was about 15 dB at the lowest frequencies tested and reduced, with increasing signal frequency, asymptoting to 2–3 dB for frequencies above about 1.5 kHz. The situation for $N_\pi S_0$, while following a similar overall pattern, was slightly different, however. BMLDs were always smaller and this deficit compared to N_0S_π became larger at frequencies below 500 Hz. The effect of frequency on the BMLD with narrowband maskers is different again and is addressed in Sect. 8.1.3.

Langford and Jeffress (1964) investigated the effect of ITDs ($N_\tau S_0$ and $N_\tau S_\pi$) or a 500-Hz tone using noise delays of up to 9 ms (Fig. 8.1b), well beyond the ecological range of ±700 μs. They found that BMLDs followed a cyclic pattern as a function of noise delay. The BMLD for $N_\tau S_\pi$ was high at zero delay, dropped to zero for $\tau = 0.5$, 1.5, 2.5 ms, etc. but returned to positive values at 1, 2 ms, etc. This 2-ms cycle corresponds to the period of the signal. As delay increased, however, each cycle was reduced in magnitude. The BMLD for $N_\tau S_0$ followed a reciprocal cycle, with a similar decline in magnitude with increasing delay.

Robinson and Jeffress (1963) investigated the effect of interaural correlation of the noise ($N_\rho S_0$ and $N_\rho S_\pi$ for 500-Hz tones (Fig. 8.1c). They found that the BMLD was maximal in $N_\rho S_\pi$ for $\rho = 1$ and in $N_\rho S_0$ for $\rho = -1$ (equivalent to $N_\rho S_\pi$ and $N_\rho S_0$ respectively; see Table 8.1). Away from these optimal cases, BMLD reduced steeply but only gradually approached zero, such that a BMLD of about 2 dB was still observed for $\rho = 0$ (equivalent to $N_\mu S_0$ and $N_u S_\pi$).

Rabiner et al. (1966) investigated IPDs applied to the noise ($N_\varphi S_0$) with pure tones of 200 and 400 Hz (Fig. 8.1d). The BMLDs increased steeply with small differences in the IPD and more slowly as the BMLD approached its maximum value at π radians but showed no sign of decline as the ecological range of delays was exceeded.

These four early studies represent a fairly comprehensive summary of the parametric influences on the BMLD in broadband noise. It should be noted, however, that where the same conditions were examined in more than one of these studies, there is considerable inconsistency in the levels of BMLD reported. Specifically,

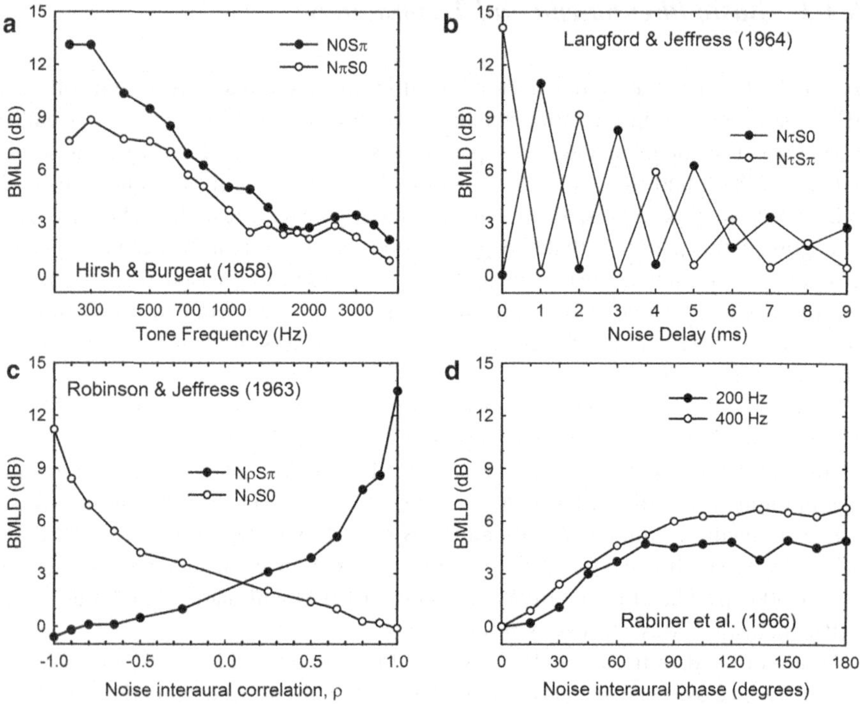

Fig. 8.1 Results from four classic binaural masking level difference (BMLD) experiments. (**a**) Hirsh and Burgeat (1958) measured N_0S_π and $N_\pi S_0$ as a function of signal frequency. (**b**) Langford and Jeffress (1964) measured $N_\tau S_0$ and $N_\tau S_\pi$ as a function of the interaural time difference (τ) of the noise for a 500-Hz signal. (**c**) Robinson and Jeffress (1963) measured $N_\rho S_0$ and $N_\rho S_\pi$ as a function of interaural noise correlation (ρ) for a 500-Hz signal. (**d**) Rabiner et al. (1966) measured $N_\varphi S_\pi$ as a function of the interaural phase difference (φ) of the noise

N_0S_π at 500 Hz is reported in three of them but is 9.5 dB in Hirsh and Burgeat (1958), 12.8 dB in Langford and Jeffress (1964), and 13.5 dB in Robinson and Jeffress (1963). Moreover, although Hirsh and Burgeat's (1958) BMLD was the lowest of those three, Rabiner et al. (1966) reported the BMLD for $N_\pi S_0$ at 400 Hz as 6.8 dB, which is even lower than the 7.7 dB reported by Hirsh and Burgeat (1958). These discrepancies probably just reflect differences between the participants, the experimental procedure, and the overall sound level as well as some degree of measurement error, but they still make it difficult to fit a model to all these basic datasets simultaneously.

8.1.2 Underlying Cues

Because the cochlea analyzes incoming sound into different frequency channels, Webster (1951) pointed out that the BU process does not need to extract the signal from the full spectrum of noise that may be present in the stimulus. Internally, a tonal signal will be masked only by the narrow band of noise that resonated at the same point in the cochlea. By considering only this narrow band, the stimulus can be analyzed using vector theory (Jeffress 1972).

The narrow band of noise that masks the signal within a frequency channel resembles an amplitude- and frequency-modulated sinusoid. When combined with the signal, a vector summation occurs at each ear (Fig. 8.2). In conditions that produce unmasking, such as N_0S_π, the vector sums differ between the ears, producing rapidly fluctuating IPDs and ILDs. One can calculate the range of vector sums that might occur at threshold for N_0S_π and thus the ILDs and IPDs that occur. Webster (1951) hypothesized that fluctuating IPDs were the dominant cue, but it proved unclear from vector theory whether one cue or the other might be more detectable at low frequencies (Jeffress 1972).

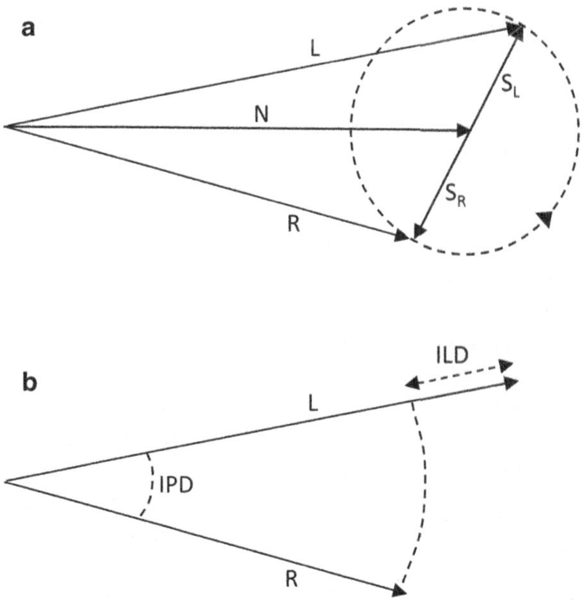

Fig. 8.2 Illustration of vector theory. (**a**) Vector summation of noise (N) and signal (S) at one moment in an N_0S_π stimulus. The noise constantly varies in amplitude and phase, but the phase is identical at the two ears and is neglected by drawing the noise vector horizontally. The noise amplitude (vector ***N*** length) changes randomly over time. The signal is fixed in amplitude so the signal vectors (S_R and S_L) are fixed in length and describe a circle (*dashed line*) over time. SR and SL are in opposite phase and so on opposites sides of the circle. The resultant vectors (L and R) represent the stimulus received at each ear. (**b**) Instantaneous interaural level difference (ILD) and IPD that would be produced by these resultant vectors

However, vector theory is revealing in other ways. Introducing a fixed ILD to an N_0S_0 tone by reducing its level at one ear will produce unmasking, even though the SNR is reduced at one ear and unchanged at the other (e.g., in N_0S_m, the ILD is infinite). Vector theory shows that even when a fixed ILD has been deliberately introduced to the signal, the vectors at the ears still produce fluctuating ILDs and IPDs. This can be seen from Fig. 8.2. For N_0S_m, the vector for the ear with no signal will simply be the vector $\boldsymbol{N}$ in Fig. 8.2a. The vector for the other ear will still rotate around the end of this vector, creating vectors for each ear that differ in both amplitude and phase.

Detection of IPDs depends on the temporal correspondence ("phase locking") of action potentials on the auditory nerve to individual pressure cycles (the "fine structure") of the stimulating waveform. At high frequencies, it is known that the auditory system is insensitive to IPDs for tonal stimuli due to a breakdown of phase locking to the fine structure, which could explain why there is a corresponding decline in the BMLD with increasing frequency. However, action potentials are still elicited by rapid temporal fluctuations in sound level in high-frequency channels. Consequently, Durlach (1964) proposed that fluctuating ILDs, encoded by phase locking to the different temporal fluctuations at each ear, probably underpin the smaller BMLD that is observed above 1500 Hz.

Empirical support for this suggestion took some decades to arrive because it is difficult to present the IPD and ILD cues separately. A Hilbert transform can be used to cleanly separate the phase and amplitude fluctuations from an N_0S_π stimulus for purposes of analysis, but the method does not facilitate manipulation of the cues; resynthesis of the waveform, with one or the other cue removed, invariably results in clearly audible artifacts. The resulting waveforms are consequently unusable as experimental stimuli.

The dominance of IPD cues at low frequencies was eventually established empirically by van der Heijden and Joris (2010). They used a technique for applying quasi-frequency modulation (QFM) to any waveform. Like conventional frequency modulation, if QFM is applied out-of-phase between the ears, the IPDs of individual frequency components will modulate. Similarly, interaurally out of phase amplitude modulation (AM) will produce ILD modulation. van der Heijden and Joris applied QFM or AM interaurally out of phase to all the stimuli (both with and without embedded signals) in a BU experiment. Because thresholds for N_0S_π were elevated only when the QFM was out of phase, they inferred that the IPD was the dominant cue; the resulting IPD modulation of the whole stimulus had obscured IPD modulations introduced by the signal.

Van der Heijden and Joris (2010) did not repeat the test at high frequencies. Had they done so, the outcome might have been reversed, with only ILD fluctuations elevating the thresholds. Culling (2011) found such a reversal but in a different way. He used AM and QFM to *simulate* the individual unmasking cues rather than to obscure them. The AM and QFM were applied to a narrow band of noise that represented the signal and masker mixture in the frequency band of the signal. Flanking bands of diotic noise were then added to represent the rest of the spectrum in a

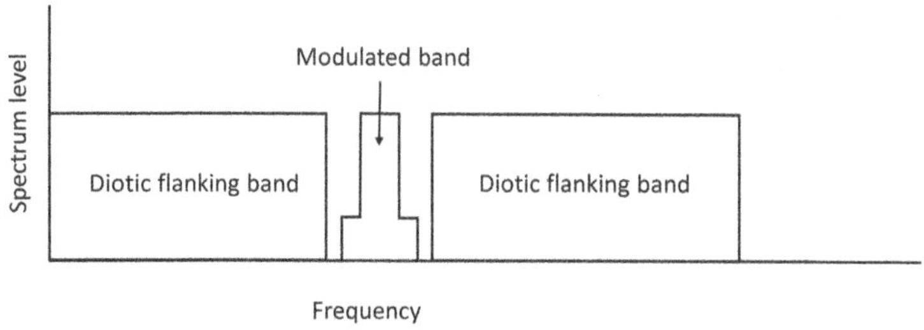

Fig. 8.3 Schematic illustration of the stimuli used by Culling (2011) to simulate a broadband NOS_π stimulus. Notches in the spectrum around the modulated band prevented the flanking bands from directly interfering with the modulation sidebands (shoulders on the modulated band)

broadband BU stimulus (Fig. 8.3). The modulation threshold was measured. Up to 750 Hz, QFM thresholds were lowest, indicating that listeners were most sensitive to IPD modulations. This result corroborated van der Heijden and Joris (2010) in emphasizing IPDs at low frequencies. Consistent with the progressive loss of phase locking with increasing frequency, QFM thresholds increased steeply with frequency. By contrast, AM thresholds (ILD fluctuations) were somewhat higher than QFM thresholds at low frequencies but scarcely increased with frequency, becoming lower than the QFM thresholds for frequencies of 1250 and 1500 Hz. These findings support the hypotheses advanced by Webster (1951) and Durlach (1964) that fluctuations in the IPD are dominant at low frequencies and fluctuations in the ILD are dominant at high frequencies. Moreover, consistent with the smaller size of the BMLD at high frequencies, the threshold for AM at high frequencies was higher than the threshold for QFM at low frequencies.

Although fluctuations in the interaural parameters of the stimulus are clearly involved, the mechanism by which they are detected is a separate issue. ILDs and ITDs are cues that listeners use to judge sound location, but it is unlikely that the listeners hear the stimulus changing location when these cues change over time. The rate of fluctuation is high, and the binaural system is notoriously "sluggish" in responding to rapid changes in binaural cues (Grantham and Wightman 1978). Consequently, the percept of an unmasked tone in broadband noise appears to be of a sound at a fixed location, usually one that differs from that of the masking noise. Alternative ideas have been introduced, arguing that ILDs and IPDs are not processed separately but are embodied in other variables such as interaural correlation (Osman 1971) or the residue from an interaural subtraction process (Durlach 1963). Most models of BU (see Sect. 8.1.4) assume that these are the real decision variables and that the relative dominance of IPD and ILD fluctuations only reflects the reduction in phase locking with increasing frequency in the auditory periphery.

8.1.3 Narrowband Unmasking and Across-Frequency Binaural Interference

Because the effective masker is only a narrow band of noise, it might seem that the stimulus could be simplified by presenting only that narrow band. It has been found, however, that the BU of tones in narrowband noise produces a pattern of data very different from that for broadband noise. In particular, Metz et al. (1966) found that the BMLD in narrowband noise can be larger than in broadband noise (up to 30 dB), and McFadden and Pasanen (1978) found that it is also sustained at higher frequencies rather than reducing to 2–3 dB like the BMLD in broadband noise. Moreover, McFadden and Pasanen found that the narrowband BMLD at 4 kHz was very variable across listeners, whereas most listeners have very consistent BMLDs in broadband noise. Bernstein et al. (1998) reported similar individual variability for narrowband masking at 500 Hz, making it a general property of data from narrowband maskers rather than of high-frequency unmasking. The increased variability arose from the N_0S_π thresholds rather than the N_0S_0 thresholds. It is quite difficult to generalize from unmasking results found with narrowband maskers to those found using broadband maskers. Nonetheless, narrowband unmasking is of interest in itself.

The effectiveness of unmasking at high frequencies when the stimulus is restricted to a narrow band is reminiscent of the ability to lateralize at high frequencies using stimuli that are strongly amplitude modulated. For instance, Harris (1960) found that ITDs could still be detected in high-pass-filtered clicks, even though the ITDs of pure tones at frequencies above 1500 Hz had long been known to be indiscriminable. This ability is generally explained in terms of phase locking of auditory nerve fibers to the amplitude envelope of the stimulus rather than to its fine structure (Colburn and Esquissaud 1976).

van de Par and Kohlrausch (1997) examined this idea experimentally by attempting to reproduce the auditory nerve response to low-frequency, narrowband N_0S_π stimuli in nerve fibers tuned to 4 kHz. The idea was that high-frequency BMLDs could be the same size as low-frequency BMLDs if the same information was encoded on the auditory nerve. The stimuli at 125 or 250 Hz were passed through a simple model of auditory transduction to capture the expected auditory nerve responses. This waveform was then multiplied by a 4-kHz carrier to produce a "transposed" stimulus that would excite auditory nerve fibers tuned to 4 kHz when presented to the listeners (see Stecker, Bernstein, and Brown, Chap. 6). For stimuli originally generated at 125 Hz, BMLDs were almost identical for the low-frequency and transposed stimuli, suggesting that the same mechanism might operate across frequency. At 250 Hz, BMLDs for the transposed stimuli were larger than for stimuli generated at 4 kHz but fell somewhat short of those for the original low-frequency stimuli.

Although phase locking to the envelope may be the mechanism by which ILD fluctuations are detected in unmasking, this does not explain the contrast between results with narrowband maskers and those with broadband maskers. However, if phase locking to the envelope is combined with a second feature of sound lateralization, that of across-frequency interference (see Best, Goupell, and Colburn, Chap. 7), a plausible account can be framed. In across-frequency binaural interference, the ability to lateralize a narrowband stimulus using either ILDs or ITDs is disrupted by the presence of additional energy in remote frequency regions. Interference also occurs in narrowband BU (Bernstein 1991). Such interference would be an integral property of a broadband unmasking stimulus due to all the off-frequency noise. Interestingly, the interference was asymmetrical in frequency, with the BMLD at 800 Hz being disrupted by noise at 400 Hz but the BMLD at 400 Hz being relatively immune to the presence of an 800-Hz noise band. Bernstein et al. (1995) later observed similar interference from an interferer at 500 Hz on BU at 4 kHz but did not examine the reverse arrangement using these frequencies.

Interference effects in BU have not received much attention since, but Culling (2011) encountered a similar effect when testing the roles of ILD and ITD fluctuations (see Sect. 8.1.2). He presented narrowband interaurally modulated noise bands both with and without flanking bands, mimicking broadband and narrowband N_0S_π stimuli, respectively. Although thresholds for QFM (ITD modulation) were largely unaffected by the presence or absence of flanking bands, thresholds for interaural AM (ILD modulation) were markedly poorer when flanking bands were present regardless of frequency.

These findings provide support to the idea that two distinct mechanisms of BU exist; one responds to IPD fluctuations and is immune to across-frequency interference, whereas the other responds to ILD fluctuations and is sensitive to across-frequency interference. This suggestion stands in contrast to the idea that changes with frequency are caused entirely by reductions in phase locking at higher frequencies in the auditory periphery. Instead, it suggests that due to changes in the peripheral encoding of IPDs with increasing frequency, there is a shift in dominance between these two mechanisms. At low frequencies, detection of IPD fluctuation dominates and listeners are very sensitive, but as the frequency increases, IPD encoding degrades and there is a shift to detection of ILD fluctuation, whose effectiveness is relatively independent of frequency. Due to the influence of interference in processing of ILD fluctuation, there is a sharper decline in sensitivity with increasing frequency in broadband unmasking than in narrowband unmasking. The results can also explain the existence of an asymmetry in binaural interference for the BMLD because the BMLD up to about 800 Hz is dominated by detection of IPD modulation, which is immune to interference, whereas the BMLD above around 800 Hz is dominated by ILD modulation, which is subject to interference. Thus, low-frequency interferers can affect high-frequency unmasking but not vice versa (see Best, Goupell, and Colburn, Chap. 7).

A likely candidate for the second mechanism is the sound localization system. Narrowband BU provides the listener with very different perceptual cues to the presence of the tone. Rather than hearing a tone stand out from the noise, the noise itself sounds broader in extent. The greater variability among listeners in narrowband unmasking may be explained by greater difficulty in picking up on this breadth cue. Moreover, the intervention of across-frequency interference would simply reflect the existence of the same effect in sound localization.

8.1.4 Modeling Binaural Unmasking

Various models of BU have been proposed (see Dietz and Ashida, Chap. 10), but the focus here is on the two general frameworks that continue to enjoy significant support. Moreover, the focus is on unmasking of low-frequency tones in broadband noise. Neither theoretical framework suggests that ILDs and IPDs are processed separately, but as explained in Sect. 8.1.2, IPD fluctuations play a dominant role at a low frequency in broadband noise. In general, the two theoretical accounts are very difficult to differentiate experimentally (Domnitz and Colburn 1976).

The correlation theory (Colburn 1977) proposes a network of internal delays and coincidence detectors similar to those suggested by Jeffress (1948) to explain sound lateralization by ITDs (Fig. 8.4). The network introduces delays through the transmission of action potentials along axons that vary systematically in length. The pattern of activity across an array of such coincidence detectors resembles a cross-correlation of the stimulus waveform. When no signal is present, the noise

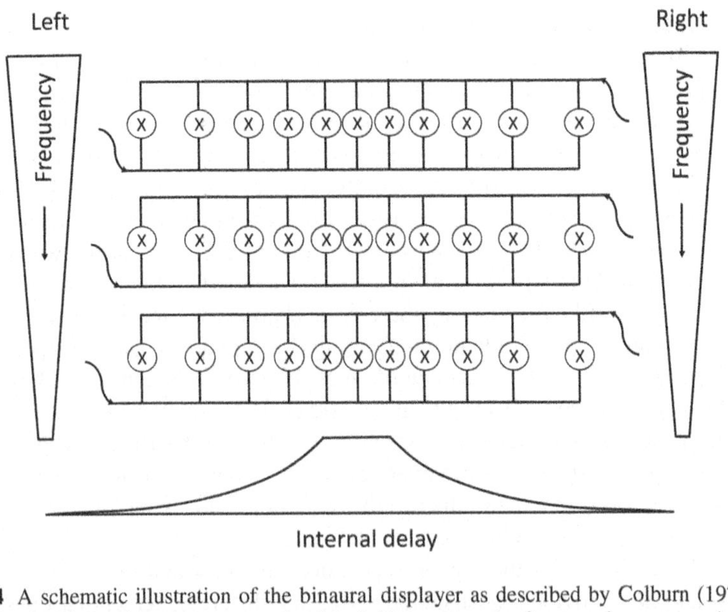

Fig. 8.4 A schematic illustration of the binaural displayer as described by Colburn (1977). An array of coincidence-detecting neurons (*top*) is fed by a network of converging axons originating from corresponding locations in each cochlea (*left* and *right*). Differences in the lengths of the axons from each ear introduce conduction delays that can compensate for externally applied ITDs. The distribution of delays among the coincidence detectors is uniform across frequency and up to ±150 μs but decays exponentially outside that range (*bottom*)

correlates perfectly when the internal delay feeding a particular coincidence detector compensates for the (external) ITD of the noise. This coincidence detector is consequently very active. In Colburn's (1977) model, detection of a signal in the noise occurs primarily by virtue of a *reduction* in activity at this coincidence detector because the signal reduces the maximum of the cross-correlation function (also known as the "coherence") of the combined waveform.

A potential problem with the correlation theory is that detection of a reduction in neuronal activity logically requires some comparator. This might be achieved in several ways. For a two-interval psychophysical task, one can simply compare activity in one interval with that in the other. However, some tasks, notably speech perception, require signal identification rather than detection, and there is no comparison interval. One could also compare activity in one frequency channel with neighboring channels that (are presumed to) contain no signal. Finally, one could assess the sound level at each ear and predict the activity level they would produce in the coincidence-detector network if they were perfectly correlated. It is not unreasonable to think that different strategies are used in tasks with different constraints.

To test these ideas, van de Par et al. (2001) deliberately designed a task in which activity comparisons across intervals or across frequency would be very problematic. They roved the overall sound level of their stimuli over a 30 dB range from one presentation interval to the other with the intention of disrupting comparisons of activity levels between intervals. They measured the listener's ability to detect reductions in the interaural correlation of noise, thus directly manipulating the presumed cue across the whole bandwidth of their stimuli so that across-frequency comparisons would also be useless. This left only the prediction-from-sound-level strategy. Detection of correlation changes were only mildly impaired compared with conditions with no roving of sound level. van de Par et al. (2001) argued that the level of performance the listeners displayed could not plausibly be attributed to the prediction-from-sound-level strategy because it would have required much finer discrimination of absolute sound levels than has ever been observed psychophysically, casting doubt on the sufficiency of a correlation mechanism.

The equalization-cancellation (EC) theory (Durlach 1963) proposes that the auditory system applies internal transformations that "equalize" the waveforms at each ear before they are subtracted one from the other. In appropriate circumstances, the output of this process has an improved SNR because the noise has been canceled out in the subtraction process. It is assumed that the system can select the optimal transformation to produce an improvement in the SNR. Finally, the system can select between this output and the two waveforms at each of the individual ears, the channel that has the best SNR. Because noise and signal are mixed, the same transformations must be applied to both together. If the SNR is improved after the cancellation, then the EC process predicts a BMLD. Otherwise, it is assumed that threshold is unchanged because monaural signal detection processes will still operate. In the case of N_0S_π, it is simple to see how this scheme would work. Because the noise is identical at the two ears, no transformation is needed; the model would simply subtract the waveforms arriving at the two ears in order to remove the noise.

Because the signal is inverted at one ear, the subtractive process also increases the signal amplitude. This combination of optimal subtraction with summation of the signal provides an explanation for the fact that the N_0S_π configuration always provides the largest BMLDs.

Various versions of the EC theory have been published over the years. The earliest, suggested by Kock (1950), could apply delays, followed by subtraction of the signals from the two ears. The model was only presented at a conceptual level. Durlach (1963) described an equalization process that could apply phase shifts and level adjustments as well as delays and then subtraction. The model contained only two parameters (σ_δ and σ_δ), which were fitted by eye to Hirsh and Burgeat's (1958) data; these control the efficiency of the equalization process, limiting the maximum BMLD to 15 dB and reducing its effectiveness with increasing signal frequency. Durlach (1972) revised his model, removing the option of phase shifts, limiting the time delays to half a period of the channel center frequency. The two original parameters were also kept, but since the revised model incorporated a more explicit frequency analysis stage, the bandwidth of each frequency channel required definition. These were derived by Rabiner et al. (1966) from an additional set of data using the assumed delay limitation. Although Durlach (1972) always insisted that this was a "black-box" model, the revised model is easy to conceive as a network of converging axons (e.g., Breebaart et al. 2001). Durlach (1972) applied his model to a very wide range of data from the literature, including all the data in Fig. 8.1. The revised version of the model is taken here as the recognized version and the benchmark against which any further revisions should be compared.

Abandoning the phase shifts has a key role in explaining the difference between N_0S_π and $N_\pi S_0$ thresholds (Fig. 8.1a). To understand this difference, one must appreciate that the revised model was conceived as operating within a given frequency channel. We may think of this frequency channel arising from the frequency selectivity of the basilar membrane in the two cochleas. Within this channel, the noise becomes narrowband, and within that band a π IPD would be approximately, but only approximately, equivalent to an ITD. For instance, at 500 Hz, the band might include frequencies from 460 to 540 Hz. At 480 Hz, the π IPD is equivalent to an ITD of $0.5/480 = 1042$ μs, whereas at 520 Hz, the equivalent ITD is 962 μs. Thus, it is impossible for the system to apply a single time delay that exactly equalizes all of the noise in the band, but a single time delay will approximately equalize them all, and so the subsequent subtraction process will be partially successful. The N_0S_π configuration would have no such problem because the noise is identical at the two ears.

Although Durlach's (1972) revised model produced satisfying fits to a wide range of data, it also had some weaknesses. Durlach (1972, pp. 431–437) identified a number of remaining shortcomings in the quality of fit that it gave to those data, but there are also some more fundamental problems. One issue is that the theory assumes that the auditory system knows a priori which equalization produces the most improved SNR. It is not obvious how this should be achieved. Culling and Summerfield (1995) suggested the possibility that the system selects the delay at which cancellation of the stimulus (signal and noise) is most complete, but this strategy will only work if

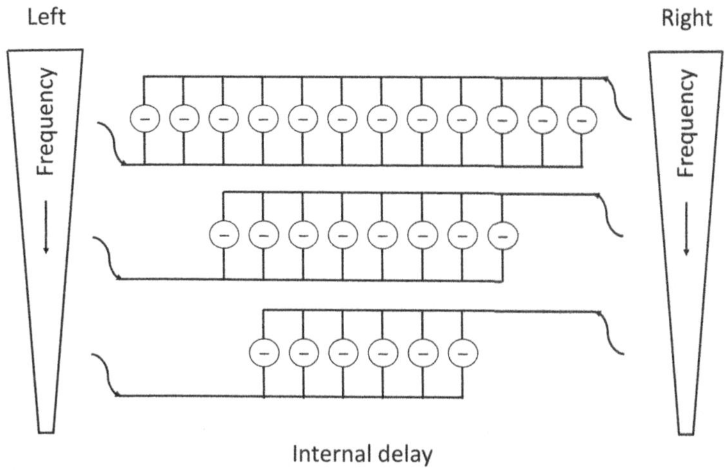

Fig. 8.5 Schematic representation of equalization-cancellation (EC) processing in a format similar to Fig. 8.3. The range of internal delays is, in effect, evenly distributed up to a fixed limit that varies with frequency

the stimulus is dominated by the noise; it may fail or be suboptimal when the target level is above threshold. This is because the most complete cancellation may, in that case, be a cancellation that has successfully eliminated part of the signal. Another major shortcoming in the EC theory is the absence of an appropriate physiological substrate of the type outlined in Fig. 8.5. Although there are neurons in the lateral superior olive that receive timed inhibition from one side, these tend to be tuned only to high frequencies at which unmasking is relatively ineffective. In contrast, there is considerable evidence for the existence of the coincidence detection neurons employed by correlation theories at low frequencies (Fig. 8.4).

Despite these issues, Culling (2007) made an empirical comparison between the correlation and EC principles and identified a case that is only explained by the EC theory. Because the two theories produce almost identical predictions of thresholds (e.g., those summarized in Fig. 8.1), Culling compared their predictions for a suprathreshold task. Once a binaural cue has been detected, further increases in its strength should produce increases in the loudness of the resulting percept, and such increases in loudness have been measured as equivalent to up to 6 dB of increased spectrum level (Culling and Lewis 2010). One can therefore ask the question of whether the loudness that listeners experience is more closely related to the correlation of the stimulus in the frequency channel of the target or to the residue from EC. Again, Culling (2007) used simulations of an N_0S_π stimulus based on manipulated bands of noise. In Expt. 3 of that paper, listeners compared the loudness of two such stimuli while the spectrum level of the target bands was increased by a few decibels in the target interval (Fig. 8.6). In separate conditions, each of the two putative cues was held constant while the other was allowed to increase or decrease in the target interval. To understand the results, it is important to note that when

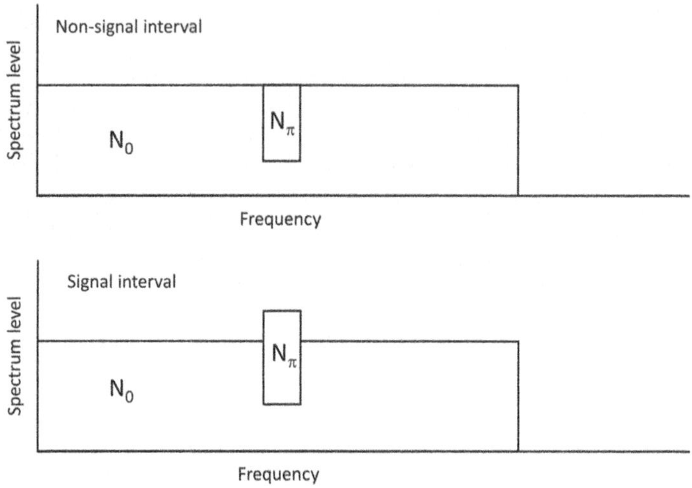

Fig. 8.6 Schematic illustration of the stimuli used by Culling (2007). In the signal interval, there is an increase in spectrum level at the target frequency for listeners to detect. The results showed that detection was facilitated if the amount of N_π noise, represented by the size of the box, also increased. Such an increase is required to keep the interaural correlation of the target band constant with increasing level, whereas EC theory predicts the increase in detection because the N_π noise is the component of the stimulus that will be left after cancellation

interaural correlation is held constant, the residue from cancellation increases with increasing spectrum level. The results showed that if the cancellation residue was fixed across intervals, then detection of the loudness change was consistent only with the spectrum-level change. If the interaural correlation was fixed, detection of the loudness change grew disproportionately to the spectrum level. The latter effect can only be explained by the EC theory because it predicts an increase in loudness from *both* the spectrum level and the residue from cancellation.

8.1.5 Binaural Unmasking of Speech

Demonstration of the BU of speech followed that of tones by only a few months (Licklider 1948). The results were essentially similar, although, Licklider measured percent correct word identification at two SNRs rather than detection thresholds. When Schubert (1956) took full psychometric functions so that threshold changes could be measured, the benefit for speech of N_0S_π compared with N_0S_0 was about 5 dB, which seems small compared with the BMLDs of 10–15 dB that occur for tones in broadband noise. This smaller effect can be explained by assuming that the audibility of the speech is improved in each frequency band by the size of the BMLD for tones at that frequency. Because the band of frequencies below 1000 Hz where the BU is most effective are less important than some higher frequencies for speech perception (ANSI 1997), the effect of their improved audibility is limited (Levitt and Rabiner 1967).

Measurement of the intelligibility of speech rather than the detectability of tones also raises further issues with respect to ecological validity. Speech is a broadband stimulus that carries information at many different frequencies. Consequently, the phase shifts used in many tone-detection experiments produce highly unrealistic spatial cues when applied to speech; not only can the ITDs they produce be larger than those generated by the separation of the ears but also they are different at every frequency. For instance, a π-radian IPD is equivalent to a 2-ms ITD at 250 Hz, a 1-ms ITD at 500 Hz, and a 500-μs ITD at 1 kHz. By contrast, real ITDs decrease by only 20–30% as a function of frequency (Kuhn 1977; see also Hartmann, Chap. 2). Moreover, the ITDs produced by a π-radian IPD are ambiguous; the 1-ms ITD at 500 Hz could equally be +1 ms or −1 ms, potentially placing the speech on either side of the head. Licklider (1948) noted this issue and explored the idea that the perceived locations of the speech and noise needed to be different for unmasking to occur but ultimately rejected the idea.

The question of whether distinct perceived locations are needed for speech to be unmasked has since been tested more directly. Schubert (1956) showed that differences in ITD between speech and noise were less effective than differences in IPD, even though the ITDs provide a much more sharply localized, ecologically plausible, and unambiguous sound image. Edmonds and Culling (2005) divided speech and noise into two different frequency bands and gave them different ITDs in each band. The BU was unimpaired when the lower speech bands had an equal but opposite sign ITDs, potentially muddling their distinct locations (Edmonds and Culling 2005, Fig. 8.6).

These and other results suggest that the optimal situation in any given frequency band is that the speech and noise differ in IPD by π radians. It does not matter whether these IPDs or ITDs match those in other channels or where the listener perceives the speech and noise to be. BU appears to be a low-level process that recovers signals from noise in each frequency channel individually and then offers up the results to higher level processes for interpretation. The irrelevance of perceived position is also supported by modeling work that relies only on ITD and ILD calculations but which accurately predicts target-speech intelligibility in multiple interferer scenarios, even if the overall interfering noise is generally not perceived as coming from a given position (Jelfs et al. 2011). However, as discussed in Sect. 8.2.3, when performance is limited by informational masking, the crucial cue becomes the difference in perceived position rather than in interaural differences.

8.2 Spatial Release from Masking and Speech Intelligibility

In the real world, having two ears helps us better understand speech in noise when the two sources have different spatial locations: a competing sound source causes less masking when it is spatially separated from the target speech compared with when the sources are co-located (Plomp 1976). This SRM results in part from the BU mechanism described in Sect. 8.1. However, in real-world listening, ITDs and ILDs co-occur and the presence of the head between our two ears creates large and stable head-related ILDs at high frequencies (see Hartmann, Chap. 2). These stable ILDs can influence the SNR at each ear, particularly when the interfering sound comes predominantly from one side, creating a head-shadow effect, and listeners can exploit these improvements in SNR.

The first thing to consider when investigating speech intelligibility in noise is the type of noise involved. The situations in question relate closely to the "cocktail party problem" introduced by Cherry (1953) but can be broadly defined as situations where a listener attempts to understand speech among localized interferers. These interferers would produce sounds ranging from stationary noises to competing voices. A "localized" interferer is a discrete competing source with a spatial position, as opposed to a diffuse, ambient noise. A cocktail party is the quintessential example of a situation where multiple people are talking. Other examples include open-plan offices, classrooms, bars, restaurants, or street scenes where interferers can be other people talking or any other sound source that might mask the target (e.g., an air conditioner or road noise). The term "noise" is used specifically to describe a signal with random phase variations (e.g., white, pink, or speech-spectrum noise), whereas "interferers" describe any type of sound, including noise, that competes with the speech that the listener is trying to understand.

Depending on the nature of the interferer (e.g., speech or noise), distinct types of interference can degrade speech intelligibility. Energetic masking, typically generated by signals that overlap in the temporal and spectral domains, has been described as interference that occurs when the sounds compete for representation at the auditory periphery (Culling and Stone 2017). Because of this competition, the target speech becomes less audible and intelligibility is reduced. By contrast, informational masking refers to additional degradations to intelligibility that occur when a speech target is masked by similar competing talkers (Kidd and Colburn 2017). There can be difficulties in segregating the competing voices (i.e., determining which signal parts belong to the target) and difficulties in attending to the right source in the mixture (i.e., overcoming confusion or distraction). Finally, room reverberation can disrupt speech intelligibility and SRM (see Zahorik, Chap. 9). Energetic masking is relatively well understood, including the effects of reverberation, because it can largely be explained in terms of variations in the SNR.

8.2.1 Stationary Noise Interferers

The advantage produced by spatial unmasking can be quantified by computing the decrease of the speech reception threshold (the SNR at which 50% of the target is understood) between a colocated condition (used as a reference) and the spatially separated condition of interest. Figure 8.7 shows data from a range of studies for a speech target in front and a single speech-spectrum noise from different azimuths.

The SRM increases monotonically when the noise is moved from 0° (front) to about 60°, dips at 90°, and then reaches a second maximum at 120° before decreasing when the noise is moved further to the back of the listener and the interaural differences decrease again. This pattern can be explained through a combination of two mechanisms (Zurek 1993): BU and better-ear listening.

Better-ear listening results from the difference in the sound level produced by each source at the two ears, their ILDs. A target and interferer at different locations produce different ILDs so that one ear will offer a better SNR than the other, and listeners can simply use the information coming from whichever ear offers the better SNR. For the situation illustrated in Fig. 8.7, as the noise source is moved further to one side of the head, the far ear is progressively more shadowed from this noise by the head, providing an improved SNR at that ear. The effect reduces once more as the sound moves to the back and the head is no longer directly between the noise source and that ear.

BU is described in detail in Sect. 8.1. In the speech-intelligibility literature, it is sometimes called "binaural interaction," "binaural advantage," or "binaural squelch." In the context of the SRM, an ITD difference between the target signal and the noise at the ears allows the auditory system to cancel the noise to some extent, improving the internal SNR. For the situation illustrated in Fig. 8.7, as the noise moves away from the target source, there is a difference in their ITDs. Within each

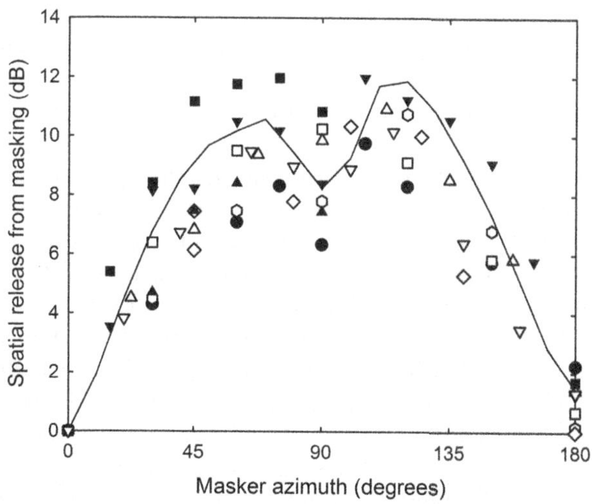

Fig. 8.7 Spatial release from masking (SRM) for speech in front as a function of masker (stationary noise) azimuth from nine studies. Azimuths to left and right are folded into the same plot. When the SRM was measured on both sides, it was averaged and plotted (*solid symbols*). Studies were by Peissig and Kollmeier (1997; *solid circles*), Muller (1992; *solid inverted triangles*), Bronkhorst and Plomp (1988; *open squares*), Plomp and Mimpen (1981; *open upright triangles*), Platte and vom Hövel (1980; *open hexagons*), Beutelmann and Brand (2006; *open diamonds*), Cosentino et al. (2014; *solid upright triangles*), Ozimek et al. (2013; *solid squares*), and Andersen et al. (2016; *open inverted triangles*). The *solid line* is from a predictive model (Jelfs et al. 2011)

frequency band, this results in a difference in the IPD, which creates the unmasking effect. The magnitude of the unmasking component of the SRM is largely dependent on the coherence of the noise (Lavandier and Culling 2010) and the differences between the IPDs of the speech and the noise in each frequency channel (Edmonds and Culling 2005).

Like ILDs and ITDs, better-ear listening and BU are frequency dependent. Although BU produces most of its benefit below 1500 Hz, better-ear listening is more effective above this frequency when the ILDs produced by the acoustic shadow of the head become substantial (Kuhn 1977).

It may seem surprising that the SRM dips when the noise is at 90°. Indeed, this spatial configuration is used in the apparent presumption that it should maximize the effect. The ILD is reduced for a noise placed at this particular azimuth due to constructive interferences of the sound waves traveling around both sides of the head, which sum at the ear opposite to the noise (for a review, see Jelfs et al. 2011). The head shadow is weakened; hence better-ear listening is diminished and the SRM reduced.

To assess the relative contributions of better-ear listening and BU to the overall SRM, Bronkhorst and Plomp (1988) processed the speech and noise stimuli to set either the ILDs or ITDs to zero and isolate the remaining cue. For a frontal target and a noise source at different azimuths, BU (ITDs) produced a release of 4–5 dB, relatively independent of the noise azimuth as soon as it was different from that of the target (30° was the smallest azimuth difference tested). On the other hand, better-ear listening (ILDs) was strongly dependent on the noise azimuth, varying between 3.5 and 8 dB. Zurek (1993) measured a slightly smaller better-ear effect, so that on average across studies, the release associated with BU is about 4 dB whereas better-ear listening produces a release between 4 and 8 dB depending on the noise azimuth.

Some studies evaluated these relative contributions by assuming that they are additive to produce the full spatial release (e.g., Hawley et al. 2004), and others (e.g., Jelfs et al. 2011) have been quite successful in predicting a range of datasets from the literature using a model that assumed this additivity. However, the full SRM is generally a little smaller than the addition of the two contributions measured individually (Bronkhorst and Plomp 1988). Figure 8.8 elaborates the predictions of the Jelfs et al. (2011) model for the case in Fig. 8.7 by separating out the BU and better-ear listening components. The model reproduces the smaller effect of unmasking and the dip at 90° for the better-ear listening component.

The contributions of better-ear listening and BU can change in the presence of multiple noise interferers. When the number of interferers increases (e.g., Hawley et al. 2004), better-ear listening is severely reduced when there are interferers on both sides of the target. The overall interferer level can then be similar at each ear so that no ear provides a better SNR than the other. On the other hand, BU can remain relatively unaffected by the number and configuration of the interferers. BU can still be effective against multiple interferers because it then acts on the overall interfering sound; the combined interferers can have a resultant IPD in each frequency channel that is still different from that of the target, even if this overall

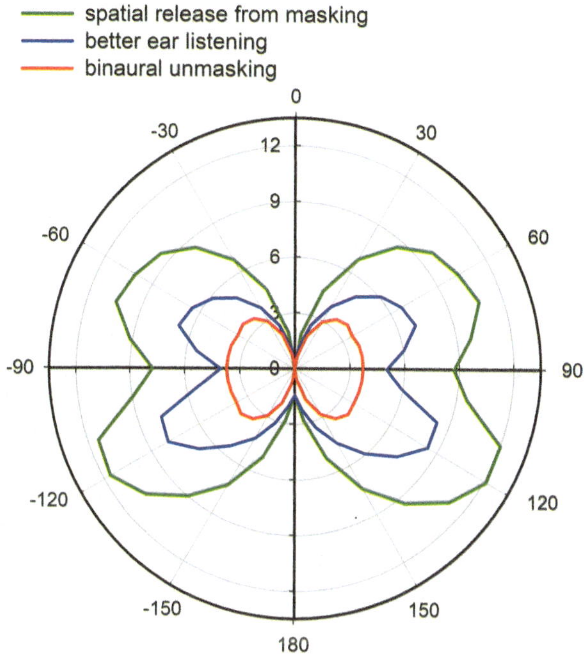

Fig. 8.8 Predicted SRM (*green*) in decibels as a function of noise azimuth for target speech in front. Also shown are the component effects of better-ear listening (*blue*) and BU (*red*). Predictions were generated by the Jelfs et al. (2011) model using head-related impulse responses from Gardner and Martin (1995)

interferer IPD may not correspond to any of the ITDs of the individual interferers. Of course, when the number of interferers increases further such that the overall interferer becomes diffuse and uncorrelated at the ears, BU is expected to be severely impaired. BU is also susceptible to room reverberation (see Sect. 8.2.4).

8.2.2 Models of Spatial Release from Masking in Stationary Noise

Beutelmann and Brand (2006) developed a computational model of SRM in stationary noise that directly implements an EC process. The target and interferer signals at each ear need to be available separately. They are first processed through a gammatone filterbank, to simulate cochlear frequency analysis, and an EC mechanism and then resynthesized with the binaurally enhanced SNR. The speech intelligibility index (Kryter 1962) is computed from the resynthesized speech to evaluate intelligibility. For the EC mechanism, the left and right ear signals are attenuated and delayed with respect to each other (equalization) and then the right channel is subtracted from the left channel. The delays and attenuations that maximize an effective SNR after cancellation are selected. Unlike Durlach's (1963) formulation, the efficiency of the EC mechanism is limited by adding independent internal noise to the signals from each ear. The SRM model of Wan et al. (2010) is conceptually very similar, except that efficiency of the EC process is limited by time-varying jitters in time and amplitude, just as proposed by Durlach (1963).

Another group of models does not directly implement an EC process, but instead BU is taken into account by assuming that the effective SNR in each frequency band is increased by the predicted BMLD for pure-tone detection in noise at the center frequency of the band (Levitt and Rabiner 1967). The BMLD is predicted using an analytical expression. The better-ear listening and BU components are predicted independently from the signals produced by the sources at the ears. Better-ear listening is estimated from the SNR computed as a function of frequency at each ear, selecting band-by-band the ear for which the ratio is higher. BU is then taken into account by increasing the better-ear SNR by the size of the BMLD in each band. In the model proposed by Zurek (1993), this BMLD is estimated from the target and interferer ITDs and ILDs using a simplified expression proposed by Colburn (1977). The broadband intelligibility prediction is then computed as the weighted sum of the resulting SNRs. This model is limited to anechoic situations because the BMLD calculation does not take into account the interaural coherence of the interferer, which mediates the effect of reverberation on BU.

The models of Lavandier and Culling (2010) and Jelfs et al. (2011) are conceptually similar to that of Levitt and Rabiner (1967), except that different weightings (ANSI 1997) and a different BMLD equation are used. This new equation (Eq. 8.1) depends on the center frequency of the band (ω_0 in rad/s), on the target (φ_S) and interferer (φ_N) IPDs, and on the interferer coherence (ρ_N), allowing predictions in reverberant situations

$$BMLD = 10\log_{10}\left[\frac{k-\cos\left(\varphi_S-\varphi_N\right)}{k-\rho_N}\right] \tag{8.1}$$

where, $k=\left(1+\sigma_\varepsilon^2\right)\exp\left(\omega_0^2\sigma_\delta^2\right)$, $\sigma_\varepsilon = 0.25$, and $\sigma_\delta = 105\ \mu s$

(standard deviations of the time and amplitude jitters, respectively, characterizing the internal noise in the EC model; Durlach 1972).

The model assumes additive contributions of BU and better-ear listening, neglecting their known interaction (Bronkhorst and Plomp 1988). Moreover, the BMLD calculation does not take head-related ILDs into account when evaluating the BU effect, even though BU of tonal targets can be reduced when the target or masker has a large ILD (Egan 1965). The accuracy of the model predictions might indicate that these effects are very limited when realistic ILDs are involved (Lavandier et al. 2012) or that the magnitude of low-frequency ILDs are small in the cases in which the model was tested. Alternatively, they may reflect the fact that BU and better-ear listening tend to operate in different frequency regions (low frequencies for BU and high frequencies for better-ear listening) such that when they are summed, one of them is usually negligible. This model predicted the SRM even in the presence of multiple interferers and reverberation (Lavandier et al. 2012). In the modeling, the signals produced by multiple interferers are simply summed to obtain the overall interferer at the ears.

Other SRM models have been proposed in the literature. Alternative modeling approaches focus on different characteristics of the signals such as their temporal modulations or the correlation between the clean and noisy speech, but BU is usually modeled using the EC theory. All these models, their ability to explain the data, and their limitations are described in detail in a recent review chapter (Lavandier and Best 2019).

8.2.3 *Modulated Noise Interferers*

Modulations in the temporal envelope of the interferer allow one to hear the target better while the SNR is high (Festen and Plomp 1990). This ability to exploit temporal fluctuations in the level of the interferer is called (temporal) dip listening, listening in the gaps, or glimpsing (Cooke 2006). Collin and Lavandier (2013) showed that for strongly modulated interferers, listeners can take advantage of the predictability of the gaps in a modulated noise interferer. This indicates that the dip-listening advantage in realistic cocktail-party situations might be overestimated from studies that used predictable modulations such as periodic modulations or constant ("frozen") speech modulations. These predictable modulations allow the evaluation of glimpsing for a listener who could use this unmasking mechanism almost optimally. In situations where the interferer gaps are less predictable (e.g., for interfering speech), the listener might not be able to use glimpsing optimally due to the uncertainty of the timing of the gaps within the masking sound.

Masker modulation can also make SRM fluctuate moment by moment. Depending on the configuration of the maskers, two types of interactions can be observed. For configurations with maskers all on one side of the target, modulations generally tend to reduce SRM, as confirmed by Culling and Mansell (2013) using a single-noise interferer with and without artificial modulations. This interaction could be explained by the fact that a strongly modulated noise causes less masking than stationary noise to start with, so that there is less room for improvement from SRM. The interaction might be limited to a strongly modulated masker though; when using one or multiple noises modulated by natural speech envelopes, SRM was not affected by noise modulations when averaged across levels of modulation and numbers of interferers (Hawley et al. 2004; Collin and Lavandier 2013).

The interaction between interferer modulations and SRM is different when the fluctuating interferers are positioned on both sides of the target. Then, the ear offering the best SNR can change rapidly over time (Fig. 8.9). The auditory system has the opportunity to take advantage of these changes across ears to improve overall speech intelligibility and SRM, but the extent of this ability remains controversial. Schoenmaker et al. (2017) suggested that the better ear should be defined based on the number of good target "glimpses" available at each ear, and the better-ear advantage be estimated based on short-term rather than long-term SNRs. Brungart and

Iyer (2012) previously suggested that the better ear could be chosen on a moment by moment basis, independently in each frequency channel. Better-ear listening is then often referred to as better-ear glimpsing. However, it is not established yet whether better-ear glimpsing is a "true" binaural mechanism involving switching across ears to get the most relevant information (best SNR across ears) or two monaural mechanisms taking place simultaneously, providing the SNRs at both ears.

Culling and Mansell (2013) provided evidence that switching is necessary because they found that performance was highly dependent on the required switching rate. They presented speech from the front with two noise interferers in symmetrical locations on either side of the target. The interferers were modulated in perfect alternation at different rates such that the better ear switched instantaneously back and forth. The stimuli were processed to set either ILDs or ITDs to zero or retained both cues. In each case, intelligibility declined sharply with increasing modulation rate, although a residual SRM was still seen at the fastest modulation rates used. The ILD-only case seems to indicate that better-ear glimpsing involves switching across ears (otherwise, it would not have been affected by the modulation rate) and that this binaural switching is rather sluggish.

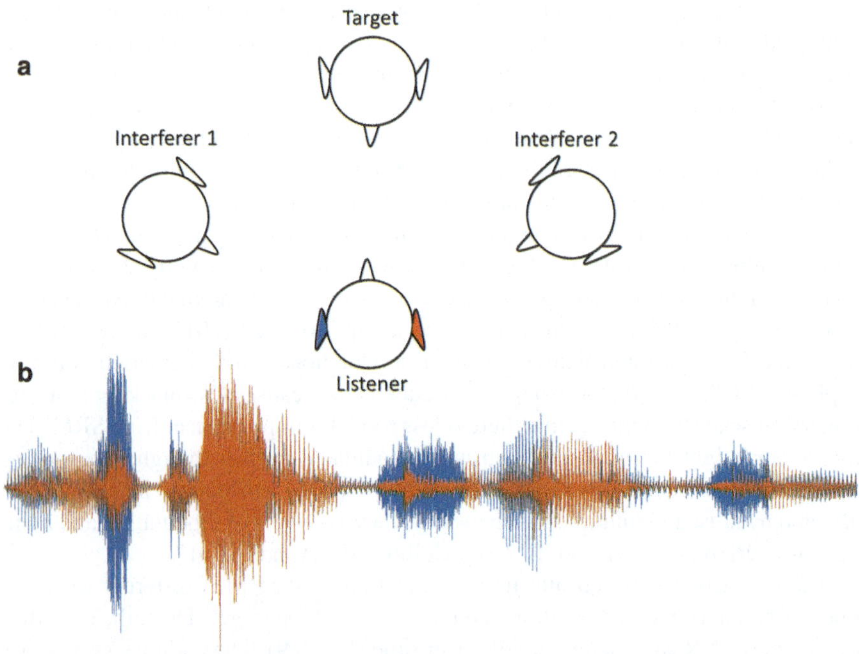

Fig. 8.9 Illustration of the potential for better-ear glimpsing. Speech interferers are at ±60 ° and a target speaker is in front (**A**). The head shadow is sufficiently strong for the combined interferers received at each ear (**B**) to differ from moment by moment in sound level. The waveforms are color coded to the ears at which they would be received

The particular case of two maskers symmetrically placed on each side of the target is often investigated because it creates no long-term better ear. Independent modulations in the maskers create asynchronous dips at the two ears so that better-ear glimpsing allows for short-term SRM. This does not occur with stationary-noise maskers. The benefit of better-ear glimpsing on the SRM should only occur when there is one masker on each side of the target; otherwise, there is a long-term better ear and modulations do not bring any advantage in terms of better ear.

Few studies have investigated the relative contributions of better-ear listening and BU in the presence of modulated noise interferers. In a spatially symmetrical masker configuration (frontal target), Ewert et al. (2017) found that the relative contribution of better-ear listening/glimpsing increased when introducing modulations in the noise maskers. However, because this configuration produces no long-term better ear for stationary noises, better-ear listening is only possible on a short-term basis once modulation is added. The contribution of BU remains similar for the two types of noise so that the increase in the SRM results from the stronger (absolute and relative) contribution of better-ear glimpsing.

To predict spatial unmasking from a modulated noise, Beutelmann et al. (2010) and Collin and Lavandier (2013) applied the (monaural) method for modeling modulated maskers (Rhebergen and Versfeld 2005) to extend their corresponding stationary-noise models (Beutelmann and Brand 2006; Lavandier and Culling 2010). Rhebergen and Versfeld (2005) used short-time frames before averaging the resulting predictions across frames. Peaks in the interferer signal induce an increase of masking, whereas pauses induce a decrease of this masking. The models consider interfering energy as a function of time. To consider the envelope modulations in the target as important information for its intelligibility, the models (at least conceptually) consider the average level of the target across time rather than its instantaneous level within short time frames.

Wan et al. (2014) revised the model of Wan et al. (2010) so that it uses EC parameters varying across short time frames along the duration of the stimuli, thus improving the possibility for the model to cancel the dominant masker over time when the direction of the dominant masker varies in time. However, this model also uses long-term SNRs calculated over the whole stimulus duration so that short-time variations in the SNR are not explicitly considered, preventing better-ear glimpsing from being explicitly predicted.

8.2.4 Speech Interferers

Speech interferers can, in some circumstances, add informational masking. Informational masking cannot be explained in terms of energy overlap and is thought to occur more centrally (Kidd and Colburn 2017). A simplified picture is that energetic masking prevents you from hearing the target, whereas informational masking prevents you from listening to the target (even if you can hear it). Listening to the target means hearing it distinctly from the competing voices (thus achieving segregation) and focusing your attention to the appropriate voice. This dichotomy is rather simplistic, however. Often, an effect that cannot be attributed to energetic masking is classified as informational masking, but, at the same time, it is not always perfectly clear how one defines energetic masking to start with or, at least, this definition seems to vary across researchers.

For instance, recent studies have shown that even "stationary" noise contains intrinsic envelope modulations and masking of speech by noise is partially due to masking of the envelope modulations in the speech by the modulations in the noise ("modulation masking"). Consequently, if the noise modulations are reduced, intelligibility of the speech improves, even though the spectral overlap of the speech and noise is unchanged (Culling and Stone 2017). For this reason, one can differentiate modulation masking and energetic masking, but researchers disagree on whether modulation masking should be classified as energetic masking, informational masking, or a new category. Modulation could be seen as the information content of the speech and modulation masking as the concealment of that information by additional modulations. On the other hand, one might consider that signals contain energy in both the spectral and modulation domains and that energetic masking occurs in both cases due to energy overlap in one or both of these domains (keeping in mind that the underlying perceptual mechanisms are different and that their modeling might not be equivalent).

There are many other examples of masking and unmasking phenomena that cannot easily be pigeonholed as energetic or informational (Culling and Stone 2017). Complicating this problem, the temporal dips and harmonicity of speech interferers offer opportunities for masking release that typically make speech a *less* effective masker than noise (for a review, see Culling and Stone 2017) despite its additional potential for informational masking.

In experiments with speech interferers, informational masking effects tend to occur when the target and interfering speech are very similar and/or when there is more than one speech interferer but not so many that the individual voices merge into a babble (Culling 2016). The amount of informational masking depends on the potential for confusion and the uncertainty in the listening task (Kidd et al. 2005), which depend on the task design and stimuli being used. Informational masking is generally reduced as soon as the similarity between the interferer and target is reduced (Watson 2005). Any perceived difference between the competing voices can potentially help listeners focus their attention on the target (Darwin and Hukin 2000; Shinn-Cunningham et al. 2005). A difference in source position is also a strong cue to reduce the uncertainty about how to disentangle the different voices in a mixture (David et al. 2017). As a result, a release of informational masking can increase the observed SRM when there is little other available information to separate the competing voices. However, spatial separation may become a redundant cue when these voices are not easily confused because they differ in any other perceived salient cue (e.g., pitch, timbre, or loudness) so that the observed SRM is then not increased.

Although spatial unmasking can result from a release from both energetic and informational masking, the underlying mechanism is quite different. As described in Sect. 8.1.5, energetic masking relies on better-ear listening and BU not on the differences in the perceived position of the competing sources. In the case of informational masking, it is the opposite and the crucial cue seems to be the difference in perceived position. Freyman et al. (1999) used the precedence effect to create the illusion of spatial separation between competing talkers so that a clear difference in the perceived position was introduced without creating the corresponding difference in ITD/ILD in the stimuli. They found a large SRM despite a small increase in energetic masking.

It seems very difficult to provide a quantitative prediction of SRM with speech maskers, because the release depends on the initial amount of masking present for the co-located sources. With speech maskers, this masking will depend on the amount of informational masking in addition to the energetic masking. As mentioned previously, this informational masking is strongly dependent on the presence of any other nonspatial cue available, such as a pitch difference, to differentiate the competing voices. In experimental conditions that provide no other cue, the SRM can be very large, above the 10 dB SRM mentioned for a stationary noise. In more realistic situations, however, many nonspatial cues that considerably reduce informational masking are generally available (Westermann and Buchholz 2015). A crucial (and never easy to solve) issue when considering the SRM from speech maskers is thus to quantify the relative contributions of energetic and informational masking. Currently, none of the existing models in the literature can fully predict speech intelligibility among speech interferers and the associated spatial release from both informational and energetic masking (Lavandier and Best 2019).

8.2.5 Reverberation

When communicating in rooms, reverberation (the multiple sound reflections from room boundaries) has several effects on speech intelligibility (see also Zahorik, Chap. 9). First, reverberation exerts a well-known temporal smearing on the target speech, which occurs even in quiet. When the speech signal at the ears is mixed with the multiple delayed versions of itself, it can be temporally smeared and self-masked. This smearing reduces the amplitude modulations in the target speech and impairs its intelligibility (Houtgast and Steeneken 1985). Having two ears may ameliorate the smearing effect on target intelligibility (binaural dereverberation); binaural listening has been shown to slightly improve intelligibility for reverberant speech in quiet (Nábělek and Robinson 1982) and also in the presence of a noise interferer (Lavandier and Culling 2008).

Importantly, reverberation is generally detrimental to spatial unmasking (Plomp 1976). Sound reflections traveling around the listener reduce ILDs, thus impairing better-ear listening (Plomp 1976). These reflections also impair BU by decorrelating the interfering sound at the two ears (Lavandier and Culling 2008). Lavandier and Culling (2010) showed that increasing reverberation on the interferer reduced the BU advantage. This reduction largely resulted from the decrease in interferer coherence, which evaluates the proportion of noise energy that is susceptible to cancellation. Changes in the interaural phase can produce a secondary effect.

Lavandier and Culling (2008) compared the effects of reverberation on temporal smearing and on the SRM by independently controlling the reverberation of the target and the interferer. Reverberation of the target speech needed to be relatively high for temporal smearing of the target to disrupt intelligibility, whereas the SRM was disrupted by relatively low levels of reverberation, suggesting that this latter effect occurs more readily in most of the rooms encountered in everyday life.

The models developed by Beutelmann and Brand (2006) and Lavandier and Culling (2010) also made accurate predictions of the SRM in the presence of reverberation. Increasing the effect of reverberation on the interferer coherence reduced the predicted BU advantage for all target azimuths, with a floor effect observed for azimuths close to that of the interferer. If the room reflections coming from the interferer are different at the listener's ears, then they reduce the interferer coherence, resulting in less BU. Lavandier and Culling (2010) noted that such models predict that no matter how late or energetic these reflections are, they should not impair intelligibility if they are identical at the two ears. They tested this prediction using a very reverberant room but symmetrical positioning of the interferer and listener within the space such that all reflections from the interferer would be identical at the listener's ears. They confirmed that speech intelligibility was as good as for an anechoic interferer. Of course, if the configuration is such that the reflections coming from the interferer are different at the two ears, the more energetic they are, the more deleterious the effect they should have on interferer coherence and target intelligibility. The main influence of interferer coherence generally occurred at low levels of reverberation for coherence between 1 and 0.75, and further decreases in coherence had less influence. This agrees with the results of Licklider (1948), who showed that most of the variation in intelligibility for speech in noise occurs for a noise coherence between 1 and 0.75.

Interestingly, in the special case of a target and interferer that lay in the same direction, increasing the effect of reverberation on the interferer can create an unmasking effect (Lavandier and Culling 2010). For small azimuth separations, where BMLDs are small in anechoic situations, room reflections can distribute part of the interferer energy to other directions, introducing differences in the IPD that did not exist without the reflections. In consequence, this energy may be canceled out and the BU advantage increased. This phenomenon could occur in real life when the interferer is more distant than the target and so has a lower direct-to-reverberant ratio.

Beutelmann and Brand (2006) interpreted the influence of early reflections of an interferer as responsible for the creation of a "mirror source" acting as a secondary interferer with a considerably different azimuth, thus disrupting the SRM (cafeteria condition; Fig. 8.10). This mirror source would indeed greatly reduce the head shadow if it were on the other side of the head compared with the interferer (Hawley et al. 2004), and it would also influence BU if it affected the overall interferer coherence at the ears. It should be noted, however, that mirror sources might not systematically reduce BU. When considering only the interaural phase, if the interferer azimuth is close to that of the target, then any mirror source could stand at a widely different angle, distributing interfering energy where it can be canceled so that the BU advantage might increase depending on the associated effect of reflections on coherence.

In real life, the effects of reverberation on spatial unmasking and temporal smearing of the target occur simultaneously. To model this, Rennies et al. (2011) and Leclère et al. (2015) combined their corresponding binaural models (Beutelmann and Brand 2006; Jelfs et al. 2011) with the (monaural) useful-to-detrimental ratio approach developed in room acoustics (Lochner and Burger 1964). This SNR approach regards the early reflections of the target as useful and as part of the "signal" because they reinforce the direct sound, whereas the late reflections are regarded as detrimental and effectively a part of the noise. The binaural room impulse response measured at the target position is split into early and late parts that are used to create an "early speech" signal and a "late speech" signal. The prediction process is then similar to that of the original binaural models, except that the target signal is replaced by the early speech target and the late speech target is added to the interferer(s) so that the detrimental influence of late reflections is taken into account before the binaural process. However, Leclère et al. (2015) found that a fixed early/late separation was not sufficient to predict speech intelligibility in rooms. Their best model performance was obtained by adjusting the early/late separation for each tested room. This indicates that the distinction between useful and detrimental reflections could be dependent on the room considered, and that refinements might be needed for their categorization such that it does not rely only on their delay after the direct sound (early/late). This might explain the wide range of early/late limits used in the literature.

This recent modeling work has triggered the idea that temporal smearing and binaural dereverberation could be interpreted within the framework of SRM (Leclère et al. 2015). Temporal smearing during speech transmission can be seen as masking of the direct sound and its early reflections by the late reflections from this same sound. Binaural dereverberation can then be understood simply in terms of spatial unmasking from these late reflections. The late reflections would thus be an additional masker, treated like any other interfering source by the binaural system. Reverberation spreads part of the late energy to different ITDs from that of the early target so that an EC mechanism can eliminate part of this late target (its coherence determining the level of cancellation). It should be noted that early and late targets might have different ILDs so that better-ear listening could also contribute to dereverberation.

It should be emphasized that the effect of reverberation on SRM cannot be predicted by the classical measurements used in room acoustics: the reverberation time (T60, duration for a 60 dB decrease in sound level after turning off the source), the clarity (C80, ratio of early-to-late impulse response energy; ISO 3382 1997), the speech transmission index (Houtgast and Steeneken 1985), or the useful-to-detrimental ratio (Lochner and Burger 1964). These monaural measures mainly depend on the characteristics of the target and the room but lack most of the inputs relative to the interferers' properties (e.g., positions, envelope modulations) that are

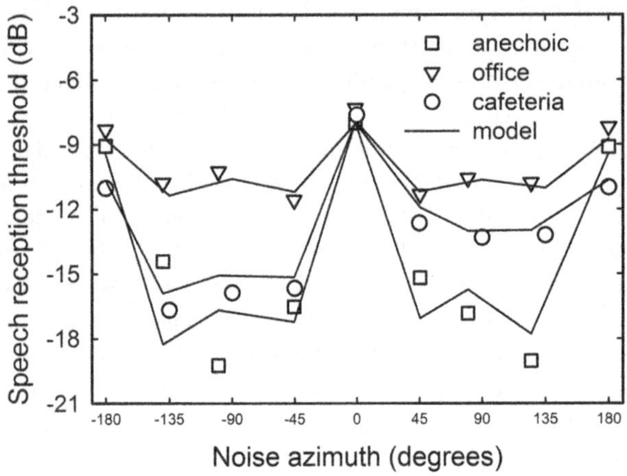

Fig. 8.10 Speech reception thresholds measured by Beutelmann and Brand (2006) for noise at various azimuths in three listening environments (*open symbols*). The speech target came from 0°. Also shown are the predictions from the Jelfs et al. (2011) model derived from the same binaural room impulse responses (*black line*)

crucial to predict segregation. For example, Beutelmann and Brand (2006) measured spatial unmasking in three different rooms: an anechoic room, a small office, and a large cafeteria. They showed that the cafeteria, which had the longest reverberation time (1.3 s), led to a larger SRM than the office, which had only half the reverberation time (0.6 s; Fig. 8.10). In addition, the large asymmetry observed in the cafeteria while moving the noise interferer around the listener cannot be described by a simple reverberation time statistic, which does not change with recording position. These statistics are thus not sufficient to predict speech segregation in rooms. Monaural measures cannot predict the SRM that relies on the binaural system.

Sound reflections in rooms can also impair dip listening (Collin and Lavandier 2013). The same temporal smearing process tends to reduce the envelope modulations of the masker, producing more masking by filling in the gaps through which the target could be heard. Finally, the modification of source spectra at the ears of the listener by room "coloration" can affect intelligibility. Room coloration results from both the constructive/destructive interference of sound reflections and the frequency-dependent absorption characteristics of the room materials. The spectrum produced by each source at each ear depends on the ear and source positions within the room. Thus, coloration influences intelligibility by determining the frequency-dependent SNR at the ears and potentially creating frequency-dependent ILDs that affect better-ear listening (Lavandier and Culling 2010). This influence varies with the positions of the listener, target, and interferers.

It is currently unclear whether reverberation can affect a release from informational masking, and very limited data are available to answer this question. Kidd et al. (2005) controlled the amount of energetic/informational masking by varying the spectral overlap between the target and masker. They found that in the presence of a primarily energetic interferer, the SRM obtained by moving the interferer from a 0° (front, target position) to a 90° azimuth was reduced by increasing reverberation but not when a primarily informational interferer was used. This robustness to reverberation may occur because the release from information masking relies on the differences in perceived location, which are largely preserved in rooms thanks to the precedence effect (Wallach et al. 1949). Reverberation and room coloration have even been shown to be cues that promote sequential segregation of competing sources (David et al. 2014). By contrast, Deroche et al. (2017) found that the release from informational masking obtained by moving the target from a 0° (front, speech masker position) to a 60° azimuth was reduced by increasing reverberation despite the conditions being carefully designed not to affect energetic masking (as also verified using a noise masker). Increased cognitive load associated with reverberation was speculated to be potentially responsible for the decrease of SRM.

8.3 Summary

The SRM is now well understood for the case of stationary noise interferers. Models of the process that incorporate better-ear listening and BU, with the latter implemented through an EC mechanism, have been very successful at predicting speech reception thresholds for any number and distribution of interferers and any room geometry. Although stationary noise may seem a rather limited case, many real-life situations where multiple interferers are combined will approximate the statistics of stationary noise. In this case, the effect of BU is quite predictable, even though the neural basis of interaural cancellation remains obscure. For maskers that retain some temporal modulation, the effects are harder to predict. There is a clear tension between the concepts of dip listening and modulation masking, and no theory yet predicts the relative dominance of these antagonistic mechanisms. Where the temporal dips are asynchronous at the two ears, the role of better-ear glimpsing is also poorly understood, and the extent of its role in everyday listening is a matter of contention. Similarly, spatial release from informational masking is a strong effect when substantial information masking is present, but, at the same time, it is currently unclear how much informational masking is present in real-life situations.

Compliance with Ethics Requirements
John Culling declares that he has no conflict of interest.
Mathieu Lavandier declares that he has no conflict of interest.

References

Andersen AH, de Haan JM, Tan Z, Jensen J (2016) Predicting the intelligibility of noisy and nonlinearly processed binaural speech. IEEE Trans. Audio Speech Lang Process 24:1908–1920
ANSI S3.5 (1997) Methods for calculation of the speech intelligibility index. American National Standards Institute, New York

Bernstein LR (1991) Spectral interference in a binaural detection task. J Acoust Soc Am 89:1306–1313

Bernstein LR, Trahiotis C, Bernstein LR (1995) Binaural interference effects measured with masking-level difference and with ITD- and liD-discrimination paradigms. J Acoust Soc Am 98:155–163

Bernstein LR, Trahiotis C, Hyde EL (1998) Inter-individual differences in binaural detection of low-frequency or high-frequency tonal signals masked by narrow-band or broadband noise. J Acoust Soc Am 103:2069–2078

Beutelmann R, Brand T (2006) Prediction of speech intelligibility in spatial noise and reverberation for normal-hearing and hearing-impaired listeners. J Acoust Soc Am 120:331–342

Beutelmann R, Brand T, Kollmeier B (2010) Revision, extension, and evaluation of a binaural speech intelligibility model. J Acoust Soc Am 127(4):2479–2497

Breebaart J, van de Par S, Kohlrausch A (2001) Binaural processing model based on contralateral inhibition. II. Dependence on spectral parameters. J Acoust Soc Am 110:1089–1104

Bronkhorst AW, Plomp R (1988) The effect of head-induced interaural time and level differences on speech intelligibility in noise. J Acoust Soc Am 83:1508–1516

Brungart DS, Iyer N (2012) Better-ear glimpsing efficiency with symmetrically placed interfering talkers. J Acoust Soc Am 132:2545–2556

Cherry EC (1953) Some experiments on the recognition of speech, with one or with two ears. J Acoust Soc Am 25:975–979

Colburn HS (1977) Theory of binaural interaction based on auditory-nerve data. II. Detection of tones in noise. J Acoust Soc Am 61:525–533

Colburn HS, Esquissaud P (1976) An auditory-nerve model for interaural time discrimination of high-frequency complex stimuli. J Acoust Soc Am 59(Suppl. 1):S23

Collin B, Lavandier M (2013) Binaural speech intelligibility in rooms with variations in spatial location of sources and modulation depth of noise interferers. J Acoust Soc Am 134:1146–1159

Cooke M (2006) A glimpsing model of speech perception in noise. J Acoust Soc Am 119:1562–1573

Cosentino S, Marquardt T, McAlpine D, Culling JF, Falk TH (2014) A model that predicts the binaural advantage to speech intelligibility from the mixed target and interferer signals. J Acoust Soc Am 135:796–807

Culling JF (2007) Evidence specifically favoring the equalization-cancellation theory of binaural unmasking. J Acoust Soc Am 122:2803–2813

Culling JF (2011) Subcomponent cues in binaural unmasking. J Acoust Soc Am 129:3846–3855

Culling JF (2016) Speech intelligibility in virtual restaurants. J Acoust Soc Am 140:2418–2426

Culling JF, Lewis HG (2010) Trading of intensity and interaural coherence in dichotic pitch stimuli. J Acoust Soc Am 128:1908–1914

Culling JF, Mansell ER (2013) Speech intelligibility among modulated and spatially distributed noise sources. J Acoust Soc Am 133:2254–2261

Culling JF, Stone MA (2017) Energetic masking and masking release. In: Middlebrooks J, Simon J, Popper AN, Fay RR (eds) The auditory system at the cocktail party: Springer handbook of auditory research. Springer, Cham, pp 41–73

Culling JF, Summerfield Q (1995) Perceptual separation of concurrent speech sounds: absence of across-frequency grouping by common interaural delay. J Acoust Soc Am 98:785–797

Darwin CJ, Hukin RW (2000) Effects of reverberation on spatial, prosodic, and vocal-tract size cues to selective attention. J Acoust Soc Am 108:335–342

David M, Lavandier M, Grimault N (2014) Room and head coloration can induce obligatory stream segregation (L). J Acoust Soc Am 136:5–8

David M, Lavandier M, Grimault N, Oxenham AJ (2017) Discrimination and streaming of speech sounds based on differences in interaural and spectral spatial cues. J Acoust Soc Am 142:1674–1685

Deroche MLD, Culling JF, Lavandier M, Gracco VL (2017) Reverberation limits the release from informational masking obtained in the harmonic and binaural domains. Atten Percept Psychophys 79:363–379

Dolan TR (1968) Effect of masker spectrum level on masking-level differences at low signal frequencies. J Acoust Soc Am 44:1507–1512
Domnitz RH, Colburn HS (1976) Analysis of binaural detection models for dependence on interaural target parameters. J Acoust Soc Am 59:598–601
Durlach NI (1963) Equalization and cancellation theory of binaural masking-level differences. J Acoust Soc Am 35:416–426
Durlach NI (1964) Note on binaural masking-level differences at high frequencies. J Acoust Soc Am 36:576–581
Durlach NI (1972) Binaural signal detection: equalization and cancellation theory. In: Tobias JV (ed) Foundations of modern auditory theory, vol II. Academic, New York
Edmonds BA, Culling JF (2005) The spatial unmasking of speech: evidence for within-channel processing of interaural time delay. J Acoust Soc Am 117:3069–3078
Egan JP (1965) Masking-level differences as a function of interaural disparities in intensity of signal and noise. J Acoust Soc Am 38:1043–1049
Ewert S, Schubotz W, Brand T, Kollmeier B (2017) Binaural masking release in symmetric listening conditions with spectro-temporally modulated maskers. J Acoust Soc Am 142:12–28
Festen JM, Plomp R (1990) Effects of fluctuating noise and interfering speech on the speech reception threshold for impaired and normal hearing. J Acoust Soc Am 88:1725–1736
Freyman RL, Helfer KS, McCall DD, Clifton RK (1999) The role of perceived spatial separation in the unmasking of speech. J Acoust Soc Am 106:3578–3588
Gardner WG, Martin KD (1995) HRTF measurements of a KEMAR. J Acoust Soc Am 97:3907–3908
Grantham DW, Wightman FL (1978) Detectability of varying interaural temporal differences. J Acoust Soc Am 63:511–523
Harris G (1960) Binaural interactions of impulsive stimuli and pure tones. J Acoust Soc Am 32:685–692
Hawley ML, Litovsky RY, Culling JF (2004) The benefit of binaural hearing in a cocktail party: effect of location and type of interferer. J Acoust Soc Am 115:833–843
Hirsh IJ (1948) The influence of interaural phase on interaural summation and inhibition. J Acoust Soc Am 20:536–544
Hirsh IJ, Burgeat M (1958) Binaural effects in remote masking. J Acoust Soc Am 30:827–832
Houtgast T, Steeneken HJM (1985) A review of the MTF concept in room acoustics and its use for estimating speech intelligibility in auditoria. J Acoust Soc Am 77:1069–1077
ISO 3382 (1997) Acoustics – measurement of the reverberation time of rooms with reference to other acoustical parameters. International Organization for Standardization, Geneva
Jeffress LA (1948) A place theory of sound localization. J Comp Physiol Psychol 41:35–39
Jeffress LA (1972) Binaural signal detection: vector theory. In: Tobias JV (ed) Foundations of modern auditory theory, vol II. Academic, New York, pp 349–368
Jelfs S, Culling JF, Lavandier M (2011) Revision and validation of a binaural model for speech intelligibility in noise. Hear Res 275:96–104
Kidd G, Colburn HS (2017) Informational masking in speech recognition. In: Middlebrooks J, Simon J, Popper AN, Fay RR (eds) The auditory system at the cocktail party. Springer, Cham, pp 75–109
Kidd GJ, Mason CR, Brughera A, Hartmann WM (2005) The role of reverberation in release from masking due to spatial separation of sources for speech identification. Acta Acust United Acust 91:526–535
Kock WE (1950) Binaural localization and masking. J Acoust Soc Am 22:801–804
Kryter KD (1962) Methods for the calculation and use of the articulation index. J Acoust Soc Am 34:1689–1697
Kuhn GF (1977) Model for the interaural time differences in the azimuthal plane for the interaural time differences in the azimuthal plane. J Acoust Soc Am 62:157–167
Langford TL, Jeffress LA (1964) Effect of noise cross-correlation on binaural signal detection. J Acoust Soc Am 36:1455–1458

Lavandier M, Best V (2019) Modeling binaural speech understanding in complex situations. In: Blauert J, Braasch J (eds) The technology of binaural understanding. Springer, Berlin-Heidelberg/New York

Lavandier M, Culling JF (2008) Speech segregation in rooms: monaural, binaural, and interacting effects of reverberation on target and interferer. J Acoust Soc Am 123:2237–2248

Lavandier M, Culling JF (2010) Prediction of binaural speech intelligibility against noise in rooms. J Acoust Soc Am 127:387–399

Lavandier M, Jelfs S, Culling JF, Watkins AJ, Raimond AP, Makin SJ (2012) Binaural prediction of speech intelligibility in reverberant rooms with multiple noise sources. J Acoust Soc Am 131:218–231

Leclère T, Lavandier M, Culling JF (2015) Speech intelligibility prediction in reverberation: towards an integrated model of speech transmission, spatial unmasking and binaural de-reverberation. J Acoust Soc Am 137:3335–3345

Levitt H, Rabiner LR (1967) Predicting binaural gain in intelligibility and release from masking for speech. J Acoust Soc Am 42:820–829

Licklider JCR (1948) The influence of interaural phase relations upon the masking of speech by white noise. J Acoust Soc Am 20:150–159

Lochner JPA, Burger JF (1964) The influence of reflections on auditorium acoustics. J Sound Vib 1:426–454

McFadden D, Pasanen EG (1978) Binaural detection at high frequencies with time-delayed waveforms. J Acoust Soc Am 63:1120–1131

Metz PJ, von Bismarck G, Durlach NI (1966) Further results on binaural unmasking and the EC model. II noise bandwidth and interaural phase. J Acoust Soc Am 43:1085–1091

Müller C (1992) Perzeptive Analyse und Weiterentwicklung eines Reimtestverfahrens für die Sprachaudiometrie, Dissertation, Universität Göttingen, Germany

Nábělek AK, Robinson PK (1982) Monaural and binaural speech perception in reverberation for listeners of various ages. J Acoust Soc Am 71:1242–1248

Osman E (1971) A correlation model of binaural masking level differences. J Acoust Soc Am 50:1494–1511

Ozimek E, Jedrzej K, Kutzner D, Sek A, Wicher A (2013) Speech intelligibility for different spatial configurations of target speech and competing noise source in a horizontal and median plane. Speech Comm 55:1021–1032

Peissig J, Kollmeier B (1997) Directivity of binaural noise reduction in spatial multiple noise-source arrangements for normal and impaired listeners. J Acoust Soc Am 101:1660–1670

Platte HJ, vom Hövel H (1980) Zur Deutung der Ergebnisse von Sprachverständlichkeitsmessungen mit Störschall im Freifeld. Acta Acust United Acust 45:139–150

Plomp R (1976) Binaural and monaural speech intelligibility of connected discourse in reverberation as a function of azimuth of a single competing sound source (speech or noise). Acta Acust United Acust 34:200–211

Plomp R, Mimpen AM (1981) Effect of the orientation of the speaker's head and the azimuth of a noise source on the speech-reception threshold for sentences. Acta Acust United Acust 48:325–328

Rabiner LR, Laurence CL, Durlach NI (1966) Further results on binaural unmasking and the EC model. J Acoust Soc Am 40:62–70

Rennies J, Brand T, Kollmeier B (2011) Prediction of the influence of reverberation on binaural speech intelligibility in noise and in quiet. J Acoust Soc Am 130:2999–3012

Rhebergen KS, Versfeld NJ (2005) A speech intelligibility index-based approach to predict the speech reception threshold for sentences in fluctuating noise for normal-hearing listeners. J Acoust Soc Am 117:2181–2192

Robinson DE, Jeffress LA (1963) Effect of varying the interaural noise correlation on the detectability of tonal signals. J Acoust Soc Am 35:523–527

Schoenmaker E, Sutojo S, van de Par S (2017) Better-ear rating based on glimpsing. J Acoust Soc Am 142:1466–1481

Schubert ED (1956) Some preliminary experiments on binaural time delay and intelligibility. J Acoust Soc Am 28:895–901

Shinn-Cunningham BG, Ihlefeld A, Satyavarta LE (2005) Bottom-up and top-down influences on spatial unmasking. Acta Acust United Acust 91:967–979

van de Par S, Kohlrausch A (1997) A new approach to comparing binaural masking level differences at low and high frequencies. J Acoust Soc Am 101:1671–1680

van de Par S, Trahiotis C, Bernstein LR (2001) A consideration of the normalization that is typically included in correlation-based models of binaural detection. J Acoust Soc Am 109:830–833

van der Heijden M, Joris PX (2010) Interaural correlation fails to account for detection in a classic binaural task: dynamic ITDs dominate N0Sπ detection. J Assoc Res Otolaryngol 11:113–131

Wallach H, Newman EB, Rosenzweig MR (1949) The precedence effect in sound localization. Am J Psychol 62:315–336

Wan R, Durlach NI, Colburn HS (2010) Application of an extended equalization-cancellation model to speech intelligibility with spatially distributed maskers. J Acoust Soc Am 128:3678–3690

Wan R, Durlach NI, Colburn HS (2014) Application of a short-time version of the equalization cancellation model to speech intelligibility experiments with speech maskers. J Acoust Soc Am 136:768–776

Watson CS (2005) Some comments on informational masking. Acta Acust United Acust 91:502–512

Webster FA (1951) The influence of interaural phase on masked thresholds I. The role of interaural time deviation. J Acoust Soc Am 23:452–462

Westermann A, Buchholz JM (2015) The influence of informational masking in reverberant, multitalker environments. J Acoust Soc Am 138:584–593

Zurek PM (1993) Binaural advantages and directional effects in speech intelligibility. In: Studebaker G, Hochberg I (eds) Acoustical factors affecting hearing aid performance. Allyn and Bacon, Needham Heights, pp 255–276

Chapter 9
Spatial Hearing in Rooms and Effects of Reverberation

Pavel Zahorik

9.1 Introduction

Everyday listening environments are filled with surfaces that reflect sound. As a result, the sounds we hear are almost always complex combinations of both direct and reflected sounds. From a scientific standpoint, this can be a problem for the valid study of direct sound effects, where indirect sound is unequivocally viewed as a contaminant. For this reason, a large portion of all psychoacoustical research has been conducted under conditions that seek to minimize the impact of indirect sound. Although careful control of extraneous variables is a hallmark of quality science, the elimination of indirect sound arguably leads to unnatural listening situations where ecological validity may be questioned. This point is dramatically demonstrated by the historical work of Stevens and Newman (1936) on sound localization. Figure 9.1 shows the experimental setup designed to minimize indirect sound. Clearly, this environment is quite different from the environments in which we spend most of our time! More modern research has used headphone presentation methods or anechoic chambers to minimize the impact of indirect sound, but questions of ecological validity can remain. It is therefore a goal of this chapter to embrace indirect sound and to describe how and to what extent it affects what we hear.

Another potential explanation for the relatively small literature on the psychophysical effects of reflected sound is that it often goes unnoticed perceptually. A classic demonstration of this compares tape recordings of impact sounds and speech in environments that range in their reflectivity from anechoic to highly reverberant (Houtsma et al. 1987). When the recordings are played in the forward direction, the

P. Zahorik (✉)
Department of Otolaryngology-Head and Neck Surgery and Communicative Disorders and Department of Psychological and Brain Sciences, University of Louisville, Louisville, KY, USA
e-mail: pavel.zahorik@louisville.edu

© Springer Nature Switzerland AG 2021

R. Y. Litovsky et al. (eds.), *Binaural Hearing*, Springer Handbook of Auditory Research 73, https://doi.org/10.1007/978-3-030-57100-9_9

FIG. 1. THE EXPERIMENTAL SET-UP

Fig. 9.1 From Stevens (*standing*) and Newman (*seated*) on the roof of the Biological Laboratories at Harvard University (Cambridge, MA) with their experimental setup designed to minimize sound-reflecting surfaces for measurement of directional sound localization. (From Stevens and Newman 1936, with permission of the University of Illinois Press (Champaign). Copyright 1936 by the Board of Trustees of the University of Illinois (Champaign-Urbana))

prominent effects of indirect sound are only heard for the most reverberant environment. When the recordings are played in reverse, however, the indirect sounds are much more audible for all reflective environments, and a recreation of this demonstration confirms that speech is much less intelligible (Longworth-Reed et al. 2009). These results show that under normal situations, the human auditory system is surprisingly insensitive to clearly extant reflected sound. This suggests that the effortless nature of perceiving source information from signals at the ears contaminated by reflected sound (Picou et al. 2016) may result from unconscious processes in the both the peripheral and central auditory system that suppress the physical contributions of indirect sound.

There are, of course, many situations where reflected sound is audible and therefore not entirely suppressed. In these cases, it can substantially affect what is heard, where it is heard, and how good it sounds. This chapter seeks to introduce concepts and summarize current knowledge that relates to all of these issues. Concepts related to the physical aspects of reflected sound are described first, followed by psychophysical effects. These range from basic sensitivities to indirect sound to its impacts on higher-level listening tasks, such as localizing sound, understanding speech, and enjoying music. The chapter is not a comprehensive review of these areas but rather a conceptual and, in some cases, historical introduction and overview intended to facilitate and stimulate further research.

9.2 Acoustical Aspects

Room acoustics can have a significant impact on the sound reaching a listener. When physical measurements are compared between the source of a sound and that registered at a distant receiving point in a room, distortions along a number of primary acoustic dimensions may be observed. These include distortions of sound spectrum, temporal properties, spatial properties, and sound level. Fortunately, the distortions are all linear time invariant (LTI; see Table 9.1) distortions. As a result, standard system identification techniques for LTI systems can be used to characterize the acoustics of any room.

9.2.1 The Impulse Response of a Room

The response of the room to an ideal impulse (infinite energy, infinitesimal duration) is a complete acoustic description of the room. This is known as the impulse response (IR) of the room or room impulse response (RIR). For all LTI systems, the IR is a complete representation of how system output relates to system input so that

Table 9.1 List of abbreviations

Abbreviation	Definition
AM	Amplitude modulation
BRIR	Binaural room impulse response
D/R	Direct-to-reverberant energy ratio
IACC	Interaural cross-correlation coefficient
ILD	Interaural level difference
IR	Impulse response
ITD	Interaural time difference
JND	Just-noticeable difference
LTI	Linear time invariant
MTF	Modulation transfer function
RIR	Room impulse response
SNR	Signal-to-noise ratio
SRM	Spatial release from masking
STI	Speech transmission index
VR	Virtual reality

in the case of room acoustics, when the input source signal and the RIR based on a particular listening point in the room are known, the output signal for that listening point in the room can be determined exactly. The RIR is therefore a powerful tool for description and analysis of room acoustics.

Historically, effective measurement of a RIR required the production of an intense impulsive sound. More modern methods derive the IR from the response to known steady-state signals, and averaging is used to improve the signal-to-noise ratio of the measurement. Figure 9.2a displays an example of a RIR measured using a maximum-length sequence technique (Rife and Vanderkooy 1989). The measurement was made using an omnidirectional microphone and a point-source loudspeaker at a distance of 4.9 m on the stage of a small concert hall, as shown in Fig. 9.2b. Key features of the RIR, which include the direct-path response, individual early reflections, and later arriving reverberation, are indicated in Fig. 9.2a. The direct-path response occurs first in time and here contains primarily the electroacoustic effects of the equipment in the measurement chain plus a pure delay related to the distance between the sound source and receiving point. In this case, the pure delay was approximately 15 ms. To facilitate time referencing relative to the direct path, time zero was set to the point where the direct-path response was initiated. Following the direct-path response in time, the responses to individual reflections from surfaces in the room may be observed as "spikes" in the RIR, out to delay times of approximately 100 ms in this situation. These reflections result from the floor, side walls, and ceiling "clouds" visible in Fig. 9.2b. At longer delays, the reflections become increasingly dense in time, and their levels decay exponentially as a function of time. Collectively, these later arriving reflections are known as reverberation.

When the sound receiving point is a listener instead of a microphone, additional effects are introduced related to the acoustical contributions of the listener's head and external ears. Figure 9.2c displays a binaural RIR (BRIR) for the same situation depicted in Fig. 9.2b but using a KEMAR (Burkhard and Sachs 1975) measurement manikin with microphones at the two ears. The sound source was located 90 ° to the left of the KEMAR. Direct path, early reflections, and reverberation can be observed in Fig. 9.2c, just as in Fig. 9.2a; however, there are now substantial differences in the responses between the two ears. These differences are most prominent in the direct-path and early reflections where substantial interaural level (amplitude) differences (ILDs) and interaural time differences (ITDs) can be observed. For the direct path, the ILD and ITD result from the spatial location of the sound source relative to the listener. The source at 90 ° to the listener's left produces a slightly delayed (~600 μs) and decreased level to the right ear relative to the left. Early reflections also exhibit strong ILD and ITD patterns. For example, the reflection in Fig. 9.2c at approximately 50 ms has a much greater level in the right ear than the left ear. This results from the reflecting surface location to the listener's right side. In contrast to the clear directional components of the direct-path and early reflections, reverberation is spatially diffuse. This means that reverberation reaches the listener not from a single direction but from all directions. As a result, the average levels of reverberation at the two ears are relatively equal (see reverberation at time >160 ms; Fig. 9.2c), but the instantaneous levels at the two ears are uncorrelated.

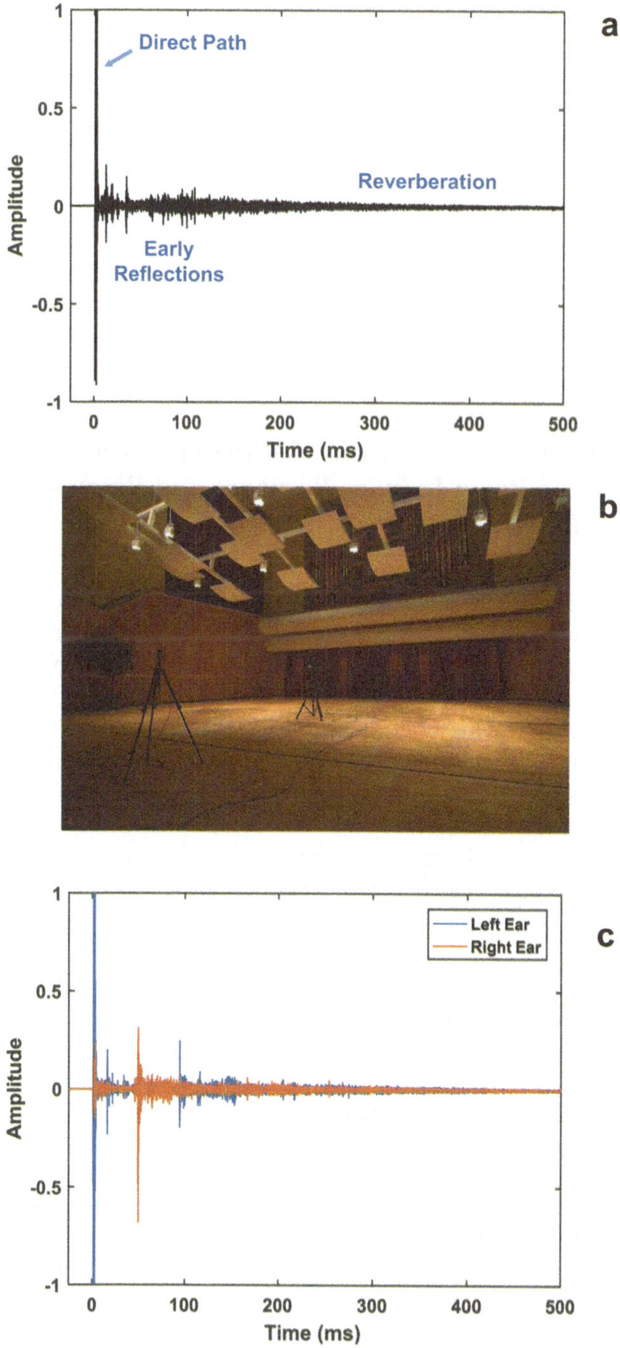

Fig. 9.2 Example of a room impulse response (RIR; **a**) and a measurement setup/environment (**b**) on the stage of a 558-seat auditorium (Comstock Hall, University of Louisville School of Music, KY). A binaural RIR (BRIR) is shown (**c**) for the same environment using a KEMAR manikin oriented at −90 ° azimuth such that the left ear faced the sound source. See text for a detailed explanation

It must be emphasized that the example RIR shown in Fig. 9.2a and the BRIR in Fig. 9.2c are specific to the particular room environment and source/receiver configuration within the room. Different rooms and different source/receiver configurations will produce different RIRs/BRIRs. This, of course, is one of the strengths of the RIR/BRIR method; it represents a complete characterization of a specific listening situation. Comparison of RIRs/BRIRs across different listening environments or source/receiver configurations allows for accurate physical quantification of environment or configuration effects. The BRIR can also be used to "auralize" a particular room listening situation using virtual auditory space techniques (Vorländer 2007). Thus, the BRIR is also critically important to enable detailed study of the perceptual effects of different listening room environments or configurations within the room.

9.2.2 Common Room Acoustic Measures that Can be Derived from the Room Impulse Response and Binaural Room Impulse Response

Although RIRs and BRIRs represent complete descriptions of the acoustics of the room both without and with the listener present, it is often difficult to easily summarize the key aspects of room acoustics from these measurements. Specific summary measures are therefore widely used. Sections 9.2.2.1, 9.2.2.2, 9.2.2.3, 9.2.2.4, 9.2.2.5, and 9.2.2.6 provide brief descriptions of selected common room acoustic summary measures, grouped by category. This list is by no means exhaustive, but it does cover the primary categories, which are based, in part, on the perceptual attributes of a sound in rooms that are relatively independent of one another. Because most measures can vary as a function of frequency in rooms, the measures are typically conducted in octave or suboctave bands, covering between 100 and 5000 Hz. ISO-3382-1 (2009) provides additional details on many of the measures.

9.2.2.1 Reverberation Time

The reverberation time represents the time it takes for sound energy to decay by a specified amount following the cessation of sound production at the source. Typically, the value that is used is 60 dB; hence, T_{60} is the most common measurement. Developed by Sabine (1922), reverberation time is the oldest and most common quantitative summary measure of room acoustics. It relates strongly to perceived reverberation, known as reverberance. Because it can often be difficult to set up situations in which the measurement signal is at least 60 dB above the background noise level of the environment, signal averaging and extrapolation methods are often used to estimate T_{60}. An alternative to T_{60} is a measure called early decay time (EDT). It is defined as six times the time it takes for sound energy to decay from 0 to −10 dB (Jordan 1970). Because this measure is based on early portion of the decay function, it is more sensitive to location within the room environment.

9.2.2.2 Relative Reverberation Energy

A second common type of summary measure compares the energy of reverberant sound to that of earlier arriving sound that is dominated by the direct path. Measures of this type are thought to relate to the perception of sound clarity and, in general, take the form of an energy ratio represented in decibels, where the numerator is early arriving energy and the denominator is later arriving energy. These energies can be derived from the RIR. There are two clarity index measures, C_{50} and C_{80}, that use values of 50 and 80 ms for as the separation criteria between early and late energy, respectively. C_{50} is typically used for speech and C_{80} for music. The direct-to-reverberant energy ratio (D/R) uses a separation criterion sufficiently brief to capture the direct-path sound only, with minimal contamination by reflections. A value of 2.5 ms is often effective (Zahorik 2002b). Another measure of balance between early and late arriving sound energy is center time (T_s) or the center of mass of the squared IR. This measure avoids the issue of defining the early and late portions of the RIR. Measures of relative reverberation energy, in addition to relating to the impression of sound clarity, can also relate to the overall perceived reverberation or reverberance.

9.2.2.3 Sound Strength

Sound strength (G) is a measure thought to be related to the loudness of sound in rooms. It is defined as the difference in decibels between the level of a sound source measured in a room and the level the same source generates in anechoic space at a distance of 10 m. The anechoic response energy can be estimated from the RIR by integrating only up to end of the direct-path response, and the anechoic level at 10 m can be estimated from an arbitrary measurement distance ≥3 m, assuming the inverse square law for sound propagation in the free field. ISO-3382-1 (2009) provides additional details on many of the measures.

9.2.2.4 Spatial Impression

Aspects related to the spatial impression of a sound in a room are often characterized by the interaural cross-correlation coefficient (IACC), which defines the degree of correlation between the left and right ear signals as a function of the delay between the signals. Often, delay values between −1 and +1 ms are considered, given that this is the approximate range of ITDs experienced by humans. To describe room acoustics, larger delays are sometime considered, and early (<80 ms) or late

(>80 ms) portions of the BRIR are be selectively targeted because they are thought to relate to more specific spatial attributes of a sound in rooms. For example, $IACC_{early}$ is thought to relate to the apparent spatial width of the sound source, and $IACC_{late}$ is related to the impression of being enveloped or surrounded by sound in the room. An additional common measure for spatial impression, the lateral energy fraction (LF) requires measurement using a microphone with a figure-of-eight response pattern. The LF, therefore, cannot be derived from the RIR or BRIR.

9.2.2.5 Timbre/Tone Color

Another primary perceptual attribute of listening to sound in rooms relates to the relative balance of sound across various frequency regions. This affects the tone color or timbre of sound in the room. Frequency balance can easily be evaluated by examining T_{60} or G values as a function of frequency. Another option from the literature on timbre perception (Grey and Gordon 1978) is to compute the spectral centroid, which conceptually is the center of mass of a signal's magnitude spectrum. Spectral centroid is not described in ISO-3382-1 (2009) but has been used to characterize the timbre effects in small rooms (Zahorik 2009). Colorations due to early reflections can be particularly strong in small rooms as a result of comb-filtering effects (see Sect. 9.3.3.3).

9.2.2.6 Speech Intelligibility

The most commonly used measure to estimate speech intelligibility within a room is the speech transmission index (STI; Houtgast and Steeneken 1985). The STI is based on the concept of the modulation transfer function (MTF), which describes how the amplitude modulation (AM) characteristics of an input signal are modified by a system, in this case, the acoustics of a room. Because speech is a signal that is rich in AM characteristics and the intelligibility of speech is known to depend critically on the preservation of these AM characteristics, distortions to these AM characteristics caused by room acoustics and quantified by the MTF can negatively impact speech intelligibility. The STI is a measure of this AM distortion caused by the room in the audio-frequency and modulation-frequency regions most important for speech understanding.

Figure 9.3 displays a contour plots of MTFs derived for an anechoic space and from the RIR shown in Fig. 9.2a. In general, room reverberation successively attenuates higher modulation frequencies, and thus their MTFs exhibit a low-pass characteristic (Houtgast and Steeneken 1985). This general pattern is evident by comparing the MTF results for the environment under test (Fig. 9.3b) with those from the anechoic space (Fig. 9.3a). Individual reflections also influence the MTFs in Fig. 9.3b and contribute to the complex patterns of MTF behavior as functions of the modulation and audio frequencies, especially when compared with anechoic space (Fig. 9.3a). The STI is computed from the MTFs over the range of 125–8000 Hz

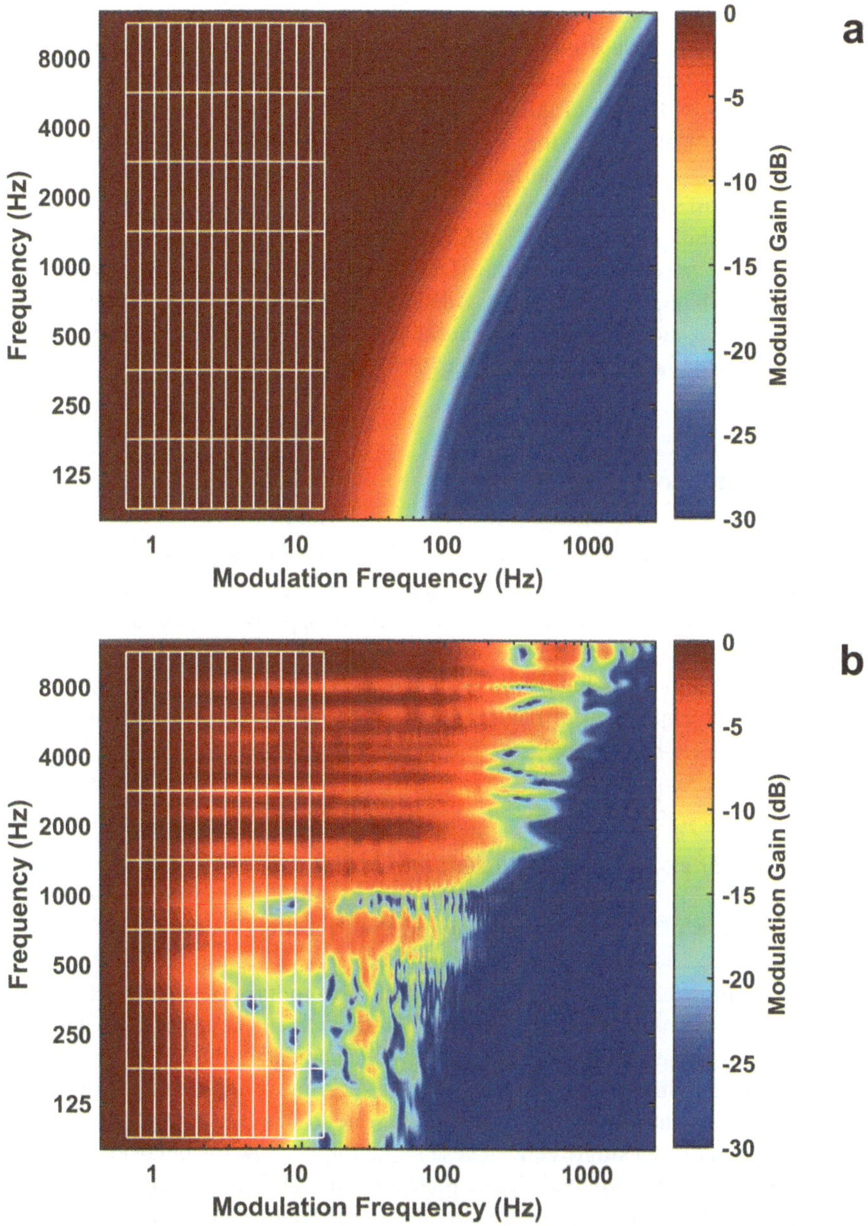

Fig. 9.3 Contour plot of modulation transfer functions (MTFs) as a function of audio frequency (80 Hz to 12.5 kHz) for (**a**) an anechoic space and (**b**) the room shown in Fig. 9.2b. Grid of audio-frequency by modulation-frequency regions used to compute the speech transmission index (STI). Here, the computed STI approaches 1 for an anechoic space and was 0.67 for the room environment

in audio frequency and 0.63–12.5 Hz in modulation frequency, in octave and one-third octave bands, respectively. A modulation gain value is then determined for each cell in this matrix of 7 audio by 14 modulation frequencies (shown superimposed in Fig. 9.3). Details of the computation of the STI are provided in Houtgast and Steeneken (1985), but the STI is essentially a weighted average of this modulation gain matrix, normalized to the range between 0 and 1. High values of STI indicate little modulation reduction by the room and therefore predict good speech intelligibility in the room. The STI for the room analyzed here is 0.67, which predicts "good" speech understanding. Details of the validation studies relating STI values to speech intelligibility are provided in Sect. 9.3.2.3.1. Although the original STI is based on monaural input, binaural versions of the STI have been implemented (van Wijngaarden and Drullman 2008).

9.3 Psychoacoustical Aspects

From Sect. 9.2 on the physical effects, it is clear that room acoustics can have a major impact on the sound reaching a listener. Separate and critical questions concern the extent to which the listener is sensitive to aspects of room acoustics and how room acoustics may affect various objective and subjective listening capabilities and applications, including the localization of sound, the understanding of speech, and the enjoyment of music. This section on psychoacoustical aspects of room acoustics explores knowledge in these areas.

9.3.1 Listener Sensitivity to Room Acoustics

9.3.1.1 Effects of a Single Reflection and the Precedence Effect

Perhaps the most basic listening scenario related to room acoustics is listening to the source of a sound with a single ideal reflection, or echo, that is simply a delayed copy of the source signal. There has been considerable scientific study of this scenario, conducted almost exclusively under simplified acoustical conditions using two loudspeakers in an anechoic space: one to represent the direct path and another to represent a single echo. These simulations allow the delay (τ) between presentation of the simulated direct path and the simulated echo to be easily manipulated. Different percepts are elicited for different values of τ. At the shortest delays ($\tau < 1$ ms for an impulse signal), the percept is one of a single fused-sound image at a phantom location somewhere between the two loudspeakers (an effect often referred to as "summing localization"). At intermediate delays (1 ms $< \tau <$ 5 ms), the percept is of a fused-sound image at the location of the direct-path loudspeaker. This dominance of the direct-path sound in specifying the spatial location is commonly referred to as the "precedence effect" (Wallach et al. 1949). Finally, for longer delays ($\tau > 5$ ms), both direct path and echo are perceived as originating from distinctly separate spatial locations. Although the exact values of τ that demarcate each of the three percept categories are affected by a number of factors, particularly source signal type, there is a range of τ that physically produces a single reflection that is not audible as a spatially separate, distinct sound. Thus, in this range of τ, the direct path takes perceptual precedence over reflected sound in terms of specifying the spatial location of the sound source.

The precedence effect has been perhaps the most widely studied psychophysical aspect of listening to sound in rooms because it offers an explanation as to why reflections that are clearly physically measurable are often not heard as separate events. Although some of the original observations of this effect were concerned exclusively with the perceptual dominance of the first-arriving wave front in specifying the spatial location of the sound source (Wallach et al. 1949), the term precedence effect now has come to be associated with a suite of phenomena all related to auditory perception in reflective acoustic environments. There are a number of reviews of the extensive precedence effect literature. Interested readers are referred to Brown et al. (2015) for review of the most recent literature, Blauert (1997) and Litovsky et al. (1999) for excellent comprehensive reviews of the classic and older literature, and Gardner (1968) for historical perspectives.

One of the most common ways to quantify the precedence effect is to determine the minimum τ required hear the source and the echo as two distinct sounds. This is known as the echo threshold, which, of course, requires an inherently subjective criterion as to what constitutes two sounds. Common criteria have included that the echo is audible as a spatially separate auditory event (Wallach et al. 1949; Blauert 1997), the echo is audible as a disturbance (Haas 1972), and the source and echo are equally loud (Lochner and Burger 1958). Note that all of these criteria for the echo threshold exclude changes in the sound quality of the source caused by the echo, such as loudness or timbre changes, which can occur at much shorter delays. For example, two-click thresholds as low as $\tau = 10$ μs have been reported (Leshowitz 1971), but this situation does not result in two separate auditory events. Instead, differences in the sound quality (i.e., timbre) of a single sound are heard for one versus two clicks, and listeners base their judgements on these differences. Subjective sound quality effects due to room acoustics are discussed in Sect. 9.3.3.

The echo threshold depends on a variety of stimulus factors. The most major factor is the type of source signal. In general, the lowest echo thresholds are obtained for impulsive signals, and longer signals with more gradual onsets/offsets yield larger thresholds. For example, the echo threshold when the independent variable is τ ranges from less than 3 ms for clicks (Babkoff and Sutton 1966) to 50 ms or more for continuous signals such as speech (Lochner and Burger 1958). Additional factors known to influence the echo threshold include sound level (Lochner and Burger 1958), frequency content (Dizon and Colburn 2006), and spatial location (Litovsky and Shinn-Cunningham 2001; Brown and Stecker 2013). There is also often considerable intersubject variability in measures of the echo threshold. For example,

Litovsky and Shinn-Cunningham (2001) reported echo thresholds for a pulse stimulus as a function of τ that range from 1.5 ms to nearly 12 ms across 6 trained listeners.

An additional factor that can substantially influence the echo threshold is the recent stimulus history experience by the listener (Clifton 1987). For example, when a repeating source signal (e.g., a train of clicks) is presented along with a correspondingly repeating echo, the echo becomes less audible over the repetitions. This "buildup" of echo suppression can increase the echo threshold by a factor of two or more compared with standard precedence experiments using a single stimulus presentation (Freyman et al. 1991) and can take tens of seconds to reach maximum suppression (Freyman et al. 1991). This dynamic change in the strength of the precedence effect based on ongoing input can be thought of as a form of perceptual adaptation.

Precedence effect buildup can also be dramatically destroyed when the auditory input indicates implausible or unnatural changes to the source and echo relationship, such as an abrupt change in the spatial locations of the source and echo (Clifton 1987). This "breakdown" in precedence results in echo thresholds that are lower than would be observed for a single source plus echo stimulus presentation (Yost and Guzman 1996) and may be a form of negative aftereffect of the perceptual adaptation. It has been suggested that this process of adaptation is perhaps one in which a model of the acoustic environment is constructed and represented in the brain (Clifton et al. 1994; Keen and Freyman 2009). Such a model would allow subsequent inputs to be evaluated in the context of the current acoustic environment, effectively suppressing the potentially misleading spatial information provided by echoes and reverberation. The buildup effects are perhaps indicative of the experience-driven nature of the environmental models. Likewise, breakdown of this adaptation results when current sensory input becomes implausible in the context of the environmental model and is manifest as a negative aftereffect. This hypothesized model-building process does not appear to be mediated by cognition, however, because it has been shown to be resistant to practice and learning (Clifton and Freyman 1997).

There has also been considerable research on the neural bases of the precedence effect. For classic demonstrations of the precedence effect with a single source/echo pair stimulus, clear physiological correlates have been observed in the responses of signal neurons within the inferior colliculus (Yin 1994). Evidence for weak echo suppression is also evident in more peripheral parts of the auditory system (see Brown et al. 2015 for a review). Other aspects of the precedence effect, including its behavioral demonstration and adaptational nature, clearly require more central brain processes. For example, precedence effect behavior in cats has been shown to be dependent on the function of auditory cortex (Cranford et al. 1971), and the behavioral demonstration of the precedence effect in humans emerges along a trajectory consistent with childhood development of cortical processing (Litovsky 1997). This latter effect is even more striking given that precedence effect processing at the level of the inferior colliculus has been demonstrated at birth (Litovsky 1998). Cortical involvement in precedence effect buildup has also been shown

(Grantham 1996; Sanders et al. 2011). See Chap. 4 by Takahashi, Kettler, Keller, and Bala for additional information on the neural bases of the precedence effect.

Hearing loss can also influence the precedence effect. Although it is clear that individuals with hearing loss still experience precedence, there is large individual variability in the size of the effect. Factors such as age (Cranford et al. 1993) and the degree of hearing loss (Akeroyd and Guy 2011) have been shown to contribute, but significant unexplained individual variability remains. The relationship to more general declines in sound localization abilities with hearing loss is unknown (Brown et al. 2015).

9.3.1.2 Audibility of Multiple Reflections

Relatively little study of the precedence effect has been conducted in situations with more than a single acoustic reflection. Early work by Seraphim (1961) showed that for ideal simulated reflections that are delivered at the same level and at the same frontal location as the direct-path sound all are similarly audible. Although the audibility criterion in this study was absolute detection (e.g., any detectable change in sound qualities, not limited to spatial location), these results suggest that precedence may operate in an ongoing fashion. That is, each successive reflection may contribute to preserving the precedence effect for subsequent reflections, out to delays of at least 70 ms (Seraphim 1961). Of course, in real rooms, reflections do not all have equal levels. Nevertheless, sound sources in real rooms (Hartmann 1983) or loudspeaker simulations of rooms (Bech 1998) are seldom localized at the locations of sound reflections, indicating that the precedence effect must be active in these situations. Similarly, Olive and Toole (1989) have shown that image shifts caused by a single strong reflection (e.g., summing localization) are remarkably unaffected by later reflections and reverberation. This is again consistent with the idea that earlier arriving sound takes perceptual precedence spatially over later arriving sound. The apparent lack of spatial influence in early reflections in rooms may also be facilitated by precedence effect buildup, which has been demonstrated to generalize to a second echo (Yost and Guzman 1996; Goupell et al. 2012) and to a room simulation with 12 echoes delayed as much as 40 ms from the direct path (Djelani and Blauert 2001).

9.3.1.3 Differential Sensitivity to Room Acoustic Measures

A number of studies have examined listeners' differential sensitivities to aspects of room acoustics. Such studies seek to determine the minimum change in a single-room acoustic parameter that a listener can discriminate. In psychophysics, this is known as the just-noticeable difference (JND). Table 9.2 provides a summary of measured JNDs for common room acoustic parameters described in Sect. 9.2.2.

Table 9.2 Just-noticeable difference values for common room acoustic measures

Measure	JND	Reference
T_{60}	24 ms	Seraphim (1958)
D/R	2–6 dB	Reichardt and Schmidt (1966) and Zahorik (2002b)
C_{50}	1 dB	Bradley et al. (1999)
C_{80}	0.5–1 dB	Cox et al. (1993)
T_s	6–11 ms	Cox et al. (1993)
G	1 dB	Okano (2002)
IACC	5%	Okano (2002)
f_c	7%	Emiroglu and Kollmeier (2008)

JND just-noticeable difference, T_{60} reverberation time, *D/R* direct-to-reverberant energy ratio, C_{50} clarity index measure with value of 50 ms, C_{80} clarity index measure with value of 80 ms, T_s center time, *G* sound strength, *IACC* interaural cross-correlation coefficient, f_c spectral centroid

One on the primary uses of these JND measures is in the comparison of different physical rooms (Bradley 2011) or virtual simulations of rooms (Bork 2000; Katz 2004). If two different listening situations produce physical measures of a given acoustic parameter that fall within one JND for that parameter, then the two listening situations would be indistinguishable from one another based on that parameter. Rooms can, of course, be characterized based on a number of acoustic parameters as described in Sect. 9.2.2. Therefore, if the end goal is to claim that two listening situations are indistinguishable, typically a number of room acoustic parameters are measured and compared with their corresponding JND values. This approach assumes that the set of room acoustic parameters evaluated represents a complete description of the listening environments, which, of course, is difficult to guarantee. One alternative approach is direct comparison between different listening situations that allows listeners to judge differences between the situations based on any listening aspects or parameters of their choosing. This approach, which has been used to evaluated virtual sound simulation techniques in anechoic (Zahorik et al. 1995; Kulkarni and Colburn 1998) and reverberant (Zahorik 2009) spaces, is not practical for real spaces because such comparisons involve physically moving the listener from one room to another. Preference rating scale approaches are therefore often used in real room evaluations (Beranek 2012).

An important consideration in interpreting JNDs for room acoustic measures is that the process for determining the JND requires that a given measure be isolated and then manipulated in the absence of changes to any other measures. In real listening scenarios, such change of a single parameter in isolation is uncommon.

9.3.2 *Objective Effects of Room Acoustics on Listening Performance*

9.3.2.1 Directional Sound Localization

Under many situations, reflected sound has a minimal impact on listeners' abilities to determine the directional location of a sound source. This remarkable insensitivity has been demonstrated in both large (Hartmann 1983) and small (Bech 1998) rooms and appears to result from processes that underlie the precedence effect. This is not to say that directional localization is completely unaffected by room acoustics, however. Strong early reflections and late reverberation can both distort directional localization cues and result in decreased directional localization accuracy. For example, Rakerd and Hartmann (1985) showed that single reflections, particularly lateral reflections, can distort both ITD and ILD cues to direct-path sound direction and that small but measurable degradations in directional localization performance result.

Greater degradations in localization accuracy can be observed in reverberation. This is due to that fact that late reverberation is spatially diffuse, meaning that it contains energy from all spatial locations and therefore is nondirectionally specific. Diffuse sound fields result in equal energy but decorrelated signals at the two ears. Relative increases in reverberant energy will therefore cause distortions to primary acoustic cues to direction. Specifically, the ILD will tend toward zero and the ITD will because less reliable. The ITD cue is thought to be extracted from ongoing signals through analysis of the correlation between signals at the two ears (Jeffress 1948), also known as binaural coherence (Rakerd and Hartmann 2010). Binaural coherence can be quantified by the IACC (see Sect. 9.2.2.4). Lower values of the IACC can often make the ITD more difficult to extract and, therefore, less reliable. For example, Fig. 9.4 demonstrates that the room from Fig. 9.2 causes IACC values to decrease and become more variable across frequency. This may explain why listeners have been shown to increasingly weigh biased ILD cues to direction when reverberation is increased (Ihlefeld and Shinn-Cunningham 2011), where otherwise in the absence of reverberation, the ITD is a more dominant directional cue (Wightman and Kistler 1992).

Perhaps the most dramatic example of room acoustics affecting directional localization was first described by Franssen (1960) and is now known as the Franssen illusion or Franssen effect. To produce the effect, two loudspeakers are required in typical stereophonic reproduction locations, as shown in Fig. 9.5. A brief tone is presented to one loudspeaker with abrupt onset and gradual offset. As this first tone is faded out, a second tone of the same frequency is faded in to the other loudspeaker. Surprisingly, this second tone continues to be (mis)localized at the location of the first tone. Thus, the first arriving wave front determines the perceived location of the sound source, seemingly consistent with the precedence effect, and the result is a large error in sound localization.

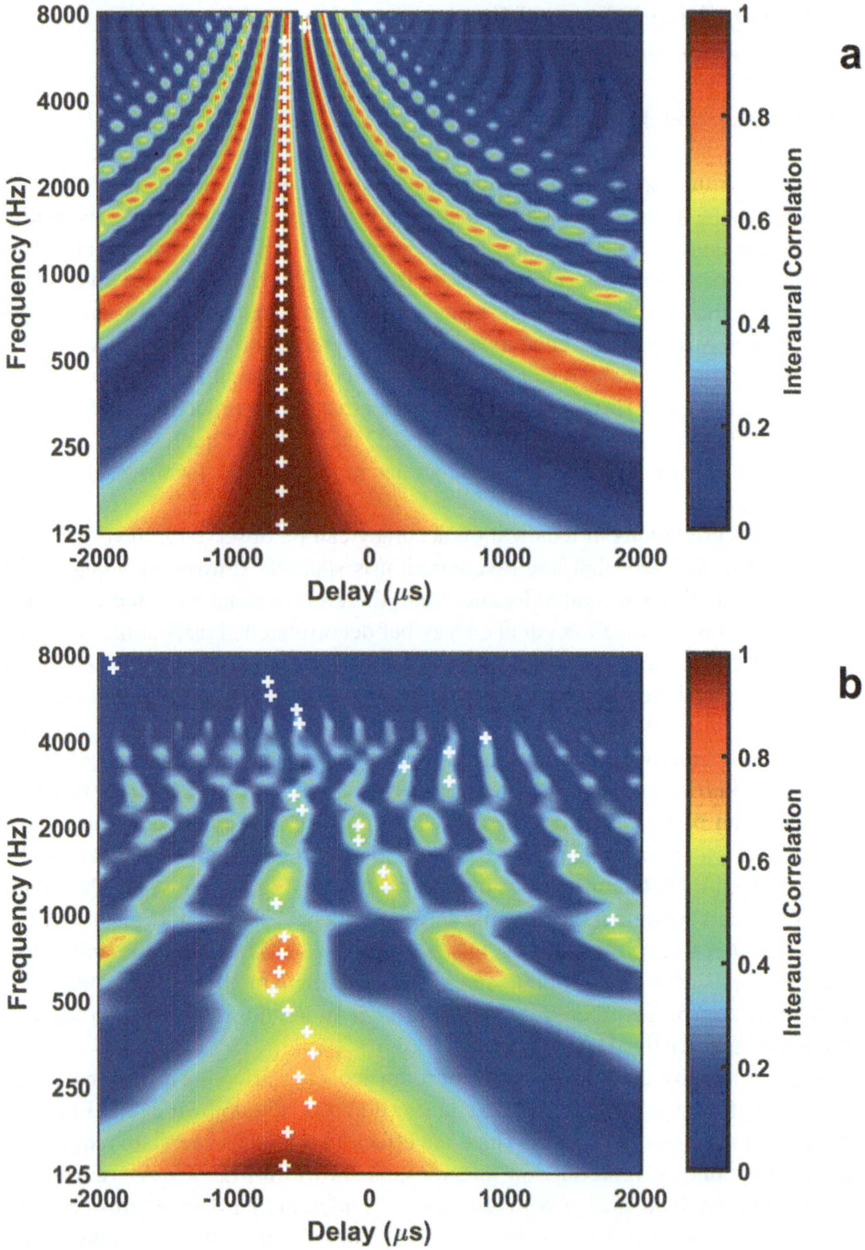

Fig. 9.4 Binaural cross-correlegrams (contour plots of interaural cross-correlation as a function of frequency) for (**a**) an anechoic space and (**b**) the room shown in Fig. 9.2b. The sound source was located at −90° azimuth, and produced the BRIRs shown in Fig. 9.3. Maximum interaural cross-correlation (IACC) for each auditory filter is shown (+ symbol). Note the across-frequency consistency of IACC in the anechoic space and inconsistency in the reverberant room

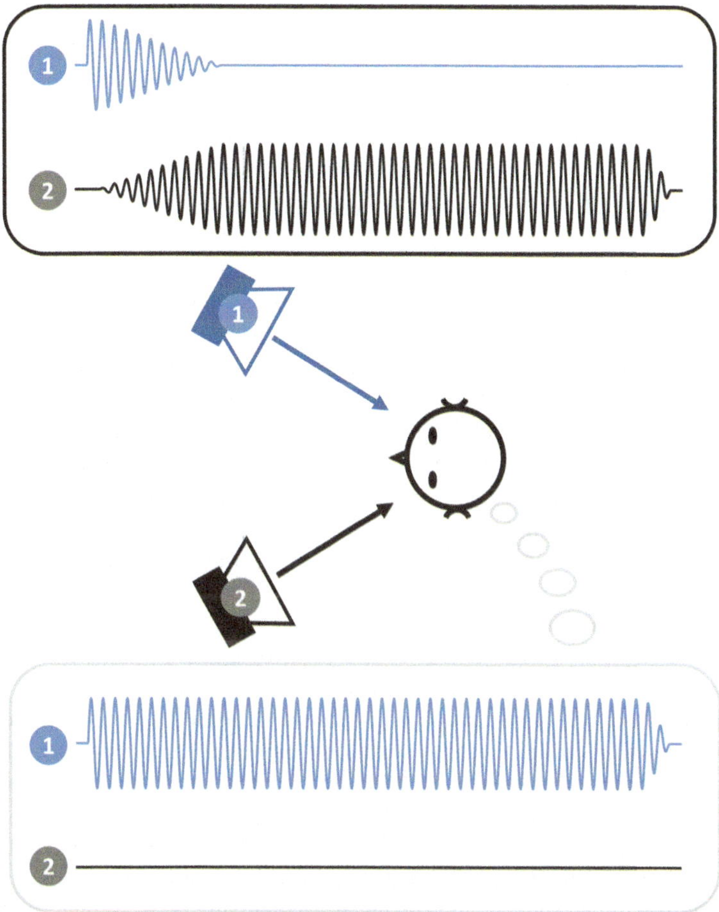

Fig. 9.5 Physical stimulus and setup for the Franssen illusion (*top* and *center*) where a tonal signal is presented initially to the right loudspeaker (1) and then faded off while the same signal is faded on to the left loudspeaker (2). The resulting spatial perception (*bottom*) is of a continuous tone at the right loudspeaker (1)

The stimulus conditions required to produce the Franssen illusion are more restrictive, however. Unlike the precedence effect, which works for a variety of source signals in a variety of acoustic spaces including an anechoic space, the Franssen illusion requires a reverberant room and tonal signals of relatively low frequency (Hartmann and Rakerd 1989). These stimulus conditions make it difficult to localize the ongoing signal due to ITD distortions caused by room reverberation but have little effect on the ITD information present in the abrupt signal onset. Thus, the apparent source location is dominated by the signal onset. Broadband signals and/or presentation in an anechoic space provide the listener with more robust localization cues in the ongoing signal and, hence, destroy the effect (Hartmann and Rakerd 1989). The fact that other species, such as birds, also experience the Franssen illusion (Dent et al. 2007) implies that illusion is driven by relatively low-level processing, although human brain imaging results suggest that the effect relates to perceptual representations of auditory space that emerge at higher-order regions of auditory cortex (Higgins et al. 2017).

9.3.2.2 Distance Localization and Externalization

Unlike directional sound localization, where reflected sound is generally antithetical to good localization accuracy, the ability to localize sound in distance depends critically on indirect sound. A primary cue to sound source distance is the D/R, which systematically decreases with increases in source distance and is independent of sound source power. Auditory distance perception in humans is known to be dependent on the D/R (Mershon and King 1975; Bronkhorst and Houtgast 1999) and can be dramatically disrupted when no indirect sound energy is present, such as in anechoic listening environments (Gardner 1969). Sensitivity to the D/R for distance localization has been demonstrated in nonhuman species such as song birds (Naguib 1995), and the D/R cue is present in a variety of listening environments in addition to rooms, ranging from forests (Sakai et al. 1998) to city streets (Wiener et al. 1965).

In addition to the D/R, other acoustic cues such as relative sound level, frequency content, and interaural differences can also contribute to the auditory distance perception, all of which are influenced by indirect sound in the listening environment. Humans appear to be flexible at utilizing distance information from multiple acoustic cues depending on the listening situation (Zahorik 2002a), however. This may help to minimize poor or misleading distance information in certain listening situations, much the way that the visual system combines and processes information across multiple and potentially conflicting visual distance cues (Landy et al. 1995). Even with these relatively sophisticated cue-processing strategies, human abilities to accurately estimate sound source distance are relatively poor compared with horizontal or vertical plane localization. High variability in the estimates and systematic biases are commonly observed, with near sources typically overestimated in distance and far sources underestimated. Interested readers are referred to two comprehensive reviews of the distance perception literature including a detailed description of the acoustic cues and human performance characteristics (Zahorik et al. 2005; Kolarik et al. 2016).

Because acoustic distance cues are affected by individual listening environments, it is natural to suppose that listeners might benefit from previous exposure to a particular environment and therefore be able to process indirect energy more effectively. There is some evidence for this in reverberant rooms with improvements in the abilities to localize the distances of nearby sounds (Mershon et al. 1989; Shinn-Cunningham 2000), which generalizes to improved localization performance at different listener positions in the same room (Kopčo et al. 2004). There is also evidence to suggest that listeners may have implicit knowledge of listening environment acoustics as it relates to sound propagation losses (Zahorik and Kelly 2007). In theory, such knowledge could facilitate the use of sound level cues in estimating the source distance, yet systematic biases (Zahorik et al. 2005) and variability (Anderson and Zahorik 2014) in distance judgments suggest that listeners may not optimally use this information.

Relatively little is known regarding the neural correlates to perceived sound source distance. On the one hand, high-level spatial processing in the posterior nonprimary auditory cortex has been implicated in the intensity-independent encoding of auditory distance cues (Kopčo et al. 2012). On the other hand, the processing of certain distance cues, such as the relative level and spectral shape, are implemented at early stages in the auditory pathway. Although the processing of reverberation is clearly central to distance perception and other important auditory capabilities, until recently, the nature of its representation within the auditory system was unknown. There is now emerging evidence that reverberation is not represented directly but rather indirectly as either monaural changes to the depth of amplitude modulation (Kim et al. 2015) or the degree of binaural coherence (Slama and Delgutte 2015) because both of these variables covary with the relative reverberation level. There is also evidence to suggest that the ability to process reverberation for source distance estimation is negatively affected by hearing loss (Akeroyd et al. 2007).

Another important perceptual role of reverberation is to convey a sense that a sound source is external to the listener's head (Plenge 1972; Laws 1973). Externalization is a central issue in sound reproduction over headphones where sound is typically perceived as internal to the head. It is also likely a multidimensional construct because multiple factors in addition to reverberation have been shown to contribute to externalization, including spectral details related to the acoustics of the head and external ears (Kulkarni and Colburn 1998), head movement (Wallach 1940; Brimijoin et al. 2013), binaural cues (Hartmann and Wittenberg 1996), visual information (Gil-Carvajal et al. 2016), and even expectation (Plenge 1974). There is evidence to suggest that reflected sound and reverberation may play primary roles in externalization (Begault et al. 2001) and that dynamic fluctuations in binaural cues caused by interaction with room reverberation are particularly important (Catic et al. 2013, 2015). Although there is not a clear consensus as to whether and how externalization might be related to distance perception and whether it is a categorical or continuous variable, externalization has a clear practical importance for realistic virtual sound reproduction applications over headphones (Begault 1992; Durlach et al. 1992). Hearing loss has also been shown to negatively impact sound externalization (Boyd et al. 2012), and sound reproduced over hearing aids or cochlear implants is often internalized (see Ricketts and Kan, Chap. 13).

9.3.2.3 Speech Intelligibility

Speech understanding has long been known to be affected by room acoustics (Knudsen 1929). Although room acoustic effects are generally thought to degrade speech understanding, this degradation is primarily caused by late reverberation. Strong early reflections often have little effect on speech perception (Haas 1972) due to the precedence effect (see Sect. 9.3.1.1) and can even enhance speech understanding (Bradley et al. 2003). This enhancement results from effective amplifica-

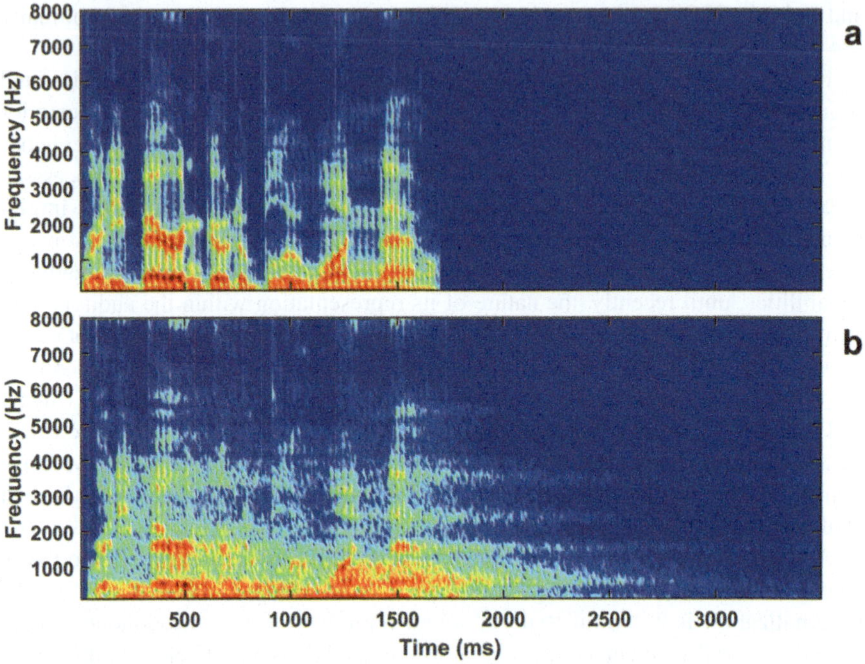

Fig. 9.6 Spectrograms of the sentence "Ready Baron go to blue one now" (CRM corpus, Bolia et al. 2000) in (**a**) an anechoic space and (**b**) the reverberant space shown in Fig. 9.2b

tion caused by integration of early reflection energy with direct-path energy (Bradley et al. 2003). Room reverberation, however, results is a variety of degradations to the speech signal that directly affect speech understanding.

9.3.2.3.1 Temporal Distortion and Validation of the Speech Transmission Index

The primary degradation to speech caused by reverberation is temporal in nature. Figure 9.6 displays spectrograms for anechoic speech and speech in the room shown in Fig. 9.2. A clear temporal "smearing" of the speech signal may be observed where, in general, the amplitude variations of the speech signal become less pronounced as a function of time. This physical effect is well characterized by the modulation transfer function concept and the STI described in Sect. 9.2.2.6. Another advantage of the STI is that it allows different acoustical attributes of rooms known to affect speech understanding, such as reverberation time (Knudsen 1929) or relative reverberation level (Bradley 1986), all represented using a single common measure.

An elegant body of work has confirmed the predictive power of the STI through validation with a variety of speech-understanding measures. In general, the validation studies related a range of STI values resulting from various room environments and signal manipulations to measures of speech understanding. Figure 9.7 summa-

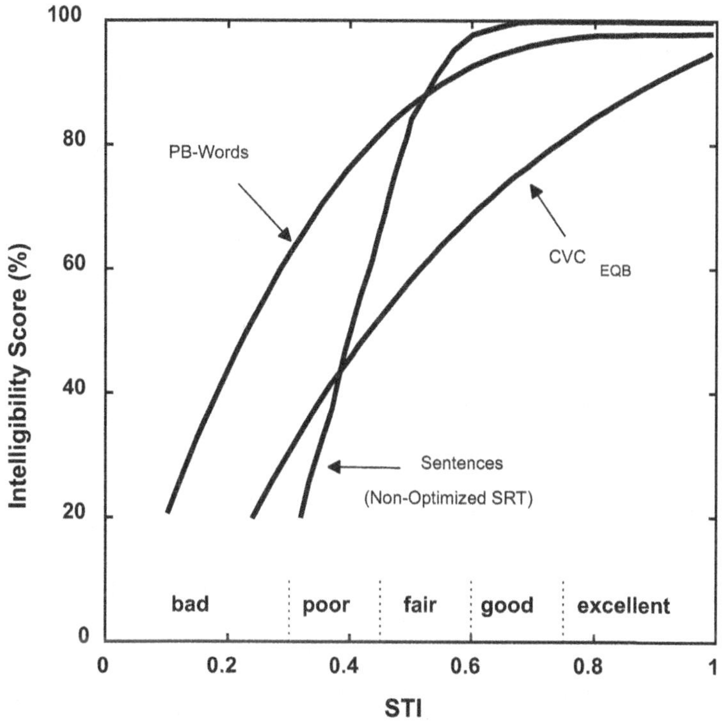

Fig. 9.7 Relationship between the STI and three measures of speech intelligibility based on consonant-vowel-consonant (CVC) nonsense words (Steeneken and Houtgast 1999), phonetically balanced (PB) words (Anderson and Kalb 1987), and sentences (Steeneken 1992). (Based on Steeneken and Houtgast 2002, with permission)

rizes the results of three validation studies conducted over many listening conditions with multiple normal-hearing listeners for three different types of speech materials: nonsense consonant-vowel-consonant (CVC) words, phonetically balanced (PB) words, and sentences. The differing relationships between STI and speech intelligibility as a function of speech material stem from at least two contributing factors. First, the different materials produce different performance levels as listening difficulty is increased. Sentence understanding benefits from context compared with isolated words, and familiar words can be guessed more easily than nonsense words. These, of course, are nonacoustical factors. A second contributing factor is acoustical and relates more specifically to how decreased temporal modulation can affect speech understanding. For isolated words, the temporal modulation characteristics required for identification are quite different (i.e., shorter timescale, faster modulation frequency) than for sentences where longer timescales and slower modulation frequencies can still aid in understanding and are less sensitive to degradations due to room acoustics.

The issue of the acoustical timescale and the STI relates to a well-known distinction for describing the effects of reverberation on speech understanding based on masking. In general, reverberant energy from preceding speech sounds can mask speech that follows, and the masking can occur on different timescales. For short timescales within a phoneme, the masking is referred to as self-masking (Bolt and MacDonald 1949), and for longer timescales across one or more phonemes, the masking is referred to as overlap masking (Nábělek et al. 1989). The general principles of energetic and temporal masking can be applied to this situation and are consistent with various vowel and consonant identification testing results in reverberation (Nábělek et al. 1989).

9.3.2.3.2 Effect of Background Noise

Although it is well-known that background noise can degrade speech understanding, there is also evidence to suggest that the combined effects of noise and reverberation are worse than either effect alone (Nábělek and Mason 1981). Subsequent work has shown that this result is predictable based on the STI (George et al. 2010). Another advantage of the STI method is that in addition to reverberation, it can be used to predict the effects of stationary background noise on speech understanding because noise also affects the modulation characteristics of the speech signal. STI measures that include the effects of noise are conducted in a different way than the impulse response based methods described in Sect. 9.2.2.6. Interested readers are referred to IEC-60268-16 (2011) for measurement methodology details.

There is also a different type of effect that occurs for nonstationary noise such as competing speech. In this situation, the signal-to-noise ratio (SNR) is time varying. As a result, there are instants in time and frequency where the SNR is more favorable such that listeners are provided with "glimpses" of relatively unmasked speech. Thus, overall speech intelligibility can be improved by glimpsing (Festen and Plomp 1990). In reverberation, however, intelligibility advantages due to glimpsing are reduced (George et al. 2008) because the reverberation causes the nonstationary noise source to become more stationary. Unfortunately, neither the STI nor the speech intelligibility index (SII; ANSI 1997), which is another common method of predicting speech intelligibility, are valid for nonstationary maskers. An extended SII (ESII) was therefore developed (Rhebergen and Versfeld 2005) and has been shown to predictive with nonstationary noise in reverberation (George et al. 2008).

A related effect is known as perceptual restoration (Warren 1970), where the intelligibility of temporally interrupted speech is improved when the interruptions are caused by a masking noise versus silence. This surprising result is attributed to top-down processes that perceptually fill in the missing speech but only when plausibly masked by a competing sound. Although such processing would clearly be advantageous in real-world situations with a fluctuating SNR, it is dramatically reversed in reverberation (Srinivasan and Zahorik 2012). This is because the silent intervals are effectively filled in by the reverberation, therefore creating restoration, and masking noise creates overlap masking that limits intelligibility.

9.3.2.3.3 Binaural and Spatial Effects

In reverberation, listening with two ears provides benefits in speech understanding relative to monaural listening. This binaural advantage, first demonstrated objectively for speech by Moncur and Dirks (1967), likely relates to more general aspects of binaural processing. For example, Koenig (1950) pointed out that binaural listening in reverberation results in a subjective decrease in reverberation level for both speech and nonspeech sounds. This "binaural squelch" of reverberation is likely related to how the coherent direct-path information and incoherent reverberation information combine from the two ears. Results from the binaural masking level difference (BMLD) literature (Jeffress et al. 1956) suggest that such a situation should result in roughly a 4 dB decrease in effective reverberation level relative to diotic presentation (see also Culling and Lavandier, Chap. 8). There is also evidence showing a binaural advantage for the detection of AM in reverberation (Danilenko 1969) that, based on the predictive strength of the STI concept, may offer an additional explanation for the binaural advantages observed for speech understanding in reverberation.

When speech is presented in an environment with one or more competing sound sources, speech understanding is improved when the target speech is spatially separated in direction from the competing sound maskers. Known as the spatial release from masking (SRM), the effect can greatly improve speech understanding (see Culling and Lavandier, Chap. 8). For example, Arbogast et al. (2002) report a SRM as large as 18 dB (see Middlebrooks et al. 2017 for a review). In reverberation, however, the SRM has been shown to be much smaller (Plomp 1976; Marrone et al. 2008). This is likely due to decreases in the precision with which the target speech and competing sound sources can be localized in reverberation and the resulting distortions to ILD and ITD cues (see Sect. 9.3.2.1). Reverberation can facilitate the SRM in the distance dimension, however (Westermann and Buchholz 2015).

9.3.2.3.4 Effects of Hearing Loss and Age

Although there is a general consensus that both hearing loss and age can be predictors of poor speech understanding in reverberation, the literature is somewhat varied in its assessment of the strength of these predictors and their interrelationship. The strong association between hearing loss and age contributes to the difficulty in assessing these effects independently (Anderson et al. 2018). For example, Nábělek and Robinson (1982) show a clear decrease in reverberant speech understanding with increasing age, but listeners with pure-tone thresholds up to 40 dB hearing loss at 4 kHz were included and relationships between speech understanding and pure-tone thresholds were not analyzed. Related confounds between age and hearing loss existed in a study by Harris and Reitz (1985), which reported poorer reverberant speech understanding in quiet for older hearing-impaired listeners (>60 years) relative to older normal-hearing and younger normal-hearing listeners, who showed similar performances. However, in a study in which age and hearing loss where found to be uncorrelated for the sample of listeners tested, both age and hearing loss were (independent) contributing factors to understanding reverberant speech (Helfer and Wilber 1990). A relationship between speech understanding in reverberation and working memory has also been shown (Reinhart and Souza 2016) as have developmental performance improvements in children (Neuman and Hochberg 1983; Neuman et al. 2010). This latter result may have particularly important implications for the acoustical design of classrooms (Yang and Bradley 2009). Overall, the fact that understanding speech in reverberation is an inherently difficult and complex listening task that is affected by a variety of both auditory and nonauditory factors clearly contributes to the research challenges in this area.

Common strategies for the treatment of hearing loss unfortunately are not entirely effective in reverberation. For example, it has been shown that even with modern hearing aids fit bilaterally, hearing-impaired individuals still do not reach speech understanding performance levels in reverberation of young normal-hearing listeners (Xia et al. 2018). This may be due, in part, to the fact that reverberation can lessen the benefit of certain hearing aid processing strategies such as wide dynamic range compression with fast attack and release times (Reinhart and Souza 2016). Regardless, it is clear that performance in reverberation is a major issue for hearing aid users because it is listed as a primary complaint based on one of the most commonly used self-report scales of hearing aid benefit (Johnson et al. 2010).

Reverberation is also a significant problem for cochlear implant users, given that the device only conveys a temporal coding of the speech signal and reverberation distorts this coding. Simulation results have confirmed this problem, although it is somewhat lessened if more frequency channels are used (Poissant et al. 2006). Bilateral implantation has also been shown to result in improved outcomes in reverberation (Kokkinakis 2018). See Ricketts and Kan, Chap. 13, for further details on hearing aid and cochlear implant users.

9.3.2.3.5 Adaptation Effects

A growing body of evidence suggests that speech understanding in rooms is dependent on recent acoustical exposure to the room. This can be thought of as a form of perceptual adaptation to room acoustics. Both changes in the listeners' abilities to perceptually categorize speech tokens (Watkins 2005a, b) and objective improvements in speech intelligibility have been demonstrated with prior room-listening exposure (Brandewie and Zahorik 2010; Srinivasan and Zahorik 2013). Only about 1 s of exposure is required to produce this facilitation (Brandewie and Zahorik 2013), which has been shown to be about an 18% improvement in intelligibility, on average, for a small room with a reverberation time of 0.4 s (Brandewie and Zahorik 2010). The effect appears to be largest for rooms with broadband reverberation times between 0.4 s and 1 s (Zahorik and Brandewie 2016) and is insensitive to listener/source locations within the room (Brandewie and Zahorik 2018). A study that examined this adaptation effect separately for the speech amplitude envelope versus temporal fine-structure cues reported observing adaptation only for speech envelope exposure in a room (Srinivasan and Zahorik 2014). This result is consistent with the important role of amplitude envelope coding in speech and its distortions caused by room acoustics (see Sects. 9.2.2.6 and 9.3.2.3.1). Other dynamic perceptual effects related to room acoustics may perhaps be considered forms of adaptation, such as precedence effect buildup/breakdown (see Sect. 9.3.1.1), distance perception improvements over time (Shinn-Cunningham 2000), and increasing insensitivity to spectral colorations caused by rooms (Olive et al. 1995). The extent to which these effects are related to the facilitation in speech understanding caused by prior room exposure is not currently known.

9.3.3 Subjective Effects of Room Acoustics

Perhaps more than any other area, the subjective effects of room acoustics have been a focus of sustained study for more than a century. Applications in architectural acoustics have been a driving force in this research. Understanding how physical acoustics relates to the subjective perceptual attributes of sound is critical for the effective design and improvement of any listening environment, ranging from home and work spaces to concert halls. Room acoustic effects are also critically important for sound quality for both live and reproduced sound. There are a number of classic reference works in these areas, including but not limited to Barron (2009), Beranek (2012), and Toole (2017).

Numerous perceptual attributes result from listening to sound in rooms. As described in Sect. 9.2.2, common room acoustic measures are designed to quantify the primary attributes, including but not limited to reverberance, clarity, strength, spatial impression, and timbre. Listener sensitivities to these measures in isolation are summarized in Sect. 9.3.1.3 and Table 9.2. Of course, under realistic listening conditions in rooms, these attributes and their related acoustical measures do not exist in isolation. Complex interactions, both acoustical and psychoacoustical, are commonplace. For example, the seminal work by Barron (1971) has demonstrated relationships between spatial and timbral aspects of sound when the delay and level of a single reflection is varied. These relationships are displayed in Fig. 9.8, where darker gray scale and increased crosshatch density represent increases in spatial impression and tone coloration, respectively. Figure 9.8 also displays various echo thresholds based on absolute detectability ("Threshold"), image shift, and subjective sound disturbance. A curve showing combinations of delay and level that produce spatial impression equal to a situation with the reflection delay fixed at 40 ms and level set to −6 dB is also displayed.

Over the past 40 years, work in architectural acoustics, particularly concert hall acoustics, has greatly expanded our understanding of the complex relationships and interrelationships between acoustical and psychoacoustical aspects of sound in rooms. Techniques from statistics and experimental psychology, such as multidimensional scaling (MDS), have been particularly useful in classifying different subjective perceptual attributes of rooms, determining their relative importance in room preference, and their relationships to various objective acoustical measures. For

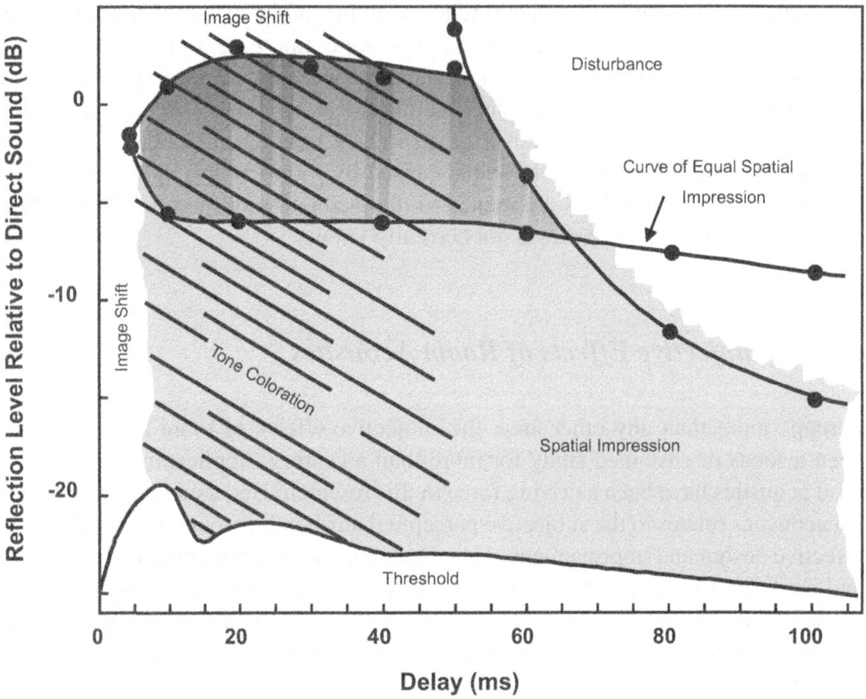

Fig. 9.8 Subjective effects of a single side reflection (horizontal plane, 40 ° to the left of midline) as a function of level and delay using music as the sound source. See text for additional details. (Based on Barron 1971, with permission)

example, Yamaguchi (1972) asked listeners to rate the similarity between pairs of musical samples recorded in different concert halls. MDS techniques were then used to estimate the perceptual space underlying the similarity ratings, which was found to have three dimensions. The primary and secondary dimensions, which accounted for most of the rating variance, were highly correlated with the sound pressure level and reverberation time. This suggests that the listeners judged the similarity between halls based primarily on attributes of loudness and reverberance. There are, of course, challenges to generalizing results such as these to different concert halls and to different listeners with potential response biases. Advances in the virtual "auralization" of concert halls have made it possible to directly compare realistic representations of many halls from a single listening location (Lokki et al. 2016), and nonbehavioral methods sensitive to emotional aspects of a musical performance may represent better metrics for success of a performance space (Pätynen and Lokki 2016). The relationships between the physical and perceptual aspects of sound in concert halls remains an active area of study today.

In addition to the collective study of subjective attributes of rooms, there has been psychoacoustical study of certain attributes individually. Sections 9.3.3.1, 9.3.3.2, and 9.3.3.3 summarize work related to loudness, spatial impression, and timbre. Apart from JND measures (Sect. 9.3.1.3), there is a notable absence of basic psychoacoustical research on reverberance and clarity. The impact of hearing loss on subjective aspects of room acoustics is also discussed.

9.3.3.1 Loudness

Loudness is a perceptual attribute of sound that can be influenced by a number of underlying physical stimulus factors. Although most strongly related to the overall sound pressure level, there are also strong dependencies on other factors such as frequency (Fletcher and Munson 1933) and time (Buus et al. 1997). Rooms and reflected sound have clear influences on sound loudness. Fundamentally, this is because they cause increases in the sound pressure level at the listener's location relative to what would have been experienced in an acoustic free field that is absent reflected sound. The room acoustic measure of G (see Sect. 9.2.2.3) is intended to capture this relationship. Notably, it does not reflect other spectral or temporal stimulus factors known to influence loudness, although there have been efforts to develop better measures of sound strength in rooms based on more detailed psychoacoustical models of loudness that consider these factors loudness (Lee et al. 2012). Another surprising aspect of loudness in rooms is that it has been shown to be insensitive to the distance to the sound source (Zahorik and Wightman 2001) even though the physical sound pressure level at the listener's location is strongly influenced by distance. This constancy of loudness in rooms is also not reflected in traditional sound strength measures such as G and underscores the importance for more detailed perceptual study of loudness in environments with reflected sound.

9.3.3.2 Spatial Impression

In concert halls, some decorrelation of the signals at the two ears, particularly from early lateral reflections, can produce a desirable broadening in the width of the sound image. This is one important subjective attribute of room spatial impression, or spaciousness, and has been shown to be an important component to concert hall preference (Schroeder et al. 1974). Subjective measures of apparent source width (ASW) have confirmed its relationship to measures of binaural coherence such as IACC, both under headphones (Blauert and Lindemann 1986) and with loudspeakers (Okano et al. 1998). The subtle increases in localization error and variability reported by Rakerd and Hartmann (1985) with a single strong lateral reflection are also perhaps consistent with increases in the ASW (see Sect. 9.3.2.1). Another aspect of spatial impression is the extent to which the listener feels surrounded or enveloped by sound. This aspect is also thought to be related to binaural coherence but specifically in the late reverberation (Beranek 2012).

9.3.3.3 Timbre

Timbre or tone color is an aspect of sound not related to pitch, loudness, duration, or spatial location. It is instead related to sound quality and is often described in terms such as "bright" or "dark." There is a considerable body of research on timbre in general (see Siedenburg et al. 2019 for a review), and rooms can clearly affect timbre due to the changes in the sound spectrum caused by reflected sound. Consider the most basic example of a single ideal impulse and a delayed copy of the impulse represent an ideal reflection (Fig. 9.9a). Compared with the perfectly flat magnitude spectrum that results from an ideal impulse alone, the addition of a reflection results in pronounced notches in the magnitude spectrum (Fig. 9.9b). This pattern, commonly known as a comb filter, causes noticeable changes to sound timbre. Comb filter notches occur at odd integer multiples of frequencies equal to half the wavelength of the reflection delay. Because the depth of the notches increases with the increasing level of the reflection, coloration effects are most prominent in small rooms with nearby reflecting surfaces and are dominated by first-order reflections (Bech 1995, 1996). Scientific research in this area has important application to sound reproduction via loudspeakers (Toole 2017), which often takes place in small rooms. Because timbral effects of the room are antithetical to accurate loudspeaker sound reproduction, it is important to fully understand listener sensitivity to these effects. Interestingly, like speech understanding, there is evidence to suggest that prior listening exposure to a room can lessen its perceptual contributions and that this is also evident in loudspeaker coloration testing in rooms (Olive et al. 1995).

9.3.3.4 Effects of Hearing Loss

Although it is certainly true that hearing loss can impact subjective aspects of room acoustics, the details and extent of the impact are largely unknown. Based on one of the only subjective self-report scales of listening ability to include assessment items related to room acoustics or reverberation, the Speech, Spatial, and Qualities Hearing Scale (SSQ), qualitative decrements in reverberant listening environments are reported for hearing-impaired listeners (Gatehouse and Noble 2004). There are also the known effects of hearing loss on objective measures such as echo threshold (Sect. 9.3.1.1), distance localization (Sect. 9.3.2.2), and speech understanding (Sect. 9.3.2.3). These objective measures have accompanying subjective attributes and therefore likely also experience decrements with hearing loss. In one of the only studies to directly assess the effect of hearing impairment on reverberation preference ratings for music, strong effects of hearing impairment were found that decreased listener sensitivity to reverberation time (Reinhart and Souza 2018). Clearly, additional work is needed in this area.

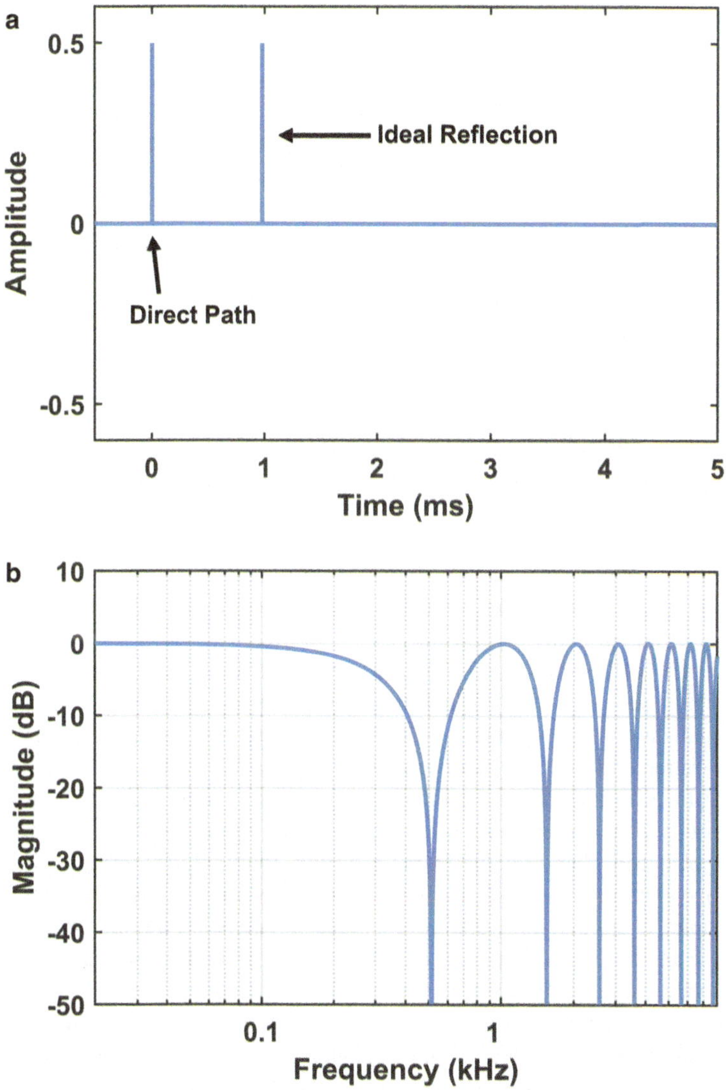

Fig. 9.9 Examples of comb filtering caused by a single ideal reflection with delay of 1 ms. (**a**) Time waveform; (**b**) magnitude spectrum

9.4 Summary

The purpose of this chapter was to provide a conceptual overview of research related to spatial hearing in rooms and the effects of reverberation. Section 9.2 summarizes the physical aspects of room acoustics through descriptions of common measurement techniques using a small auditorium as an illustrative test case. These techniques are based on the impulse response of the room measured either with a single reference microphone or binaurally. From these measurements, a variety of additional physical measures may be derived that are targeted toward certain perceptual aspects of sound in rooms. ISO-3382-1 (2009) provides further detail on these measures. Section 9.3 summarizes the psychophysical work in key areas relevant to listening in acoustically reflective environments. These areas included basic listener sensitivities to reflected sound as well as the impact of reflected sound on both objective listening performance and subjective sound quality. Sound localization and speech intelligibility were particular areas of focus in terms of objective measures. Where known, relevant neural correlates and impacts of hearing loss were also summarized in each area.

From this overview of physical and psychophysical aspects of room acoustics, at least three important takeaway points emerge.

1. In everyday listening, the physical effects of room acoustics are minimized or unnoticed, but this is not an accident. Complex processes ranging from those that support the precedence effect to perhaps more general forms of perceptual adaptation actively facilitate sound localization and speech understanding when reflected sound is present. This apparent disconnect between the physical and perceptual aspects of room acoustics underscores the need for further research on the details of their relationship.
2. There are both positive and negative aspects of room acoustics. On one hand, reflected sound provides critical information for distance estimation and sound externalization and provides major enhancements in sound quality. On the other hand, reflected sound can be detrimental to speech understanding and directional sound localization. These detriments are generally magnified with hearing loss.
3. The study of room acoustic effects is inherently multidisciplinary. Work summarized in this chapter came from disciplines that included architectural acoustics, psychological acoustics, physiological acoustics, audiology, experimental psychology, engineering, and speech/hearing science. Although integration of work from these different areas can sometimes be challenging, there is clearly a great value in the diversity of expertise when focused on common research problems. Prime examples of this include the use auralization techniques developed by architectural acousticians and engineers that enable psychoacousticians to test more natural and ecologically valid listening situations or the use of psychoacoustical concepts and models to more accurately predict practical room listening outcomes of interest to architectural acousticians. Continued collaborations of these types will advance the field as a whole.

Although considerable knowledge regarding listening in rooms has been gained over more than a century of research, there are nevertheless gaps in this knowledge. Some of the most important gaps include

1. Psychophysical understanding of reverberance and clarity. Some perceptual attributes of room listening such as loudness, timbre, and spatial effects have experienced detailed psychophysical study. Such study has resulted in computational models that can accurately predict human performance. In many cases, the

models are inspired by known physiological processes. At present, however, there are no models for reverberance and clarity, and very little psychophysical data exist on these attributes themselves.

2. Neural correlates of reverberation. Although it is clear that reverberation has important perceptual attributes and consequences, little is known about its encoding in the auditory system. What is known suggests that it may be coded indirectly at the brainstem level through monaural (Kim et al. 2015) and/or binaural (Slama and Delgutte 2015) means. Evidence related to distance perception (Kopčo et al. 2012) and emotional aspects of listening to music in concert halls (Pätynen and Lokki 2016) suggests that higher level processes are also involved. Clearly additional work is needed in this area.
3. Impact of hearing loss on listening to sound in rooms. There has been considerable work in this area, but it has been targeted almost exclusively to speech understanding. Expanded study of hearing loss to other areas of listening in rooms would be of great benefit. Areas of particular need include distance perception and externalization and subjective sound quality effects. Research in these areas will contribute to the improvement of assistive devices for hearing loss, which have historically performed poorly in reverberation.
4. The role of room effects in auditory scene analysis. There has been very little research done in this area. Basic questions such as how much information about our environment can be gained acoustically or how can a sound-producing object can be differentiated from background-reflected sound are largely unanswered. Questions such as these relate to our situational awareness through sound, and pioneering work by Traer and McDermott (2016) suggests that the statistical properties of sound in natural reverberant scenes may be critically important in this regard. Beyond the considerable basic science interests in this area, auditory scene analysis and situational awareness are of great practical importance in virtual reality (VR) applications, where such aspects are critical for high-quality simulation. A critical issue will be understanding the minimal conditions that support effective scene analysis within the constraints of the VR system. Room effects also aid in auditory distance perception and externalization, which have historically been problem areas in VR.

Compliance with Ethics Requirements
Pavel Zahorik declares that he has no conflict of interest.

References

Akeroyd MA, Guy FH (2011) The effect of hearing impairment on localization dominance for single-word stimuli. J Acoust Soc Am 130:312–323

Akeroyd MA, Gatehouse S, Blaschke J (2007) The detection of differences in the cues to distance by elderly hearing-impaired listeners. J Acoust Soc Am 121:1077–1089

Anderson BW, Kalb JT (1987) English verification of the STI method for estimating speech intelligibility of a communications channel. J Acoust Soc Am 81:1982–1985

Anderson PW, Zahorik P (2014) Auditory/visual distance estimation: accuracy and variability. Front Psychol 5:1097
Anderson S, Gordon-Salant S, Dubno JR (2018) Hearing and aging effects on speech understanding: challenges and solutions. Acoust Today 14(4):10–18
ANSI (1997) ANSI S3.5-1997: methods for calculation of the speech intelligibility index. American National Standards Institute, New York
Arbogast TL, Mason CR, Kidd G Jr (2002) The effect of spatial separation on informational and energetic masking of speech. J Acoust Soc Am 112:2086–2098
Babkoff H, Sutton S (1966) End point of lateralization for dichotic clicks. J Acoust Soc Am 39:87–102
Barron M (1971) The subject effects of first reflections in concert halls – the need for lateral reflections. J Sound Vib 15:475–494
Barron M (2009) Auditorium acoustics and architectural design, 2nd edn. Routledge, London
Bech S (1995) Timbral aspects of reproduced sound in small rooms I. J Acoust Soc Am 97:1717–1726
Bech S (1996) Timbral aspects of reproduced sound in small rooms II. J Acoust Soc Am 99:3539–3549
Bech S (1998) Spatial aspects of reproduced sound in small rooms. J Acoust Soc Am 103:434–445
Begault DR (1992) Perceptual effects of synthetic reverberation on three-dimensional audio systems. J Audio Eng Soc 40:895–904
Begault DR, Wenzel EM, Anderson MR (2001) Direct comparison of the impact of head tracking, reverberation, and individualized head-related transfer functions on the spatial perception of a virtual speech source. J Audio Eng Soc 49:904–916
Beranek L (2012) Concert halls and opera houses: music, acoustics, and architecture. Springer, New York
Blauert J (1997) Spatial hearing. Revised Edition edn. MIT Press, Cambridge
Blauert J, Lindemann W (1986) Spatial mapping of intracranial auditory events for various degrees of interaural coherence. J Acoust Soc Am 79:806–813
Bolia RS, Nelson WT, Ericson MA, Simpson BD (2000) A speech corpus for multitalker communications research. J Acoust Soc Am 107:1065–1066
Bolt RH, MacDonald AD (1949) Theory of speech masking by reverberation. J Acoust Soc Am 21:577–580
Bork I (2000) A comparison of room simulation software – the 2nd round robin on room acoustical computer simulation. Acta Acust United Acust 86:943–956
Boyd AW, Whitmer WM, Soraghan JJ, Akeroyd MA (2012) Auditory externalization in hearing-impaired listeners: the effect of pinna cues and number of talkers. J Acoust Soc Am 131:EL268–EL274
Bradley JS (1986) Predictors of speech intelligibility in rooms. J Acoust Soc Am 80:837–845
Bradley JS (2011) Review of objective room acoustics measures and future needs. Appl Acoust 72:713–720
Bradley JS, Reich R, Norcross S (1999) A just noticeable difference in C50 for speech. Appl Acoust 58:99–108
Bradley JS, Sato H, Picard M (2003) On the importance of early reflections for speech in rooms. J Acoust Soc Am 113:3233–3244
Brandewie EJ, Zahorik P (2010) Prior listening in rooms improves speech intelligibility. J Acoust Soc Am 128:291–299
Brandewie EJ, Zahorik P (2013) Time course of a perceptual enhancement effect for noise-masked speech in reverberant environments. J Acoust Soc Am 134:EL265–EL270
Brandewie EJ, Zahorik P (2018) Speech intelligibility in rooms: disrupting the effect of prior listening exposure. J Acoust Soc Am 143:3068–3078
Brimijoin WO, Boyd AW, Akeroyd MA (2013) The contribution of head movement to the externalization and internalization of sounds. PLoS One 8(12):e83068

Bronkhorst AW, Houtgast T (1999) Auditory distance perception in rooms. Nature 397(6719):517–520
Brown AD, Stecker GC (2013) The precedence effect: fusion and lateralization measures for headphone stimuli lateralized by interaural time and level differences. J Acoust Soc Am 133:2883–2898
Brown AD, Stecker GC, Tollin DJ (2015) The precedence effect in sound localization. J Assoc Res Otolaryngol 16:1–28
Burkhard MD, Sachs RM (1975) Anthropometric manikin for acoustic research. J Acoust Soc Am 58:214–222
Buus S, Florentine M, Poulsen T (1997) Temporal integration of loudness, loudness discrimination, and the form of the loudness function. J Acoust Soc Am 101:669–680
Catic J, Santurette S, Buchholz JM, Gran F, Dau T (2013) The effect of interaural-level-difference fluctuations on the externalization of sound. J Acoust Soc Am 134:1232–1241
Catic J, Santurette S, Dau T (2015) The role of reverberation-related binaural cues in the externalization of speech. J Acoust Soc Am 138:1154–1167
Clifton RK (1987) Breakdown of echo suppression in the precedence effect. J Acoust Soc Am 82:1834–1835
Clifton RK, Freyman RL (1997) The precedence effect: beyond echo suppression. In: Gilkey RH, Anderson TR (eds) Binaural and spatial hearing in real and virtual environments. Erlbaum, Mahwah, pp 233–255
Clifton RK, Freyman RL, Litovsky RY, McCall D (1994) Listeners' expectations about echoes can raise or lower echo threshold. J Acoust Soc Am 95:1525–1533
Cox TJ, Davies W, Lam YW (1993) The sensitivity of listeners to early sound field changes in auditoria. Acta Acust United Acust 79:27–41
Cranford J, Ravizza R, Diamond IT, Whitfield IC (1971) Unilateral ablation of the auditory cortex in the cat impairs complex sound localization. Science 172(3980):286–288
Cranford JL, Andres MA, Piatz KK, Reissig KL (1993) Influences of age and hearing loss on the precedence effect in sound localization. J Speech Lang Hear Res 36:437–441
Danilenko L (1969) Binaural hearing in non-stationary diffuse sound field. Kybernetik 6:50–57
Dent ML, McClaine EM, Welch TE (2007) The Franssen effect illusion in budgerigars (*Melopsittacus undulatus*) and zebra finches (*Taeniopygia guttata*). J Acoust Soc Am 122:3609–3614
Dizon RM, Colburn HS (2006) The influence of spectral, temporal, and interaural stimulus variations on the precedence effect. J Acoust Soc Am 119:2947–2964
Djelani T, Blauert J (2001) Investigations into the build-up and breakdown of the precedence effect. Acta Acust United Acust 87:253–261
Durlach NI, Rigopulos A, Pang X, Woods W, Kulkarni A, Colburn H, Wenzel E (1992) On the externalization of auditory images. Presence 1:251–257
Emiroglu S, Kollmeier B (2008) Timbre discrimination in normal-hearing and hearing-impaired listeners under different noise conditions. Brain Res 1220:199–207
Festen JM, Plomp R (1990) Effects of fluctuating noise and interfering speech on the speech-reception threshold for impaired and normal hearing. J Acoust Soc Am 88:1725–1736
Fletcher H, Munson W (1933) Loudness, its definition, measurement and calculation. J Acoust Soc Am 5:82–108
Franssen NV (1960) Some considerations on the mechanism of directional hearing. Dissertation, Delft University of Technology
Freyman RL, Clifton RK, Litovsky RY (1991) Dynamic processes in the precedence effect. J Acoust Soc Am 90:874–884
Gardner MB (1968) Historical background of the Haas and-or precedence effect. J Acoust Soc Am 43:1243–1248
Gardner MB (1969) Distance estimation of 0 degrees or apparent 0 degree-oriented speech signals in anechoic space. J Acoust Soc Am 45:47–53

Gatehouse S, Noble W (2004) The speech, spatial and qualities of hearing scale (SSQ). Int J Audiol 43:85–99

George ELJ, Festen JM, Houtgast T (2008) The combined effects of reverberation and nonstationary noise on sentence intelligibility. J Acoust Soc Am 124:1269–1277

George ELJ, Goverts ST, Festen JM, Houtgast T (2010) Measuring the effects of reverberation and noise on sentence intelligibility for hearing-impaired listeners. J Speech Lang Hear Res 53:1429–1439

Gil-Carvajal JC, Cubick J, Santurette S, Dau T (2016) Spatial hearing with incongruent visual or auditory room cues. Sci Rep 6:37342

Goupell MJ, Yu G, Litovsky RY (2012) The effect of an additional reflection in a precedence effect experiment. J Acoust Soc Am 131:2958–2967

Grantham DW (1996) Left-right asymmetry in the buildup of echo suppression in normal-hearing adults. J Acoust Soc Am 99:1118–1123

Grey JM, Gordon JW (1978) Perceptual effects of spectral modifications on musical timbres. J Acoust Soc Am 63:1493–1500

Haas H (1972) The influence of a single echo on the audibility of speech. J Audio Eng Soc 20:146–159

Harris RW, Reitz ML (1985) Effects of room reverberation and noise on speech discrimination by the elderly. Audiology 24:319–324

Hartmann WM (1983) Localization of sound in rooms. J Acoust Soc Am 74:1380–1391

Hartmann WM, Rakerd B (1989) Localization of sound in rooms. IV: The Franssen effect. J Acoust Soc Am 86:1366–1373

Hartmann WM, Wittenberg A (1996) On the externalization of sound images. J Acoust Soc Am 99:3678–3688

Helfer KS, Wilber LA (1990) Hearing loss, aging, and speech perception in reverberation and noise. J Speech Lang Hear Res 33:149–155

Higgins NC, McLaughlin SA, Da Costa S, Stecker GC (2017) Sensitivity to an illusion of sound location in human auditory cortex. Front Syst Neurosci 11:35

Houtgast T, Steeneken HJM (1985) A review of the MTF concept in room acoustics and its use for estimating speech intelligibility in auditoria. J Acoust Soc Am 77:1069–1077

Houtsma AJM, Rossing TD, Wagenaars WM (1987) Effects of echoes. Auditory demonstrations. Institute for Perception Research (IPO), Eindhoven, the Netherlands

IEC-60268-16 (2011) Sound system equipment – Part 16: Objective rating of speech intelligibility by speech transmission index. International Electrotechnical Commission, Geneva

Ihlefeld A, Shinn-Cunningham BG (2011) Effect of source spectrum on sound localization in an everyday reverberant room. J Acoust Soc Am 130:324–333

ISO-3382-1 (2009) Acoustics – measurement of the reverberation time of rooms with reference to other acoustical parameters. International Organization for Standardization, Geneva

Jeffress LA (1948) A place theory of sound localization. J Comp Physiol Psychol 41:35–39

Jeffress LA, Blodgett HC, Sandel TT, Wood CL III (1956) Masking of tonal signals. J Acoust Soc Am 28:416–426

Johnson JA, Cox RM, Alexander GC (2010) Development of APHAB norms for WDRC hearing aids and comparisons with original norms. Ear Hear 31:47–55

Jordan VL (1970) Acoustical criteria for auditoriums and their relation to model techniques. J Acoust Soc Am 47:408–412

Katz BFG (2004) International round robin on room acoustical impulse response analysis software 2004. Acoust Res Lett 5:158–164

Keen R, Freyman RL (2009) Release and re-buildup of listeners' models of auditory space. J Acoust Soc Am 125:3243–3252

Kim DO, Zahorik P, Carney LH, Bishop BB, Kuwada S (2015) Auditory distance coding in rabbit midbrain neurons and human perception: monaural amplitude modulation depth as a cue. J Neurosci 35:5360–5372

Knudsen VO (1929) The hearing of speech in auditoriums. J Acoust Soc Am 1:56–82

Koenig W (1950) Subjective effects in binaural hearing. J Acoust Soc Am 22:61–62

Kokkinakis K (2018) Binaural speech understanding with bilateral cochlear implants in reverberation. Am J Audiol 27:85–94
Kolarik AJ, Moore BCJ, Zahorik P, Cirstea S, Pardhan S (2016) Auditory distance perception in humans: a review of cues, development, neuronal bases, and effects of sensory loss. Atten Percept Psychophys 78:373–395
Kopčo N, Schoolmaster M, Shinn-Cunningham B (2004) Learning to judge distance of nearby sounds in reverberant and anechoic environments. In: Joint Conference of CFA/DAGA, Strasbourg
Kopčo N, Huang S, Belliveau JW, Raij T, Tengshe C, Ahveninen J (2012) Neuronal representations of distance in human auditory cortex. Proc Natl Acad Sci U S A 109:11019–11024
Kulkarni A, Colburn HS (1998) Role of spectral detail in sound-source localization. Nature 396:747
Landy MS, Maloney LT, Johnston EB, Young M (1995) Measurement and modeling of depth cue combination: in defense of weak fusion. Vis Res 35:389–412
Laws P (1973) Auditory distance perception and the problem of "in-head localization" of sound images. Acta Acust United Acust 29:243–259
Lee D, Cabrera D, Martens WL (2012) The effect of loudness on the reverberance of music: Reverberance prediction using loudness models. J Acoust Soc Am 131:1194–1205
Leshowitz B (1971) Measurement of the two-click threshold. J Acoust Soc Am 49:462–466
Litovsky RY (1997) Developmental changes in the precedence effect: estimates of minimum audible angle. J Acoust Soc Am 102:1739–1745
Litovsky RY (1998) Physiological studies of the precedence effect in the inferior colliculus of the kitten. J Acoust Soc Am 103:3139–3152
Litovsky RY, Shinn-Cunningham BG (2001) Investigation of the relationship among three common measures of precedence: fusion, localization dominance, and discrimination suppression. J Acoust Soc Am 109:346–358
Litovsky RY, Colburn HS, Yost WA, Guzman SJ (1999) The precedence effect. J Acoust Soc Am 106:1633–1654
Lochner PA, Burger JF (1958) The subjective masking of short time delayed echoes by their primary sounds and their contributions on the intelligibility of speech. Acta Acust United Acust 8:1–10
Lokki T, Pätynen J, Kuusinen A, Tervo S (2016) Concert hall acoustics: repertoire, listening position, and individual taste of the listeners influence the qualitative attributes and preferences. J Acoust Soc Am 140:551–562
Longworth-Reed L, Brandewie E, Zahorik P (2009) Time-forward speech intelligibility in time-reversed rooms. J Acoust Soc Am 125:EL13–EL19
Marrone N, Mason CR, Kidd G Jr (2008) The effects of hearing loss and age on the benefit of spatial separation between multiple talkers in reverberant rooms. J Acoust Soc Am 124:3064–3075
Mershon DH, King E (1975) Intensity and reverberation as factors in the auditory perception of egocentric distance. Percept Psychophys 18:409–415
Mershon DH, Ballenger WL, Little AD, McMurtry PL, Buchanan JL (1989) Effects of room reflectance and background noise on perceived auditory distance. Perception 18:403–416
Middlebrooks JC, Simon JZ, Popper AN, Fay RR (eds) (2017) The auditory system at the cocktail party. Springer, New York
Moncur JP, Dirks D (1967) Binaural and monaural speech intelligibility in reverberation. J Speech Lang Hear Res 10:186–195
Nábělek AK, Mason D (1981) Effect of noise and reverberation on binaural and monaural word identification by subjects with various audiograms. J Speech Lang Hear Res 24:375–383
Nábělek AK, Robinson PK (1982) Monaural and binaural speech perception in reverberation for listeners of various ages. J Acoust Soc Am 71:1242–1248
Nábělek AK, Letowski TR, Tucker FM (1989) Reverberant overlap- and self-masking in consonant identification. J Acoust Soc Am 86:1259–1265

Naguib M (1995) Auditory distance assessment of singing conspecifies in Carolina wrens: the role of reverberation and frequency-dependent attenuation. Anim Behav 50:1297–1307
Neuman AC, Hochberg I (1983) Children's perception of speech in reverberation. J Acoust Soc Am 73:2145–2149
Neuman AC, Wroblewski M, Hajicek J, Rubinstein A (2010) Combined effects of noise and reverberation on speech recognition performance of normal-hearing children and adults. Ear Hear 31:336–344
Okano T (2002) Judgments of noticeable differences in sound fields of concert halls caused by intensity variations in early reflections. J Acoust Soc Am 111:217–229
Okano T, Beranek LL, Hidaka T (1998) Relations among interaural cross-correlation coefficient (IACCE), lateral fraction (LFE), and apparent source width (ASW) in concert halls. J Acoust Soc Am 104:255–265
Olive SE, Toole FE (1989) The detection of reflections in typical rooms. J Audio Eng Soc 37:539–553
Olive SE, Schuck PL, Sally SL, Bonneville M (1995, 6–9 October) The variability of loudspeaker sound quality among four domestic-sized rooms. In: Audio engineering society convention 99, New York
Pätynen J, Lokki T (2016) Concert halls with strong and lateral sound increase the emotional impact of orchestra music. J Acoust Soc Am 139:1214–1224
Picou EM, Gordon J, Ricketts TA (2016) The effects of noise and reverberation on listening effort in adults with normal hearing. Ear Hear 37:1–13
Plenge G (1972) On the problem of "in head localization". Acta Acust United Acust 26:241–252
Plenge G (1974) On the differences between localization and lateralization. J Acoust Soc Am 56:944–951
Plomp R (1976) Binaural and monaural speech intelligibility of connected discourse in reverberation as a function of azimuth of a single competing sound source (speech or noise). Acta Acust United Acust 34:200–211
Poissant SF, Whitmal NA III, Freyman RL (2006) Effects of reverberation and masking on speech intelligibility in cochlear implant simulations. J Acoust Soc Am 119:1606–1615
Rakerd B, Hartmann WM (1985) Localization of sound in rooms, II: the effects of a single reflecting surface. J Acoust Soc Am 78:524–533
Rakerd B, Hartmann WM (2010) Localization of sound in rooms. V. Binaural coherence and human sensitivity to interaural time differences in noise. J Acoust Soc Am 128:3052–3063
Reichardt W, Schmidt W (1966) The audible steps of spatial impression in music performances. Acta Acust United Acust 17:175–179
Reinhart PN, Souza PE (2016) Intelligibility and clarity of reverberant speech: effects of wide dynamic range compression release time and working memory. J Speech Lang Hear Res 59:1543–1554
Reinhart PN, Souza PE (2018) Effects of varying reverberation on music perception for Young normal-hearing and old hearing-impaired listeners. Trends Hear 22:1–11
Rhebergen KS, Versfeld NJ (2005) A speech intelligibility index-based approach to predict the speech reception threshold for sentences in fluctuating noise for normal-hearing listeners. J Acoust Soc Am 117:2181–2192
Rife DD, Vanderkooy J (1989) Transfer-function measurement with maximum-length sequences. J Audio Eng Soc 37:419–444
Sabine WC (1922) Collected papers on acoustics. Harvard University Press, Harvard
Sakai H, Sato S-I, Ando Y (1998) Orthogonal acoustical factors of sound fields in a forest compared with those in a concert hall. J Acoust Soc Am 104:1491–1497
Sanders LD, Zobel BH, Freyman RL, Keen R (2011) Manipulations of listeners' echo perception are reflected in event-related potentials. J Acoust Soc Am 129:301–309
Schroeder MR, Gottlob D, Siebrasse KF (1974) Comparative study of European concert halls: correlation of subjective preference with geometric and acoustic parameters. J Acoust Soc Am 56:1195–1201

Seraphim H (1958) Investigations on the difference limen of exponentially decaying bandlimited noise pulses. Acta Acust United Acust 8:280–284

Seraphim HP (1961) Über die Wahrnehmbarkeit mehrerer Rückwürfe von Sprachschall. Acta Acust United Acust 11:80–91

Shinn-Cunningham BG (2000, 25 April) Learning reverberation: considerations for spatial audio displays. In: 2000 International conference on auditory display, Atlanta

Siedenburg K, Saitis C, McAdams S, Popper AN, Fay RR (eds) (2019) Timbre: acoustics, perception, and cognition. Springer, Cham

Slama MC, Delgutte B (2015) Neural coding of sound envelope in reverberant environments. J Neurosci 35:4452–4468

Srinivasan NK, Zahorik P (2012) Phonemic restoration effect reversed in a reverberant room. J Acoust Soc Am 131:EL28–EL34

Srinivasan NK, Zahorik P (2013) Prior listening exposure to a reverberant room improves open-set intelligibility of high-variability sentences. J Acoust Soc Am 133:EL33–EL39

Srinivasan NK, Zahorik P (2014) Enhancement of speech intelligibility in reverberant rooms: role of amplitude envelope and temporal fine structure. J Acoust Soc Am 135:EL239–EL245

Steeneken HJ (1992) Quality evaluation of speech processing systems. In: Digital speech processing. Springer, Boston, pp 127–160

Steeneken HJ, Houtgast T (1999) Mutual dependence of the octave-band weights in predicting speech intelligibility. Speech Commun 28:109–123

Steeneken HJ, Houtgast T (2002) Validation of the revised STIr method. Speech Commun 38:413–425

Stevens SS, Newman EB (1936) The localization of actual sources of sound. Am J Psychol 48:297–306

Toole FE (2017) Sound reproduction: the acoustics and psychoacoustics of loudspeakers and rooms. Routledge, New York

Traer J, McDermott JH (2016) Statistics of natural reverberation enable perceptual separation of sound and space. Proc Natl Acad Sci U S A 113:E7856–E7865

van Wijngaarden SJ, Drullman R (2008) Binaural intelligibility prediction based on the speech transmission index. J Acoust Soc Am 123:4514–4523

Vorländer M (2007) Auralization: fundamentals of acoustics, modelling, simulation, algorithms and acoustic virtual reality. Springer Science & Business Media, New York

Wallach H (1940) The role of head movements and vestibular and visual cues in sound localization. J Exp Psychol 27:339–368

Wallach H, Newman EB, Rosenzweig MR (1949) The precedence effect in sound localization. Am J Psychol 62:315–336

Warren RM (1970) Perceptual restoration of missing speech sounds. Science 167(3917):392–393

Watkins AJ (2005a) Perceptual compensation for effects of echo and of reverberation on speech identification. Acta Acust United Acust 91:892–901

Watkins AJ (2005b) Perceptual compensation for effects of reverberation in speech identification. J Acoust Soc Am 118:249–262

Westermann A, Buchholz JM (2015) The effect of spatial separation in distance on the intelligibility of speech in rooms. J Acoust Soc Am 137:757–767

Wiener FM, Malme CI, Gogos CM (1965) Sound propagation in urban areas. J Acoust Soc Am 37:738–747

Wightman FL, Kistler DJ (1992) The dominant role of low-frequency interaural time differences in sound localization. J Acoust Soc Am 91:1648–1661

Xia J, Xu B, Pentony S, Xu J, Swaminathan J (2018) Effects of reverberation and noise on speech intelligibility in normal-hearing and aided hearing-impaired listeners. J Acoust Soc Am 143:1523–1533

Yamaguchi K (1972) Multivariate analysis of subjective and physical measures of hall acoustics. J Acoust Soc Am 52:1271–1279

Yang W, Bradley J (2009) Effects of room acoustics on the intelligibility of speech in classrooms for young children. J Acoust Soc Am 125:922–933
Yin TC (1994) Physiological correlates of the precedence effect and summing localization in the inferior colliculus of the cat. J Neurosci 14:5170–5186
Yost WA, Guzman SJ (1996) Auditory processing of sound sources: is there an echo in here? Curr Dir Psychol Sci 5(4):125–131
Zahorik P (2002a) Assessing auditory distance perception using virtual acoustics. J Acoust Soc Am 111:1832–1846
Zahorik P (2002b) Direct-to-reverberant energy ratio sensitivity. J Acoust Soc Am 112:2110–2117
Zahorik P (2009) Perceptually relevant parameters for virtual listening simulation of small room acoustics. J Acoust Soc Am 126:776–791
Zahorik P, Brandewie EJ (2016) Speech intelligibility in rooms: effect of prior listening exposure interacts with room acoustics. J Acoust Soc Am 140:74–86
Zahorik P, Kelly JW (2007) Accurate vocal compensation for sound intensity loss with increasing distance in natural environments. J Acoust Soc Am 122:EL143–EL150
Zahorik P, Wightman FL (2001) Loudness constancy with varying sound source distance. Nat Neurosci 4:78–83
Zahorik P, Wightman FL, Kistler DJ (1995) On the discriminability of virtual and real sound sources. In: Proc ASSP (IEEE). IEEE Press, New York
Zahorik P, Brungart DS, Bronkhorst AW (2005) Auditory distance perception in humans: a summary of past and present research. Acta Acust United Acust 91:409–420

Chapter 10
Computational Models of Binaural Processing

Mathias Dietz and Go Ashida

10.1 Introduction

What is a model? The answer to this question is undoubtedly context dependent. Here it is a set of mathematical functions or relationships, typically implemented numerically on a computer. It mimics or simplifies certain parts of the auditory system. The model functions are accompanied by a set of parameters. Ideally, model functions are fixed across subjects and parameters are used to fit the input-output relationship of a specific subject or a specific neuron. In more general terms, any research beyond pure data reporting is modeling.

At first glance, computational modeling may seem like a special discipline, potentially hard to learn. One purpose of this chapter is orientation for newcomers to the field of modeling. Thus, before embarking on the details involved in modeling, we recommend starting with obtaining the code of one or two models that appear to be useful for the respective research question. Typically, it is not clear at first why the model does what it does. Hence we recommend taking a deeper look inside the model, "zooming" into every function and testing some inputs until the behavior of the model is understood. There is a likelihood that the output of the model will deviate from the data that are being modeled; however, deep understanding of the model can lead to adjustments that better fit the data.

The original version of this chapter was revised. The correction to this chapter is available at https://doi.org/10.1007/978-3-030-57100-9_14

M. Dietz (✉)
Department of Medical Physics and Acoustics, University of Oldenburg, Oldenburg, Germany
e-mail: mathias.dietz@uni-oldenburg.de

G. Ashida
Department of Neuroscience, University of Oldenburg, Oldenburg, Germany
e-mail: go.ashida@uni-oldenburg.de

© Springer Nature Switzerland AG 2021, Corrected Publication 2021

R. Y. Litovsky et al. (eds.), *Binaural Hearing*, Springer Handbook of Auditory Research 73, https://doi.org/10.1007/978-3-030-57100-9_10

10.1.1 Purpose of Modeling

In many scientific disciplines, computational modeling approaches complement experimental methods that directly examine complex systems. A model contains a limited number of components and thus enables us to test, analyze, and ultimately understand the modeled phenomena in a simplified setting. Even though the model itself may sometimes seem to be complicated (as seen in Sects. 10.3 and 10.4), its level of complexity is generally lower than that of the real-world system, which helps us identify key factors and operational rules underlying the simulated phenomena.

One of the most important roles of modeling is to simulate the function of a complex system and predict its possible outcome. Simulation results can be used to confirm or falsify current hypotheses about a system. A second important role of modeling is to make predictions that have not been tested (but are potentially testable) in experimental studies, thus guiding future directions of the research field. Because components of a model can be easily manipulated, models enable us to study how the system may function under hypothetical conditions.

For verification, a model generally needs to be tested with experimental data that are obtained from the real-world system under study. The binaural models reviewed in this chapter have been validated against various physiological or psychological measurements. It should be noted, however, that no single model can fill the demand of all users because every model has its own applicability and limitations. Therefore, users need to choose an appropriate model depending on their goals.

10.1.2 Auditory Modeling

Auditory models, including the ones reviewed in this chapter, can largely be divided into two categories: "physiological (biophysical) models" and "functional (phenomenological) models." A physiological model is constructed from and normally calibrated with the data obtained from studies on living tissue (e.g., ears, nerves, and brains of animals or humans). Studies using a physiological model most frequently aim to reveal how and what biophysical processes underlie the specific auditory function being modeled. One such example is simulating sound-driven electrical activity of binaural neurons to study interaural time difference (ITD; see Table 10.1) coding (e.g., Ashida et al. 2017). Physiological models often (but not always) focus on a specific stage of the auditory pathway. Naturally, when the focus is specific, the desire for detail becomes relatively high. Models described in Sect. 10.3, where neuronal substrates of binaural processing are discussed, have a relatively high degree of physiological detail.

In contrast to physiological models, functional models do not necessarily include biological details of the modeled system. The validity of a functional model is confirmed when its input-output relationship is similar enough to that of the system

Table 10.1 List of abbreviations

Abbreviation	Definition
AM	Amplitude modulation or amplitude-modulated
AN	Auditory nerve
AVCN	Anteroventral cochlear nucleus
BMLD	Binaural masking level difference
EC	Equalization-cancellation
EE	Excitatory-excitatory
EI	Excitatory-inhibitory
HH	Hodgkin-Huxley
IC	Inferior colliculus
IF	Integrate-and-fire
ILD	Interaural level difference
IPD	Interaural phase difference
ITD	Interaural time difference
LNTB	Lateral nucleus of the trapezoid body
LSO	Lateral superior olive
MNTB	Medial nucleus of the trapezoid body
MSO	Medial superior olive
TFS	Temporal fine structure

being modeled. Functional models are preferred over physiological models when the detailed biophysics of the phenomena is not the primary focus of the study or when the experimental knowledge of the biological processes behind the auditory function under study is not sufficient. Furthermore, auditory pathway models that aim to relate the sound reaching the ear to its perception in the brain tend to be on the functional side. Section 10.4, which is about models of binaural perception, therefore includes more functional models than physiological models.

It should be noted, however, that the distinction between physiological and functional models is neither strict nor mutually exclusive. A physiological model can often possess phenomenological or functional components without sufficient biological grounds, whereas many functional models of auditory processing are actually inspired by physiological observations.

10.2 Signal and System Theory Applied to Binaural Interaction

To understand the input-output relationship of a binaural neuron or, in more general terms, of a binaural interaction process, it is helpful to take one step back and consider the theory behind such an operation (see Hartmann, Chap. 2). Note that, for the most part, the description here focuses on the output of a single interaction unit (e.g., a single neuron rather than the complete output of a multiunit binaural processor). The second key to understanding the function of the system is to quantify the information that is present in the inputs to the binaural interaction unit. Therefore, Sect. 10.2 describes some properties of sound signals at several stages along the subcortical human auditory pathway that are critical for an understanding of binaural hearing in the most general terms.

10.2.1 Peripheral Filtering and Binaural Coherence

The first important transformation is characterized by the band-pass filter bank of the basilar membrane. At any place, the output bandwidth is much smaller than the center frequency (e.g., about 78 Hz wide at 500 Hz). Only because of these relatively narrow filters can we usually identify a carrier and a slow amplitude modulation (AM) at the output level of a basilar membrane channel (Fig. 10.1a). This is not

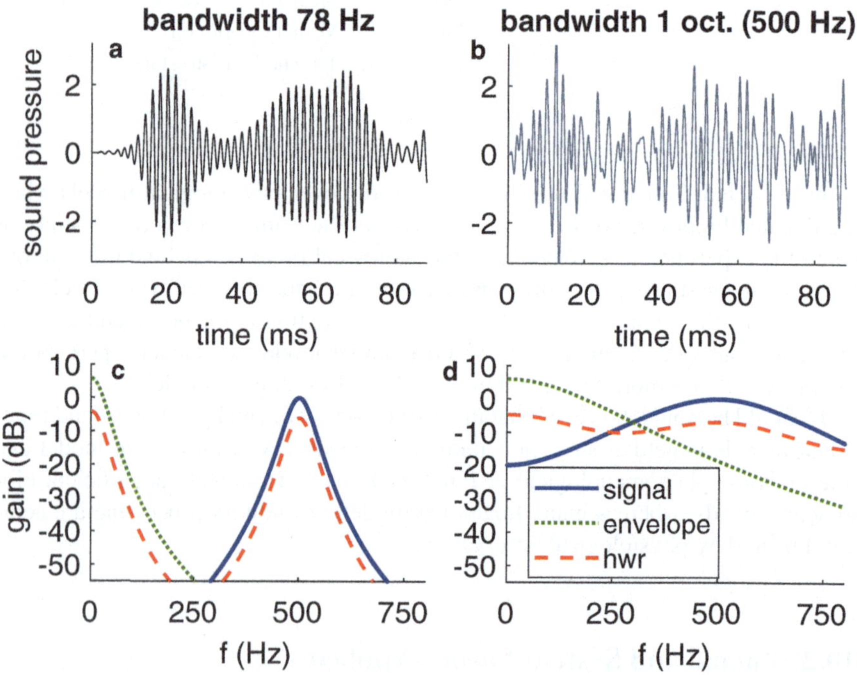

Fig. 10.1 (**a**) Gaussian white noise after passing through a gammatone filter centered at 500 Hz with a bandwidth of 78 Hz. (**b**) Same as in **a** but with a deliberately unrealistic filter bandwidth of 1 octave. Note that in contrast to **a**, it is not as easy to define a meaningful envelope for **b**. (**c**) Transfer function for the same filter used in **a** for the carrier (signal; *blue line*), for the Hilbert envelope (*red dashed line*), and for the signal after half-wave rectification (hwr; *yellow line*). (**d**) Same format as in **c** but for a hypothetical filter with a bandwidth of 1 octave (500 Hz). Note how the envelope and signal (fine-structure region) have a spectral overlap in the 300-Hz region

possible for signals wider than approximately 50% of the center frequency (Fig. 10.1b).

Following the band-pass filter bank at the level of the basilar membrane, the acoustic vibration induces movement of the basilar membrane and fluids in the scala media, which produces movement of the hair cells. Movement in one displacement direction causes the opening of ion channels located at the tip links connecting the stereocilia of hair cells. The most simplified model of this hair cell property is applying half-wave rectification to the signal. A consequence of the half-wave rectification is that any AM is demodulated, which means that the AM frequencies (the broadness of the spectrum around the center or carrier frequency) become physically present in the signal (Fig. 10.1c). A common terminology is to use "envelope" for the demodulated AM and "temporal fine structure" (TFS) for the carrier-related signal components that remain in the band-pass frequency range even after hair cell processing.

In normal-hearing listeners, the input to the hair cells has a sufficiently narrow bandwidth, and, therefore, the demodulated AM frequencies do not spectrally overlap with the original frequencies. This is a prerequisite for identifying the temporal envelope and the TFS of the filtered signal (compare separable envelope and TFS in Fig. 10.1c and overlap in Fig. 10.1d). Up to this stage, this reduced version of the model is a linear time-invariant system whose properties are fully described by the transfer functions for each location along the cochlea.

The last input stage discussed here is the synapse between the hair cell and auditory nerve (AN) fibers, leading to the generation of action potentials in the innervating AN fibers. The continuous waveform of the presynaptic signal is now transformed into discrete events, also known as spikes or action potentials. However, not all models include this stage. Instead, many models operate on the basis of a continuous signal representation (e.g., the hair cell potential). Many properties of both monaural and binaural processing can be modeled and understood without the considerations of complex neural code, whereas some system properties are often not captured correctly with nonspiking models (see Sect. 10.4.3).

After this summary of input transformations, the reader is now prepared to understand the actual binaural interaction from a signal theory point of view. Section 10.2 focuses on ITD information that can be extracted from comparing the inputs. To study the system theory behind extracting ITD information, consider the simple case where the left and right signals are identical except for a fixed ITD. This is the same experimental concept that has been widely used in optics to study temporal coherence or interferometry. Typical optical devices for such experiments are Michelson-Morley or Mach Zender interferometers. Interferometers split a light beam into two identical copies that can be delayed relative to each other (e.g., see Hecht 2014 for a comprehensive textbook). The core of binaural processing in the brain (i.e., the interaction of two oscillatory signals) is described by the same mathematical expressions that have been developed for temporal coherence in optics. The terminology is also mostly the same, but some branches and laboratories have adapted some deviations. To keep a consistent terminology in this broad chapter, conflicts have been resolved by favoring the conventions and terms used in general textbooks and across disciplines.

The most important aspects of the binaural processing stage are summarized here with accompanying terminology and examples from auditory signals, which obviously have a much lower frequency and propagation speed than light, but this discrepancy does not influence the theory at all.

Temporal coherence is a measure of how well the signal phase in one moment in time can be estimated from the phase at a different moment in time. More simplified, it can be seen as a measure of self-similarity over time. In the present example two identical copies of the signal (s) are used, so the temporal coherence [$\gamma(\tau)$] describes how the phase at time $t = \tau$ can be estimated from knowing the phase at $t = 0$. The coherence function simplifies to the normalized autocorrelation function of the signal

$$\gamma(\tau) = \frac{\langle s(t+\tau) s^*(t) \rangle}{\langle |s(t)|^2 \rangle} \tag{10.1}$$

where * denotes the complex conjugate and triangular parentheses denote the temporal average. The denominator is the signal energy, normalizing the coherence to 1 for $\tau = 0$.

The autocorrelation function is always proportional to the Fourier transform $(\mathcal{F})$ of the power spectrum (Wiener-Khinchin theorem). Therefore, the power spectrum $S(f)$ of the input to the binaural processor is sufficient to fully determine the complex-valued $\gamma(\tau)$

$$\gamma(\tau) \propto \mathcal{F}[S(f)] \tag{10.2}$$

For example, the extreme case of a pure tone has an infinitely long coherence. Generally speaking, the bandwidth of the input signal determines the decay of $|\gamma(\tau)|$: the broader the spectrum, the shorter the temporal coherence. If we assume that the bandwidth is limited predominantly by the band-pass filters of the basilar membrane, their bandwidth imposes the limit for how fast the coherence can decay. If the signal is narrower than the filter, $|\gamma(\tau)|$ will decay slower. Figure 10.2 shows an example of the two different bandwidths that were also employed in Fig. 10.1.

Note that coherence is not the same as the normalized interaural correlation coefficient (e.g., Klein-Hennig et al. 2011), which is sometimes referred to as the interaural correlation. The latter is the same as the real part of the coherence. For example, a tone with a 90° interaural phase difference (IPD) is fully coherent, but the correlation coefficient and the real part of the coherence are zero.

Langford and Jeffress (1964) measured the decay of the binaural benefit in detecting a 500-Hz tone by increasing the masking-noise ITDs from 0 to 2, 4, 6, and 8 ms (see Culling and Lavandier, Chap. 8). They accounted for the decay by calculating the noise cross correlation at $\tau = 0$ by using a 100-Hz-wide auditory filter centered at 500 Hz. With these ITDs, the noise correlation at $\tau = 0$ is identical to the

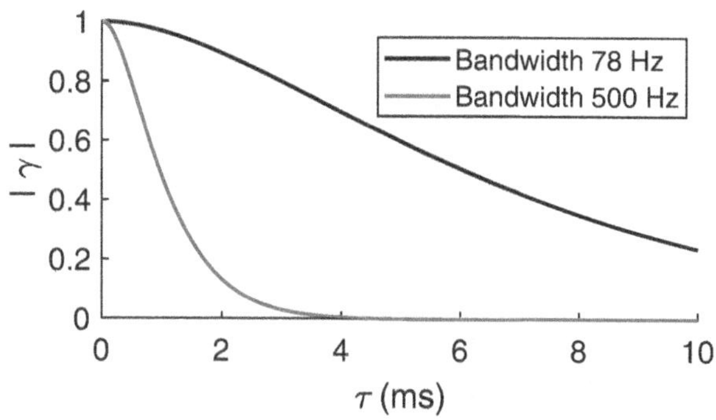

Fig. 10.2 Temporal coherence function (Eqs. 10.10 and 10.12). Note that the decay rate only depends on the bandwidth, not on the center frequency (500 Hz in this example)

abovementioned coherence because the ITDs are an integer multiple of the 2-ms cycle duration at 500 Hz. Their finding is remarkable because it essentially consists of binaural data with up to 8-ms ITDs explained by a purely monaural property, namely, the auditory filter bandwidth (see, e.g., Colburn and Durlach 1978 for more details). A related consequence is that very different binaural model concepts all have a similar outcome for binaural detection tasks (Domnitz and Colburn 1976). Stimulus inherent properties like bandwidth and frequency as well as the peripheral model stages are the crucial components that influence the output more strongly than the actual binaural interaction model.

10.2.2 Extraction of Interaural Time Difference

The aforementioned system using binaural coherence is not able to determine whether a sound comes from the left or the right. It only derives the similarity of the left and right signals at this stage. To describe the processing underlying directional hearing, however, an actual binaural model is required that maps the ITD [and the interaural level difference (ILD)] to an azimuthal direction or lateralization. To continue with the system theory approach for this part of binaural processing, the implications of two general concepts are discussed here: cross-correlation and IPD extraction. These concepts are, in fact, already simple binaural models but treating them in their purest form without any attempt to be realistic or physiological will be helpful for understanding the specific models later on.

The first concept is ITD extraction through the cross-correlation function $C(\tau)$

$$C(\tau) = \int s_L(t) s_R(t+\tau) dt \tag{10.3}$$

Here the complex conjugate * is omitted because only real-valued signals are considered. Equation 10.3 contains two simple operations: time delay and multiplication. Jeffress (1948) proposed this mechanism and suggested an array of time delay elements of different lengths, coined *delay lines*, followed by *coincidence* *detectors*. Because of its fundamental importance, the mathematical relationship between coincidence detection and cross-correlation is presented in the following paragraph. The next three equations can be skipped if this is already known or not of interest.

Assume that a coincidence detector receives inputs $s_L(t_L)$ and $s_R(t_R)$; here, two time variables (t_L and t_R) are introduced to account for the time difference between them. The coincidence detector produces an output when it receives synchronized inputs within a time window $w(t)$. The average output of this coincidence detector, called A_{out}, over a time period is thus a function of delay τ, which compensates the timing difference of the two inputs, and is calculated as

$$\begin{aligned} A_{out}(\tau) &\propto \iint s_L(t_L) s_R(t_R) w(t_R - t_L - \tau) dt_L dt_R \\ &= \iint s_L(t_L) s_R(t_L + x + \tau) w(x) dt_L dx \end{aligned} \tag{10.4}$$

where $x = t_R - t_L - \tau$. It is assumed that the coincidence window is narrow [i.e., $w(x) \neq 0$ only for small $|x|$] and the input rate modulates slowly enough compared with the coincidence window [i.e., $s_R(t_L + x + \tau) \approx s_R(t_L + \tau)$ for small $|x|$]. Then we have

$$A_{out}(\tau) \propto \int s_L(t_L) s_R(t_L + \tau) w(x) dt_L dx = \int s_L(t_L) s_R(t_L + \tau) dt_L \int w(x) dx \tag{10.5}$$

Because the integration for $w(x)$ in Eq. 10.5 gives a constant that does not depend on τ, we finally have

$$A_{out}(\tau) \propto \int s_L(t_L) s_R(t_L + \tau) dt_L = C(\tau) \tag{10.6}$$

which implies that the output of coincidence detector (with delay τ) is directly related to the cross-correlation of the two inputs.

Note that the variable τ corresponds to internal delays that counteract the external ITD in Eqs. 10.3, 10.4, 10.5, and 10.6. Now the ITD can simply be determined by finding the value of τ that counteracts best (i.e., that maximizes the cross-correlation between the left and the right ear inputs)

$$\text{ITD} = \text{argmax}\{C(\tau)\} \tag{10.7}$$

Different variants of the delay line concept are introduced in Sect. 10.4. It does not necessarily have to be the multiplication or the argmax operation.

The second concept of ITD extraction is very different from cross-correlation: an IPD extractor without the need of any internal time delay mechanism. The argument φ of the complex coherence is identical to the IPD

$$\gamma(\text{ITD}) = |\gamma(\text{ITD})| \times e^{i\varphi} \tag{10.8}$$

The IPD can be understood as equivalent to the ITD of the carrier or the TFS. In practice, a single neuron cannot be expected to unambiguously code the IPD because of its circular nature. Even with only one neuron, however, it is still possible to code some directional information through its response rate (*R*). Consider the example

$$R_1(\text{ITD}) \propto |\gamma(\text{ITD})| \times \cos\varphi = Re\{\gamma(\text{ITD})\} \tag{10.9}$$

Of course, neural response rates are never negative. For simplicity, this model neuron allows for negative response rates. An offset could be added to avoid this but is not important here. Solving for the IPD we get

$$\varphi = \cos^{-1}\left(R_1 / R_1^{\max}\right) \tag{10.10}$$

Typically, Eq. 10.10 has two solutions, so a second neuron with a characteristic phase or time offset is required to resolve the ambiguity. The easiest case would be a 90 ° (i.e., π/2) offset, causing two orthogonal rate IPD functions

$$R_2(\text{ITD}) \propto |\gamma(\text{ITD})| \times \cos\left(\varphi - \frac{\pi}{2}\right) = |\gamma(\text{ITD})| \times \sin\varphi = Im\{\gamma(\text{ITD})\} \tag{10.11}$$

from which we get

$$\varphi = \sin^{-1}\left(R_2 / R_2^{\max}\right) \tag{10.12}$$

This again has two solutions, but only one solution is in common with Eq. 10.10, which gives the true IPD. Alternatively, the phase offset of +π/4 in one and −π/4 in the second neuron can be used, as observed in mammals (McAlpine et al. 2001).

In contrast to the delay-line concept, the two-channel concept (Eqs. 10.10 and 10.12) can only be applied when the TFS can be spectrally separated from the envelope, i.e., with sufficiently narrow filters (Fig. 10.1). Otherwise, a cyclic carrier phase would not even exist. Also, at frequencies above 1400 Hz where humans are not sensitive to TFS IPD (Brughera et al. 2013), this method is no longer applicable. In the important midfrequency range (300–1400 Hz in humans), however, the TFS and the demodulated envelope are both physically present in the stimulus at this stage of processing (Fig. 10.1c). Fortunately, they are in different frequency regions due to the narrow auditory filters. This separation does not exist for an unrealistic one-octave-wide filter (Fig. 10.1d). Finally, temporal differences in the demodulated envelope from left and right provide ITD information, which is particularly important for carrier frequencies above 1400 Hz where the TFS IPD cannot be exploited.

To summarize, it is critically important to understand the implications of cochlear processing and to have a basic knowledge of systems theory for a deep understanding of binaural modeling. Most importantly, the narrow bandwidth of the basilar membrane filters allows for a spectral separation of envelope and fine structure after the hair cell nonlinearity at low and intermediate frequencies. It further provides outputs with a high coherence up to large ITDs (e.g., $|\gamma(1\ \text{ms})| > 0.97$). In theory, this enables a pair of ideal binaural units (neurons) to determine the IPD (Eqs. 10.10 and 10.12). At high frequencies, the hair cell nonlinearity is even more important for extracting the envelope from an AM signal, a prerequisite for high-frequency envelope ITD sensitivity.

10.3 Modeling Physiological Data

10.3.1 Background

Section 10.3 covers the physiological modeling of binaural interaction. Most frequently, the aim of physiological modeling is to explain how binaural information (such as ITD and ILD) is represented or "encoded" in the physiological responses of binaural neurons (e.g., spiking patterns). The discussion on how to transfer and "decode" binaural information to form the perception of the sound source is found in Sect. 10.4.1. As an introduction to physiological models, Sect. 10.3 first reviews the underlying neuronal circuits for binaural processing and the classification of physiological models.

10.3.1.1 Binaural Pathways

The first auditory stage where the information from the two ears converges is the superior olivary complex located in the mammalian brainstem (Grothe et al. 2010; see also Owrutsky, Benichoux, and Tollin, Chap. 5). It is subdivided into several nuclei, including the medial superior olive (MSO; Fig. 10.3a) and the lateral

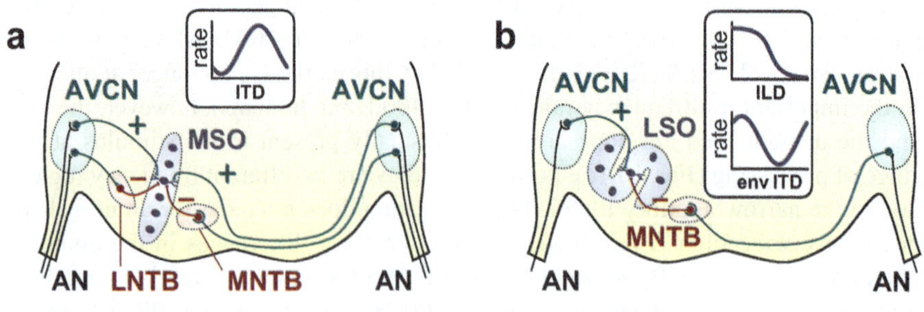

Fig. 10.3 Binaural circuits in the auditory brainstem. (**a**) Interaural time difference (ITD) coding circuit in the medial superior olive (MSO). (**b**) Interaural level difference (ILD) coding circuit in the lateral superior olive (LSO), where the envelope (env) ITD is also encoded. *Insets:* typical binaural tuning curves. +, Excitatory synaptic transmission (mediated by glutamate); –, inhibitory synaptic transmission (mediated by glycine). AVCN anteroventral cochlear nucleus, LNTB lateral nucleus of the trapezoid body, MNTB medial nucleus of the trapezoid body, AN auditory nerve

superior olive (LSO; Fig. 10.3b). Principal neurons in the MSO are sensitive to ITDs of low-frequency sounds (typically below 2 kHz), whereas those in the LSO respond to ILDs of high-frequency sounds (typically above 2 kHz) as well as envelope ITDs of high-frequency amplitude-modulated sounds. Note that here only models of the mammalian brainstem pathways for sound localization are explained. Birds and reptiles have comparable yet different structures (Ashida and Carr 2011; see Takahashi, Kettler, Keller, and Bala, Chap. 4).

The mammalian MSO neuron receives binaural excitatory inputs from spherical bushy cells in the anteroventral cochlear nucleus (AVCN) of the ipsilateral and contralateral sides (Fig. 10.3a; see also Owrutsky, Benichoux, and Tollin, Chap. 5). Because of this bilateral excitation, the response type of an MSO neuron is referred to as excitatory-excitatory (EE). The MSO neuron acts as a coincidence detector that varies its output spiking rate according to the degree of synchrony of bilateral excitatory inputs. Consistent with Eq. 10.6, the output spike rate of an MSO neuron becomes high when ipsilateral and contralateral excitatory inputs arrive in time. Because the ITD reflects to the relative timing of the bilateral synaptic inputs, the MSO output rate consequently modulates with the ITD. A number of cellular and synaptic specializations underlie this submillisecond coincidence detection in MSO (Golding and Oertel 2012). There are, however, additional sources that affect the relative input timing, including the cochlear delay, geometry of the bushy cell axon (total length, axonal diameter, internodal length), and fast inhibitory inputs (Grothe et al. 2010; Vonderschen and Wagner 2014). The primary source of inhibition in the MSO is the medial nucleus of the trapezoid body (MNTB), which is driven by the contralateral AVCN (Fig. 10.3a). MSO neurons receive additional inhibitory inputs from the lateral nucleus of the trapezoid body (LNTB), but this ipsilateral inhibition is usually not included in existing MSO models because characterization of its physiological properties was performed only recently (e.g., Roberts et al. 2013; Franken et al. 2016) and its role in ITD coding is still unclear.

The LSO neuron receives excitatory synaptic inputs from spherical bushy cells in the ipsilateral AVCN and inhibitory synaptic inputs from the MNTB driven by contralateral sound stimuli (Fig. 10.3b; see Owrutsky, Benichoux, and Tollin, Chap. 5). This excitatory-inhibitory interaction (referred to as EI) makes the LSO neuron sensitive to ILDs. Namely, the higher the sound level at the ipsilateral ear relative to the contralateral ear, the higher the output spike rate of an LSO neuron. In many earlier models of ILD coding (e.g., those reviewed in Colburn 1996; Jennings and Colburn 2010), the LSO neurons were thus assumed to be a simple comparator of ipsilaterally driven excitatory inputs with contralaterally driven inhibitory inputs. In addition to ILD coding, LSO neurons vary their spike rates according to the ITDs of envelopes of amplitude-modulated sounds (Joris and Yin 1995). Furthermore, LSO neurons with low characteristic frequencies are sensitive to ITDs of low-frequency tones below 1 kHz (Tollin and Yin 2005). Early anatomical investigations failed to identify the cluster of MNTB neurons in humans; only recently was the existence of the human MNTB confirmed, suggesting a possible role of timed inhibition for binaural coding both in laboratory animals and in humans (Kulesza and Grothe 2015).

10.3.1.2 Classification of Physiological Models

As seen in Sect. 10.1, binaural models can be largely divided into two families: physiological models and functional models. Physiological models of binaural neurons can be further categorized into several groups (Table 10.2). Figure 10.4 illustrates the fundamental properties of different physiological models. Models with membrane potentials (e.g., Fig. 10.4c, d) are used to investigate, for example, how bilateral synaptic inputs are integrated into the modeled neuron or how underlying biophysical processes (such as ion channels, membrane nonlinearity, and neuronal

Table 10.2 Classification of binaural models binaural models

<table>
<tr><td colspan="6">Binaural models</td></tr>
<tr><td colspan="5">Physiological models</td><td rowspan="5">Functional models</td></tr>
<tr><td colspan="3">Models with membrane potential (conductance-based models)</td><td colspan="2">Models without membrane potential (shot-noise models)</td></tr>
<tr><td colspan="2">HH-type models</td><td rowspan="2">IF-type models</td><td rowspan="2">Stein models</td><td rowspan="2">Coincidence-counting model</td></tr>
<tr><td>Multicompartment HH-type models</td><td>Single compartment HH-type models</td></tr>
</table>

See text for an explanation
HH Hodgkin-Huxley, *IF* integrate-and-fire

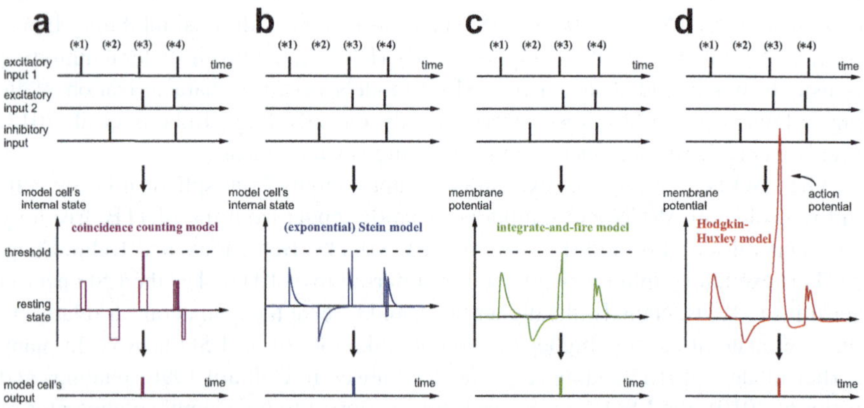

Fig. 10.4 Schematic drawing of various physiological models. To simplify the illustration, all the models shown here are assumed to receive two excitatory inputs and one inhibitory input. Excitatory (*1) and inhibitory (*2) inputs (*top row*), respectively, make positive and negative impacts on the model's internal state or the membrane potential (*center row*). The shape of the synaptic inputs differs between models. The coincidence-counting model (**a**) uses rectangles, whereas the Stein model (**b**) uses an exponentially decaying function. Synaptic inputs in conductance-based models (**c** and **d**) usually have curved shapes similar to those observed in physiological experiments. When there are coincident excitatory inputs (*3), the model produces an output spike (*bottom row*). The internal mechanism of coincidence detection may either be threshold-crossing detection (in **a–c**) or more naturalistic action potential generation (**d**). An inhibitory input cancels the accumulation of excitatory inputs (*4) and suppresses the output spike generation. See Table 10.2 and the text for categorization of the models

morphology) may play a role in binaural processing. In contrast, models without a membrane potential (e.g., Fig. 10.4a, b) usually focus on the input-output relationship of the modeled neuron without paying much attention in its subcellular processes.

Within the category of Hodgkin-Huxley (HH)-type conductance-based models, the model proposed by Rothman and Manis (2003) has been most frequently used in auditory simulation studies. They characterized various ionic conductances of the guinea pig ventral cochlear nucleus neurons in vitro, including the fast-sodium, high-voltage-activated (delayed rectifier) potassium (K_{HVA}), low-voltage-activated potassium (K_{LVA}), transient (inactivating) potassium (K_A), and hyperpolarization-activated cation (I_h) channels. By changing the amounts of these conductances, a wide variety of spiking behavior can be simulated (Rothman and Manis 2003; Manis and Campagnola 2018). The Rothman-Manis model (usually with some modifications) has been applied to various auditory neurons in the AVCN, MSO, and LSO.

The items in Table 10.2 are organized by the relative complexity of the models, with the most complex category on the left. Complex models are used for investigating the detailed relationship between underlying biophysical processes and spiking responses. Simple models are preferred when mathematical simplicity is desired (e.g., for calculating analytical solutions), when computational efficiency is required (e.g., for engineering applications and large-scale simulations), when a minimal description of the phenomenon is investigated, and/or when they are applied to perception-oriented whole pathway models (Sect. 10.4). There is a general trade-off between simplicity and biological plausibility in that no model would be appropriate for all purposes. Users of a model are therefore advised to carefully consider which model may fit the goals of their intended applications (Ashida et al. 2017).

10.3.2 Medial Superior Olive Neuron Models

10.3.2.1 Hodgkin-Huxley-Type Medial Superior Olive Models

MSO neurons vary their spike rates to ITD changes of less than 100 μs (Grothe et al. 2010; Ashida and Carr 2011). This remarkable precision has attracted the attention of many theoretical and experimental studies that investigated the underlying synaptic and cellular specializations (Golding and Oertel 2012). Studies using single-compartment HH-type models (e.g., Fig. 10.4d) revealed that the large amount of K_{LVA} conductance found in the MSO plays a major role for this computational precision. The dendrotoxin-sensitive K_{LVA} conductance that is activated in the subthreshold regimen (typically around −50 mV) reduces the membrane time constant, enhances integration of subthreshold signals (Svirskis et al. 2002), defines a narrow temporal window for synaptic integration (Svirskis and Rinzel 2003), makes the neuron sensitive to the rising phase of inputs (Gai et al. 2014; Dietz et al. 2016), and sets a frequency-dependent threshold for excitatory synaptic inputs (Mikiel-Hunter et al. 2016).

The I_h current, which is mediated by hyperpolarization-activated cyclic nucleotide-gated channels and activated also in the subthreshold regimen, plays a complementary role to K_{LVA} in temporal coding. Simulations demonstrated that I_h current counteracts the inactivation of K_{LVA} to maintain the size of synaptic inputs over time (Khurana et al. 2011). The tonotopic organization of I_h conductance and its dynamics contribute to the fine temporal filtering of inhibitory synaptic inputs in the MSO (Baumann et al. 2013). The frequency-dependent responses characterized with these subthreshold currents were further simulated with a simplified model, which showed that resonant properties of the MSO (and low-frequency LSO) neurons are suitable for extracting ITDs of the TFS (Remme et al. 2014).

Anatomical studies revealed that a typical MSO neuron has a bipolar morphology; ipsilateral excitatory inputs converge to one dendritic branch while axons from the contralateral AVCN projects to the other branch (Grothe et al. 2010). The effects of having bipolar dendrites in auditory coincidence detectors (such as the MSO) were examined with multicompartment conductance-based models. Sublinear summation of synaptic inputs at each dendritic tree enhances the coincidence detection of the bilateral EE inputs (Agmon-Snir et al. 1998; Remme and Rinzel 2011). Spatial distribution of K_{LVA} conductance along the dendrite underlies the stable transmission of synaptic potentials (Mathews et al. 2010). Asymmetrical distribution of inhibitory synapses and axons may contribute to the observed time shift in the ITD-tuning curve (Brughera et al. 2013). A similar bipolar dendritic model was also used to examine the origin of the exceptionally large field potential in the MSO (Goldwyn et al. 2014; Goldwyn and Rinzel 2016). A multicompartment HH-type model suggested that inactivation of the sodium channels distributed along the MSO axon plays a major role in enabling high-frequency spiking to widen the dynamic range of the rate code for ITDs (Lehnert et al. 2014). To summarize, HH-type models have contributed (and will keep contributing) to revealing various biophysical mechanisms underlying submillisecond coincidence detection in the MSO.

10.3.2.2 Integrate-and-Fire-Type Medial Superior Olive Models

Because the HH model is a set of nonlinear differential equations, it is difficult to directly analyze its dynamics using mathematical techniques. For the mathematical examination of neuron models, the integrate-and-fire (IF) model and its variations are often preferred (Fig. 10.4c; Jolivet et al. 2004; Burkitt 2006). An IF model with a K_{LVA} current was used as a simple model to study how small noisy signals are integrated in the MSO (Svirskis and Rinzel 2003). To simulate the "onset-spiking" behavior of MSO, an IF model with a dynamic (nonstationary, voltage-dependent) threshold was developed (Meng et al. 2012). Analyses of this model indicated that the phasic response (i.e., spiking preferably at the stimulus onset) is important for precise temporal processing. In addition, IF-type models with synaptic learning mechanisms have been used to study how spike-timing-dependent plasticity plays a role in tuning binaural information processing in auditory coincidence detectors such as the MSO (Glackin et al. 2010; Fontaine and Brette 2011).

10.3.2.3 Medial Superior Olive Models Without Membrane Potential

Physiological neuron models that lack the membrane potential but still receive (modeled) synaptic inputs are called shot-noise models (Colburn 1996). This term was borrowed from electrical engineering and indicates that each impulse input is regarded as a delta function and that the set of inputs follows the Poisson (or other random) process. In these models, each synaptic input is converted into, for example, an exponential decay function (the so-called Stein model; Fig. 10.4b) or a rectangular function (the coincidence counting model; Fig. 10.4a) and then summed to form a "virtual membrane potential" that represents the internal state of the model. When the summed input reaches the preset threshold (Fig. 10.4a–c, *dashed lines*), an output spike is generated. Shot-noise models are even simpler than the IF model; its response can be approximated by stochastic processes (such as an Ornstein-Uhlenbeck process) under some mathematical assumptions (Burkitt 2006).

Despite its simplicity, a shot-noise model can simulate the coincidence detection in the MSO reasonably well (Colburn et al. 1990). A similar model was used to analytically calculate the input-output relationship of MSO-like EE-type coincidence detector neurons (Toth and Marsalek 2015). Franken et al. (2014) created a simple shot-noise-type model with only a few parameters to simulate the nonlinear binaural interaction in the MSO (an example of the simulated ITD tuning curve is shown in Fig. 10.5). Their model used different thresholds for monaural and binaural coincidence detection and was fed with spike sequences recorded from putative input fibers in vivo. Their series of simulations demonstrated that the binaural threshold should be lower than the monaural threshold to reproduce realistic MSO responses. This finding is consistent with the nonlinear summation of synaptic inputs along the bipolar dendrite (Agmon-Snir et al. 1998).

10.3.3 Lateral Superior Olive Neuron Models

10.3.3.1 Hodgkin-Huxley-Type Lateral Superior Olive Models

In contrast to the abundance of MSO models, the LSO has attracted somewhat less attention in computational studies. An early study using a multicompartment HH-type model of the LSO showed that the distributions of excitatory synapses on dendrites and inhibitory synapses on the cell body underlie the divisive effects of the EI interaction (Zacksenhouse et al. 1998). Single-compartment models with various ionic conductances demonstrated the importance of hyperpolarization-activated currents (Szalisznyó 2006) and the K_{LVA} current (Wang and Colburn 2012) in

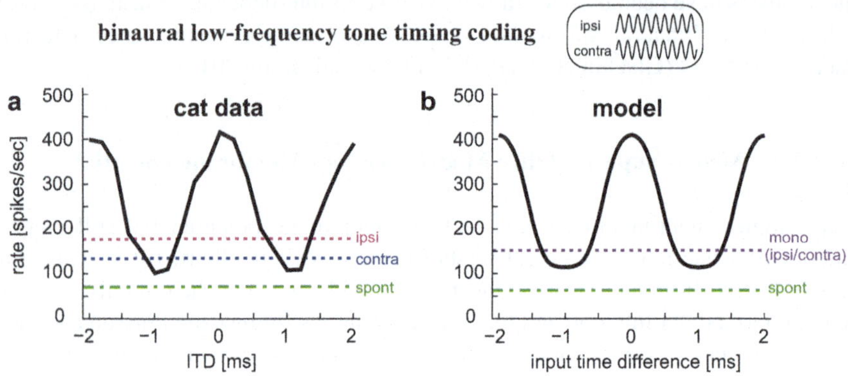

Fig. 10.5 Comparison of cat MSO in vivo recording data (replotted data from Yin and Chan 1990; **a**) and corresponding simulation results with the coincidence-counting model (**b**). Binaural tuning curves (*black curves*) as well as monaural rates (contralateral [contra], *blue*; ipsilateral [ipsi], *pink*; monoaural [mono], *purple*) and spontaneous (spont) rates (*green*) are shown. The horizontal axis for the modeled MSO response represents the time difference of bilateral synaptic inputs, which reflects the ITD (and possibly other delays) in the corresponding experimental data. For the model, the following parameters were used: sound frequency = 500 Hz; average sound-driven rate of input fibers = 300 spikes/s; sound-driven vector strength of input fibers = 0.9; spontaneous rate of input fibers 100 spikes/s; number of excitatory inputs = 4 fibers/side; coincidence window = 0.25 ms; monaural threshold = 4 inputs; binaural threshold = 2 inputs; and refractory period = 1.6 ms

characterizing the spiking properties of LSO neurons, including the subthreshold oscillation, onset response, and entrainment. Tonotopic variation of the membrane properties and their effects on temporal coding were also demonstrated in a study that combined in vitro slice recordings with an electrical circuit membrane model (Remme et al. 2014).

10.3.3.2 Integrate-and-Fire-Type Lateral Superior Olive Models

When detailed ionic descriptions of the membrane are not required, models simpler than the HH type are selected. A recent LSO modeling effort with IF-type models investigated how spike timing-dependent plasticity plays a role in ILD tuning (Fontaine and Peremans 2007), how spike after hyperpolarization characterizes the interspike-interval statistics (Zhou and Colburn 2010), and how temporal jitter in the input may affect the ILD coding in the LSO (Karcz et al. 2011). Simulations using a modified IF model with potassium conductance revealed that a variety of discharge patterns seen in the LSO can be replicated by changing the timing and strength of ipsilateral inhibition from the LNTB (Greene and Davis 2012). This ipsilateral LNTB inhibition has rarely been considered in other existing LSO models (and is therefore omitted in Fig. 10.3b).

10.3.3.3 Lateral Superior Olive Models Without Membrane Potential

Shot-noise models have also been applied to LSO neurons (Colburn 1996). For EI-type neurons, such as the LSO, excitatory and inhibitory inputs are modeled to make a positive and a negative impact, respectively, to the internal state of the shot-noise model. In a simple setting, for example, the counts of inhibitory inputs within a preset time window is weighted and subtracted from the counts of summed excitatory inputs (Fig. 10.4a, coincidence-counting model). Even this simple model can simulate representative responses of LSO neurons, including monaural modulation frequency tuning as well as binaural ILD and envelope ITD tunings (Fig. 10.6). The model spike rate varies periodically with the envelope ITD (Fig. 10.6a_1, a_2) and monotonically with the ILD (Fig. 10.6b_1, b_2). These modeling results suggest that the LSO neuron is not just a comparator of excitatory and inhibitory inputs over a long time period but rather operates as a temporally precise anticoincidence detector, whose spike rate becomes maximal when its excitatory and inhibitory inputs arrive asynchronously (Ashida et al. 2016, 2017). The term "anti" indicates the opposing effects of excitatory and inhibitory inputs, which make a contrast with MSO neurons where the synchronized arrival of bilateral excitatory inputs is detected.

Anatomically, both MSO and LSO neurons have bipolar dendrites. Excitatory synaptic terminals are found primarily on their dendrites, whereas inhibitory terminals are located at or near the cell body (Grothe et al. 2010). Physiologically, MSO and LSO neurons have short membrane time constants of the (sub)millisecond order mediated by various ion channels, as mentioned in Sect. 10.3.1. On average, the response timescale of MSO neurons appears to be faster than that of LSO neurons, suitable for extracting the temporal information of the TFS (i.e., carrier) of the sound stimulus (Remme et al. 2014). Corresponding to these anatomical and physiological characteristics, MSO and LSO models often share the fundamental structures, whereas their parameters are tuned differently to account for the distinction in their known response characteristics. The structure and parameters of a model need

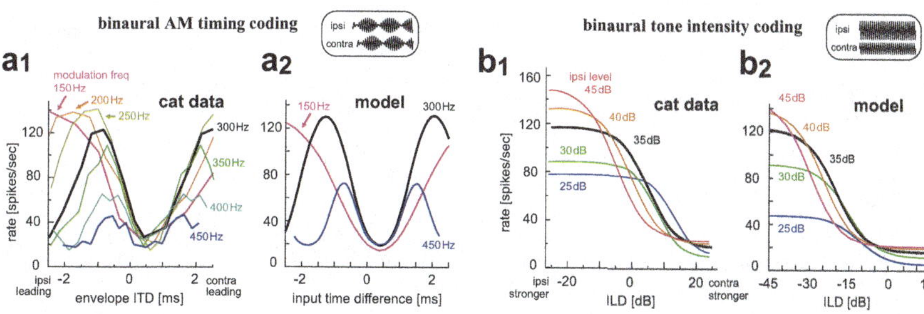

Fig. 10.6 Comparison of cat LSO in vivo recording data and corresponding simulation results of the coincidence-counting model. $\mathbf{a_1}$ and $\mathbf{a_2}$ Binaural envelope ITD tuning curves with different modulation frequencies (freq). AM amplitude modulated. $\mathbf{b_1}$ and $\mathbf{b_2}$ Binaural-tone ILD tuning curves with different ipsilateral sound levels. Original data from Joris and Yin (1998) and Tsai et al. (2010); figures adapted from Ashida et al. (2017)

to be continually revised according to available empirical data. For example, a recent intracellular recording study in vivo revealed that principal neurons in the LSO have much finer temporal properties than previously estimated (Franken et al. 2018). This finding further supports the role of the LSO in precise temporal coding and necessitates a revision of current LSO models.

10.3.4 Applications of Medial Superior Olive/Lateral Superior Olive Models

10.3.4.1 Model Comparison

Applications of models have diverse goals. Depending on the goal of an application, different constraints can be imposed on the model. For the investigations of the biological mechanisms of binaural sound localization, for example, relevant anatomical and physiological restrictions should be considered. Such restrictions include the head size, which influences the maximum ITD in free-field listening, or the temporal precision of the spike generation of binaural neurons.

To date, most physiological MSO and LSO models aim to replicate a certain small set of empirical data. It is often unclear how similarly or how differently each model responds to the same input. A comparative study of single-compartment LSO models demonstrated that, when calibrated with common criteria, both shot-noise models and conductance-based models show more or less similar tunings for ILD and envelope ITD tuning (Fig. 10.7). For monaural AM tones, different models share a peak response at a modulation frequency of about 200 Hz (Fig. 10.7a). For binaural stimulation, all models present periodic ITD tunings (Fig. 10.7b), whereas they show monotonic decrease of spike rates with ILD (Fig. 10.7c). These results

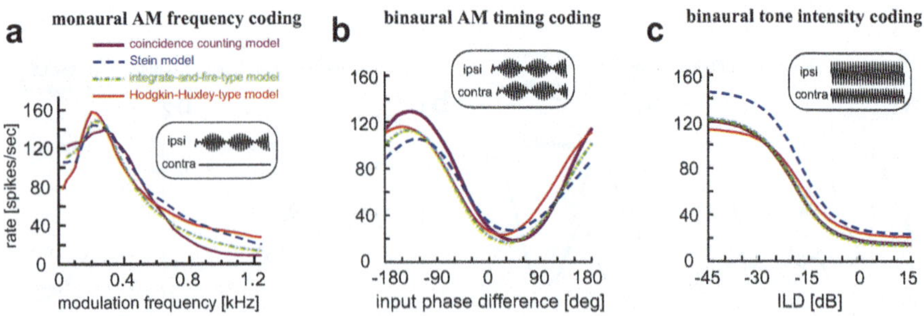

Fig. 10.7 Comparisons of physiological LSO models. Four models (coincidence counting, Stein, integrate-and-fire-type, and Hodgkin-Huxley type) are calibrated with the common dataset and their resulting tuning curves are plotted. (**a**) Frequency tuning of monaural AM sounds. (**b**) ITD tuning of binaural AM sounds. (**c**) ILD tuning of binaural tones. Note that all the models show similar tuning curves, although the structures and complexity of the models are considerably different (see Fig. 10.4). (Adapted from Ashida et al. 2017)

indicate that a model user can (and should) select an appropriate LSO model according to the desired level of biological complexity and computational efficiency of the model (Ashida et al. 2017). Systematic comparisons of MSO models have not been carried out yet.

10.3.4.2 Simulations with Arbitrary Sound Stimuli

In most of the neurophysiological experiments, where the fundamental response characteristics of auditory neurons are investigated, relatively simple acoustic stimuli are used, such as pure tones, broadband noise, sinusoidal AM tones, and clicks. Hence, physiological models of binaural neurons are usually calibrated and tested only with these simple sounds. In more naturalistic situations as well as in psychophysical experiments, animal or human subjects experience more complex acoustical stimuli, including a combination of simple sounds (such as a tone in noise), reverberation, and vocalizations or speech that is rich in spectral and temporal components. To adapt a binaural neuron model to such complex stimulations, auditory periphery models that can handle arbitrary acoustic stimuli are employed. These models receive a sound waveform as their inputs and produce spike trains of modeled ANs at each frequency band (Meddis 2006; Bruce et al. 2018). This AN output is used to drive model neurons located along the central auditory pathways.

In the last decade, an increasing number of simulation studies have combined an auditory periphery model with a binaural neuron model. Examples include the simulation of binaural responses driven by pure and AM tones (LSO: Wang and Colburn 2012), by transposed tones and speech tokens (MSO: Gai et al. 2014), by click trains with varied levels of time lags to simulate the precedence effect (MSO: Xia et al. 2010; Gai et al. 2015), by animal vocalization (MSO and LSO: Remme et al. 2014), by sound waveforms filtered with a measured head-related transfer function (LSO: Wall et al. 2012), and by AM tones with various envelope shapes (LSO: Dietz et al. 2016).

A previous modeling study (Brughera et al. 1996) investigated how the spiking pattern of AVCN bushy cells may affect the binaural tuning properties of a modeled MSO neuron. Other than this example, studies combining an AN model with a binaural neuron model often assume the AVCN and MNTB to be a relay that merely transfers AN output to the target MSO or LSO neuron without making any modification to the temporal spiking patterns. Although this assumption helps to simplify the entire model structure, it may constitute an oversimplification of the biological reality because physiological recordings showed an improvement in phase locking, especially at low frequencies, between ANs and AVCN bushy cells (Joris et al. 1994). It is yet to be investigated how this enhancement of temporal features at the AVCN and MNTB may contribute to the binaural processing in MSO and LSO.

10.3.4.3 Simulations of Hearing Impairment and Electrical Hearing

By modifying the auditory periphery model, impaired hearing can be simulated. For example, the effects of damaged hair cells on the spiking behavior of ANs were investigated by changing the frequency tuning parameters of a periphery model (Zilany and Bruce 2007). To date, binaural models that include hearing impairment use an impaired periphery but do not impair the binaural interaction per se. Not many studies exist on LSO or MSO pathophysiology (see, e.g., Lee 2013). Also, on the other end of the spectrum, only a few behavioral studies appear to require an impairment specific to binaural interaction. The unusually high binaural loudness summation of some hearing-impaired individuals (Oetting et al. 2016) is arguably the clearest example of such effects, but no binaural model exists that tries to explain this effect. In many cases, a reduction in binaural sensitivity can be explained simply by factors preceding binaural interaction (Bernstein and Trahiotis 2018).

In stark contrast to the periphery focus of impaired acoustic hearing, modeling studies of binaural hearing with cochlear implants focus on the binaural interaction (Chung et al. 2015; Kelvasa and Dietz 2015). This is likely motivated by discrepancies between prominent temporal precision of electrically stimulated AN fibers and very low rate limits of ITD perception, which cannot be explained by “normal” binaural models. For a further review of electrical hearing binaural models, see Dietz (2016).

10.3.5 Models of the Inferior Colliculus

The inferior colliculus (IC) is a central hub in the auditory midbrain where almost all ascending information from the lower auditory stations converges. The IC receives inputs directly and indirectly (via the lateral lemniscal nuclei) from the MSO, LSO, and other brainstem nuclei (Malmierca 2004; Cant and Benson 2006). Reflecting this convergence, neurons in the IC show diverse response characteristics to acoustic stimuli (Malmierca 2004; Graña et al. 2017). Some IC neurons are sensitive to the ITD and/or ILD, whereas the response of other IC neurons is independent of these interaural cues and rather depends on other temporal and spectral patterns of the sound (reviewed in Davis 2005). IC neurons can be categorized into several physiological and morphological groups (Malmierca 2004; Ono and Ito 2015), but the relationship between binaural coding properties and these cellular categories remains to be investigated. Furthermore, the IC is anatomically divided into several subdivisions, and the fractions of binaural units are different between the areas (Graña et al. 2017). It is still uncovered what kind of physiologically distinct roles in binaural coding these subdivisions may have.

IC neurons receive inputs from multiple sources, and not all of the sources are physiologically well characterized (Malmierca 2004; Cant and Benson 2006). Furthermore, as noted above, the relationships among the anatomical, physiological, and functional characteristics of the IC neurons are largely unclear. Previous IC models thus focused mostly on replicating or explaining a specific small set of data instead of simulating the entire functions of the IC.

A few binaural models have been used to simulate ITD/ILD coding in the IC. A study using an IC model that receives inputs from the MSO on both sides suggested that the cellular adaptation and EI interaction may underlie the history-dependent IPD sensitivity in the IC (Borisyuk et al. 2002). More recently, an IF-type IC neuron model was combined with HH-type MSO and LSO models (Wang et al. 2014) to investigate how the converging inputs from the MSO and LSO may characterize the binaural envelope tuning curves of the IC neurons. A model of the IC is particularly relevant for physiological modeling studies that aim also to simulate behavioral or psychophysical data. Aharonson and Furst (2001) developed a neural network model to examine possible effects of neuronal lesions on coding in the IC and consequences for binaural perception in stroke patients (see also Sect. 10.4.4). In this model, an IC neuron was assumed to receive inputs from multiple sources, including monaurally driven inputs from the cochlear nucleus on both sides as well as binaurally driven inputs from an array of ipsilateral MSO and LSO neurons tuned to different ITDs and ILDs. The IC is the last nucleus of the ascending binaural auditory pathway studied with the physiological modeling approaches discussed in Sect. 10.3.

10.4 Simulating Psychoacoustic Data

The physiological modeling in Sect. 10.3 focused mostly on how the ITD, the ILD, and other important information are represented in the spiking patterns of binaural neurons. To connect the neuronal activity to binaural perception, however, a fundamental question needs to be addressed: "How do higher auditory stages 'read out' the relevant information encoded by the binaural neurons?" Section 10.4.1 discusses two prevailing model-based methods for the decoding stage that operates on the spiking outputs of binaural neurons.

In contrast to physiological models, a fair number of binaural models that aim to predict psychoacoustic data do not directly employ simulations of spiking neurons in the first place. These functional models operate on internal quantities that may, for example, be interpreted as spiking probability. If set up properly, these models can be functionally very similar to spiking models and are often computationally much more efficient. The cross-correlation function from Sect. 10.2 would be an example of such a model. This class of models often use further mathematical operations as decision variables and thus circumvents the decoding stage discussed in Sect. 10.4.1.

Irrespective of whether the model employs spikes or not, the two central goals of modeling should be (1) no functional contradiction to physiological data and (2) no significant deviations from behavioral data. To date, there seems to be some dichotomy between these goals. Models that are physiologically most plausible are not able to account for some of the classic psychoacoustic data, and delay-line-based models appear to account best for the classic psychoacoustic data, but the existence of delay lines in mammals is inconsistent with recent anatomical and physiological observations (Grothe et al. 2010). In light of the two goals, Sects. 10.4.2, 10.4.3, and 10.4.4 focus on the modeling of those behavioral datasets that help to contrast between the different model concepts.

10.4.1 Decoding the Output of Binaural Nuclei

Roughly speaking, there are two types of approaches to neuronal decoding of binaural information. The first approach aims to estimate the optimal performance of the system from the spiking pattern of neuronal elements without considering the detailed mechanisms of decoding. The second approach is to predict the behavioral performance from the underlying neuronal activity using various (hypothetical) decoding methods. Section 10.4.1 briefly reviews these two approaches.

The first type of approach employs an information-theoretic measure to estimate the theoretical lower bound of errors in behavioral experiments. Earlier studies took the stochastic spiking pattern of ANs and calculated the Cramér-Rao bound for the variance of estimated monaural (Siebert 1965) and binaural (Colburn 1973) quantities such as frequency, intensity, and interaural differences. More recent studies calculated the theoretical error bounds in ITD/ILD estimation from the activity of real IC neurons (Brown and Tollin 2016) to account for the just-noticeable differences measured in behavioral experiments. A similar but slightly different approach is to apply the Bayesian inference to evaluate the performance of the system. In the field of visual neuroscience, the idea of "Bayesian ideal observer" is broadly used to estimate how much information is available to the sensory system and how much information is preserved or lost at each stage of sensory processing (Geisler 2003). The ideal observer method was recently used in combination with the measured head-related transfer function to estimate the optimal sound localization ability in human listeners (Reijniers et al. 2014). Bayesian inference was also applied to modeled midbrain neurons to explain the measured behavioral accuracy of sound localization in owls (Fischer and Peña 2011).

In contrast to this first approach where the decoder is assumed to act as an optimal processor (in terms of the Fischer information or Bayesian inference), the second approach aims to extract the ITD/ILD information from the spiking activity of binaural neurons. There are several methods proposed for decoding (see Dietz et al. 2018 for a review of these concepts). For example, a place code, in which ITDs are encoded through the label of the most active neuron (Eq. 10.7), requires a broad range of neurons with different internal delays. On the other hand, for a rate code similar to that found in Eqs. 10.10 and 10.12, binaural neurons should ideally have very distinct properties from each other. This codependence of encoding and decoding offers fertile grounds for binaural models because they can translate experimental evidence from one stage into predictions about the other stage. However, it is also a common pitfall to entangle the two stages within the analysis or description of the processing mechanisms.

A hybrid of the two approaches is matching the activity pattern of binaural neurons to previously learned patterns (Day and Delgutte 2013; Goodman et al. 2013). It is unknown, however, whether and how pattern matching is achieved in the actual auditory system. Existing computational studies often resort to artificial neural network models to perform such operations; some of these models employ an additional hidden layer (Wall et al. 2012; Encke and Hemmert 2018) and others do not (Day and Delgutte 2013).

10.4.2 Binaural Unmasking

The phenomenon of binaural unmasking is arguably among the most described dichotic listening peculiarities (see Culling and Lavandier, Chap. 8, for an introduction and description of binaural unmasking). Despite much attention on binaural unmasking, there is no physiologically plausible modeling approach available that can account for the binaural unmasking of speech (i.e., binaural speech intelligibility level difference).

The most commonly used model concept to account for binaural unmasking of speech is the equalization-cancellation (EC) mechanism (Kock 1950). The left and right inputs are altered in level and timing so that the noise is best equalized. Then the channels are subtracted from each other, canceling out the noise but leaving any signal components that have other interaural differences (see Culling and Lavandier, Chap. 8, for details and variations). The unique aspect of models following the EC mechanism is that their output is a noise-reduced signal. One can use the EC model output in the same way as an unprocessed audio signal, presenting it to a listener or automatic speech recognizer. By contrast, all other models provide just a function of interaural differences over time, derive some measure related to interaural coherence, or directly provide spike trains. Despite the practical benefit, there is doubt from both physiology and psychoacoustics if an EC mechanism is operational in humans. First, time equalization requires a delay line, although this has been challenged by mammalian physiology (Grothe et al. 2010). However, EC-type models have been shown to account for a variety of data based on temporal coherence alone, i.e., without delay lines (e.g., Rabiner et al. 1966), so this does not impose an argument against the EC idea. Problematic is level equalization orthogonal to the delay line, which is far from physiological observations in the LSO. Second, the absence of harmonic features for signals presented below diotic but above dichotic detection threshold (Krumbholz et al. 2009; Klein-Hennig et al. 2018) indicates that a noise-reduced signal generated by the EC model is not psychoacoustically assessable.

Binaural unmasking of tones, the so-called binaural masking level difference (BMLD), can be modeled without the need for an output in the same format as the input. The most comprehensive model for BMLD to date is a delay-line-based framework, which detects the target tone trough a reduction in interaural correlation

(Bernstein and Trahiotis 2017, 2018). Another concept that is closely related to interaural coherence is tone detection by means of temporal fluctuations in the ITD and ILD (Goupell and Hartmann 2007). As a third concept, EC-type models can also account for a vast range of BMLD data (e.g., Breebaart et al. 2001a, b). The main problem here is that BMLD data are often not able to discern competing model concepts. Domnitz and Colburn (1976) identified that much of the general data and model trends directly follow from the stimulus itself. More precisely, BMLD magnitudes follow from the stimulus properties after peripheral processing (Trahiotis and Bernstein 2014). They speculate that datasets with varying spectral and interaural configurations of the masker should allow for a better disambiguation of model outcomes. Indeed, so-called double-delayed noise maskers (van der Heijden and Trahiotis 1999) and noise with different IPDs in its flanking bands (Marquardt and McAlpine 2009) appear to be the best candidates for testing the existing models.

There is no comprehensive neural model accounting for binaural unmasking, presumably because physiological studies have not yet provided the type of data required for addressing the modelers' questions, including where and how the information processing relevant to binaural unmasking is actually performed in the brain. The gap between physiology and modeling is most apparent in the fact that there is currently no alternative to EC models for predicting binaural unmasking of speech.

10.4.3 Perceptual Discrimination

Under ideal conditions, the difference in ITD between two noise intervals that can be discriminated by highly trained normal hearing listeners with a 75% correct rate is on average 6.9 μs (Thavam and Dietz 2019). This threshold ITD is orders of magnitude below any other temporal discrimination thresholds in the human sensory system and is henceforth subject to detailed and broad neuroscientific and psychophysical investigations.

Models that are similar to those described in Sect. 10.4.2 have been developed to account for a variety of ITD discrimination phenomena. The model concept to account for most of the data in a quantitative manner is the delay-line model where an ITD is detected by means of a reduced correlation coefficient, i.e., with a single delay-line unit at zero ITD (Bernstein and Trahiotis 2002). This is virtually identical to detecting an out-of-phase tone in diotic noise (Bernstein and Trahiotis 1997). The "delay-line and coincidence detection"-based model approach can explain discrimination data for a variety of different AM stimuli (Bernstein and Trahiotis 2009). Stimuli with steeper AM slopes cause a stronger ITD dependence on the cross-correlation function. On this basis and after peripheral preprocessing with low-pass filters and compression, the model can quantitatively account for the threshold ITD dependence on AM depth, AM frequency, and AM exponent of a sinusoidal modulator. The data and the model predictions are also in-line with rate ITD functions recorded from neurons in the guinea pig IC (Griffin et al. 2005). Despite these model successes, the argument by Domnitz and Colburn (1976) about BMLD data and models also applies here. Not all of the experiments can shed light on the processing within the binaural interaction stage itself.

One fact that many of the functional nonspiking models neglect is that temporal information is carried by spike trains in the brain. Especially for the abovementioned AM stimuli, many models represent the input to the binaural neurons as a transformed stimulus envelope (see Trahiotis and Bernstein 2014 for a review). For these models, an envelope ITD is detected by cross-correlating the left and the right inputs to the binaural interaction stage. This captures temporal differences in spike rates, but it omits most of the information from the timing of the first spike after stimulus onset or AM cycle onset. These first spike timings can be very precise, with a jitter as small as 0.2 ms for AN fibers (see Fig. 10.8c for a model fiber response to a 30-ms tone burst), similar to behavioral discrimination thresholds. Neural models (see Sect. 10.3.3) routinely use spike timing for envelope ITD sensitivity. Nonspiking models, on the other hand, need exceptionally high correlation reduction sensitivity to reach down to submillisecond thresholds only with the temporal rate-correlation

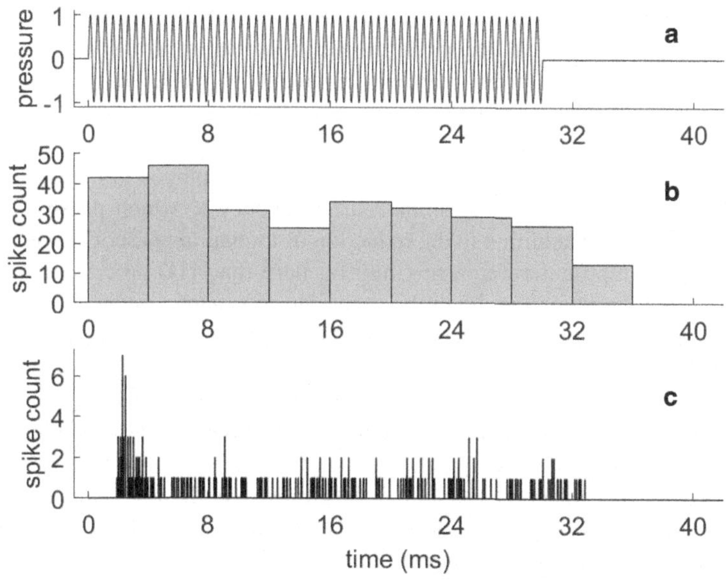

Fig. 10.8 Processing of a high-frequency tone burst simulated with the AN model from Bruce et al. (2018). Primarily, the first spike to the stimulus (or AM cycle) carries the temporal information useful for ITD extraction. (**a**) 4-kHz tone with a 30-ms duration. (**b**) Peristimulus-time histogram of an example AN model fiber response to the stimulus (38 repetitions) shown in **a**. The coarse bin width (4 ms) helps to ignore the precise timing and indicates a weak spike rate adaptation. After a high onset rate, a small decline is observed starting 8 ms after onset. (**c**) Analysis of the same spike train as in **b** but with a bin width reduced by a factor of 100 to 40 μs. What now looks like a very strong spike rate adaptation is primarily a difference in spike time precision. The small bin width emphasizes that (16 of 38) onset spikes fall within four adjacent 40-μs bins (i.e., in 160-μs windows). The standard deviation of all 40 first spike latencies is 0.3 ms, whereas the standard deviations of the second and last spike latencies are 2.6 and 4.2 ms, respectively

cue. Klein-Hennig et al. (2011) required a correlation detection threshold close to 0.999 (from a 1.0 reference) to account for the behavioral envelope threshold ITD. They also highlighted that classical (rate- or amplitude-) correlation models fail to predict ITD detection thresholds for some envelope shapes, particularly those that were designed to test the differences between AM onset and AM offset sensitivity. Despite a good agreement of model LSO neurons (e.g., Dietz et al. 2016) with both behavioral (Klein-Hennig et al. 2011) and neural data (Greenberg et al. 2017), an appropriate back-end model relating the LSO or IC model output to ITD perception does not exist yet. Such models are expected to exploit modulation onset spike ITDs and automatically ignore the timing of sustained activity.

Even more fundamental, but no less complex, is the modeling of pure-tone ITD sensitivity. The rapid decline of TFS ITD sensitivity within less than 1 octave (from maximal at 800 Hz to completely insensitive at 1500 Hz) is often attributed to the low-pass characteristic of the hair cell membrane potential. This results in a decline in phase locking (i.e., of TFS phase encoding in the AN response timing) with increasing frequency. This has been traditionally modeled with steep low-pass filters after half-wave rectification, such as a fourth order (Bernstein and Trahiotis 2002) or even a fifth order low-pass filter (Breebaart et al. 2001a). When employing more detailed physiological peripheral models, however, several studies report that ITD sensitivity of their models can exist above 1400–1600 Hz (Brughera et al. 2013; Moncada-Torres et al. 2018; Bouse et al. 2019). Due to missing direct observations, the degree of phase locking in humans remains subject to speculation (Verschooten et al. 2019), but the peripheral models employed in these studies are generally not suspected to overestimate the frequency to which phase locking is limited. Therefore, in addition to the reduction of monaural phase locking, binaural stages seem to impose the frequency limit for pure-tone ITD processing. Potential sources of this limitation can lie in the timescale of synaptic inputs to the MSO or in the band-pass filter characteristics of MSO neurons (see Sect. 10.3; Remme et al. 2014).

Given that both the TFS and envelope contribute to ITD sensitivity and that the additional effects of ILDs need to be considered, modelers face the challenge to disambiguate single pathway hypotheses (i.e., two cues are processed by the same neurons) from the independent parallel pathways hypothesis. Hafter and Carrier (1972) reported at least some "trading" between TFS ITD and ILD, suggesting a single pathway model. On the contrary, Furukawa (2008) improved the experimental paradigm and found that TFS ITD and ILD appear to be processed in two rather independent channels, presumably the MSO and LSO, respectively. Furukawa further showed that at high frequencies, envelope ITD and ILD combine as predicted by a single channel model, hinting at a LSO processing of envelope ITD, again in line with physiology (e.g., Joris 1996). Moore et al. (2018) used different methods and tested the third possible hypothesis of a pairwise comparison between TFS ITD and envelope ITD. Their "TFS ITD versus envelope ITD trading" data indeed hints at two independent ITD channels. Although some of these data can likely be modeled by separating the cues (e.g., Dietz et al. 2009), the temporal envelope does in fact have a strong influence on TFS ITD sensitivity. In other words, the auditory system is most sensitive to ITDs near modulation onset. ITDs applied to a 600-Hz carrier during segments of decreasing amplitude (envelope) can barely be detected (Hu et al. 2017). These phenomena, which are further reviewed by Stecker, Bernstein, and Brown (Chap. 6), cannot be modeled easily by nonspiking correlation-based models and are thus expected to be important for future model refinement.

In summary, ITD discrimination models can account for a wide range of psychoacoustic data. In particular, cue-trading data and threshold ITD dependence on envelope shape can help to distinguish competing model concepts. So far, however, this potential has rarely been exploited.

10.4.4 Laterality and Localization

A broadband sound presented over a loudspeaker is received by the two ears and creates an externalized perception of the sound at the position of the loudspeaker. This is referred to as *localization*. By contrast, in most of the above-described headphone experiments, the ITDs and ILDs cause a *laterality* of the intracranial percept (see Hartmann, Chap. 2; Stecker, Bernstein, and Brown, Chap. 6). For models of binaural interaction, which are in the focus of this chapter, measurements of laterality are usually more informative than localization performance.

As discussed in Sect. 10.2, models need to be sufficiently complex for predicting the extent of laterality or even localization of a sound. Many of the models reviewed in Sects. 10.4.2 and 10.4.3 operate (or can operate) on a single coincidence-detecting unit per frequency. For example, a unit without a time delay between its inputs can account for several datasets of BMLD (Rabiner et al. 1966; Bernstein and Trahiotis 2017). For laterality, on the contrary, most models rely on an array of differently tuned units and a complex mechanism to combine the information (e.g., Stern and Shear 1996). This model family is able to account for challenging behavioral data of both normal-hearing subjects (e.g., Bernstein and Trahiotis 2003, 2012), and subjects with lesions or other neural impairments along the auditory pathway (Aharonson and Furst 2001). However, especially for high-frequency stimuli, the cross-correlation approach of this model family is not in accordance with the EI domination of the most envelope ITD sensitive neurons (see Sect. 10.3). Trying to overcome this discrepancy, Klug et al. (2020) demonstrated that the response rate difference between a left and a right EI model neuron is proportionally related to the extent of laterality for a wide range of high-frequency stimuli. No delay lines were required and the same model neurons simulated both ILD and envelope ITD-based laterality.

At low frequencies, where the TFS ITD can be exploited, the most common test stimulus is band-pass-filtered noise with an ITD corresponding to a more than 180 ° carrier IPD at its center frequency (Trahiotis and Stern 1989). Therefore, the TFS ITD (or IPD) gives a cue toward the "wrong" lagging side that can be resolved either by envelope ITDs or by comparing ITDs across frequencies. Best, Goupell, and Colburn (Chap. 7) describe how such stimuli are consistently perceived at the lagging side for narrow and at the leading side for broad bandwidths. Different strategies have been suggested to model this perception. Apart from our unpublished work in progress, as of 2020, only models with an array of coincidence-detecting neurons with delays of 1.5 ms can quantitatively account for the behavioral data (Stern and Shear 1996). However, such long delays are likely not employed in the human binaural system (Grothe et al. 2010). This is arguably the most famous dichotomy between the two main model goals set at the beginning of Sect. 10.4 and is therefore an important question for future binaural models.

10.5 Summary and Future Directions

This chapter has addressed a variety of challenging questions that have been investigated with binaural models. Much has been learned about the auditory system from modeling approaches, but many open questions remain. Here, we briefly summarize the main advances of binaural models and then connect the open questions with an outlook.

Current physiological modeling of binaural coding and decoding (reviewed in Sects. 10.3.2 and 10.3.3) faces multiple challenges and limitations. Models of binaural neurons are normally verified with physiological findings from MSO and LSO recordings. The most limiting factor here, however, is that physiological datasets from different laboratories are not in full agreement with each other, prohibiting a universally acceptable model in the first place. Functional consequences are fortunately minor, and MSO/LSO model responses are in many cases very similar to the data (e.g., Wang and Colburn 2012; Ashida et al. 2017).

The same generally holds for models of binaural perception. Computational models for the auditory processing stages before binaural interaction have become more accurate and can now account for a vast range of experimental data. However, this happened irrespective of the details of the binaural interaction stage itself. The three central classes of behavioral experiments highlighted in Sects. 10.4.2, 10.4.3, and 10.4.4 are designed to understand the mechanisms of binaural interaction in humans.

Limitations arise mostly from the interface between the (physiological) front end and the (rather functional) decoding stage that relates the output of the model LSO/MSO neurons to a behavioral response. As mentioned in Sect. 10.3.5, it is not clear which neuronal population in the IC actually encodes the relevant information for binaural sound localization. This makes it difficult for a modeler to properly choose the set of neurons whose output is used for decoding. Furthermore, most prior modeling studies (if not all) on binaural coding and decoding simply ignored known physiological and anatomical variations, including tonotopic specializations and neuron-to-neuron variability, which might nevertheless be essential for precise binaural processing. Most current models that account for perceptual data circumvent these difficulties by not using a spiking front-end model in the first place. This comes, however, at the risk of missing out, misinferring, or misrepresenting the information carried by the neural code.

One could expect that the combination of physiological front-end models with perception predicting back-end models is the next big milestone for the field. However, it is unlikely that there is a single model that satisfies the demands of all model users. Different laboratories develop models for different task classes. The current wave of open science and data/model sharing now offers a good selection of models (e.g., Ashida et al. 2017; Manis and Campagnola 2018; see Dietz et al. 2018 for a review of available auditory toolboxes). The future will show if these attempts are broadly in-line with each other and generate synergies toward a new standard model or if they are more in conflict with each other providing novel research questions.

In any case, physiological models that can account for binaural perception are expected to open a whole new dimension of possibilities. For example, hearing-impaired patients will benefit from individualizing the model to diagnose the individual pathology and, subsequently, using the patient-specific model to optimize the individual hearing aid preprocessing.

Equally evident is that new physiological insights reduce the number of free-model parameters. Psychophysics and physiology together bring a good number of constraints for models. In fact, they are so restrictive that, to date, no binaural model that operates on physiologically plausible mechanisms can account quantitatively for key behavioral data. The cases discussed in this chapter can be considered the tip of the iceberg. A further modeling effort is awaited by a variety of other stimulus or task dimensions.

Acknowledgments This work was supported by the European Research Council (ERC) under the European Union Horizon 2020 Research and Innovation Programme under Grant Agreement 716800 (ERC Starting Grant to Mathias Dietz) and the Cluster of Excellence Hearing4all (EXC2177/1) financed by the Deutsche Forschungsgemeinschaft (DFG; German Research Foundation; Project ID 390895286) at the University of Oldenburg, Germany.

Compliance with Ethics Requirements

Mathias Dietz declares that he has no conflict of interest.

Go Ashida declares that he has no conflict of interest.

References

Agmon-Snir H, Carr CE, Rinzel J (1998) The role of dendrites in auditory coincidence detection. Nature 393:268–272. https://doi.org/10.1038/30505

Aharonson V, Furst M (2001) A model for sound lateralization. J Acoust Soc Am 109:2840–2851. https://doi.org/10.1121/1.1371756

Ashida G, Carr CE (2011) Sound localization: Jeffress and beyond. Curr Opin Neurobiol 21:745–751. https://doi.org/10.1016/j.conb.2011.05.008

Ashida G, Kretzberg J, Tollin DJ (2016) Roles for coincidence detection in coding amplitude-modulated sounds. PLoS Comput Biol 12:e100499. https://doi.org/10.1371/journal.pcbi.1004997

Ashida G, Tollin DJ, Kretzberg J (2017) Physiological models of the lateral superior olive. PLoS Comput Biol 13:e1005903. https://doi.org/10.1371/journal.pcbi.1005903

Baumann VJ, Lehnert S, Leibold C, Koch U (2013) Tonotopic organization of the hyperpolarization-activated current (I_h) in the mammalian medial superior olive. Front Neural Circuits 7:117. https://doi.org/10.3389/fncir.2013.00117

Bernstein LR, Trahiotis C (1997) The effects of randomizing values of interaural disparities on binaural detection and on discrimination of interaural correlation. J Acoust Soc Am 102:1113–1120. https://doi.org/10.1121/1.419863

Bernstein LR, Trahiotis C (2002) Enhancing sensitivity to interaural delays at high frequencies by using "transposed stimuli". J Acoust Soc Am 112:1026–1036. https://doi.org/10.1121/1.1497620

Bernstein LR, Trahiotis C (2003) Enhancing interaural-delay-based extents of laterality at high frequencies by using "transposed stimuli". J Acoust Soc Am 113: 3335–3347. https://doi.org/10.1121/1.1570431

Bernstein LR, Trahiotis C (2009) How sensitivity to ongoing interaural temporal disparities is affected by manipulations of temporal features of the envelopes of high-frequency stimuli. J Acoust Soc Am 125:3234–3242. https://doi.org/10.1121/1.3101454

Bernstein LR, Trahiotis C (2012) Lateralization produced by interaural temporal and intensitive disparities of high-frequency, raised-sine stimuli: data and modeling. J Acoust Soc Am 131:409–415. https://doi.org/10.1121/1.3662056

Bernstein LR, Trahiotis C (2017) An interaural-correlation-based approach that accounts for a wide variety of binaural detection data. J Acoust Soc Am 141:1150–1160. https://doi.org/10.1121/1.4976098

Bernstein LR, Trahiotis C (2018) Effects of interaural delay, center frequency, and no more than "slight" hearing loss on precision of binaural processing: empirical data and quantitative modeling. J Acoust Soc Am 144:292–307. https://doi.org/10.1121/1.5046515

Borisyuk A, Semple MN, Rinzel J (2002) Adaptation and inhibition underlie responses to time-varying interaural phase cues in a model of inferior colliculus neurons. J Neurophysiol 88:2134–2146. https://doi.org/10.1152/jn.00073.2002

Bouse J, Vencovsky V, Rund F, Marsalek P (2019) Functional rate-code models of the auditory brainstem for predicting lateralization and discrimination data of human binaural perception. J Acoust Soc Am 145:1–15. https://doi.org/10.1121/1.5084264

Breebaart J, van de Par S, Kohlrausch A (2001a) Binaural processing model based on contralateral inhibition. I. Model structure. J Acoust Soc Am 110:1074–1088. https://doi.org/10.1121/1.1383297

Breebaart J, van de Par S, Kohlrausch A (2001b) Binaural processing model based on contralateral inhibition. II. Dependence on spectral parameters. J Acoust Soc Am 110:1089–1104. https://doi.org/10.1121/1.1383298

Brown AD, Tollin DJ (2016) Slow temporal integration enables robust neural coding and perception of a cue to sound source location. J Neurosci 36:9908–9921. https://doi.org/10.1523/JNEUROSCI.1421-16.2016

Bruce IC, Erfani Y, Zilany MSA (2018) A phenomenological model of the synapse between the inner hair cell and auditory nerve: implications of limited neurotransmitter release sites. Hear Res 360:40–54. https://doi.org/10.1016/j.heares.2017.12.016

Brughera AR, Stutman ER, Carney LH, Colburn HS (1996) A model with excitation and inhibition for cells in the medial superior olive. Aud Neurosci 2:219–233

Brughera A, Dunai L, Hartmann WM (2013) Human interaural time difference thresholds for sine tones: the high-frequency limit. J Acoust Soc Am 133:2839–2855. https://doi.org/10.1121/1.4795778

Burkitt AN (2006) A review of the integrate-and-fire neuron model: I. Homogeneous synaptic input. Biol Cybern 95:1–19. https://doi.org/10.1007/s00422-006-0082-8

Cant NB, Benson CG (2006) Organization of the inferior colliculus of the gerbil (*Meriones unguiculatus*): differences in distribution of projections from the cochlear nuclei and the superior olivary complex. J Comp Neurol 495:511–528. https://doi.org/10.1002/cne.20888

Chung Y, Delgutte B, Colburn HS (2015) Modeling binaural responses in the auditory brainstem to electric stimulation of the auditory nerve. J Assoc Res Otolaryngol 16:135–158. https://doi.org/10.1007/s10162-014-0492-6

Colburn HS (1973) Theory of binaural interaction based on auditory-nerve data. I. General strategy and preliminary results on interaural discrimination. J Acoust Soc Am 54:1458–1470. https://doi.org/10.1121/1.1914445

Colburn HS, Han Y, Culotta CP (1990) Coincidence model of MSO responses. Hear Res 49:335–346. https://doi.org/10.1016/0378-5955(90)90112-3

Colburn HS (1996) Computational models of binaural processing. In: Hawkins HL, McMullen TA, Popper AN, Fay RR (eds) Auditory computation. Springer, New York, pp p332–p400. https://doi.org/10.1007/978-1-4612-4070-9_8

Colburn HS, Durlach NI (1978) Models of binaural interaction. In: Carterette EC, Friedman MP (eds) Handbook of perception. Vol. 4, Hearing. Academic Press, New York, pp 467–518

Davis KA (2005) Spectral processing in the inferior colliculus. Int Rev Neurobiol 70:169–205. https://doi.org/10.1016/S0074-7742(05)70006-4

Day ML, Delgutte B (2013) Decoding sound source location and separation using neural population activity patterns. J Neurosci 33:15837–15847. https://doi.org/10.1523/JNEUROSCI.2034-13.2013

Dietz M (2016) Models of the electrically stimulated binaural system: a review. Network 27:186–211. https://doi.org/10.1080/0954898X.2016.1219411

Dietz M, Ewert SD, Hohmann V (2009) Lateralization of stimuli with independent fine-structure and envelope-based temporal disparities. J Acoust Soc Am 125:1622–1635. https://doi.org/10.1121/1.3076045

Dietz M, Wang L, Greenberg D, McAlpine D (2016) Sensitivity to interaural time differences conveyed in the stimulus envelope: estimating inputs of binaural neurons through the temporal analysis of spike trains. J Assoc Res Otolaryngol 17:313–330. https://doi.org/10.1007/s10162-016-0573-9

Dietz M, Lestang J-H, Majdak P, Stern RM, Marquardt T, Ewert SD, Hartmann WM, Goodman DFM (2018) A framework for testing and comparing binaural models. Hear Res 360:92–106. https://doi.org/10.1016/j.heares.2017.11.010

Domnitz R, Colburn H (1976) Analysis of binaural detection models for dependence on interaural target parameters. J Acoust Soc Am 59:598–601. https://doi.org/10.1121/1.380904

Encke J, Hemmert W (2018) Extraction of inter-aural time differences using a spiking neuron network model of the medial superior olive. Front Neurosci 12:140. https://doi.org/10.3389/fnins.2018.00140

Fischer BJ, Peña JL (2011) Owl's behavior and neural representation predicted by Bayesian inference. Nat Neurosci 14:1061–1066. https://doi.org/10.1038/nn.2872

Fontaine B, Brette R (2011) Neural development of binaural tuning through Hebbian learning predicts frequency-dependent best delays. J Neurosci 31:11692–11696. https://doi.org/10.1523/JNEUROSCI.0237-11.2011

Fontaine B, Peremans H (2007) Tuning bat LSO neurons to interaural intensity differences through spike-timing dependent plasticity. Biol Cybern 97:261–267. https://doi.org/10.1007/s00422-007-0178-9

Franken TP, Bremen P, Joris PX (2014) Coincidence detection in the medial superior olive: mechanistic implications of an analysis of input spiking patterns. Front Neural Circuits 8:42. https://doi.org/10.3389/fncir.2014.00042

Franken TP, Smith PH, Joris PX (2016) *In vivo* whole-cell recordings combined with electron microscopy reveal unexpected morphological and physiological properties in the lateral

nucleus of the trapezoid body in the auditory brainstem. Front Neural Circuits 10:69. https://doi.org/10.3389/fncir.2016.00069
Franken TP, Joris PX, Smith PH (2018) Principal cells of the brainstem's interaural sound level detector are temporal differentiators rather than integrators. eLife 7:e33854. https://doi.org/10.7554/eLife.33854
Furukawa S (2008) Detection of combined changes in interaural time and intensity differences: segregated mechanisms in cue type and in operating frequency range? J Acoust Soc Am 123:1602–1617. https://doi.org/10.1121/1.2835226
Gai Y, Kotak VC, Sanes DH, Rinzel J (2014) On the localization of complex sounds: temporal encoding based on input-slope coincidence detection of envelopes. J Neurophysiol 112:802–813. https://doi.org/10.1152/jn.00044.2013
Gai Y, Ruhland JL, Yin TC (2015) Behavior and modeling of two-dimensional precedence effect in head-unrestrained cats. J Neurophysiol 114:1272–1285. https://doi.org/10.1152/jn.00214.2015
Geisler WS (2003) Ideal observer analysis. In: Chalupa LM, Werner JS (eds) The visual neurosciences. MIT Press, Cambridge, pp 825–837
Glackin B, Wall JA, McGinnity TM, Maguire LP, McDaid LJ (2010) A spiking neural network model of the medial superior olive using spike timing dependent plasticity for sound localization. Front Comput Neurosci 4:18. https://doi.org/10.3389/fncom.2010.00018
Golding NL, Oertel D (2012) Synaptic integration in dendrites: exceptional need for speed. J Physiol 590:5563–5569. https://doi.org/10.1113/jphysiol.2012.229328
Goldwyn JH, Rinzel J (2016) Neuronal coupling by endogenous electric fields: cable theory and applications to coincidence detector neurons in the auditory brain stem. J Neurophysiol 115:2033–2051. https://doi.org/10.1152/jn.00780.2015
Goldwyn JH, Mc Laughlin M, Verschooten E, Joris PX, Rinzel J (2014) A model of the medial superior olive explains spatiotemporal features of the local field potentials. J Neurosci 34:11705–11722. https://doi.org/10.1523/JNEUROSCI.0175-14.2014
Goodman DFM, Benichoux V, Brette R (2013) Decoding neural responses to temporal cues for sound localization. eLife 2:e01312. https://doi.org/10.7554/eLife.01312
Goupell MJ, Hartmann WM (2007) Interaural fluctuations and the detection of interaural incoherence III. Narrowband experiments and binaural models. J Acoust Soc Am 122:1029–1045. https://doi.org/10.1121/1.2734489
Graña GD, Hutson KA, Badea A, Pappa A, Scott W, Fitzpatrick DC (2017) The organization of frequency and binaural cues in the gerbil inferior colliculus. J Comp Neurol 525:2050–2074. https://doi.org/10.1002/cne.24155
Greenberg D, Monaghan JJM, Dietz M, Marquardt T, McAlpine D (2017) Influence of envelope waveform on ITD sensitivity of neurons in the auditory midbrain. J Neurophysiol 118:2358–2370. https://doi.org/10.1152/jn.01048.2015
Greene NT, Davis KA (2012) Discharge patterns in the lateral superior olive of decerebrate cats. J Neurophysiol 108:1942–1953. https://doi.org/10.1152/jn.00908.2011
Griffin SJ, Bernstein LR, Ingham NJ, McAlpine D (2005) Neural sensitivity to interaural envelope delays in the inferior colliculus of the Guinea pig. J Neurophysiol 93:3463–3478. https://doi.org/10.1152/jn.00794.2004
Grothe B, Pecka M, McAlpine D (2010) Mechanisms of sound localization in mammals. Physiol Rev 90:983–1012. https://doi.org/10.1152/physrev.00026.2009
Hafter ER, Carrier SC (1972) Binaural interaction in low-frequency stimuli: the inability to trade time and intensity completely. J Acoust Soc Am 51:1852–1862. https://doi.org/10.1121/1.1913044
Hecht E (2014) Optics, 5th edn. Pearson Education (ISBN 0133977226)
Hu H, Ewert SD, McAlpine D, Dietz M (2017) Differences in the temporal course of interaural time difference sensitivity between acoustic and electric hearing in amplitude modulated stimuli. J Acoust Soc Am 141:1862–1873. https://doi.org/10.1121/1.4977014
Jeffress LA (1948) A place theory of sound localization. J Comp Physiol Psychol 41:35–39. https://doi.org/10.1037/h0061495

Jennings TR, Colburn HS (2010) Models of the superior olivary complex. In: Meddis R et al (eds) Computational models of the auditory system (Springer handbook of auditory research 35). Springer, New York, pp 65–96. https://doi.org/10.1007/978-1-4419-5934-8_4

Jolivet R, Lewis TJ, Gerstner W (2004) Generalized integrate-and-fire models of neuronal activity approximate spike trains of a detailed model to a high degree of accuracy. J Neurophysiol 92:959–976. https://doi.org/10.1152/jn.00190.2004

Joris PX (1996) Envelope coding in the lateral superior olive. II. Characteristic delays and comparison with responses in the medial superior olive. J Neurophysiol 76:2137–2156. https://doi.org/10.1152/jn.1996.76.4.2137

Joris PX, Yin TCT (1995) Envelope coding in the lateral superior olive. I. Sensitivity to interaural time differences. J Neurophysiol 73:1043–1062. https://doi.org/10.1152/jn.1995.73.3.1043

Joris PX, Yin TCT (1998) Envelope coding in the lateral superior olive. III. Comparison with afferent pathways. J Neurophysiol 79:253–269. https://doi.org/10.1152/jn.1998.79.1.253

Joris PX, Carney LH, Smith PH, Yin TCT (1994) Enhancement of neural synchronization in the anteroventral cochlear nucleus. I. Responses to tones at the characteristic frequency. J Neurophysiol 71:1022–1036. https://doi.org/10.1152/jn.1994.71.3.1022

Karcz A, Hennig MH, Robbins CA, Tempel BL, Rübsamen R, Kopp-Scheinpflug C (2011) Low-voltage activated Kv1.1 subunits are crucial for the processing of sound source location in the lateral superior olive in mice. J Physiol 589:1143–1157. https://doi.org/10.1113/jphysiol.2010.203331

Kelvasa D, Dietz M (2015) Auditory model-based sound direction estimation with bilateral cochlear implants. Trends Hear 19:1–16. https://doi.org/10.1177/2331216515616378

Khurana S, Remme MWH, Rinzel J, Golding NL (2011) Dynamic interaction of I_h and $I_{K\text{-}LVA}$ during trains of synaptic potentials in principal neurons of the medial superior olive. J Neurosci 31:8936–8947. https://doi.org/10.1523/JNEUROSCI.1079-11.2011

Klein-Hennig M, Dietz M, Hohmann V, Ewert SD (2011) The influence of different segments of the ongoing envelope on sensitivity to interaural time delays. J Acoust Soc Am 129:3856–3872. https://doi.org/10.1121/1.3585847

Klein-Hennig M, Dietz M, Hohmann V (2018) Combination of binaural and harmonic masking release effects in the detection of a single component in complex tones. Hear Res 359:23–31. https://doi.org/10.1016/j.heares.2017.12.007

Klug J, Schmors L, Ashida G, Dietz M (2020) Neural rate difference model can account for lateralization of high-frequency stimuli. J Acoust Soc Am 148: 678–691. https://doi.org/10.1121/10.0001602

Kock WE (1950) Binaural localization and masking. J Acoust Soc Am 22:801–804. https://doi.org/10.1121/1.1906692

Krumbholz K, Magezi DA, Moore RC, Patterson RD (2009) Binaural sluggishness precludes temporal pitch processing based on envelope cues in conditions of binaural unmasking. J Acoust Soc Am 125:1067–1074. https://doi.org/10.1121/1.3056557

Kulesza RJ, Grothe B (2015) Yes, there is a medial nucleus of the trapezoid body in humans. Front Neuroanat 9:35. https://doi.org/10.3389/fnana.2015.00035

Langford TL, Jeffress LA (1964) Effect of noise crosscorrelation on binaural signal detection. J Acoust Soc Am 36:1455–1458. https://doi.org/10.1121/1.1919224

Lee KY (2013) Pathophysiology of age-related hearing loss (peripheral and central). Korean J Audiol 17:45–49. https://doi.org/10.7874/kja.2013.17.2.45

Lehnert S, Ford MC, Alexandrova O, Hellmundt F, Felmy F, Grothe B, Leibold C (2014) Action potential generation in an anatomically constrained model of medial superior olive axons. J Neurosci 34:5370–5384. https://doi.org/10.1523/JNEUROSCI.4038-13.2014

Malmierca MS (2004) The inferior colliculus: a center for convergence of ascending and descending auditory information. Neuroembryol Aging 3:215–229. https://doi.org/10.1159/000096799

Manis PB, Campagnola L (2018) A biophysical modelling platform of the cochlear nucleus and other auditory circuits: from channels to networks. Hear Res 360:76–91. https://doi.org/10.1016/j.heares.2017.12.017

Marquardt T, McAlpine D (2009) Masking with interaurally "double-delayed" stimuli: the range of internal delays in the human brain. J Acoust Soc Am 126:EL177–EL182. https://doi.org/10.1121/1.3253689
Mathews PJ, Jercog PE, Rinzel J, Scott LL, Golding NL (2010) Control of submillisecond synaptic timing in binaural coincidence detectors by K_v1 channels. Nat Neurosci 13:601–609. https://doi.org/10.1038/nn.2530
McAlpine D, Jiang D, Palmer AR (2001) A neural code for low-frequency sound localization in mammals. Nat Neurosci 4:396–401. https://doi.org/10.1038/86049
Meddis R (2006) Auditory-nerve first-spike latency and auditory absolute threshold: a computer model. J Acoust Soc Am 119:406–417. https://doi.org/10.1121/1.2139628
Meng X, Huguet G, Rinzel J (2012) Type III excitability, slope sensitivity and coincidence detection. Discrete Contin Dyn Syst Ser A 32:2729–2757. https://doi.org/10.3934/dcds.2012.32.2729
Mikiel-Hunter J, Kotak V, Rinzel J (2016) High-frequency resonance in the gerbil medial superior olive. PLoS Comput Biol 12:e1005166. https://doi.org/10.1371/journal.pcbi.1005166
Moncada-Torres A, Joshi SN, Prokopiou A, Wouters J, Epp B, Francart T (2018) A framework for computational modelling of interaural time difference discrimination of normal and hearing-impaired listeners. J Acoust Soc Am 144:940–954. https://doi.org/10.1121/1.5051322
Moore BCJ, Heinz MG, Braida LD, Léger AC (2018) Effects of age on sensitivity to interaural time differences in envelope and fine structure, individually and in combination. J Acoust Soc Am 143:1287–1296. https://doi.org/10.1121/1.5025845
Oetting D, Hohmann V, Appell JE, Kollmeier B, Ewert SD (2016) Spectral and binaural loudness summation for hearing-impaired listeners. Hear Res 335:179–192. https://doi.org/10.1016/j.heares.2016.03.010
Ono M, Ito T (2015) Functional organization of the mammalian auditory midbrain. J Physiol Sci 65:499–506. https://doi.org/10.1007/s12576-015-0394-3
Rabiner LR, Laurence CL, Durlach NI (1966) Further Results on Binaural Unmasking and the EC Model. J Acoust Soc Am 40:62–70. https://doi.org/10.1121/1.1910065
Reijniers J, Vanderelst D, Jin C, Carlile S, Peremans H (2014) An ideal-observer model of human sound localization. Biol Cybern 108:169–181. https://doi.org/10.1007/s00422-014-0588-4
Remme MWH, Rinzel J (2011) Role of active dendritic conductances in subthreshold input integration. J Comput Neurosci 31:13–30. https://doi.org/10.1007/s10827-010-0295-7
Remme MWH, Donato R, Mikiel-Hunter J, Ballestero JA, Foster S, Rinzel J, McAlpine D (2014) Subthreshold resonance properties contribute to the efficient coding of auditory spatial cues. Proc Natl Acad Sci U S A 111:E2339–E2348. https://doi.org/10.1073/pnas.1316216111
Roberts MT, Seeman SC, Golding NL (2013) A mechanistic understanding of the role of feedforward inhibition in the mammalian sound localization circuitry. Neuron 78:923–935. https://doi.org/10.1016/j.neuron.2013.04.022
Rothman JS, Manis PB (2003) The roles potassium currents play in regulating the electrical activity of ventral cochlear nucleus neurons. J Neurophysiol 89:3097–3113. https://doi.org/10.1152/jn.00127.2002
Siebert WM (1965) Some implications of the stochastic behavior of primary auditory neurons. Kybernetik 2:206–215. https://doi.org/10.1007/BF00306416
Stern RM, Shear GD (1996) Lateralization and detection of low-frequency binaural stimuli: effects of distribution of internal delay. J Acoust Soc Am 100:2278–2288. https://doi.org/10.1121/1.417937
Svirskis G, Rinzel J (2003) Influence of subthreshold nonlinearities on signal-to-noise ratio and timing precision for small signals in neurons: minimal model analysis. Network 14:137–150
Svirskis G, Kotak V, Sanes D, Rinzel J (2002) Enhancement of signal-to-noise ratio and phase locking for small inputs by a low-threshold outward current in auditory neurons. J Neurosci 22:11019–11025. https://doi.org/10.1523/JNEUROSCI.22-24-11019.2002
Szalisznyó K (2006) Role of hyperpolarization-activated conductances in the lateral superior olive: a modeling study. J Comput Neurosci 20:137–152. https://doi.org/10.1007/s10827-005-5637-5

Thavam S, Dietz M (2019) Smallest perceivable interaural time differences. J Acoust Soc Am 145:458–468. https://doi.org/10.1121/1.5087566
Tollin DJ, Yin TCT (2005) Interaural phase and level difference sensitivity in low-frequency neurons in the lateral superior olive. J Neurosci 25:10648–10657. https://doi.org/10.1523/JNEUROSCI.1609-05.2005
Toth P, Marsalek P (2015) Analytical description of coincidence detection synaptic mechanisms in the auditory pathway. Biosystems 136:90–98. https://doi.org/10.1016/j.biosystems.2015.07.006
Trahiotis C, Bernstein LR (2014) Advances in the understanding of binaural information processing: consideration of the stimulus as processed. In: Perspectives on auditory research, vol 50. Springer, New York, pp 585–600. https://doi.org/10.1007/978-1-4614-9102-6
Trahiotis C, Stern RM (1989) Lateralization of bands of noise: effects of bandwidth and differences of interaural time and phase. J Acoust Soc Am 86:1285–1293. https://doi.org/10.1121/1.398743
Tsai JJ, Koka K, Tollin DJ (2010) Varying overall sound intensity to the two ears impacts interaural level difference discrimination thresholds by single neurons in the lateral superior olive. J Neurophysiol 103:875–886. https://doi.org/10.1152/jn.00911.2009
van der Heijden M, Trahiotis C (1999) Masking with interaurally delayed stimuli: the use of "internal" delays in binaural detection. J Acoust Soc Am 105:388–399. https://doi.org/10.1121/1.424628
Verschooten E, Shamma S, Oxenham AJ, Moore BCJ, Joris PX, Heinz MG, Plack CJ (2019) The upper frequency limit for the use of phase locking to code temporal fine structure in humans: a compilation of viewpoints. Hear Res 377:109–121. https://doi.org/10.1016/j.heares.2019.03.011
Vonderschen K, Wagner H (2014) Detecting interaural time differences and remodeling their representation. Trends Neurosci 37:289–300. https://doi.org/10.1016/j.tins.2014.03.002
Wall JA, McDaid LJ, Maguire LP, McGinnity TM (2012) Spiking neural network model of sound localization using the interaural intensity difference. IEEE Trans Neural Netw Learn Syst 23:574–586. https://doi.org/10.1109/TNNLS.2011.2178317
Wang L, Colburn HS (2012) A modeling study of the responses of the lateral superior olive to ipsilateral sinusoidally amplitude-modulated tones. J Assoc Res Otolaryngol 13:249–267. https://doi.org/10.1007/s10162-011-0300-5
Wang L, Devore S, Delgutte B, Colburn HS (2014) Dual sensitivity of inferior colliculus neurons to ITD in the envelopes of high-frequency sounds: experimental and modeling study. J Neurophysiol 111:164–181. https://doi.org/10.1152/jn.00450.2013
Xia J, Brughera A, Colburn HS, Shin-Cunningham B (2010) Physiological and psychophysical modeling of the precedence effect. J Assoc Res Otolaryngol 11:495–513. https://doi.org/10.1007/s10162-010-0212-9
Yin TCT, Chan JCK (1990) Interaural time sensitivity in medial superior olive in cat. J Neurophysiol 64:465–488. https://doi.org/10.1152/jn.1990.64.2.465
Zacksenhouse M, Johnson DH, Williams J, Tsuchitani C (1998) Single-neuron modeling of LSO unit responses. J Neurophysiol 79:3098–3110. https://doi.org/10.1152/jn.1998.79.6.3098
Zhou Y, Colburn HS (2010) A modeling study of the effects of membrane afterhyperpolarization on spike interval statistics and on ILD encoding in the lateral superior olive. J Neurophysiol 103:2355–2371. https://doi.org/10.1152/jn.00385.2009
Zilany MSA, Bruce IC (2007) Representation of the vowel/ε/in normal and impaired auditory nerve fibers: model predictions of responses in cats. J Acoust Soc Am 122:402–417. https://doi.org/10.1121/1.2735117

Chapter 11
Clinical Ramifications of the Effects of Hearing Impairment and Aging on Spatial and Binaural Hearing

Frederick J. Gallun, Nirmal K. Srinivasan, and Anna C. Diedesch

11.1 Introduction

The binaural system relies on the exquisite timing of the auditory nerve and the precise architecture of the cochlear nucleus, trapezoidal body, and superior olive to create a neural code that represents extremely small differences in the timing and intensity of pressure waves arriving at each ear (see Takahashi, Kettler, Keller, and Bala, Chap. 4; Owrutsky, Benichoux, and Tollin, Chap. 5). Once this neural signal has been created, it then must be accurately transmitted to and recoded by the remainder of the auditory brainstem and the auditory cortical areas (see Takahashi, Kettler, Keller, and Bala, Chap. 4; Owrutsky, Benichoux, and Tollin, Chap. 5; Town and Bizley, Chap. 12). The effects of distortions on the creation and transmission of the spatial and binaural code have been studied for over 100 years. Durlach et al. (1981) reviewed the early literature and concluded that despite a large number of studies, it was difficult to draw firm conclusions on the effects of hearing

F. J. Gallun (✉)
VA RR&D National Center for Rehabilitative Auditory Research, VA Portland Health Care System, Portland, OR, USA

Department of Otolaryngology, Head and Neck Surgery, Oregon Health and Science University, Portland, OR, USA
e-mail: gallunf@ohsu.edu

N. K. Srinivasan
Department of Speech-Language Pathology & Audiology, Towson University, Towson, MD, USA
e-mail: nsrinivasan@towson.edu

A. C. Diedesch
Department of Communication Sciences and Disorders, Western Washington University, Bellingham, WA, USA
e-mail: Anna.Diedesch@wwu.edu

© Springer Nature Switzerland AG 2021
R. Y. Litovsky et al. (eds.), *Binaural Hearing*, Springer Handbook of Auditory Research 73, https://doi.org/10.1007/978-3-030-57100-9_11

impairment and the distortions they cause in binaural processing. Here, an updated review of the literature in this area is provided, highlighting both the progress that has been made and the areas in which important questions remain.

11.2 Impairments of the Auditory System

There are three well-established classes of hearing loss (Moore 2007), each associated with a different aspect of the hearing mechanisms: conductive, sensorineural, and retrocochlear. Each of these can be unilateral or bilateral and caused by a wide range of factors, including viral infection, side effects of other diseases and health events such as diabetes and strokes, high fevers, exposure to chemicals and/or drugs with ototoxic properties, metabolic imbalance, allergic reaction, genetic disease, noise exposure, and/or presbycusis.

Conductive hearing loss, which represents the most peripheral mechanism of impairment, involves reductions in the efficiency of the transduction of sound by the outer or middle ear. The most common causes are cerumen buildup in the outer ear, damage to the tympanic membrane ("eardrum"), presence of fluid in the middle ear, and damage to or stiffening of the middle ear bones ("ossicles").

Sensorineural hearing loss (SNHL) involves the cochlear structures (basilar membrane, cochlear fluids, organ of Corti, inner and outer hair cells, and stria vascularis) and the auditory nerve. Damage to the central auditory system, including the auditory nerve but excluding the cochlea, is known as retrocochlear hearing loss. The most common example is the growth of a benign tumor on the vestibular portion of the eighth nerve that fills the cavity through which the nerve passes and damages the nerve by pressing on it. This tumor is known as an acoustic neuroma or, more properly, a vestibular schwannoma because it is caused by the Schwann cells that are responsible for myelinating the vestibular portion of the eighth nerve. It is also the case that damage to the cochlear nucleus, other central structures of the auditory pathway, or the auditory portions of the cortex can result in impaired hearing.

The characterization of hearing impairment, including definitions of severity and an explanation of the audiogram, which is the most common clinical measure of monaural pure-tone detection ability, was reviewed by Katz (2014). Curhan and Curhan (2016) reviewed the epidemiology of hearing impairment. Although the term presbycusis is often used to refer to age-related hearing loss, this is description of a disease rather than of a mechanism because the physiological determinants of the loss are to be found in specific parts of the hearing system and are likely very similar to other types of hearing loss. See Gordon-Salant et al. (2010) for a review of the mechanisms and results of presbycusis.

11.2.1 Studying the Impacts of Impairments of the Auditory System on Binaural and Spatial Hearing

Historically, experiments on spatial hearing have only occasionally been conducted on hearing-impaired (HI) individuals, and most have focused on SNHL and presbycusis among people with symmetrical losses at the two ears. There have also been a number of studies that have examined the role of unilateral hearing loss (normal hearing in one ear and impaired hearing in the other) or asymmetrical losses (different degrees and/or etiologies of loss at the two ears) on the ways in which listeners make use of spatial sounds in the environment. Recent work has also examined how monaural conductive loss during development affects spatial perception.

It is important to consider the type of the hearing loss when designing and interpreting these studies. Most importantly, the frequency ranges in which thresholds are abnormal are hypothesized to have different implications for how hearing loss will affect an individual's ability to utilize spatial cues. For instance, a low-frequency hearing loss is hypothesized to affect spatial hearing mostly due to its impact on binaural sensitivity. Specifically, sensitivity to interaural time differences (ITDs) is expected to be reduced due to the fact that ITD cues are the largest and most detectable for low-frequency sounds (see Hartmann, Chap. 2). On the other hand, the much more common high-frequency losses (due to their association with age-related hearing loss and ototoxicity) might be presumed to have a greater effect on the listener's ability to utilize interaural level differences (ILDs) for horizontal plane localization and monaural spectral cues that are useful for vertical plane localization and the disambiguation of front/back confusions. This chapter focuses primarily on binaural cues, but it is important to recall that spatial awareness also includes monaural cues, which can be affected by both peripheral hearing loss and, potentially, damage to the central brain areas responsible for vertical plane localization.

There is great interest in understanding the ways in which various clinical conditions impact the spatial awareness, both for the scientific knowledge to be gained and to better help the patients with these conditions. As mentioned at the beginning of this section, the most commonly studied clinical patients are older people, mostly with impaired hearing, but a number of studies have also studied people with diseases of the central nervous system, such as multiple sclerosis, cortical lesions due to strokes or penetrating head wounds, and tumors (e.g., Sanchez-Longo and Forster 1958; Häusler and Levine 1980; Jabbari et al. 1987). Studying patients, either individually or in groups, presents a number of difficulties that are not often encountered when working on those with normal hearing (NH).

For a review of the early literature, Durlach et al. (1981) is the most comprehensive place to start. Their review sets out a number of the problems associated with studying the binaural hearing of patients that are still being confronted today. In their review of studies involving people with all three of the types of hearing loss, evidence is presented showing that all of these groups varied in their pure-tone threshold detection ability across frequencies as well as their ability to understand speech, as would be expected from the losses studied. It is perhaps no surprise, then, that there was great diversity across listeners in terms of binaural function as well. What is striking about these early studies (e.g., Bergman 1957; Jongkees and Veer 1958) is the repeated finding that it is very difficult to predict binaural performance on a particular task from only the audiometric thresholds of the listener. This is a finding that has continued to be observed even as methods have been further refined in terms of both testing patient populations and characterizing binaural function (e.g., Gabriel et al. 1992; Spencer et al. 2016).

Durlach et al. (1981) discussed several reasons why this might be the case. The first issue is the diversity of conditions described as "hearing loss." As noted in Sect. 11.2, there are several areas along the auditory pathway where hearing loss (peripheral or conductive, sensorineural or cochlear, and central or retrocochlear) and different audiometric configurations could occur. In addition to the inability to aggregate data across different conditions (such and bilateral and unilateral losses), it is also important to note that the exact configuration of the hearing loss could easily influence the effects on the binaural system, making it difficult to combine data across patients.

One difficulty that is present in a large number of clinical studies and that is not always considered in sufficient detail, especially in the older work, is the comorbidities between multiple conditions. Comorbidity refers to the presence of multiple disease states in the same patients, either because one causes the other or because they simply co-occur in the population or sample of interest. The most obvious example is the high correlation between age and hearing loss. Although it is possible to recruit samples in which the correlation is very low (e.g., Füllgrabe 2013), there remains the difficulty that in the actual individuals with the conditions, it is likely that two or more conditions will co-occur. Furthermore, it may be the case that those with only one of the conditions are not representative of those with multiple conditions. In the case of hearing impairment and aging, for example, young people with severe hearing loss (often referred to as "younger HI") are likely to have a different mechanism of loss than older people with a similar audiometric configuration ("older HI"). A related concern is that there may be some nonauditory factor that is leading to good performance on multiple measures. For example, older people with very good audiometric thresholds ("older NH") may have made lifestyle choices or have genetic profiles that are quite different from those whose hearing has deteriorated more rapidly with age. These genetic and lifestyle choices can create complexity in the attempts to understand the mechanisms behind why some older individuals with normal pure-tone hearing, similar to a younger listener with normal hearing ("younger NH"), may still report some difficulties of hearing, especially in noisy settings.

One of the ways to address these correlated factors in human patient populations is through animal models. One very influential model (Kujawa and Liberman 2009) suggests that when a listener experiences temporary threshold elevation due to noise exposure, the auditory nerve could be damaged even if audiometric thresholds return to normal. These findings make it possible that even listeners with normal audiometric thresholds could have degraded input to the binaural system. The finding of variability on binaural tasks across NH listeners or those with only slight losses (Ruggles et al. 2011; Bernstein and Trahiotis 2016) are consistent with this hypothesis.

Finally, the methods used by researchers can also be an important source of variability across studies. Durlach et al. (1981) noted that some researchers studied the localization of 1-kHz tones from a set of closely spaced loudspeakers, others studied the localization of speech from four loudspeakers separated by 90 °, and others focused on the sensitivity to binaural cues presented over headphones. Durlach et al. described three main difficulties associated with interpreting the clinical research literature available at the time: (1) asymmetries in hearing sensitivity were not addressed by changing the level at each ear to equate audibility; (2) sensitivity was often confounded with response bias (especially in localization studies); and (3) in most cases, only a single task was used. With these issues unresolved, they found it very difficult to draw any but the broadest conclusions.

Based on the Durlach et al. (1981) review, Häusler et al. (1983) attempted to address some of the methodological deficits by carefully controlling the stimuli and presenting them over headphones in addition to free-field testing. Figure 11.1 shows an example of the impact of sensation level (SL) on thresholds for a range of binaural tasks. These data are all from one young NH listener and show clearly that if

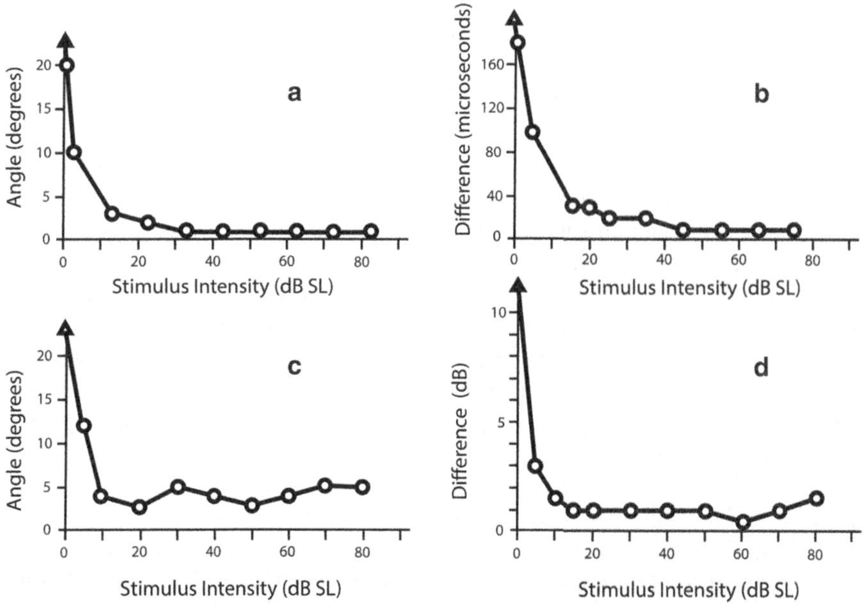

Fig. 11.1 Discrimination of spatial and binaural differences for a single young normal-hearing (NH) listener as a function of stimulus level. (**a**) Smallest discriminable difference in stimulus angle between a 0° reference sound and a horizontally displaced comparison stimulus (the minimum audible angle [MAA]). (**b**) Smallest interaural time difference (ITD) that could be detected. (**c**) vertical MAA. (**d**) Smallest interaural level difference (ILD) that could be detected. SL sensation level. (Replotted from Häusler et al. 1983, with permission)

binaural function is tested at levels near threshold, performance is worse than those at well-above threshold. Therefore, it is potentially problematic to compare thresholds from a near-detection threshold for a HI listener with thresholds from a well-above detection threshold for a NH listener.

This chapter discusses many of the ways in which researchers have made improvements in the design of experiments and the interpretation of data when it comes to examining the binaural system of patients with impairment of the peripheral and/or central auditory systems. In some areas, substantial progress has been made since Durlach et al. (1981) wrote their review, but in other areas, there is still significant work to be done. The general finding is that binaural function is more likely to be degraded when other impairments are present than when the rest of the system is functioning normally. Unfortunately, however, the repeated attempts by researchers to predict binaural and spatial hearing from other measures, especially those favored by audiologists, have been largely unsuccessful. After reviewing this literature, the implications of this disconnect between the audiological and binaural measures for the clinician and for the patient are discussed. Finally, Sect. 11.12 contains a few comments on the implications of the current state of the art for researchers interested in understanding the binaural system in normal and impaired ears. Additional information not covered here is available in the recent reviews by Freigang et al. (2015) and Akeroyd and Whitmer (2016).

11.3 Evidence from Self-Report

In addition to the measurement of the ability to detect and use binaural and spatial cues, it is also helpful to get direct reports from the patients about their experiences. One of the most common surveys developed for obtaining self-report measures of spatial hearing is the Speech, Spatial and Qualities of Hearing Scale (SSQ) developed by Gatehouse and Noble (2004). Figure 11.2 shows the data for younger and older NH listeners, for listeners with moderate hearing impairment (with and without their hearing aids), and for listeners with severe hearing impairment. Patients in the group with severe hearing impairment were unable to perform any of the tasks without hearing aids; therefore, they only answered for the aided condition. The data are mean values on a scale from 0 to 10 for 5 of the spatial items of the SSQ, where 0 indicates that the task can be performed "Not at all" and 10 indicates "Perfectly." The data show that for four of the five tasks, those with severe loss express more difficulty than all of the other listeners. In terms of types of localization abilities, it is worth noting that those in the severe-loss group and those in the moderate-loss group report equal difficulty determining the horizontal plane localization of a person speaking. On the other hand, NH listeners all report better ability to do this task than do HI listeners. The data also show that hearing aids improve the experiences of the listeners with moderate loss in most cases but that even when using hearing aids, those with severe loss only report doing as well as the moderate group without their aids. For more details on hearing aids and binaural function, see

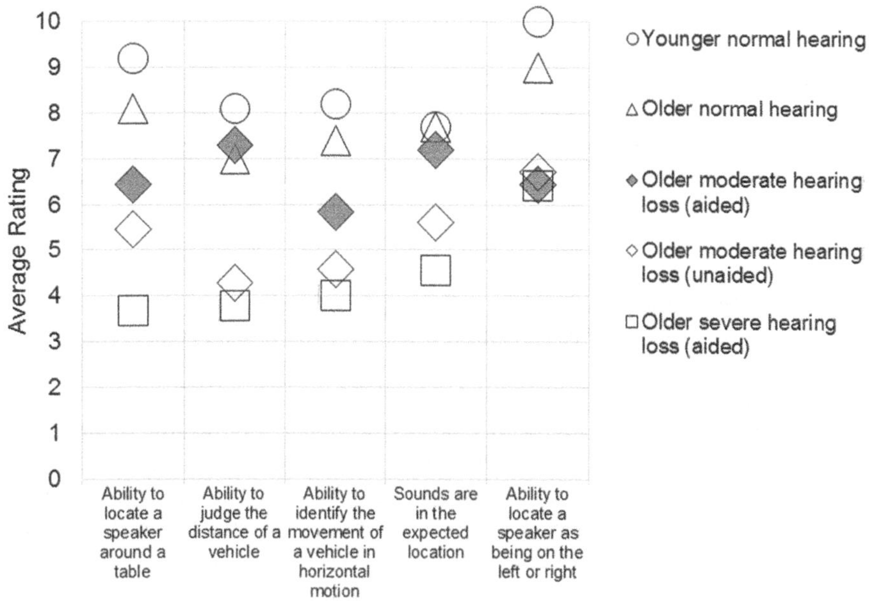

Fig. 11.2 Ratings of perceived spatial abilities in five groups of people varying in age and hearing abilities. Each rating corresponds to the degree to which listeners would endorse the statements listed along the *x*-axis, where 0 indicates that the task can be performed "Not at all" and 10 indicates "Perfectly." (Younger NH and older NH listener data are plotted based on the data of Demeester et al. 2012. The data of older moderate hearing-impaired (HI) listeners reporting their aided and unaided experiences are based on the data described by Noble and Gatehouse (2006). The data of older severe HI listeners reporting their aided experiences are based on the data from Souza et al. 2018)

Chap. 13 by Ricketts and Kan. Finally, it is interesting to note that the older NH listeners report worse localization abilities than do the younger NH listeners for all of the questions except for whether or not sounds are perceived in the expected locations.

11.4 Evidence from Physiology

Data from animal models to study the impacts of hearing loss on binaural and spatial sensitivity are relatively sparse. By contrast, there is a growing literature on the effects of hearing impairment and aging on a variety of noninvasive physiological measures of binaural hearing using human listeners. Although animal models have been developed that focus on hearing loss, few have been used to examine binaural sensitivity and those that have are primarily concerned with the effects of aging and monaural deprivation (e.g., Popescu and Polley 2010; Engle and Recanzone 2013) on the ability of bilateral cochlear implants to provide binaural cues. For more information on cochlear implants and binaural hearing, see Chap. 13 by Ricketts and Kan.

The limited literature that examines binaural coding in bilaterally HI animals has only used animals with age-related hearing loss. Thus, as in humans, it is hard to determine the degree to which the results reflect aging and the degree to which they reflect hearing loss. McFadden and Willott (1994) explored the neural discharges of inferior colliculus units in mice varying in age and hearing loss. They presented tone bursts from varying locations and found no evidence of changes in azimuthal tuning functions with age-related hearing loss. However, there was a substantial increase in the degree to which the units responded to ipsilateral presentation of sound, suggesting a reduction in inhibition. A loss of inhibition in older animals and its impact on binaural function were also reported by Engle and Recanzone (2013) in their study of aged macaques (*Macaca*), who also experienced some hearing loss. Juarez-Salinas et al. (2010) found degraded spatial tuning of neuronal units in the auditory cortex of these macaques, but Engle and Recanzone (2013) were able to further show that the timing of the neural spikes is altered by aging and that the changes in the binaural responses were due to a loss of inhibition. This loss of inhibition in older mammalian auditory systems is thought to reflect a homeostatic response to a loss of input (reviewed by Caspary et al. 2008) and thus is likely to occur in response to both aging and cochlear loss.

The animal literature on monaural deprivation has shown that both the ITDs and ILDs are altered by filling the middle ear with fluid (Thornton et al. 2012) and that plugging an ear causes both behavioral (Moore et al. 1999) and physiological (Brugge et al. 1985) alterations to binaural sensitivity. Moore et al. (1999) showed that even when the plug is removed, the binaural sensitivity of the animal improves but not to normal levels. In animal models of permanent (Tillein et al. 2016) and temporary (Popescu and Polley 2010) unilateral deafness, it has been shown that binaural stimulation during development is essential for the appropriate organization of the binaural system.

The human physiological literature is generally consistent with the suggestions of degraded spatial tuning found in the animal literature reviewed in the previous paragraph, although the results appear to be more consistent for cortical than for brainstem measures. One of the common measures used in these experiments is the binaural masking level difference (BMLD; see Culling and Lavandier, Chap. 8), which is the difference in detection threshold for a tone in noise presented in phase at the two ears and out of phase at the two ears. Jerger et al. (1984) conducted a study on nearly 1000 listeners without and with hearing impairment across the age range and found only a small, but not a significant, effect of age in the NH listeners. For hearing impairment where the losses were symmetrical (0–9 dB difference between the ears) and confined to frequencies above 2000 Hz, the BMLD was within 1 dB of the NH control subjects. However, when the loss was at 1000 Hz or below, the BMLD was reduced by more than 2 dB. As the amount of symmetrical loss increased, the average BMLD decreased. Koehnke et al. (1995) also found a difference in binaural performance for NH and HI listeners. To explore the BMLD using physiological measures, Eddins and Eddins (2018) measured the cortical auditory evoked potential (CAEP) in response to tones in phase and out of phase at the two ears in the presence of in phase narrowband noise. To distinguish aging effects from hearing loss, all of their listeners were NH but varied in age. They found that older listeners have both reduced behavioral thresholds and reduced CAEPs. The differences were only present for noises and tones at 500 Hz as opposed to 4000 Hz, leading them to suggest that the difference was in sensitivity to temporal fine structure (TFS) rather than envelope timing. This conclusion is at odds with that of Anderson et al. (2018) who used the frequency-following response to test stimulus encoding at the brainstem in older versus younger listeners, both with normal hearing. Neural encoding of both in phase and out of phase signals was reduced by the same amount for both younger and older groups, resulting in no group difference in encoding of out-of-phase signals. In the same listeners, however, the behavioral difference (measured as the BMLD) was reduced in older listeners. This difference between the older and younger listeners behaviorally but not electrophysiologically resulted in no correlation between the behavioral and electrophysiological thresholds.

Work by several other groups all support the finding of Eddins and Eddins (2018) that cortical responses are indeed related to behavior even though measures of brainstem encoding may not be as strongly related to behavior. Ross et al. (2007), Papesh et al. (2017), and Vercammen et al. (2018) each used slightly different physiological measures of cortical responses to changes in binaural configuration of a tone and each related the cortical response to a behavioral measure of binaural sensitivity other than the BMLD. Despite the different methods, all three found age- and hearing loss-related declines in binaural sensitivity using far-field measures of cortical activity that correlated with behavioral performance. Currently, it is unclear what is responsible for the lack of a relationship for the brainstem measures. One possibility is that the brainstem contains a wide range of responses to binaural stimuli, only some of which are degraded by aging and hearing loss and that an average response such as the frequency-following response is not capable of capturing this heterogeneity of responding. An even more significant problem is that the frequency-following response reflects the responses of units both with binaural sensitivity and without binaural sensitivity. It may be the case that the signal is so dominated by the responses of those units without binaural sensitivity that the responses of the binaural units are very difficult to detect.

Examples of the cortical response relationships with behavioral measures are shown in Figs. 11.3 and 11.4. Figure 11.3 shows that the BMLD measured behaviorally was a good binaural predictor of the BMLD based on the difference between the responses to the same signals using CAEPs. The best-fit lines for the older listeners are lower than the best-fit lines for the younger listeners, which represents the effect of aging on the BMLD. Figure 11.4 shows that a statistical linear regression model based on CAEP amplitude and latency was able to predict spatial release from speech on speech masking using the methods of Gallun et al. (2013). For details on the behavioral measure of spatial release used by Papesh et al. (2017), see Sect. 11.6. The data shown in Fig. 11.4 are for younger to middle-aged NH listeners, but similar results were obtained with a different group of older listeners with

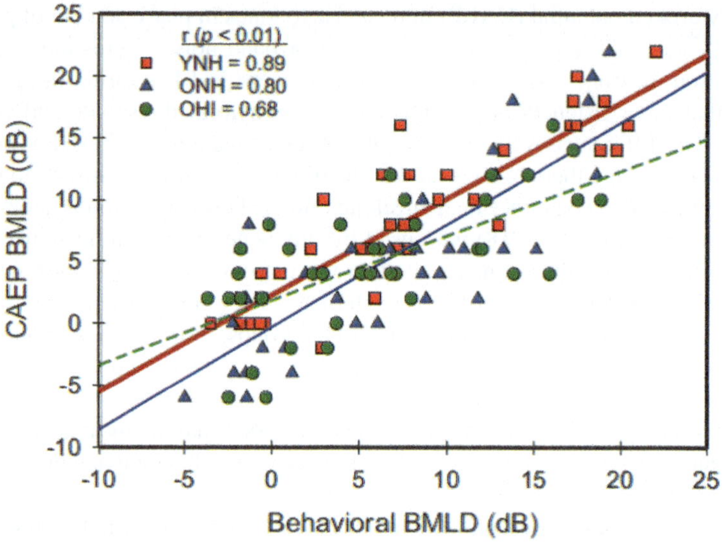

Fig. 11.3 The relationship between electrophysiological (cortical auditory evoked potential [CAEP]) and behavioral measures of the binaural masking level difference (BMLD) for younger NH (YNH), older NH (ONH), and older HI (OHI) listeners. (Reprinted from Eddins and Eddins 2018, with permission)

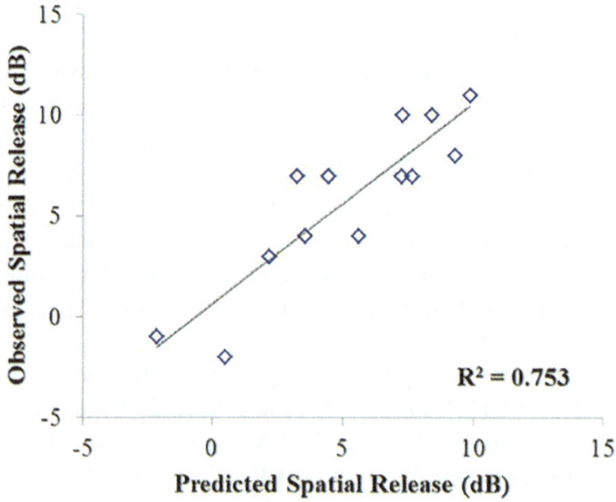

Fig. 11.4 The predicted and observed spatial release from speech on speech masking for a group of younger to middle-aged NH listeners. Model predictions are based on the latency and amplitudes of electrophysiological responses to a change in binaural phase. (Reprinted from Papesh et al. 2017, with permission)

hearing impairment. In both groups, the statistical model based on the electrophysiological measures was not improved by adding age or hearing loss as predictors.

11.5 Evidence from Behavioral Measures

11.5.1 Localization

Most of the data on the effects of age and hearing loss on binaural function comes from behavioral measures, largely using measures similar to those described by Durlach et al. (1981). A recent review of 29 studies evaluating horizontal plane localization in HI listeners was conducted by Akeroyd and Whitmer (2016). Consistent with the data reviewed by Durlach et al. (1981), Akeroyd and Whitmer (2016) concluded that although there was a small-to-moderate correlation between hearing impairment and localization ability, there was significant variability among listeners. Specifically, although HI listeners have a poorer ability to make left/right discriminations for free-field presentation than NH listeners, the worsening of threshold with hearing impairment, when averaged across studies, is only 5°. Similarly, when asked to discriminate whether two sounds were presented from the same loudspeaker or from two different loudspeakers, the minimum separation of distinguishable loudspeakers (called the "minimum audible angle" [MAA]) can be as low as 1° in hearing impairment, which is as good a threshold that can be achieved by young NH listeners.

When the distributions of thresholds are considered, however, it is clear that there are more examples of people with very poor thresholds in the HI groups, where left/right discrimination thresholds can be as high as 16° compared with the NH listeners where thresholds are all 4° or better. They also found that HI listeners are more prone to front/back errors in localization, which is consistent with a reduction in high-frequency sensitivity. Indeed, it is well known that high-frequency cues are important for vertical localization, of which front/back discrimination is an example. Akeroyd and Whitmer (2016), like Durlach et al. (1981), also pointed out a number of areas in which methodological differences make comparing studies difficult. They discuss the importance that listeners remain stationary until after the stimulus ends to ensure that ITD and ILD cues do not change with head movement as well as emphasizing the benefits of a pointing response rather than a labeling response to increase the resolution of the measure.

There are several key studies that illustrate the conclusions drawn by Akeroyd and Whitmer (2016). Abel et al. (2000) conducted an experiment across the age span to evaluate localization performance in a semireverberant chamber meant to reflect the listening situation of a small office. Reduced sound localization ability was seen as early as the third decade of life, with a decrease in performance of 12–15% over the age ranges tested. Van den Bogaert et al. (2006) measured sound localization over a range of signals, including high- and low-frequency noise bursts, a ringing telephone, and a ringing telephone in noise, and found differences in root-mean-square errors between groups of no more than 10°. Dobreva et al. (2011) investigated the effects of age and found that younger NH listeners showed horizontal overestimation and vertical underestimation of the target location, whereas older HI listeners with no more than moderate hearing loss showed pronounced horizontal plane localization deficiencies for narrowband 1250- to 1575-Hz targets. It was not possible to distinguish the aging effects from the loss of peripheral sensitivity due to the high correlation between age and hearing loss in the listeners tested.

As mentioned in Sect. 11.2, although this chapter is largely focused on binaural function, there are also monaural cues to spatial awareness that can be impaired by hearing loss and other clinical conditions. To address this, some studies have examined vertical plane localization either alone or in a two-dimensional spherical array. For example, Häusler et al. (1983) found that some older HI listeners could distinguish speakers displaced vertically using the MAA task while others could not and that a speech discrimination task predicted vertical localization performance. This is consistent with the idea that vertical localization is based primarily on monaural spectral cues, degradation of which should also influence speech performance. Brungart et al. (2017) found that NH listeners performed better than HI listeners on a localization task involving vertical and horizontal localization. The authors concluded that the poorer performance of HI listeners could be due to reduced sensitivity in both the spectral and temporal information needed for horizontal and vertical plane localization.

11.5.2 Interaural Time Differences

When testing ITD thresholds over headphones, Häusler et al. (1983) found that those with bilateral SNHL all had ITD thresholds that were slightly elevated, although over half had thresholds of 30 μs or better when tested with an 85 dB sound pressure level (SPL) noise burst. Smoski and Trahiotis (1986) measured ITD thresholds using either 80 dB SPL signals at each ear ("Equal SPL" condition) or 25 dB SL signals at each ear (25 dB above detection threshold for a monaural signal in quiet; "Equal SL" condition; see Fig. 11.5 for a subset of the data). In the Equal SPL condition with a narrowband noise as the carrier, ITD thresholds were higher for HI listeners. The difference was much greater when the stimuli were centered at 4000 Hz as opposed to 500 Hz, with one listener being unable to detect the ITD at all. When the signals were presented at 25 dB SL, thresholds for both groups were poorer for the 500-Hz signals. The detrimental effect was greater for the NH listeners. At 4000 Hz, the thresholds were much worse for NH listeners, whereas the thresholds of HI listeners were similar to the Equal SPL condition. This is to some extent explainable by the fact that at 4000 Hz, 80 dB SPL was fairly close to 25 dB SL for the HI listeners. These results are also consistent with the results of others who have performed similar studies (e.g., Hawkins and Wightman 1980; Koehnke et al. 1995). The differential effect of SL and carrier frequency are especially

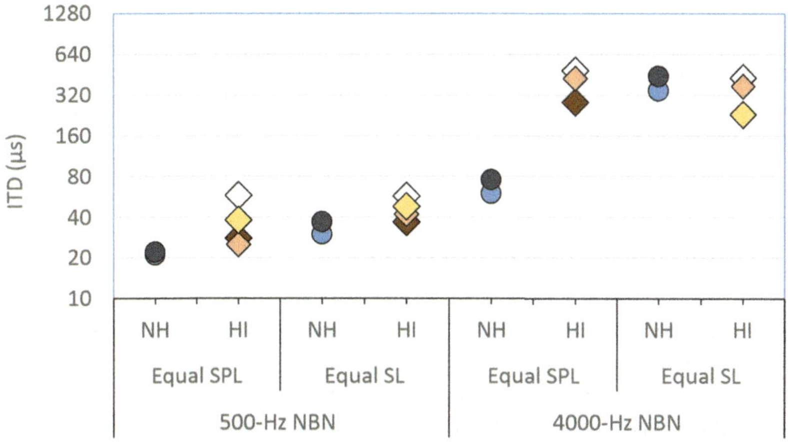

Fig. 11.5 ITD thresholds for two NH listeners and four HI listeners as a function of the center frequencies of the narrowband noise (NBN) carriers (500 Hz vs. 4000 Hz) and the manner in which the carriers were equalized at the two ears (sound pressure level [SPL]) or SL. *Circles* thresholds of NH listeners, *diamonds* thresholds of HI listeners. Each listener is represented by the same colored symbol across conditions. Missing symbols corresponded to unmeasurable thresholds. (Plots based on the data reported in Smoski and Trahiotis 1986)

important to consider when interpreting data collected with only high- or low-frequency stimuli, such as studies comparing envelope ITD imposed on a high-frequency carrier with ongoing or carrier ITD imposed on a low-frequency carrier.

Other studies (e.g., Füllgrabe 2013; Gallun et al. 2014) have examined the sensitivity to ITDs and found that both aging and hearing loss can lead to elevated thresholds on tasks in which clicks or pure tones were used as stimuli. Strouse et al. (1998) and Gallun et al. (2014) measured monaural temporal processing using gap detection thresholds and binaural sensitivity using ITD thresholds and found that older listeners had higher temporal gap detection thresholds and higher ITD thresholds compared to the younger listeners. All of these studies have also revealed substantial variation in performance, which suggests that temporal encoding may be impaired independently of pure-tone detection (as measured by the audiogram). Another important point to keep in mind, however, is that much of the earlier literature relied on expert listeners who may have been drawn from the group of either those with good temporal encoding or those who may have learned to use cues that are not readily apparent to a naive listener presented with artificial binaural stimuli and/or unfamiliar tasks.

Bharadwaj et al. (2015) provide a comprehensive discussion of the possible mechanisms by which both aging and noise exposure can lead to reduced precision of temporal coding and present evidence consistent with the mechanisms proposed. The growing literature on the degree to which listeners vary in temporal structure encoding and the relationships between temporal encoding, aging, and noise exposure (e.g., Bramhall et al. 2017) provides a potential explanation for the variability observed among HI listeners. Future work will need to provide a more systematic examination of the ways that potential auditory nerve damage could be responsible for binaural and spatial hearing deficits in older and HI listeners. It is not yet clear why some listeners with poor audiometric thresholds retain their spatial hearing abilities and some do not but retaining precision of temporal encoding may provide some of the missing explanation.

11.5.3 Interaural Level Differences

Häusler et al. (1983) found no differences between ILD sensitivity in their NH and HI listeners. Koehnke et al. (1995) found that although about half of the HI listeners they tested had ILD sensitivity in the normal range at 500 or 4000 Hz, the other half did not. Gabriel et al. (1992) and Smith-Olinde et al. (2004) also found that although some listeners with hearing loss had normal ILD thresholds, others had thresholds outside the normal range.

Currently, there is not much known about exactly why ILDs might be elevated, but it is likely related to the need to base decisions on activity in auditory nerve fibers with best frequencies outside of the region of hearing loss, as described for monaural intensity discrimination (e.g., Schroder et al. 1994). Specifically, if decisions are made based on the spread of excitation to regions of the cochlea where nerve fibers are intact (or at least less damaged), then it is likely that a more intense stimulus is needed to produce the same perceived intensity and thus the same perceived intensity difference across ears. Eddins and Hall (2010) proposed that given the greater reduction in ITD relative to ILD sensitivity, it may be the case that when both cues are present, the ILD is weighted more heavily. As of this writing, no work has been published further examining this hypothesis.

One of the few studies (Simon and Aleksandrovsky 1997) to examine lateralization based on ILDs in listeners with SNHL found that the perceived location was more consistent across listeners when a fixed SPL value was used than when a fixed SL value was used. Specifically, those with asymmetrical hearing losses lateralized an equal SL stimulus toward their ear with the poorer threshold. An Equal SPL stimulus was not shifted away from midline. These results suggest that listeners with asymmetrical hearing losses adapt to their loss and perceive ILDs that are more consistent with the levels at the eardrum than with the monaural levels they experience.

11.5.4 Auditory Source Width Sensitivity

Whitmer et al. (2012) examined auditory source width sensitivity through headphones in older HI listeners by changing the interaural coherence of the presented sound. The researchers concluded that older HI listeners exhibited decreased sensitivity to changes in the width of the source. The authors related this finding to the

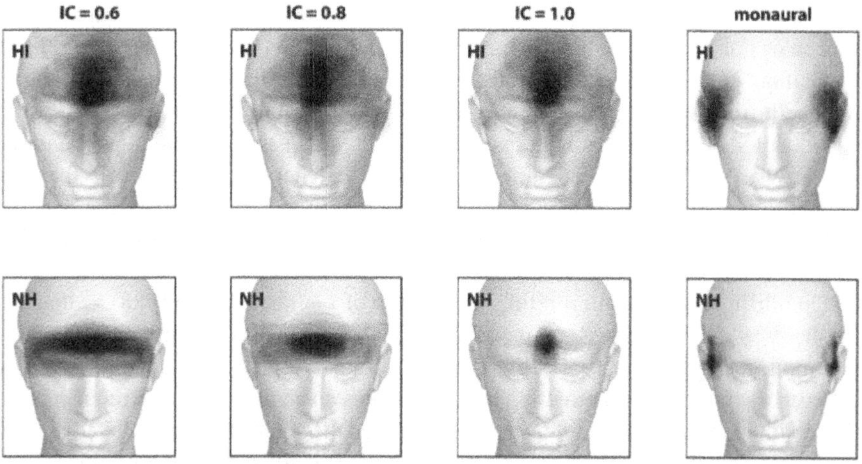

Fig. 11.6 Internal representations of broadband noise presented monaurally or binaurally. Interaural correlation (IC) of the binaurally presented noise varied between 0.6, 0.8, and 1.0, with 1.0 being perfectly correlated at the two ears (i.e., the same signals) and 0.0 being completely independent noises. After listening to these noises, listeners drew the perceived location and size of the sounds. *Top:* overlaid drawings created by HI listeners; *bottom:* those created by NH listeners. (Reprinted from Whitmer et al. 2013, with permission)

inability of older HI listeners to accurately process the timing information for representing sound source positions precisely.

Whitmer et al. (2014) measured the perceived auditory source width in NH and HI individuals by instructing them to sketch the width of the auditory image or select the closest one from a set of predetermined visual sketches. NH listeners sketched smaller and narrower widths compared with HI listeners given the same stimuli. Whitmer et al. (2013) also instructed adult NH and HI listeners to sketch the perceived width of acoustic stimuli in a simulated reverberant environment. The average results are reproduced in Fig. 11.6. Although the correlations between sketched widths and interaural coherence were higher for NH than for older HI listeners, it is noteworthy that NH listeners were both more accurate at sketching narrow widths for high coherence and sketching wide widths for low coherence. Older HI listeners sketched similar widths regardless of coherence, suggesting that the width was not perceived to vary much with interaural coherence.

11.6 Spatial Release from Masking

Although the detection and localization of tones and noises can reveal many things about binaural and spatial hearing, it is also of use to consider the tasks and stimuli that are more similar to those listeners that perform in natural environments. The most common task used to explore the use of binaural cues with natural stimuli is the identification of target speech in the presence of masking sounds, which can be speech or noise (see Culling and Lavandier, Chap. 8).

Gelfand et al. (1988) examined the effects of hearing loss and aging on the ability to benefit from spatial cues by presenting speech in 12-talker babble and compared a front colocated condition with one in which the target speech was in front and the masker was 90 ° to the side. Although the older listeners did more poorly than the younger ones in the colocated condition, the benefit of spatial separation was the same regardless of age. The older HI listeners, however, received substantially less spatial benefit.

One difficulty with interpreting the results of Gelfand et al. (1988) is that the masker was only presented on one side of the listener. This configuration, which is common in much of the early literature, confounds effects of monaural intensity differences with binaural effects. The head creates an acoustical interference pattern that results in variations in intensity at each ear as the sound source is moved around the head (Algazi et al. 2002). For a single asymmetrically offset masker, this means that the target-to-masker ratio (TMR) at each ear varies with the angle of offset.

To address the potential confound of head-related monaural intensity differences with spatial separations, Marrone et al. (2008) made an important modification to the testing methods that had been used previously by switching from an asymmetrical masker offset to using two maskers, which were symmetrically displaced to the left and right of the target location. This resulted in a reduction in the "better ear" listening strategy, which may have allowed listeners in previous studies to use the better TMR at the ear on the far side from the masker to perform the task. Thus, in a task with a single masker separated from the target to the side, the spatial benefit is not necessarily measuring binaural ability but rather the ability to take advantage of the difference in TMR at the two ears. Marrone et al. (2008) tested 40 listeners (10 in each group: younger NH, younger HI, older NH, and older HI) and measured the spatial release from masking (SRM) using three colocated speech sentences drawn from a corpus of very similar sentences or symmetrical separation of the maskers from the target by 45 °. The SRM was negatively correlated with both age and hearing loss, but the correlation with age was not significant after correcting for hearing loss. Gallun et al. (2013) addressed this issue by using the same design and recruiting listeners with the same range of ages but with a smaller range of hearing impairments. This allowed the effects of age and hearing loss to be statistically uncoupled in a mixed linear regression model, which revealed that both were meaningfully influencing performance.

Srinivasan et al. (2016) used the same stimuli and similar statistical techniques to separate the individual contributions of age from hearing loss on the SRM for a smaller range of separations, based on the hypothesis that 45° was a very large separation and that the relative effects of age and hearing loss might be more fully revealed at smaller separations. Figure 11.7 shows that the SRM was reduced by age and hearing loss, with the younger NH group achieving a SRM greater than 1 dB at 4°, the older NH group achieving a SRM greater than 1 dB at 8°, and the older HI group achieving a SRM greater than 1 dB only at 30°. The results of regression analyses predicting a SRM based on age and hearing loss supported the group

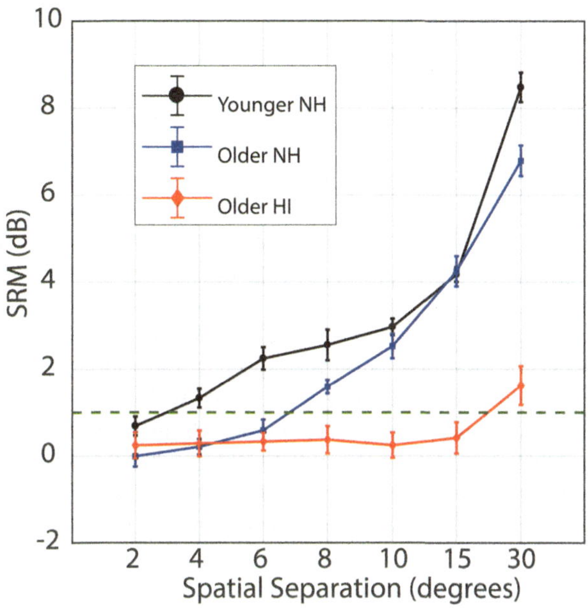

Fig. 11.7 Spatial release from masking (SRM) for younger NH, older NH, and older HI listeners as a function of spatial separation of the target and two symmetrically located maskers. *Dashed line* 1 dB SRM. (Plot derived from the data in Srinivasan et al. 2016)

results, suggesting that age was a significant predictor of the SRM at the lowest spatial separations (4° and 8°), whereas hearing loss was a significant predictor at the larger separations. These results support the hypothesis that although large separations are useful for examining hearing loss, more subtle effects such as the influence of aging may be more readily observed by examining the relative ability to use smaller cues.

Glyde et al. (2013b) examined the effects of age and hearing loss in a group of listeners ranging from children to older adults. The study used the Listening in Spatialized Noise-Sentences Test (LISN-S; Cameron and Dillon 2007), which presents target sentences at a center location in the presence of two competing stories. Virtual spatial cues were presented to all three talkers from the same location or maskers were presented from 90° to the left and right of the target. The results indicated that the advantage of spatially separating the target from the maskers deteriorated significantly with increasing hearing loss. However, no significant relationship between aging and spatial-processing capabilities were found. One limitation of the Glyde et al. (2013b) study is that the age range was so large that performance was nonmonotonically related to age, with the youngest and oldest listeners performing worse than the young adults. The dominance of hearing loss is consistent with the results of Srinivasan et al. (2016), but unlike Glyde et al. (2013b), Srinivasan et al. (2016) found significant age effects. As shown in Fig. 11.7, however, the largest age effects were found at the smallest separations. This suggests that the 90 ° spatial separations used by Glyde et al. (2013b) may have limited the ability of that study to find age effects.

Another area in which there is still uncertainty in the literature involves the relative importance of ITD and ILD cues for the SRM. Early work on this by Colburn and Durlach (1965) showed that BMLD thresholds were unaffected for NH listeners by creating a "conflicting cue" stimulus in which the ITDs were consistent with a signal on one side of the head and the ILDs were consistent with a signal on the other side of the head. Gallun et al. (2008) replicated these results and showed that they held both for tones masked by noise and for tones masked by multitone complexes.

To explore the question of consistent and inconsistent cues, Glyde et al. (2013a) used the LISN-Ss but modified the virtual space to vary only ITDs or ILDs. Thus, when ITDs alone were presented, they were paired with an ILD of 0 dB, and when ILDs alone were presented, they were paired with an ITD of 0 µs. NH listeners repeated the test three times (ITD alone, ILD alone, or both), and the results showed that the colocated thresholds were very similar. The spatially separated thresholds for ILDs alone were similar to those obtained with both cues, but both were better than for ITDs alone.

Ellinger et al. (2017) performed a similar experiment using a processed version of the stimuli from Gallun et al. (2013) but actually presented conflicting or consistent ITDs and ILDs as in Colburn and Durlach (1965). Unlike in the Glyde et al. (2013a) study, the presence of ITDs and ILDs resulted in better performance than ITDs alone, which, in turn, was better than ILDs alone. This pattern appeared in all three of the listener groups they tested (younger NH, older NH, and older HI), and there were no interactions between the group factor and the stimulus conditions. One important difference between the methods is that in Glyde et al. (2013a), the maskers were substantially more intense (resulting in a −12 dB signal-to-noise ratio [SNR] in the colocated condition) than those in Ellinger et al. (2017) where the colocated thresholds for the NH listeners were around a −3 dB SNR.

Best et al. (2013) showed that monaural factors limit the lowest SNR at which speech is intelligible regardless of binaural cue availability. Their data suggest that this "energetic limit" for noise maskers, which is about an −11 dB SNR for those with mild hearing loss (pure-tone average of 30 dB hearing loss), is lower than for speech maskers by about 4 dB and that increased hearing loss leads to an increased energetic limit. They estimated the function relating pure-tone average to the energetic limit as having a slope of 0.22 dB/dB for speech and 0.21 dB/dB for noise, resulting in energetic limits for those with moderate-to-severe hearing loss (pure-tone average of 60 dB) of around a −4 dB SNR for noise and around a 0 dB SNR for speech maskers.

The hypothesized energetic limit means that findings in which those with more hearing loss achieve a smaller SRM could reflect an energetic masking effect due to a very low SNR rather than a lack of access to binaural cues. A similar explanation could account for the importance of ILD in the LISN-S task, where the energetic limit is more likely to be a factor than for the stimuli and configurations used by Ellinger et al. (2017). The reason for this is that an ILD corresponding to a 90°

spatial separation imposed on the masker would substantially improve the SNR at the far ear, whereas the same ITD would not. This agrees with the fact that Ellinger et al. (2017) observed better performance when the ITD and ILD were consistent (more intense signal arrives first) than when they were inconsistent (more intense signal is delayed). The improved performance for the consistent cues suggests that listeners, regardless of age or hearing loss, were making use of the consistency of the cues to separate the targets from the maskers. A reasonable inference is that the perceived spatial location of the targets and maskers improved performance.

One way to reduce the chance that the energetic limit will make it difficult to compare the thresholds between groups is to use smaller spatial separations, as illustrated in Fig. 11.7. By measuring the smallest separation at which the SRM exceeds 1 dB (Fig. 11.7, *dashed line*), the limitation on the amount of SRM is decoupled from the question of whether or not binaural information is being used similarly between groups. Although the older HI group achieves less SRM than the other two groups shown in Fig. 11.7, the finding that they achieve no SRM at all for separations below 30° would still be true even if one concluded that the SRM achieved at 30° was at the energetic limit for those listeners.

The consideration of which spatial separation allows the SRM to be achieved for a given listener brings up the point that in almost all of these studies, the data reveal considerable between-subject variability. Although it is useful to use group analyses and multiple regression to distinguish the influence of various factors, the clinical utility of these results is primarily in how they can be used to improve the lives of individual patients. For example, it would be useful to know if a given person will be able to benefit from spatial separation if they were fit with a hearing aid that preserved or even enhanced binaural cues. The variability that is observed in these studies makes it clear that the answer to this question can only be obtained by testing the individual to be treated. For this reason, there is a great need to move toward individualized measures, based on what has been learned from group designs.

11.7 Central Auditory Processing of Binaural and Spatial Cues

Durlach et al. (1981) described a variety of cases of central auditory dysfunction, all of which have been confirmed in more modern studies. Jerger et al. (1986) studied patients with multiple sclerosis and found that 45% had abnormal results on the BMLD relative to control subjects. Przewoźny et al. (2015) found that 81% of their patients (9 of 11) with brainstem lesions due to a stroke, multiple sclerosis, or tumors had abnormal horizontal localization abilities compared with the control subjects without lesions or hearing loss. Only occasional small reductions in the BMLD were reported by Mueller and Beck (1987) for a group of patients with penetrating head wounds acquired during military combat. All had wounds that led to visible damage to the auditory cortex, and many had severely impacted scores on tests of dichotic speech perception. These results for listeners with penetrating head wounds suggest that the BMLD is specifically related to the integrity of the brainstem, given normal or near-normal input via the auditory nerve. In the future, it would be helpful to use direct measures of the responses of the brain to auditory stimuli (see Town and Bizley, Chap. 12) to better understand the ways in which the BMLD relates to physiology for patients with such injuries and diseases.

A more recent patient population with auditory difficulties is those who have suffered mild traumatic brain injury (mTBI) in either military combat or civilian pursuits. Although Gallun et al. (2016) found no reduction in the BMLD with mTBI, Saunders et al. (2015) found that nearly 50% of those with mTBI from blast exposure and auditory complaints performed abnormally on the spatial component of the LISN-S. Together, the findings of these two studies support the results from Mueller and Beck (1987) in which the BMLD was normal, but dichotic speech was difficult to understand. Kubli et al. (2018) found similar results when they measured localization ability in active duty military service members who had been exposed to high-intensity blasts. Although the listeners' ability to localize single speech utterances in quiet or in the presence of a single interfering talker was similar to that of age- and hearing-matched control subjects, the two groups diverged when two competing speech streams were presented and the task was to identify the location of a sentence related to a specific topic. This task resulted in substantially larger errors for the blast-exposed group. Hoover et al. (2017) examined civilians with mTBI with a set of tests of TFS sensitivity and speech, including several involving spatial and binaural cues, and found that the proportion of those with mTBI who did poorly on multiple TFS tests was much higher than among the control subjects. Overall, these results suggest that the binaural system is both a sensitive and informative way to explore dysfunction in parts of the auditory system beyond the cochlea.

11.8 Modeling Binaural and Spatial Hearing in Older and Hearing-Impaired Listeners

Historically, there have been three key aspects of auditory research in clinical populations. The first is behavioral work, establishing the phenomena to be explained and the differences between the patients and the normal population. The second is animal models of the disorder that allow the neurophysiology to be understood and the pathologies to be described anatomically and physiologically. The third is the computational modeling needed to explain how the physiological changes lead to the behavioral outcomes. At this time, there is limited progress on the modeling of impaired binaural function.

One of the most physiologically specific models is the model of the BMLD proposed by Mao et al. (2015). Based on the auditory nerve modeling of Zilany et al. (2009, 2014) and the phase-opponency model of Carney et al. (2002), Mao et al. (2015) developed a model in which binaural cues are encoded in the TFS, the envelope, or the energy of the auditory nerve-firing patterns. The model suggests that for young NH listeners detecting tones in noise, the diotic threshold is based on a combination of TFS, envelope, and energy, whereas in older NH and mildly HI listeners, the diotic threshold is based on envelope and energy but not TFS. Finally, in those older listeners with more significant hearing loss, the diotic threshold is based on energy cues alone. For the dichotic task, threshold was best predicted by a model based only on interaural envelope cues rather than models that included ITDs or ILDs alone or in combination. The model suggests that it is the redundancy of these energy, envelope, and TFS cues that leads to a good performance in the diotic condition but that only envelope is important for the dichotic condition.

A descriptive model of the SRM has been developed by Jones and Litovsky (2011) and computational models by Wan et al. (2014) and Mi et al. (2017). The descriptive model is a refined version of the model proposed by Bronkhorst (2000) and predicts a reduction in speech reception threshold when the speech target is in front of the listener and one or more interferers are placed around the listener at different azimuth angles. The reduction in the SRM predicted by the model is a sum of three separate contributions: (1) the contribution due to the angular separation between the maskers, (2) the contribution due to the asymmetry of the location of the maskers, and (3) the contribution due to front/back differences. When the maskers are symmetrically separated around the target in the frontal hemifield, the model predicts that the SRM will be based on only one of these factors: the angular separation between the maskers. This model is computationally simple in approach and is able to model the SRM for up to three maskers. Although the model is descriptive rather than mechanistic, it has been shown to capture much of the relevant data for younger NH in anechoic situations (Jones and Litovsky 2011). So far, neither the descriptive nor the computational models have been extended to predict results for HI listeners.

The promise of modeling for organizing behavioral and physiological data of older and impaired auditory systems is substantial, but the task becomes increasingly complex as more factors are included. For example, although Le Goff et al. (2013) proposed a model of horizontal localization that incorporates changes to audiometric thresholds, basilar membrane compression, and auditory filter width with hearing loss, it has not been extended to include changes in TFS processing. Future work would benefit from increased collaborations across groups and among researchers working on modeling, behavior, and neurophysiology to ensure that all of the most important factors are being explored in all three domains.

11.9 Multifactorial Approaches

The studies described up to this point have largely focused on single aspects of binaural and spatial hearing, although many studies have examined multiple measures. An alternative approach that has been only applied in a few instances involves the development of statistical models that relate hearing impairment and aging to multiple measures, with the goal of determining which binaural measures are related to each other and attempting to identify one or more factors predictive of a key outcome measure. One of the earliest studies to use this approach (Hall et al. 1984) explored the relationships among various monaural measures and binaural detection (BMLD) and discrimination (ITD) for HI listeners. It was determined that although hearing loss was strongly correlated with all of the measures, the only relationship that remained when hearing loss was taken into account was the relationship between the ITD and BMLD. Similarly, when Gabriel et al. (1992) examined four HI listeners on a range of binaural measures, they concluded that there was no interpretable relationship between the audiometric thresholds and any of the measures. As with many of the studies in this area, however, the lack of statistical power associated with only testing four listeners who varied on so many dimensions is a major obstacle to drawing strong conclusions.

Strelcyk and Dau (2009) tried to find monaural and binaural measures that were predictive of speech understanding in noise for 10 listeners with SNHL. They were fairly successful in building a model that showed that frequency selectivity was an independent predictor of speech understanding in noise and that there was a second factor involving both binaural and monaural sensitivity to TFS. Neher et al. (2011) extended the approach by adding several measures of attention in their study of HI listeners and found that binaural sensitivity predicted speech understanding in spatialized noise but that attention was an additional important factor. Most recently, Spencer et al. (2016) tested 11 HI listeners varying over a wide range of ages and found that with extensive training, there was substantial variability among listeners but good consistency between ITD and ILD and good reliability across multiple test sessions, once training was complete. Sensitivity to interaural correlation was independent of ITD and ILD thresholds, however, and none of the measures correlated with audiometric thresholds.

Future work focused on identifying common factors underlying or related to binaural hearing should pay close attention to designing a study with power sufficient to find relationships among measures that involve substantial variability. Of the studies reviewed in the previous paragraph, the one with the largest sample of HI ($n = 23$; Neher et al. 2011) was also the one in which the limitations of small sample sizes were most carefully considered. The authors specifically limited their multiple regression analyses to models with no more than two factors to respect the statistical limits imposed by the sample size with which they were working. To make progress in this area, it will be necessary to develop methods by which large numbers of listeners can be examined on a variety of measures that have already been shown to capture unique sources of variance.

11.10 Current Evaluation Techniques

The methods employed by clinical researchers are constantly being updated in accordance with new understandings of the systems under examination and the questions that are driving the research. Two of the current topics receiving substantial attention, as has been evident throughout this chapter, are the role of TFS sensitivity (Moore 2014) and the damage to auditory nerve fibers after noise exposure, especially in combination with aging (Fernandez et al. 2015; Liberman 2017). Consequently, there is an increasing interest in methods by which sensitivity to TFS can be evaluated quickly and reliably and great enthusiasm for finding evidence that behavioral or electrophysiological tests can reveal auditory nerve damage. In both of these areas, sensitivity to the ITD has been identified as potentially one of the most sensitive measures. The reason for this is that the binaural system leads to a percept that is monotonically related to the timing of the spikes on the auditory nerve. Consequently, if the listener can detect and discriminate differences in the ITD similar to those detectable by younger NH listeners, then it must be the case that the auditory nerve is functioning well. Unfortunately, it is not possible to conclude from good auditory nerve function, especially as measured by the audiogram, that the binaural system is intact. To date, far too few clinicians are even capable of testing binaural abilities, let alone incorporating such tests into their routine exams.

The methods that have been used to test binaural function in clinical populations have been evolving rapidly, however. Much of this chapter has focused on the localization of sounds in free-field environments and the lateralization of brief clicks and noise bursts. Clinically, however, these tests are difficult to administer without expensive and cumbersome equipment and facilities and substantial technical expertise on the part of the test administrator. The methods that are replacing them in the clinical research arena are more like the BMLD, which has been used clinically for over 35 years (Jerger et al. 1984), although much more often in clinical research than in clinical practice. This test and others like it are based on forced-choice detection in which no introspection is required of the listener. All that is needed is a response as to whether the target was in interval one or interval two. Bernstein and Trahiotis (2016) were able to use this paradigm to show that listeners with "slight" hearing loss, defined as monaural detection thresholds for a 4000-Hz tone presented in quiet between 7.5 dB hearing loss and 25 dB hearing loss, had BMLDs that were reduced relative to those with quiet thresholds better than 7.5 dB hearing loss. This was true for 500-Hz tones as well as for "transposed" stimuli presented at 4000 Hz but with a 125-Hz periodicity imposed on the envelope. It is not obvious why reduced BMLDs for those with losses at 4000 Hz were not found in the much larger study by Jerger et al. (1984), but it is likely a result of the more time-consuming, and thus substantially more precise, measurement procedure used by Bernstein and Trahiotis (2016).

The second approach is the TFS testing developed by Hopkins and Moore (2010) and Füllgrabe et al. (2017) and the binaural frequency modulation test (Witton et al. 2000; Grose and Mamo 2012). In both of these tests, the listener is presented with a choice between an interval in which there is a binaural difference in the phase of the tones at the two ears and one in which there is no binaural difference. Adaptive tracking is used to determine the smallest phase difference that can be detected or, in the case of Füllgrabe et al. (2017), the highest frequency at which a phase shift of a fixed value can be detected. These studies have shown that older listeners with and without hearing loss are more likely to have elevated interaural phase difference (IPD) thresholds and that IPD discrimination performance declines at relatively lower frequencies compared with that in younger listeners. Moore et al. (2012) used the binaural TFS test to show that, for older NH listeners (NH up to 1000 Hz), there was a strong correlation between age and IPD sensitivity but no correlation with audiometric thresholds. These measures are also similar to, and consistent with the data from, electrophysiological and magnetoencephalographic studies of IPD sensitivity (Ross et al. 2007; Papesh et al. 2017). These data are clear evidence that normal audiometric thresholds are not a guarantee of normal binaural function.

To better understand the relative effects of age and SNHL on TFS, King et al. (2014) conducted a study on HI listeners who varied in age using the binaural TFS test. There was, however, the important modification that instead of always adding the IPD to the carrier and not the envelope, the IPD was added to the carrier in one condition and it was added to the envelope in the other. Results suggested that age and SNHL both reduced TFS sensitivity but that increasing age reduced envelope IPD sensitivity, whereas increasing audiometric thresholds do not. Future work will need to examine in more detail the relationships among these tests and, in particular, the apparent discrepancies between the BMLD and tests of IPD sensitivity that do not use a noise masker.

One area that holds a particular promise for clinical testing is the use of virtual auditory spatial arrays. The study of Brungart et al. (2017) described previously found similar effects of hearing loss both in a free-field condition and in a virtual auditory spatial array created using measurements of the listener's head-related transfer functions. Although few clinics had access to free-field testing facilities, the technology for simulating such facilities is already available in many inexpensive consumer devices for creating virtual reality environments. It is much more feasible to turn these environments into virtual testing facilities than it is to create actual spatial arrays to test clinical populations. Future work involving binaural and spatial hearing should focus on harnessing this technology for clinical testing.

11.11 Clinical Ramifications

Although many audiologists are aware of the importance of binaural hearing, tests evaluating binaural or spatial hearing are rarely investigated clinically. One reason for this may be time constraints on the clinicians. It is typical for a busy clinician to have fewer than 15 minutes to devote to diagnostic testing, largely due to the emphasis on the fitting of hearing aids rather than a detailed diagnostic evaluation. Just like speech-in-noise testing, results of a binaural sensitivity test do not alter the way hearing aids are fit or prescribed to patients. Currently, pure-tone hearing thresholds are used to prescribe and fit hearing aids. Some tools are being developed to allow for adjustments based on the preferred sound quality of the patient, but the benefits of this technique may not translate from the clinic to the patient's real-world environments.

It is interesting that hearing aids are still being fit primarily based on pure-tone thresholds because it is well-known that two individuals with the same pure-tone hearing may have differences in spectral and temporal resolution, binaural hearing sensitivity, weighting of binaural cues, and speech understanding in noise. With advancing technology (see Ricketts and Kan, Chap. 13), hearing aid and cochlear implant manufacturers are including technology that has the ability to restore binaural cues that may be lost due to device features such as microphone placement. On the other hand, some hearing aid settings may eliminate binaural cues with bilateral beamforming technology. If hearing aid fittings are individualized on the basis of the ability to use binaural cues, it is possible that this could lead to an improvement in speech recognition performance and overall hearing aid satisfaction.

King et al. (2017) have proposed an SRM task that uses signals composed of amplitude-modulated pure tones (Arbogast et al. 2002, 2005) as a way to identify candidates for hearing aids that preserve ITDs. By basing the modulation on speech envelopes, these signals can be interpreted as speech, and yet, the ITDs in each frequency region can be carefully controlled. The difficulty with this proposal is that clinicians already have many tests to complete in minimal amounts of time. If brief tests evaluating binaural and spatial hearing are developed, the results of these tests may help individualize hearing aid fittings. With improvement in portable technology, it may be possible to collect extra tests and questionnaires (e.g., on a tablet computer) while patients wait in the lobby for their appointment. This may be important because if an individual can make use of binaural cues to achieve SRM, for example, it may be detrimental to them to use a binaural beamformer (eliminating binaural cues) or to wear a behind-the-ear style of hearing aid that alters pinna cue information. This is an exciting technological era with great flexibility in hearing aids. It would be very interesting to see if clinicians could use additional tests to better individualize hearing aid fitting and the selection of appropriate devices and technology.

11.12 Conclusions and Future Directions

This chapter can be best summarized by saying that there is no simple way to characterize the abilities of patients with auditory dysfunction to make use of binaural and spatial cues. Figure 11.1, for example, serves as a reminder of how much of the greatly diminished spatial experience of patients with severe SNHL can be attributed solely to the reduction in sensation level of the sounds they receive. The task, then, is to find ways to capture these individual experiences and understand the causes, with the ultimate goal of providing effective rehabilitation. It is illuminating and encouraging to find so many examples of listeners with good binaural function in the literature on hearing loss. It suggests that spatial hearing is capable of being preserved and/or restored in listeners with hearing loss. Future work should focus on seeking to uncover what it is that can cause two listeners with similar audiograms and no apparent retrocochlear loss to have such different binaural sensitivities.

The heterogeneity within and among disorders and methods is an important consideration to keep in mind while reviewing the data, but it should not be seen as a reason to turn away from the clinical research endeavor. In the end, much of the justification for studying the human body, especially through the use of animal models, is to someday improve the lives of those who live with these conditions. Readers of this chapter are encouraged to examine these results in this light and perhaps even be inspired to join what can be argued to be one of the most challenging and yet gratifying scientific endeavors to be found.

Future work in the area of impaired binaural hearing has many areas in need of further study.

1. Examining systematically the ways binaural and spatial hearing abilities in older and HI listeners are affected by auditory nerve damage. Designing studies with such multiple factors that have sufficient power to find such relationships;
2. Using physiological measures to better understand the ways in which the behavioral thresholds relate to physiology for patients with injuries to and diseases of the central auditory system;
3. Developing tools that will allow clinicians and clinical researchers to test large numbers of patients with a minimum of resources; and
4. Collecting evidence that supports the use of binaural test measures in the diagnosis of hearing difficulties and the fitting of hearing aids and cochlear implants.

Acknowledgments This work was supported by Grant R01-DC-011828 from the National Institute on Deafness and Other Communication Disorders, National Institutes of Health, Bethesda, MD, and the VA RR&D National Center for Rehabilitative Auditory Research, VA Portland Health Care System, Portland, OR. The opinions expressed are the private views of the authors and do not represent the official position of the United States, the National Institutes of Health, or the Department of Veterans Affairs.

Compliance with Ethics Statement

Frederick J. Gallun declares that he has no conflict of interest.
Nirmal K. Srinivasan declares that he has no conflict of interest.
Anna C. Diedesch declares that she has no conflict of interest.

References

Abel SM, Giguère C, Consoli A, Papsin BC (2000) The effect of aging on horizontal plane sound localization. J Acoust Soc Am 108:743–752

Akeroyd MA, Whitmer WM (2016) Spatial hearing and hearing aids. In: Popelka GR, Moore BCJ, Fay RR, Popper AN (eds) Hearing aids. Springer International Publishing, Cham, pp 181–215

Algazi VR, Duda RO, Duraiswami R et al (2002) Approximating the head-related transfer function using simple geometric models of the head and torso. J Acoust Soc Am 112:2053–2064. https://doi.org/10.1121/1.1508780
Anderson S, Ellis R, Mehta J, Goupell MJ (2018) Age-related differences in binaural masking level differences: behavioral and electrophysiological evidence. J Neurophysiol 120:2939–2952. https://doi.org/10.1152/jn.00255.2018
Arbogast TL, Mason CR, Kidd G (2002) The effect of spatial separation on informational and energetic masking of speech. J Acoust Soc Am 112:2086–2098. https://doi.org/10.1121/1.1510141
Arbogast TL, Mason CR, Kidd G (2005) The effect of spatial separation on informational masking of speech in normal-hearing and hearing-impaired listeners. J Acoust Soc Am 117:2169–2180. https://doi.org/10.1121/1.1861598
Bergman M (1957) Binaural hearing. AMA Arch Otolaryngol 66:572–578. https://doi.org/10.1001/archotol.1957.03830290078009
Bernstein LR, Trahiotis C (2016) Behavioral manifestations of audiometrically-defined "slight" or "hidden" hearing loss revealed by measures of binaural detection. J Acoust Soc Am 140:3540–3548. https://doi.org/10.1121/1.4966113
Best V, Thompson ER, Mason CR, Kidd G (2013) An energetic limit on spatial release from masking. J Assoc Res Otolaryngol 14:603–610. https://doi.org/10.1007/s10162-013-0392-1
Bharadwaj HM, Masud S, Mehraei G et al (2015) Individual differences reveal correlates of hidden hearing deficits. J Neurosci 35:2161–2172. https://doi.org/10.1523/JNEUROSCI.3915-14.2015
Bramhall NF, Konrad-Martin D, McMillan GP, Griest SE (2017) Auditory brainstem response altered in humans with noise exposure despite normal outer hair cell function. Ear Hear 38:e1. https://doi.org/10.1097/AUD.0000000000000370
Bronkhorst AW (2000) The cocktail party phenomenon: a review of research on speech intelligibility in multiple-talker conditions. Acta Acust United Acust 86:117–128
Brugge JF, Orman SS, Coleman JR et al (1985) Binaural interactions in cortical area AI of cats reared with unilateral atresia of the external ear canal. Hear Res 20:275–287. https://doi.org/10.1016/0378-5955(85)90032-2
Brungart DS, Cohen JI, Zion D, Romigh G (2017) The localization of non-individualized virtual sounds by hearing impaired listeners. J Acoust Soc Am 141:2870–2881. https://doi.org/10.1121/1.4979462
Cameron S, Dillon H (2007) Development of the listening in spatialized noise-sentences test (LISN-S). Ear Hear 28:196. https://doi.org/10.1097/AUD.0b013e318031267f
Carney LH, Heinzy MG, Evilsizer ME et al (2002) Auditory phase opponency: a temporal model for masked detection at low frequencies. Acta Acust United Acust 88:334. https://www.ingentaconnect.com/content/dav/aaua/2002/00000088/00000003/art00005. Accessed 21 Jun 2019
Caspary DM, Ling L, Turner JG, Hughes LF (2008) Inhibitory neurotransmission, plasticity and aging in the mammalian central auditory system. J Exp Biol 211:1781–1791. https://doi.org/10.1242/jeb.013581
Colburn HS, Durlach NI (1965) Time-intensity relations in binaural unmasking. J Acoust Soc Am 38:93–103. https://doi.org/10.1121/1.1909625
Curhan G, Curhan S (2016) Epidemiology of hearing impairment. In: Popelka GR, Moore BCJ, Fay RR, Popper AN (eds) Hearing aids. Springer International Publishing, Cham, pp 21–58
Demeester K, Topsakal V, Hendrickx J-J et al (2012) Hearing disability measured by the speech, spatial, and qualities of hearing scale in clinically normal-hearing and hearing-impaired middle-aged persons, and disability screening by means of a reduced SSQ (the SSQ5). Ear Hear 33:615. https://doi.org/10.1097/AUD.0b013e31824e0ba7
Dobreva MS, O'Neill WE, Paige GD (2011) Influence of aging on human sound localization. J Neurophysiol 105:2471–2486
Durlach NI, Thompson CL, Colburn HS (1981) Binaural interaction in impaired listeners: a review of past research. Audiology 20:181–211. https://doi.org/10.3109/00206098109072694
Eddins AC, Eddins DA (2018) Cortical correlates of binaural temporal processing deficits in older adults. Ear Hear 39:594–604

Eddins DA, Hall JW (2010) Binaural processing and auditory asymmetries. In: Gordon-Salant S, Frisina RD, Popper AN, Fay RR (eds) The aging auditory system. Springer, New York, pp 135–165

Ellinger RL, Jakien KM, Gallun FJ (2017) The role of interaural differences on speech intelligibility in complex multi-talker environments. J Acoust Soc Am 141:EL170–EL176. https://doi.org/10.1121/1.4976113

Engle J, Recanzone GH (2013) Characterizing spatial tuning functions of neurons in the auditory cortex of young and aged monkeys: a new perspective on old data. Front Aging Neurosci 4:36. https://doi.org/10.3389/fnagi.2012.00036

Fernandez KA, Jeffers PWC, Lall K et al (2015) Aging after noise exposure: acceleration of cochlear synaptopathy in "recovered" ears. J Neurosci 35:7509–7520. https://doi.org/10.1523/JNEUROSCI.5138-14.2015

Freigang C, Richter N, Rübsamen R, Ludwig AA (2015) Age-related changes in sound localisation ability. Cell Tissue Res 361:371–386. https://doi.org/10.1007/s00441-015-2230-8

Füllgrabe C (2013) Age-dependent changes in temporal-fine-structure processing in the absence of peripheral hearing loss. Am J Audiol 22:313–315. https://doi.org/10.1044/1059-0889(2013/12-0070)

Füllgrabe C, Harland AJ, Sęk AP, Moore BCJ (2017) Development of a method for determining binaural sensitivity to temporal fine structure. Int J Audiol 0:1–10. https://doi.org/10.1080/14992027.2017.1366078

Gabriel KJ, Koehnke J, Colburn HS (1992) Frequency dependence of binaural performance in listeners with impaired binaural hearing. J Acoust Soc Am 91:336–347

Gallun FJ, Durlach NI, Colburn HS et al (2008) The extent to which a position-based explanation accounts for binaural release from informational masking. J Acoust Soc Am 124:439–449. https://doi.org/10.1121/1.2924127

Gallun FJ, Diedesch AC, Kampel SD, Jakien KM (2013) Independent impacts of age and hearing loss on spatial release in a complex auditory environment. Front Neurosci 7:252. https://doi.org/10.3389/fnins.2013.00252

Gallun FJ, McMillan GP, Molis MR et al (2014) Relating age and hearing loss to monaural, bilateral, and binaural temporal sensitivity1. Front Neurosci 8:172. https://doi.org/10.3389/fnins.2014.00172

Gallun FJ, Lewis MS, Folmer RL et al (2016) Chronic effects of exposure to high-intensity blasts: results of tests of central auditory processing. J Rehabil Res Dev 53:705

Gatehouse S, Noble W (2004) The speech, spatial and qualities of hearing scale (SSQ). Int J Audiol 43:85–99

Gelfand SA, Ross L, Miller S (1988) Sentence reception in noise from one versus two sources: effects of aging and hearing loss. J Acoust Soc Am 83:248–256. https://doi.org/10.1121/1.396426

Glyde H, Buchholz JM, Dillon H et al (2013a) The importance of interaural time differences and level differences in spatial release from masking. J Acoust Soc Am 134:EL147–EL152. https://doi.org/10.1121/1.4812441

Glyde H, Cameron S, Dillon H et al (2013b) The effects of hearing impairment and aging on spatial processing. Ear Hear 34:15. https://doi.org/10.1097/AUD.0b013e3182617f94

Gordon-Salant S, Frisina RD, Fay RR, Popper A (2010) The aging auditory system. Springer Science & Business Media, New York

Grose JH, Mamo SK (2012) Frequency modulation detection as a measure of temporal processing: age-related monaural and binaural effects. Hear Res 294:49–54. https://doi.org/10.1016/j.heares.2012.09.007

Hall JW, Tyler RS, Fernandes MA (1984) Factors influencing the masking level difference in cochlear hearing-impaired and normal-hearing listeners. J Speech Lang Hear Res 27:145–154. https://doi.org/10.1044/jshr.2701.145

Häusler R, Levine RA (1980) Brain stem auditory evoked potentials are related to interaural time discrimination in patients with multiple sclerosis. Brain Res 191:589–594. https://doi.org/10.1016/0006-8993(80)91312-8

Häusler R, Colburn S, Marr E (1983) Sound localization in subjects with impaired hearing: spatial-discrimination and interaural-discrimination tests. Acta Otolaryngol 96:1–62

Hawkins DB, Wightman FL (1980) Interaural time discrimination ability of listeners with sensori-neural hearing loss. Audiology 19:495–507

Hoover EC, Souza PE, Gallun FJ (2017) Auditory and cognitive factors associated with speech-in-noise complaints following mild traumatic brain injury. J Am Acad Audiol 28:325–339

Hopkins K, Moore BCJ (2010) Development of a fast method for measuring sensitivity to temporal fine structure information at low frequencies. Int J Audiol 49:940–946. https://doi.org/10.3109/14992027.2010.512613

Jabbari B, Mueller HG, Howard DK (1987) Auditory brainstem response findings in the late phase of head injury. In: Seminars in hearing. Copyright© 1987 by Thieme Medical Publishers, Inc, New York, pp 261–265

Jerger J, Brown D, Smith S (1984) Effect of peripheral hearing loss on the masking level difference. Arch Otolaryngol 110:290–296. https://doi.org/10.1001/archotol.1984.00800310014003

Jerger JF, Oliver TA, Chmiel RA, Rivera VM (1986) Patterns of auditory abnormality in multiple sclerosis: Exemples de cas auditifs anormaux dans les scléroses en plaques. Audiology 25:193–209. https://doi.org/10.3109/00206098609078386

Jones GL, Litovsky RY (2011) A cocktail party model of spatial release from masking by both noise and speech interferers. J Acoust Soc Am 130:1463–1474. https://doi.org/10.1121/1.3613928

Jongkees LBW, Veer RAD (1958) On directional sound localization in unilateral deafness and its explanation. Acta Otolaryngol 49:119–131. https://doi.org/10.3109/00016485809134735

Juarez-Salinas DL, Engle JR, Navarro XO, Recanzone GH (2010) Hierarchical and serial processing in the spatial auditory cortical pathway is degraded by natural aging. J Neurosci 30:14795–14804. https://doi.org/10.1523/JNEUROSCI.3393-10.2010

Katz J (2014) Handbook of clinical audiology, 7th edn. Wolters Kluwer, Lippincott William & Wilkins, Philadelphia

King A, Hopkins K, Plack CJ (2014) The effects of age and hearing loss on interaural phase difference discrimination. J Acoust Soc Am 135:342–351. https://doi.org/10.1121/1.4838995

King A, Hopkins K, Plack CJ et al (2017) The effect of tone-vocoding on spatial release from masking for old, hearing-impaired listeners. J Acoust Soc Am 141:2591–2603. https://doi.org/10.1121/1.4979593

Koehnke J, Culotta CP, Hawley ML, Colburn HS (1995) Effects of reference interaural time and intensity differences on binaural performance in listeners with normal and impaired hearing. Ear Hear 16:331–353

Kubli LR, Brungart D, Northern J (2018) Effect of blast injury on auditory localization in military service members. Ear Hear 39:457. https://doi.org/10.1097/AUD.0000000000000517

Kujawa SG, Liberman MC (2009) Adding insult to injury: cochlear nerve degeneration after "temporary" noise-induced hearing loss. J Neurosci 29:14077–14085. https://doi.org/10.1523/JNEUROSCI.2845-09.2009

Le Goff N, Buchholz JM, Dau T (2013) Modeling horizontal localization of complex sounds in the impaired and aided impaired auditory system. In: Blauert J (ed) The technology of binaural listening. Springer Berlin Heidelberg, Berlin, Heidelberg, pp 121–144

Liberman MC (2017) Noise-induced and age-related hearing loss: new perspectives and potential therapies. F1000Res 6. https://doi.org/10.12688/f1000research.11310.1

Mao J, Koch K-J, Doherty KA, Carney LH (2015) Cues for diotic and dichotic detection of a 500-Hz tone in noise vary with hearing loss. J Assoc Res Otolaryngol 16:507–521. https://doi.org/10.1007/s10162-015-0518-8

Marrone N, Mason CR, Kidd G (2008) The effects of hearing loss and age on the benefit of spatial separation between multiple talkers in reverberant rooms. J Acoust Soc Am 124:3064–3075. https://doi.org/10.1121/1.2980441

McFadden SL, Willott JF (1994) Responses of inferior colliculus neurons in C57BL/6J mice with and without sensorineural hearing loss: effects of changing the azimuthal location of an unmasked pure-tone stimulus. Hear Res 78:115–131. https://doi.org/10.1016/0378-5955(94)90018-3

Mi J, Groll M, Colburn HS (2017) Comparison of a target-equalization-cancellation approach and a localization approach to source separation. J Acoust Soc Am 142:2933–2941. https://doi.org/10.1121/1.5009763

Moore BC (2007) Cochlear hearing loss: physiological, psychological and technical issues. John Wiley & Sons, London

Moore BC (2014) Auditory processing of temporal fine structure: effects of age and hearing loss. World Scientific, Singapore

Moore DR, Hine JE, Jiang ZD et al (1999) Conductive hearing loss produces a reversible binaural hearing impairment. J Neurosci 19:8704–8711. https://doi.org/10.1523/JNEUROSCI.19-19-08704.1999

Moore BCJ, Vickers DA, Mehta A (2012) The effects of age on temporal fine structure sensitivity in monaural and binaural conditions. Int J Audiol 51:715–721. https://doi.org/10.3109/14992027.2012.690079

Mueller HG, Beck WG (1987) Brainstem level test results following head injury. Semin Hear 8:253–260

Neher T, Laugesen S, Søgaard Jensen N, Kragelund L (2011) Can basic auditory and cognitive measures predict hearing-impaired listeners' localization and spatial speech recognition abilities? J Acoust Soc Am 130:1542–1558. https://doi.org/10.1121/1.3608122

Noble W, Gatehouse S (2006) Effects of bilateral versus unilateral hearing aid fitting on abilities measured by the speech, spatial, and qualities of hearing scale (SSQ). Int J Audiol 45:172–181. https://doi.org/10.1080/14992020500376933

Papesh MA, Folmer RL, Gallun FJ (2017) Cortical measures of binaural processing predict spatial release from masking performance. Front Hum Neurosci 11:124

Popescu MV, Polley DB (2010) Monaural deprivation disrupts development of binaural selectivity in auditory midbrain and cortex. Neuron 65:718–731. https://doi.org/10.1016/j.neuron.2010.02.019

Przewoźny T, Gójska-Grymajło A, Szmuda T, Markiet K (2015) Auditory spatial deficits in brainstem disorders. Neurol Neurochir Pol 49:401–411. https://doi.org/10.1016/j.pjnns.2015.10.001

Ross B, Fujioka T, Tremblay KL, Picton TW (2007) Aging in binaural hearing begins in mid-life: evidence from cortical auditory-evoked responses to changes in interaural phase. J Neurosci 27:11172–11178. https://doi.org/10.1523/JNEUROSCI.1813-07.2007

Ruggles D, Bharadwaj H, Shinn-Cunningham BG (2011) Normal hearing is not enough to guarantee robust encoding of suprathreshold features important in everyday communication. Proc Natl Acad Sci U S A 108:15516–15521. https://doi.org/10.1073/pnas.1108912108

Saunders GH, Frederick MT, Arnold M, Silverman S, Chisolm TH, & Myers P (2015) Auditory difficulties in blast-exposed Veterans with clinically normal hearing. Journal of Rehabilitation Research & Development, 52(3).

Sanchez-Longo LP, Forster FM (1958) Clinical significance of impairment of sound localization. Neurology 8:119–125. https://doi.org/10.1212/WNL.8.2.119

Schroder AC, Viemeister NF, Nelson DA (1994) Intensity discrimination in normal-hearing and hearing-impaired listeners. J Acoust Soc Am 96:2683–2693

Simon HJ, Aleksandrovsky I (1997) Perceived lateral position of narrow-band noise in hearing-impaired and normal-hearing listeners under conditions of equal sensation level and sound-pressure level. J Acoust Soc Am 102:1821–1826. https://doi.org/10.1121/1.420089

Smith-Olinde L, Besing J, Koehnke J (2004) Interference and enhancement effects on interaural time discrimination and level discrimination in listeners with normal hearing and those with hearing loss. Am J Audiol 13:80–95. https://doi.org/10.1044/1059-0889(2004/011)

Smoski WJ, Trahiotis C (1986) Discrimination of interaural temporal disparities by normal-hearing listeners and listeners with high-frequency sensorineural hearing loss. J Acoust Soc Am 79:1541–1547. https://doi.org/10.1121/1.393680

Souza P, Hoover E, Blackburn M, Gallun F (2018) The characteristics of adults with severe hearing loss. J Am Acad Audiol 29:764

Spencer NJ, Hawley ML, Colburn HS (2016) Relating interaural difference sensitivities for several parameters measured in normal-hearing and hearing-impaired listeners. J Acoust Soc Am 140:1783–1799. https://doi.org/10.1121/1.4962444

Srinivasan NK, Jakien KM, Gallun FJ (2016) Release from masking for small spatial separations: effects of age and hearing loss. J Acoust Soc Am 140:EL73–EL78. https://doi.org/10.1121/1.4954386

Strelcyk O, Dau T (2009) Relations between frequency selectivity, temporal fine-structure processing, and speech reception in impaired hearing. J Acoust Soc Am 125:3328–3345. https://doi.org/10.1121/1.3097469

Strouse A, Ashmead DH, Ohde RN, Grantham DW (1998) Temporal processing in the aging auditory system. J Acoust Soc Am 104:2385–2399

Thornton JL, Chevallier KM, Koka K et al (2012) The conductive hearing loss due to an experimentally induced middle ear effusion alters the interaural level and time difference cues to sound location. J Assoc Res Otolaryngol 13:641–654. https://doi.org/10.1007/s10162-012-0335-2

Tillein J, Hubka P, Kral A (2016) Monaural congenital deafness affects aural dominance and degrades binaural processing. Cereb Cortex 26:1762–1777. https://doi.org/10.1093/cercor/bhv351

Van den Bogaert T, Klasen TJ, Moonen M et al (2006) Horizontal localization with bilateral hearing aids: without is better than with. J Acoust Soc Am 119:515–526

Vercammen C, Goossens T, Undurraga J et al (2018) Electrophysiological and behavioral evidence of reduced binaural temporal processing in the aging and hearing impaired human auditory system. Trends Hear 22:2331216518785733. https://doi.org/10.1177/2331216518785733

Wan R, Durlach NI, Colburn HS (2014) Application of a short-time version of the equalization-cancellation model to speech intelligibility experiments with speech maskers. J Acoust Soc Am 136:768–776. https://doi.org/10.1121/1.4884767

Whitmer WM, Seeber BU, Akeroyd MA (2012) Apparent auditory source width insensitivity in older hearing-impaired individuals. J Acoust Soc Am 132:369–379. https://doi.org/10.1121/1.4728200

Whitmer WM, Seeber BU, Akeroyd MA (2013) Measuring the apparent width of auditory sources in normal and impaired hearing. In: Moore BCJ, Patterson RD, Winter IM et al (eds) Basic aspects of hearing. Springer, New York, pp 303–310

Whitmer WM, Seeber BU, Akeroyd MA (2014) The perception of apparent auditory source width in hearing-impaired adults. J Acoust Soc Am 135:3548–3559

Witton C, Green GGR, Rees A, Henning GB (2000) Monaural and binaural detection of sinusoidal phase modulation of a 500-Hz tone. J Acoust Soc Am 108:1826–1833. https://doi.org/10.1121/1.1310195

Zilany MSA, Bruce IC, Nelson PC, Carney LH (2009) A phenomenological model of the synapse between the inner hair cell and auditory nerve: long-term adaptation with power-law dynamics. J Acoust Soc Am 126:2390–2412. https://doi.org/10.1121/1.3238250

Zilany MSA, Bruce IC, Carney LH (2014) Updated parameters and expanded simulation options for a model of the auditory periphery. J Acoust Soc Am 135:283–286. https://doi.org/10.1121/1.4837815

[illegible]

Chapter 12
Physiology of Higher Central Auditory Processing and Plasticity

Stephen M. Town and Jennifer K. Bizley

12.1 Introduction

Sound waves are encoded within the cochlea according to their frequency, meaning that the spatial location of sound sources must be computed from acoustic features that are encoded as binaural (two ears) and monaural (one ear) localization cues (see Hartmann, Chap. 2). Although sound localization cues are extracted by specialized brainstem nuclei, no single cue defines a sound's location unambiguously; instead, information must be integrated across cue types and frequencies to robustly compute spatial position. Although the precise role of auditory cortex in sound perception has remained somewhat elusive, an observation that is highly consistent across methods and species is that a deficit in sound localization results from removal or silencing of auditory cortical activity (e.g., Whitfield et al. 1972; Heffner and Heffner 1990). Animals lacking an intact auditory cortex can detect a change in source location and can still accurately discriminate left from right (i.e., lateralize) sound locations; however, such subjects show impairments in their ability to compute the precise source location in the side of space contralateral to the lesion (e.g., sounds to the right of the head following a lesion to the left hemisphere of the brain). These observations from animal models mirror deficits in humans that are observed after brain damage in clinical case studies. However, despite work spanning several decades demonstrating a critical requirement for the auditory cortex in sound localization, our understanding of how neural circuits in (and beyond) the auditory cortex generate spatial perception remains limited.

This chapter begins by reviewing the cortical representation of space and our current understanding of spatial processing within auditory cortex (see Sect. 12.2). The role of auditory cortex is then discussed within the context of larger scale brain networks where regions such as the auditory, parietal, and frontal cortices may form

S. M. Town · J. K. Bizley (✉)
University College London Ear Institute, London, UK
e-mail: s.town@ucl.ac.uk; j.bizley@ucl.ac.uk

© Springer Nature Switzerland AG 2021

R. Y. Litovsky et al. (eds.), *Binaural Hearing*, Springer Handbook of Auditory Research 73, https://doi.org/10.1007/978-3-030-57100-9_12

a specialized spatial processing stream (see Sect. 12.3). This section also examines how task demands shape neural processing so that listeners can actively and dynamically sense their environment (see Sect. 12.3.3). Flexible sound localization is then considered within the context of plasticity and reorganization of cortical function, with a particular focus on the role of descending connections in facilitating this plasticity (see Sect. 12.4). Finally, the chapter discusses some of the open questions that remain regarding central processing of binaural cues. These include whether neurons in auditory cortex represent acoustic localization cues or an integrated representation of space (see Sect. 12.5.1) and how the brain represents sound location in multiple coordinate systems (see Sect. 12.5.2). To answer these and other important questions about central processing of binaural cues, significant technical and theoretical challenges must be overcome (see Sect. 12.6).

12.2 Auditory Cortex

The auditory cortex encompasses a number of distinct cortical fields, with core (also known as primary) areas innervated directly by the lemniscal thalamus (medial geniculate body [MGB]; see Table 12.1 for a complete list of abbreviations), belt (secondary) areas innervated directly by core areas (and also by the MGB), and, finally, parabelt (tertiary) areas that receive cortical input but do not receive direct connections from the auditory thalamus (Hackett 2015). In humans, auditory cortex lies along the superior temporal gyrus (Fig. 12.1a) with the core area located in the caudal two-thirds of Heschl's gyrus (HG; Hackett et al. 2001). The HG is surrounded by belt areas that extend anteriorly to the planum polare and caudally to the planum temporale (PT). Finally, parabelt areas lie beyond the belt regions, eventually emerging on the superior temporal gyrus (Hackett 2015). Nonhuman primates, carnivores, and rodents share an analogous core-belt-parabelt organization (Fig. 12.1b; Hackett 2015). In addition to the classical auditory cortex, a number of further cortical areas are critical for auditory processing and sound localization in particular, including parietal and frontal regions such as the dorsolateral prefrontal cortex and posterior parietal cortex (see Sect. 12.3.2).

The auditory cortex is divided into six layers that differ in their neuronal morphology and connectivity, and potentially, in their contribution to spatial hearing. Inputs from the ventral division of the thalamus terminate in layers III and IV. From there, information is relayed to the supragranular layers (I and II) where information is combined with inputs from other cortical fields. Finally, infragranular layers (V and VI) receive input from layers I to IV and act as the major output of auditory cortex. Within the infragranular layers, large pyramidal neurons send long-range connections across the cortex, thalamus, midbrain, and striatum (Linden and Schreiner 2003; Hackett 2015).

Table 12.1 List of abbreviations

Abbreviation	Definition
A2	Secondary auditory cortex
AAF	Anterior auditory field
AES	Anterior ectosylvian sulcus
AGF	Anterior gaze fields
dlPFC	Dorsolateral prefrontal cortex
DZ	Dorsal zone
EEG	Electroencephalography
FEF	Frontal eye field
fMRI	Functional magnetic resonance imaging
GABA	γ-Aminobutyric acid
HG	Heschl's gyrus
IC	Inferior colliculus
ILD	Interaural level difference
IPD	Interaural phase difference
IPL	Inferior parietal lobule
ITD	Interaural time difference
LIP	Lateral intraparietal cortex
MEG	Magenetoencephalography
MGB	Medial geniculate body
mPFC	Medial prefrontal cortex
OT	Optic tectum
PAF	Posterior auditory field
PET	Positron emission tomography
PFC	Prefrontal cortex
PPC	Posterior parietal cortex
PRR	Parietal reach region
PT	Planum temporale
SOC	Superior olivary complex
SPL	Superior parietal lobule
SRF	Spatial receptive field
STG	Superior temporal gyrus
VAS	Virtual acoustic space
VIP	Ventral intraparietal cortex
vlPFC	Ventrolateral prefrontal cortex
vPM	Ventral premotor area

12.2.1 Cortical Areas Necessary for Sound Localization

To identify cortical areas with a causal role in spatial hearing, deficits in sound localization must be identified either through targeted inactivation using animal models in the laboratory or in studies of patients with brain injuries. Profound and lasting sound localization deficits in primates (Heffner and Heffner 1990) and

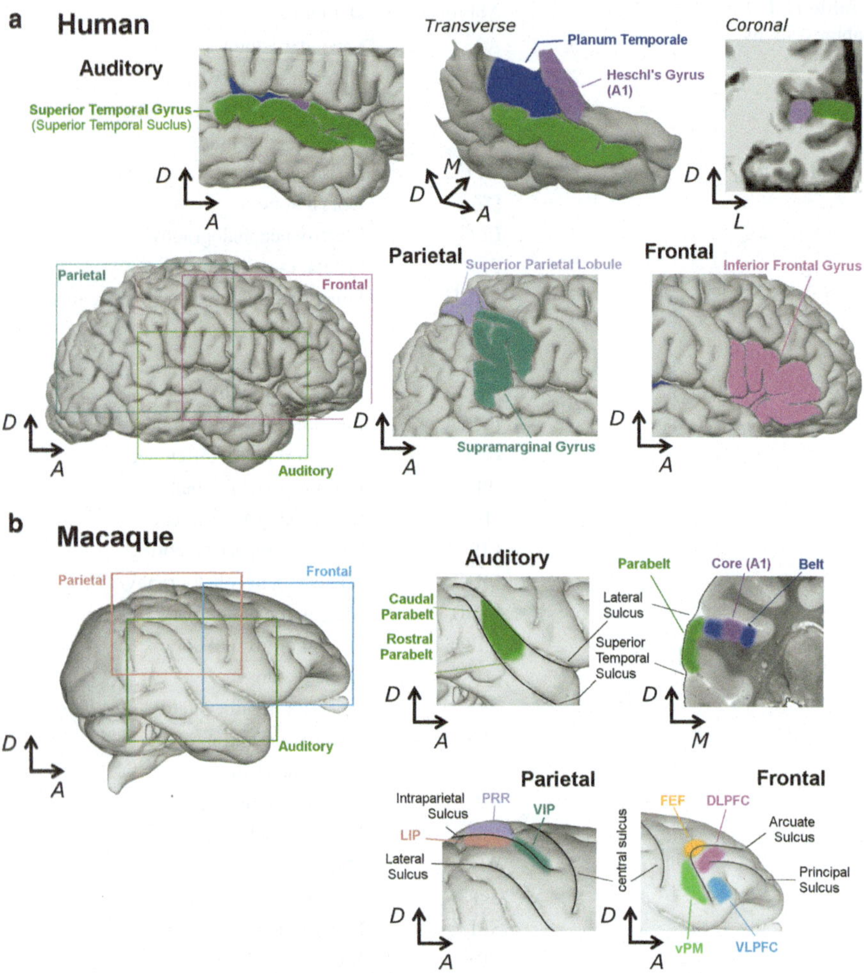

Fig. 12.1 Anatomical organization of auditory cortex. Human (**a**) and macaque (**b**) brains. Insets: lateral views of auditory cortical (lateral *[left]*; transverse [*center*], and coronal [*right*] views in A, *top row*; lateral [*left*] and coronal [*right*] views in B, *top row*), parietal, and frontal (A and B, *bottom rows*) cortical areas associated with sound localization and binaural cue processing. Specific cortical areas are highlighted in color. D, dorsal; A, anterior; M, medial; L, lateral; A1, primary auditory cortex; LIP, lateral intraparietal sulcus; PRR, parietal reach region; VIP, ventral intraparietal area; FEF, frontal eye field; DLPFC, dorsolateral prefrontal cortex; vPM, ventral premotor area; VLPFC, ventrolateral prefrontal cortex. Note that the precise boundaries for the human auditory cortex are unclear and the example shown represents one potential organization. The example provided by the macaque brain is broadly representative of auditory cortical organization across primates, carnivores, and rodents. Human brain images are modified from Society for Neuroscience (2017); macaque data from Rohlfing et al. (2012)

carnivores (Whitfield et al. 1972) result from lesioning auditory cortex. Later studies using thermal or pharmacological methods to reversibly silence auditory cortex have also shown that primary and specific non-primary auditory cortical fields are critical for normal localization in both horizontal (Malhotra et al. 2004; Wood et al. 2017) and vertical (Bizley et al. 2007) dimensions. Here, thermal inactivation is achieved by passing chilled ethanol through a steel "cooling loop" on the cortical surface to reduce the temperature of the neurons, temporarily disabling their spiking ability. Pharmacological inactivation may involve the application of muscimol to the brain; muscimol mimics the action of the inhibitory neurotransmitter γ-aminobutyric acid (GABA) and therefore silences the region in which it has been applied. Lesion, thermal, and pharmacological inactivation data all point to a role for auditory cortex in integrating binaural and spectral cues. Animals with primary auditory cortex inactivation show decreased localization accuracy, with front-back confusions accounting for a large proportion of the errors (Nodal et al. 2010). Such front-back confusions are thought to result from the ambiguity known as the "cone of confusion," whereby sound sources at multiple locations generate the same binaural cues (see Hartmann, Chap. 2; Takahashi, Kettler, Keller, and Bala, Chap. 4). Resolving front-back confusions relies on integrating binaural and monaural spectral cues. Therefore, the inability of animals with auditory cortical inactivation to resolve these front-back confusions suggests that integration of different localization cues relies on the auditory cortex. Beyond auditory cortex, a network of brain regions process sound and, in particular, sound location. Although laboratory studies that perturb activity in these areas during sound localization are limited, sound localization deficits have also been reported after lesions of the prefrontal cortex in monkeys (Wegener 1973).

Spatial listening deficits in human patients are also associated with damage to the frontal, inferior parietal, and superior temporal areas (Clarke et al. 2002), supporting the conclusion from animal studies that a cortical network contributes to sound localization. In particular, damage to the parietal cortex produces spatial deficits, including impairments in the judgment of the relative location of two sounds sources (Griffiths et al. 1997; Pavani et al. 2001) and distorted sensitivity to interaural timing cues (Tanaka et al. 1999).

The study of behavioral deficits after stroke and other forms of traumatic brain injury in patients has played a vital role in developing models of neural processing and cognitive function. Such studies offer vital insights into how the human brain works; by studying the consequences of long-term neurological damage in small clinical populations, the brain regions critical for a given function may be elucidated. However, recovery of function after trauma may mask functional links between the brain and behavior in listeners without brain damage. To complement studies in patients, approaches such as transcranial stimulation allow the study of causal mechanisms in large cohorts of healthy subjects over a relatively rapid time course, where the perturbation is under experimenter control. Using such approaches, investigators have demonstrated that repetitive focal transcranial magnetic stimulation of the posterior parietal cortex (PPC) can induce systematic errors in a subject's estimation of sound azimuth and elevation. In particular, stimulation of the right PPC shifts a subject's estimates of sound location down in elevation and to the left (i.e., towards the contralateral hemisphere; Lewald et al. 2004). Broadly speaking, therefore, studies in patients and healthy listeners highlight a role for cerebral cortex in spatial perception.

12.2.2 *How Is Sound Location Represented by Neurons in Auditory Cortex?*

Typically, experiments investigating the representation of sound location in auditory cortex use either a free-field stimulus presentation from a loudspeaker, a series of loudspeakers, or a dichotic presentation over headphones. Dichotic presentation introduces spatial cues by altering the relative timing or intensity of the signals in each ear to create interaural timing differences (ITDs) or interaural level differences (ILDs; see Hartmann, Chap. 2). An alternative is to present sounds over headphones using virtual acoustic space (VAS) stimuli that are constructed from recordings of free-field sounds made with a microphone placed in the ear canal. VAS recreates both the binaural and spectral cues present in real-world sounds and mostly reproduces a perception of a localized sound source in the world (Wightman and Kistler 1989; Middlebrooks 1999). This is in contrast to studies that only manipulate binaural cues over headphones where sounds appear to move left to right within the head (i.e., sounds differ in lateralization). As well as offering practical advantages by allowing stimulus presentation over headphones, VAS offers the potential to independently manipulate (Wightman and Kistler 1992) or eliminate (Macpherson and Middlebrooks 2002; Campbell et al. 2006a) variation in particular spatial cues. In contrast, free-field stimulation can only constrain the type of acoustic cue presented by limiting the frequency spectrum (e.g., Wood et al. 2019).

12.2.2.1 Encoding Interaural Level Differences

Historically, one of the earliest phenomena to be investigated at the neuronal level within auditory cortex was the difference in responses of the neurons to sounds presented over headphones (e.g., Imig and Adrian 1977). In these animal studies, neural responses to sounds coming from one ear were compared with neural responses to sounds delivered to both ears. Neurons in auditory cortex were then classified based on whether their responses to monaural and binaural sounds were excitatory (E) or inhibitory (I). "EE" neurons produce excitatory responses to sounds at both ears; "EI" neurons produce excitatory and inhibitory responses to sounds at the contralateral and ipsilateral ear, respectively; and "EO" neurons only produce excitatory responses to sounds at the contralateral ear and do not respond to sounds at the ipsilateral ear. Over time, the stimulus parameter space has increased such that auditory cortical neurons can be tested with sounds containing the entire physiological range of ILDs (±35 dB in humans [Middlebrooks et al. 1989]; ±25 dB in ferrets [Carlile 1990]) at a range of sound levels or "average binaural levels". Expanding the stimulus space has led to a concomitant increase in the complexity of reported receptive fields, with individual units tuned (i.e., most strongly responsive) to different ILDs. The majority of units in primary auditory cortex are tuned to ILDs associated with contralateral sounds, although across the neural population, the full range of physiological ILDs are represented (Campbell et al. 2006b). In some cases, neural responses have been reported to have a topographic distribution, with tuning to many different ILDs observed along isofrequency contours (Nakamoto et al. 2004). Two-photon imaging studies that allow visualization of neural activity through the use of voltage- or ion-sensitive dyes have provided an unprecedented ability to observe the spatial structure in the representation of ILDs, but (at least in mice) have failed to reveal any systematic organization of ILD preference across primary auditory cortex (Panniello et al. 2018).

12.2.2.2 Encoding Interaural Time Differences

Although more studies have focused on the processing of ILDs than ITDs, many auditory cortical neurons are tuned to ITDs presented through earphones, as first shown by Brugge et al. (1969). ITDs and ILDs are dependent on sound frequency, and studies investigating the physiological representation of ITDs are probably less numerous, in part, because common animal models such as the mouse do not hear the low-frequency sounds that convey ITDs (see Owrutsky, Benichoux, and Tollin, Chap. 5). Although midbrain responses to sounds varying in ITD (or interaural phase difference [IPD] for ongoing tones) are strongest to ITDs or IPDs corresponding to contralateral source locations (Shackleton et al. 2003), cortical tuning appears more heterogeneous. Most studies of cortical neurons report some degree of contralateral bias in tuning (Reale and Brugge 1990; Scott et al. 2009), although in gerbils (*Meriones unguiculatus*), each hemisphere of primary auditory cortex contains neurons tuned to ITDs that correspond to locations in both the contralateral and ipsilateral sides of space (Belliveau et al. 2014). This study recorded predominantly in layer V and raises the possibility that ipsilateral tuning might emerge through intracortical processing. Nevertheless, in humans, electroencephalography (EEG) and magnetoencephalography (MEG) adaptation studies suggest a strong bias in the neural representation for ITDs corresponding to contralateral sound locations (Magezi and Krumbholz 2010; Salminen et al. 2010). Investigations using functional magnetic resonance imaging (fMRI) also suggest that a bias of neural activity for contralateral ITDs is present but is weaker than for contralateral ILDs (Higgins et al. 2017b).

12.2.3 Spatial Receptive Fields in Auditory Cortex

In everyday listening, broadband sounds with energy at many frequencies offer multiple, redundant cues (e.g., ITDs and ILDs) for computing sound location. The spatial tuning of individual neurons can be measured by varying the position of sounds (e.g., noise bursts) while recording the spiking response of neurons with microelectrodes. The sound source can be moved in space either by presenting free-field sounds from a speaker array or by presenting sounds in a VAS. A spatial receptive field (SRF) can then be constructed by comparing the strength of the response of the neuron (i.e., the number of action potentials elicited) to sounds at different positions (Fig.12.2a; Brugge et al. 1969; Eisenman 1974). SRFs in primary auditory cortex are relatively broad and typically tuned to contralateral locations, with a minority of neurons tuned to the ipsilateral space or the midline (Harrington et al. 2008).

Spatial tuning is refined as sound signals ascend the auditory pathway from primary to non-primary auditory cortex. When presented with complex sounds varying in position or identity, neurons in the caudolateral belt show strong tuning to sound location but weak tuning to identity, whereas neurons in the anterolateral belt are sensitive to sound identity but are weakly modulated by location (Tian et al. 2001). Spatial tuning of neurons in caudolateral belt regions of the secondary auditory cortex is enhanced relative to primary auditory cortex, and only in the caudolateral belt do firing rates of populations of neurons convey sufficient spatial information to account for sound localization ability (Miller and Recanzone 2009).

Further evidence that the processing of sound location and identity are separated within the auditory cortex is provided by reversible inactivation studies in animals (Lomber and Malhotra 2008). Transiently cooling some auditory cortical fields (the primary auditory cortex, posterior auditory field [PAF], and anterior ectosylvian sulcus [AES]) of the domestic cat during a sound localization task results in behavioral deficits. In contrast, inactivation of other areas (the anterior auditory field [AAF] and secondary auditory field [A2]) leaves performance intact (Malhotra et al. 2004). When cooling either anterior or posterior cortical fields during a localization task or a (non-spatial) temporal pattern discrimination task, a double dissociation was described: cooling the posterior (but not anterior) auditory cortex resulted in only impaired localization, whereas cooling the anterior (but not posterior) cortex impaired only temporal discrimination (Lomber and Malhotra 2008). Thus, spatial tuning is refined from the primary to the non-primary auditory cortex, with specific subfields of secondary auditory cortex playing a key role.

12.2.4 Spatial Receptive Fields During Behavior

Studies of the neural representation of space in auditory cortex have typically recorded neural responses in anesthetized or passively listening animals. This leaves open the possibility that the broad spatial tuning often observed in individual neurons, which stands in contrast to the fine discrimination of sound location

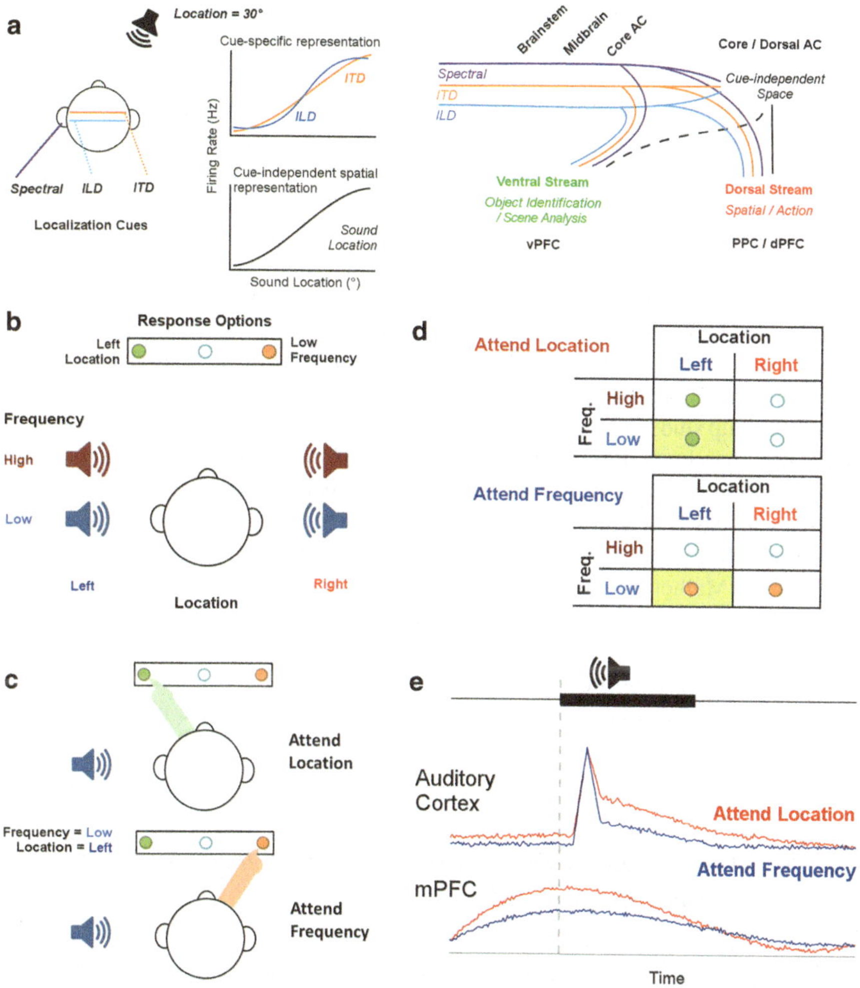

Fig. 12.2 Attentional modulation of cortical processing. (**a**), *left:* a sound source at 30° generates interaural timing differences (ITDs), interaural level differences (ILDs), and monaural spectral cues. Within the brain, neurons may be sensitive to specific cues (e.g., ITD or ILD) that covary with spatial position (*top center*) or may integrate these cues into a cue-independent representation of sound location (*bottom center*). *Right:* schematic showing the flow of localization cue information through the central auditory system and into dorsal and ventral processing streams. Within auditory cortex (AC), some neurons encode a cue-independent representation of space. vPFC, ventral prefrontal cortex; PPC, posterior parietal cortex; dPFC, dorsal prefrontal cortex. (**b–d**) studying attentional modulation using task switching. (**b**) example experimental design in which subjects are presented with high- or low-frequency sounds from either the left or right. On different trials, subjects are instructed to report either the location or frequency of the sound (**c**) and to respond differently to the same stimulus depending on attentional condition (**d**). *Yellow cell,* example stimulus shown in **c**. (**e**) attention modulates neural activity across the central auditory system before and during sound onset. Shown are hypothetical responses of two neurons recorded from auditory cortex and medial prefrontal cortex (mPFC) to the same sounds depending on whether the animal is required to report the location or frequency of the sound

behaviorally (see Hartmann, Chap. 2), might reflect the limits of the anesthetized state rather than the true capabilities of the auditory system. To address this issue, Lee and Middlebrooks (2011) recorded spatial receptive fields in primary auditory cortex of cats that were either listening passively, performing a non-spatial task, or performing a task that required attention toward sound location. Spatial tuning of individual neurons narrowed when the animals actively listened to sounds, with the greatest contrast seen between passive listening and engagement in the localization task. Thus, even at the earliest stages of auditory cortical processing, ongoing task demands shape the way that neurons process spatial information. One might expect such changes to arise from enhanced selectivity of spatially tuned neurons for their preferred sound locations; in contrast, improvements in spatial tuning came principally from omnidirectional units (i.e., those with little to no spatial preference) becoming broadly tuned to contralateral space.

A follow-up study demonstrated that the spatial tuning of neurons in non-primary fields also sharpened during behavior (Lee and Middlebrooks 2013). Across subfields, neurons in the dorsal zone (area DZ) appeared mostly tuned for midline sounds; whereas PAF neurons were mostly tuned for more lateral locations, including those behind the cat. Both these anatomical regions are necessary for sound localization (Malhotra et al. 2004) and so might play complementary roles in spatial hearing. Receptive fields of spatially tuned units in primary and non-primary auditory cortices were not dramatically sharper during task performance, and single-unit tuning was still substantially broader than behavioral thresholds. This suggests that behavioral sensitivity to sound location is most likely to arise from the activity of neuronal populations.

12.2.5 How Is Spatial Information Extracted from Auditory Cortical Responses?

If listeners can accurately localize sounds with a precision of a few degrees, whereas the SRFs in auditory cortex are broadly tuned to the contralateral space, then information must be combined across populations of neurons to yield a precise neural representation of sound location (Middlebrooks et al. 1998; Miller and Recanzone 2009). Inspired by studies of binaural cue encoding in the midbrain (see Owrutsky, Benichoux, and Tollin, Chap. 5), at least two models have been proposed to account for integration of spatial information across neurons. A *distributed model* would represent the location of a sound source through the activation of a particular neural subpopulation, with sounds originating from different locations eliciting distinct patterns of neural activity across the cortical population (Stecker and Middlebrooks 2003; Day and Delgutte 2013). In contrast, a two-channel or opponency model proposes that the relative activity of two broadly tuned neural populations determines the perceived location (Stecker et al. 2005b).

Recordings from single neurons provide support for a distributed model. Although the SRFs are typically tuned to the contralateral space, their best positions are distributed throughout space rather than being restricted to two contralateral positions as predicted by the two-channel model (Woods et al. 2006; Wood et al. 2019). Neural activity can also be decoded using two-channel models (Stecker et al. 2005a), but such models are outperformed by distributed models (Day and Delgutte 2013; Wood et al. 2019).

In contrast to the support for distributed systems provided by single-neuron recordings, functional imaging studies in primates (including humans) support a two-channel model when neural activity is measured with high temporal and low spatial resolution using MEG (Salminen et al. 2009) or at higher spatial resolution and lower temporal resolution using fMRI. These fMRI investigations reveal a strong bias for contralateral locations indicative of a two-channel representation, which is observed across cortical fields and hemispheres in both macaques (Ortiz-Rios et al. 2017) and humans (Higgins et al. 2017b).

The conflicting support for the two-channel and distributed models, from primate and nonprimate studies, respectively, suggests a difference across species or methodologies. Most neural recording studies report that individual neurons have strong contralateral biases, but there is heterogeneous spatial tuning that shows little systematic organization of sound location across the cortical surface (also known as a topographic map, observable in primary visual and somatosensory cortices). Therefore, it is possible that measuring signals over the thousands of neurons within a voxel averages out variability in individual SRFs to produce the appearance of a two-channel representation, even when distributed codes exist. This would be particularly likely if each hemisphere represents a contralateral space using a distributed code (Wood et al. 2019). Last, these models only attempt to account for azimuthal localization when a sound can be accurately pinpointed in at least two dimensions (Blauert 1996). In the future, expanded models of spatial tuning that account for our behavioral ability to perform multidimensional localization will be required.

12.3 Spatial Processing in Brain Networks

12.3.1 The Dual-Stream Theory

Insights from animal studies (see Sect. 12.2) suggest that spatial processing is refined in ascending pathways from the primary to the non-primary auditory cortex. This implies that spatial processing may be further enhanced higher up the auditory system to ultimately give rise to spatial perception. Inspired by studies in vision (Ungerleider and Mishkin 1982), the dual-stream model formalizes this idea, proposing that information about sound location is processed independently to the non-spatial features (such as pitch and timbre) that allow us to identify a sound source.

This has led to the two streams being described as the "what" and "where" pathways, with significant discussion about the nature and independence of spatial and non-spatial processing. Within the model, spatial and non-spatial information streams are processed in parallel pathways originating in auditory cortex, with the spatial stream then extending to parietal and prefrontal cortices (Fig. 12.2a; Rauschecker and Tian 2000; Rauschecker and Scott 2009).

Evidence for the dual-stream model comes from converging anatomical, neurophysiological, and neuropsychological studies. In addition to the early evidence that spatial processing was refined in the primate caudal auditory cortex (see Sect. 12.2.4), key evidence for this model was provided by anatomical experiments. Romanski et al. (1999) injected neural tracers into physiologically identified regions of the belt auditory cortex in macaque monkeys. They found that the anterior belt cortex was reciprocally connected with several areas of frontal cortex, including ventral prefrontal regions involved in processing non-spatial attributes of sound. In contrast, the caudal belt was mainly connected with the caudal principal sulcus and frontal eye fields, both areas involved in spatial processing. Thus distinct pathways from auditory to frontal cortex were found, indicating that information about sound travels in parallel spatial and non-spatial processing streams.

Functional imaging studies in humans also support the dual-stream model. Posterior auditory cortical fields are consistently activated during spatial processing in humans (Maeder et al. 2001) and only in these areas is activity correlated with spatial perception (Higgins et al. 2017a). Additional evidence for the dual-stream model has included reports of functionally independent brain networks for spatial and non-spatial processing (Ahveninen et al. 2006; Retsa et al. 2018). Ahveninen et al. (2006) compared neural responses with sound location and identity when subjects attended to either the spatial or non-spatial features of sound. Consistent with dual-stream models, the posterior superior temporal gyrus (STG) showed greater sensitivity to changing sound location, whereas the anterior STG and planum polare were more sensitive to sound identity. When listeners attended to changes in sound location (that were present in all stimuli but could be attended or ignored), activity was enhanced in posterior but not anterior regions. In contrast, attention to sound identity modulated activity in anterior but not posterior areas.

12.3.2 Cortical Processing of Space Outside Auditory Cortex

The spatial and non-spatial streams originating in auditory cortex (see Sect. 12.3.1) each target distinct brain networks that include the frontal and parietal cortices. In humans and other primates, the frontal cortex is divided into the prefrontal cortex (PFC) and the precentral gyrus that contains the motor cortex. The PFC has several subregions including the dorsolateral (dlPFC) and ventrolateral (vlPFC) PFCs as well as the frontal eye fields (FEFs; Fig. 12.1). The dlPFC and vlPFC receive distinct inputs from the spatial and non-spatial auditory pathways (Romanski et al. 1999), although there are also connections between the ventrolateral and dorsolateral regions within the PFC (Petrides and Pandya 2001). Similarly, the motor cortex has been divided into the primary motor cortex, the supplementary motor area, and the premotor cortex, all of which are important in coordinating a subject's behavioral response to sounds, with the ventral premotor area (vPM) sensitive to sound location (Graziano et al. 1999).

In PPC, several areas have been associated with spatial hearing, including the lateral intraparietal sulcus (LIP) and the parietal reach region (PRR). The parietal cortex receives auditory input via connections from the caudal temporoparietal cortex to the ventral intraparietal area (VIP) and the ventral LIP (Fig. 12.1; Lewis and Van Essen 2000; Rozzi et al. 2006). Auditory signals processed within parietal areas may then be sent to the prefrontal cortex via direct connections from the PPC (Pandya and Kuypers 1969; Divac et al. 1977).

Human neuroimaging and neurophysiology complement the evidence from animal work for a role for the prefrontal and parietal cortices in spatial auditory processing. Although studies of individual neurons in animals are often limited to a single brain region, neuroimaging allows investigators to study neural processing across the entire brain. Thus, approaches like positron emission tomography (PET) imaging and fMRI revealed that spatial auditory processing activates not only specific regions of auditory cortex but also widespread networks, including other cortical areas.

Using PET imaging while the subjects localized sounds, Bushara et al. (1999) found bilateral activation of the superior and inferior parietal lobules, bilateral middle frontal gyrus, and right medial frontal gyrus. When localizing free-field stimuli, PET activation in the right inferior parietal cortex specifically was predictive of sound localization performance (Zatorre et al. 2002). Similar results were found using fMRI, when the subjects detected changes in the virtual location of sounds (Alain et al. 2001) or identified sound location using ITDs (Maeder et al. 2001). An important question is whether the activated areas are involved in hearing generally or in spatial hearing specifically. Maeder and colleagues addressed this by contrasting activation when subjects localized or identified the same test sounds. Localization produced greater activation of the inferior parietal lobule, posterior middle frontal gyri, and inferior frontal gyri, whereas identification activated the middle temporal gyrus, precuneus, and left inferior frontal gyrus. Thus separate dorsal and ventral auditory pathways appear to be specialized for the analysis of sound location and identity, respectively.

One constraint of fMRI and PET imaging is the relatively slow temporal resolution of the measurements, which limits insight into the time course of auditory processing. This is particularly problematic for studies of audition, where signals unfold rapidly over time. Other approaches such as EEG and MEG can track neural signals with millisecond resolution, although it is much harder to pinpoint the source of these signals. These techniques have also shown the importance of the parietal and frontal cortices. EEG studies where subjects listened passively to noise bursts found that binaural cues modulated activity in the temporal-parietal and inferior frontal cortices (Tardif et al. 2006). Using MEG, Kaiser et al. (2000, 2002) showed that the posterior parietotemporal and supratemporal areas respond to shifts in sound location. These effects differed across hemispheres, with the right cortex responding to shifts in both the ipsilateral and contralateral directions, whereas corresponding regions in the left hemisphere were responsive only to contralateral shifts in sound location (Kaiser et al. 2000). This observation fits with a model of spatial attention in which the right hemisphere controls shifts in attentional focus in both directions, whereas the left hemisphere only controls shifts to the right (Mesulam 1981). Other studies using similar approaches, but focusing on early sensory responses, have found an opposite asymmetry (Salminen et al. 2010), so it remains unclear to what extent auditory spatial processing is asymmetrically distributed in humans. What is clear across these studies, however, is that a broad network of parietal and frontal areas is active during sound localization.

12.3.2.1 Frontal Cortex

A role for the PFC in sound localization was first indicated by lesion studies that demonstrated deficits in task performance following bilateral PFC lesions (Wegener 1973). Neurophysiological recordings in PFC of awake monkeys also demonstrated that neurons in the dlPFC were responsive to sounds and often tuned to contralateral locations (Azuma and Suzuki 1984). Sound-driven responses in dlPFC occurred later in time than in auditory cortex and were less tightly locked to the onset of sounds. Contralateral tuning suggests that auditory cortex and dlPFC may share functional properties; however, later results indicated such similarities were limited (Vaadia et al. 1986; Vaadia 1989). In dlPFC, relatively few units responded to unisensory auditory stimuli, and those that were sound responsive were also highly sensitive to the animal's behavior. When monkeys were engaged in a task (either sound detection or localization), a larger proportion of PFC units were responsive. Although the largest proportion of responsive cells were reported during sound localization, these cells werè also active during a visual localization task. This suggests that behaviors that required analysis of stimulus position in either modality engage these neurons rather than only localization of sounds. In contrast to the dlPFC, most units in the superior temporal plane around auditory cortex were sound responsive and relatively few of these cells were modulated by behavior (Benson et al. 1981).

Neuroimaging results in humans also indicate that spatial behaviors engage frontal cortex: both visual and auditory localization tasks activate medial frontal cortex and IPL bilaterally (Bushara et al. 1999). In contrast, regions such as the superior parietal lobule (SPL) and middle frontal gyrus were activated only by visual or auditory localization. Together, these results indicate that these brain regions use cues that are relevant to achieve the subject's goal rather than any fixed physical features or modality. In the case of sound localization, binaural cues are relevant for behavior and thus selected for further representation. However, this selection is flexible, and binaural cue processing can be dynamically engaged or ignored when necessary.

Neurons encode selection rules for directing behavior toward task-relevant spatial information within the rodent medial prefrontal cortex (mPFC). For example,

Rodgers and DeWeese (2014) trained rats to flexibly switch between two tasks in which subjects reported the lateralization (i.e., whether sounds came from the left or right) or frequency of sounds (Fig. 12.2b). Stimuli contained spatial and non-spatial information on all trials, but subjects switched between decisions using (in the case of lateralization) or ignoring (in frequency discrimination) binaural cues (Fig. 12.2c, d). In mPFC (and auditory cortex), neurons showed task-dependent modulation of baseline neural activity in the anticipatory period preceding stimulus onset as well as sound-evoked activity during stimulus presentation. Thus, an example mPFC neuron might be more active before and during sound presentation when the subjects used binaural cues to guide behavior than when those cues were ignored (Fig. 12.2e).

The idea that prefrontal cortex combines binaural cues with the goals of the listening subject is also supported by causal evidence showing that disruption of mPFC impairs flexible switching between using and ignoring binaural cues. For example, electrical stimulation of mPFC impaired alternation between lateralization and frequency discrimination tasks and thus the rats' ability to select binaural cues for behavior (Rodgers and DeWeese 2014). Similarly, optogenetic stimulation of the prelimbic cortex (an area thought comparable to the primate dlPFC) prevents mice from flexibly switching between lateralization of auditory and visual targets within an audiovisual stimulus (Wimmer et al. 2015). Intriguingly, optogenetic stimulation was only effective in the anticipatory window before sound onset, suggesting that subjects' use of binaural cues is determined *before* sounds are experienced and that behavioral priorities may shape auditory processing prospectively. Mice have relatively poor spatial acuity for sounds compared with humans (Behrens and Klump 2016) and lack sensitivity to low-sound frequencies that we use to access ITDs. In future research, it will thus be critical to test how cortical modulation of spatial processing occurs in other species with low-frequency hearing and high spatial acuity.

Findings from human neuroimaging mirror results from rodent studies. Expectation of a sound produces enhanced activity in the frontal and parietal areas, even when stimuli are not physically presented (Smith et al. 2010). This suggests that incoming sensory input is indeed processed within a broader temporal context in both human and nonhuman animals.

12.3.2.2 Parietal Cortex

Within the parietal cortex, LIP neurons are responsive to sound and encode information about the location of a sound source (Mazzoni et al. 1996; Stricanne et al. 1996). Spatially sensitive auditory neurons have also been reported in the VIP, where many cells are sensitive to both sounds and visual stimuli and auditory and visual receptive fields overlap (Schlack et al. 2005). Like nonhuman primates, rodents also have neurons in the PPC that encode sound location (Nakamura 1999).

In the parietal cortex, active behavioral performance is not required to observe spatially informative LIP responses; however, auditory responses of LIP neurons are behaviorally dependent (Linden et al. 1999). Specifically, auditory responses are stronger when performing behavioral tasks such as memory-guided saccades than during fixation of static locations. The encoding of spatial information about sound locations by LIP neurons was previously thought to depend on training animals in auditory saccade tasks, because LIP neurons recoded in naive subjects were uninformative about sound location (Grunewald et al. 1999). Subsequent analysis of the visual context in which sounds were presented revealed that auditory responses in LIP can arise in the absence of visual cues, in animals performing fixation tasks (Gifford and Cohen 2004). Such responses do not require specific training with sounds or execution of saccades, indicating that spatial processing is a natural part of LIP function.

The human parietal cortex also engages in auditory spatial processing in a task-dependent manner. For example, Michalka et al. (2016) showed that regions within the SPL are recruited when subjects engaged in spatial but not temporal short-term memory. Here, subjects listened to two consecutive sequences of four complex tones and were asked to report whether the sequences matched. In all trials, the tones came from four distinct locations (determined by ITD cues) and were presented with variable intertone intervals. Both the locations and/or timing could vary between the first (probe) and second (target) sequence. In the spatial task, the subjects were asked whether the sequence of locations in the probe and target sequences matched, and in the temporal task, the subjects reported whether the timing sequence was the same or different. In both task conditions, lateral and anterior regions of the intraparietal sulcus were activated; however, only when attending to sound location was the SPL activated. In another study, Alain et al. (2010) found that inferior parietal lobule (IPL) activation was associated with demand for memory of the sound locations; specifically, activity increased as the subjects were required to remember sound locations across longer sequences. Together, these results suggest that memory for sound location involves neural processing in the human parietal cortex.

12.3.3 Beyond "What Versus Where"

As the theory that sound location and sound identity were analyzed in parallel pathways across the central auditory system (Rauschecker and Tian 2000; Rauschecker and Scott 2009) became more widely accepted, evidence emerged suggesting that a strictly binary interpretation of this view was oversimplified. For example, neurophysiological studies in animals demonstrated spatial modulation of neurons in vlPFC, a part of the putative "what" pathway (Cohen et al. 2004). Neurons in the vlPFC are more spatially selective than neurons in the anterolateral belt region of the auditory cortex from which it receives input, suggesting that the vlPFC receives spatial information from regions other than rostral auditory cortex. These areas may include regions within the "where" pathway such as the dlPFC, to which vlPFC is interconnected (Petrides and Pandya 2001). Furthermore, vlPFC neurons have spatial tuning comparable with LIP neurons associated with the where pathway, further suggesting that parallel pathways may interact and potentially converge. Thus, models of spatial tuning have evolved to emphasize the interaction between the dorsal and ventral pathways (Rauschecker and Scott 2009).

Performing a meta-analysis across fMRI studies, Arnott et al. (2004) found that areas such as the IPL and superior frontal cortex are more often active in tasks where the subjects were asked to make spatial judgements about sounds. However, non-spatial tasks still activate areas such as the IPL in 41% (11/27) of the examined studies and, in rare cases, spatial processing engaged regions of the ventral identification pathway such as the inferior frontal cortex. Neural recordings have also indicated that LIP is not exclusively specialized for the processing of sound location, and LIP neurons can encode non-spatial spectrotemporal features relevant for identifying sounds (Gifford and Cohen 2005). In humans, sound recognition produces greater bilateral activation in the precuneus within the SPL than sound localization (Maeder et al. 2001), again bringing into question how strongly spatial and non-spatial functions are separated. Thus, spatial processing beyond the auditory cortex is not exclusively limited to the dorsal pathway, nor is non-spatial processing to the ventral pathway.

Even within non-primary auditory cortex, the division of labor into discrete processing streams might not be as simple as first proposed. For example, in ferrets (*Mustela putorius furo*), neurons in primary and secondary areas are sensitive to both spatial and non-spatial sound attributes (Bizley et al. 2009; Town et al. 2018). In humans, EEG studies also reveal position-linked object identity information in the ventral processing stream (Bourquin et al. 2013). Finally, in cats, inactivation of the caudal region AES leads to sound localization deficits, whereas inactivation of the rostral region A2 does not (Malhotra et al. 2004). However, neurons in both areas show comparable spatial tuning (Middlebrooks et al. 1998), indicating that information about sound location available in neural responses may not always be causally relevant for behavior.

12.3.3.1 Using Binaural Cues in Auditory Scene Analysis

If spatial information is isolated from the neural pathways responsible for the non-spatial processing that underlies sound identification, why then do spatially tuned neural responses exist in the ventral processing stream (e.g., vlPFC and anterior regions of the secondary auditory cortex)? One possibility is that spatial information allows sound sources to be unmixed during the analysis of auditory scenes. The organization of incoming sound mixtures into perceptual "objects" is thought to play a key role in the auditory cortex (Bizley and Cohen 2013). Consistent with this idea, similar neural populations are activated in the early auditory cortex (HG and PT) when listeners separate a sound mixture into distinct perceptual streams using either pitch or binaural cues (Schadwinkel and Gutschalk 2010). Neuropsychological studies following patients with stroke damage also show that listeners with a complete loss of sound localization abilities can still use spatial information to segregate sound mixtures into distinct objects (Spierer et al. 2007). Finally, there appears to be specialization within the right planum temporale for sound localization in complex acoustic environments with multiple distracters (Zundorf et al. 2014). Lesions to this area impair only localization of sound sources within a mixture and not of a single source by itself, suggesting that this area is critical for segregating sounds based on space and their subsequent localization. Thus, encoding of spatial information may be important for behaviors that do not require subjects to explicitly report sound location.

A common limitation of many spatial coding studies is the use of isolated sound bursts presented in otherwise silent environments. This contrasts starkly with one's daily acoustic experience, where multiple sound sources compete for our attention. Studies employing competing sound sources have revealed that the spatial tuning of neurons in the auditory cortex can be much sharper than previously observed with single sources: When cells in field L, an analogue of the mammalian primary auditory cortex in the zebra finch (*Taeniopygia guttata*), are tested with single tokens of birdsong, spiking activity locks to the temporal modulations of the song irrespective of its spatial location. When competing tokens are introduced at a second location, cells become spatially selective in their responses (Maddox et al. 2012). Similar findings have also been observed in the ferret primary auditory cortex: When competing sound sources are introduced, spatial receptive fields sharpen, becoming both narrower and more strongly modulated by sound location (Wood et al. 2019).

In another study, Middlebrooks and Onsan (2012) presented sequences of sound bursts from multiple sound sources that could be perceived as either a single regular stream of sound when bursts were colocated or two streams with distinct rhythms when sources were separated by only a few degrees. Neural responses in auditory cortex to the same stimuli showed broad spatial tuning when the stimuli were colocated. However, for separated sources, neurons selectively responded to the source closest to their preferred location (Middlebrooks and Bremen 2013). These effects were only weakly present at the earlier stages of the auditory pathway, arguing in favor of an origin within auditory cortex (Yao et al. 2015). Together, these studies illustrate the importance of stimulus competition in measuring spatial tuning properties that might otherwise be missed with single suprathreshold stimuli presented in isolation and the broader role of binaural processing in auditory scene analysis.

12.3.3.2 Perception-Action Pathway

In vision, the theoretical separation between spatial and identity processing (Ungerleider and Mishkin 1982) has evolved into a distinction between perception and action (Milner and Goodale 2006). This view emphasizes the importance of behavioral output, with the ventral and dorsal visual streams responsible for perceiving stimuli (e.g., the mosquito flying nearby) and taking action (e.g., swatting it away with one's hand), respectively.

In hearing, the perception-action distinction has also gained prominence (Belin and Zatorre 2000; Rauschecker and Scott 2009), where areas such as the parietal cortex are thought to integrate predicted changes in auditory input caused by the subject's actions, with signals sampled from the environment containing information about the actual consequence. For example, if a listener facing a static sound source rotates their head to the right, the source should move to the left of the head. Auditory input allows us track the actual position of the source and can be compared with such predicted changes to detect unexpected events and update predictive models. These "forward models" may originate within the premotor cortex where inverse models (the actions needed to reach a certain goal) and expectations from frontal areas are integrated with information about sounds from the parietal and auditory cortices (Rauschecker and Scott 2009).

12.3.3.3 Descending Connections

The central auditory system has been described as a hierarchical structure in which auditory input is transduced in the cochlea and conveyed via the auditory nerve through the brainstem and midbrain to the thalamic and cortical circuits that include both primary and non-primary fields (Kaas and Hackett 2000; Hackett 2015). This hierarchical perspective could be expanded to large-scale cortical networks extending from auditory to parietal and frontal cortices. However, the view that the auditory system (and thus central processing of binaural cues) is hierarchically organized overlooks the many descending connections from higher areas back to lower stages (Ryugo et al. 2011). These descending connections are critical for shaping sensory processing and determining how binaural cues sampled at the periphery are represented in the central auditory system.

In addition to many ascending connections from the auditory to the prefrontal cortex, the PFC also sends dense reciprocal connections back to auditory cortex (Hackett et al. 1999; Romanski et al. 1999). Similarly, auditory cortical neurons send corticothalamic projections back to the medial geniculate body (Hackett et al. 1998) and corticocollicular projections to the inferior colliculus (IC; Bajo et al. 2007). The thalamus and IC also send their own projections to brainstem nuclei such as the superior olivary complex (SOC) where binaural cues are initially extracted (see Stecker, Bernstein, and Brown, Chap. 6). IC and SOC neurons project directly to the cochlea to innervate the inner hair cells through which initial transduction occurs (Schofield and Cant 1999). Thus, the auditory system is structured as a highly interconnected cortical and subcortical network through which information can travel in multiple directions.

At a computational level, the auditory system is therefore actually a heterarchy, a form of network organization in which nodes (e.g., brain regions) are widely interconnected and multiple, complex, and potentially cyclical chains of command exist. In the auditory system, the range of descending connections offers multiple pathways for areas such as the prefrontal and parietal cortices to sculpt the neural processing of incoming signals. This may serve the ongoing behavioral goals of the organism, such as selectively attending to a particular sound source or spatial location (Ryugo et al. 2011).

Attention provides an example of descending modulation of sensory function where neural responses in auditory cortex (Fritz et al. 2003) are shaped by the features of sounds that subjects attend. In auditory cortex, attentional modulation is thought to originate from descending connections from frontal cortex (Fritz et al. 2010). Neurons in frontal cortex are highly selective for behaviorally relevant sounds, and measures of functional connectivity such as coherence between frontal and auditory cortices are enhanced when animals actively discriminate sounds. Similar effects have been observed in spatial working-memory tasks in which human subjects are required to detect whether a target sound differed in location from a reference (Lutzenberger et al. 2002). Here, activity in the γ-frequency range of EEG signals (30–80 Hz) was enhanced during the delay period of the task when the subjects were required to remember the reference location. Interestingly, γ-band coherence between left parietal and right frontal regions was also enhanced during this phase, indicating a possible coupling between frontal and parietal cortices that might preserve spatial information about binaural cues after sound presentation has ended. One of the limitations of measuring interactions between brain regions using functional connectivity metrics is that connectivity is only implied rather than causally demonstrated. However, by optogenetically stimulating neurons in mice, it has been shown that the orbitofrontal cortex directly excites neurons in the primary auditory cortex and shapes sound-evoked responses (Winkowski et al. 2018).

An example of descending control of spatial processing comes from the study of gaze direction circuits in barn owls (*Tyto alba*). In the owl, neurons in the arcopallial gaze fields (AGFs) and optic tectum (OT) are necessary for orienting to sound sources (Knudsen and Knudsen 1996). The AGF is analogous to the FEFs in mammals, and AGF neurons show tuning to localization cues such as the ITD and ILD (Cohen and Knudsen 1995). However, unlike other structures such as the OT in which ITD and ILD tuning is organized to create a spatial map, the receptive fields across AGF are not topographically arranged (Cohen and Knudsen 1995). Microstimulation of AGF sharpens the spatial selectivity and enhances the responses of neurons in OT (Winkowski and Knudsen 2006, 2007). Enhanced selectivity is specific for those OT neurons that share similar spatial tuning to the site of stimulation in the AGF, whereas the responses of OT neurons tuned to distant sound locations are suppressed. Enhanced selectivity is achieved by specific improvements in tuning consistency and spatial resolution, whereas suppression is driven by global modulations in neural gain within the OT (Winkowski and Knudsen 2008). Together, this work offers a potential mechanism by which frontal areas like the AGF may shape spatial tuning elsewhere in the auditory system.

Use of mouse models has also allowed detailed dissection of the neural circuits through which spatial attention may route binaural signals toward or away from decision making in mammals (Schmitt et al. 2017). This work suggests that the drive for the PFC to select a particular rule (i.e., prioritize or ignore binaural cues) is sustained by inputs from the mediodorsal thalamus. This study highlights that subcortical processing is critical for coordinating cognitive function in cortical circuits as well as providing ascending auditory input into auditory cortex. Thus, the use of binaural information in auditory behaviors requires large networks, including multiple brain structures that undermine strong functional divisions between cortical and subcortical processing of the auditory space.

12.4 Plasticity: Adaptive Encoding of Auditory Space

Attention and behavioral goals shape the neural processing of binaural cues on very short timescales, but the use of binaural cues in spatial perception changes over longer timescales. Such plasticity is critical for dealing with the changes in head size that occur through development, and for adapting to changes in the balance of inputs between the two ears occurring with unilateral hearing loss (Keating and King 2013).

One common approach to studying binaural plasticity is to reversibly alter the relationship between binaural cues and sound location by inserting an earplug into one ear to delay and attenuate monaural input. When juvenile owls (Knudsen et al. 1984) or ferrets (Keating et al. 2015) are raised with a unilateral earplug, they adapt to altered binaural cues to maintain accurate sound localization. Correspondingly, neurons in primary auditory cortex show a parallel ability to maintain normal spatial receptive fields by "remapping" altered binaural cues to sound source location (Fig. 12.3a, b). The ability to relearn sound localization after earplugging is not restricted to juvenile animals and can be elicited (and accelerated via training) in adult animals (Kacelnik et al. 2006), including humans (Bauer et al. 1966).

Relearning the relationship between localization cues and spatial position in the world is known as remapping. Remapping is complemented by *reweighting*, where an altered or unreliable cue is given less influence on perceptual judgments, while greater emphasis is placed on other unaltered cues (Fig. 12.3c). Reweighting is particularly relevant for sound localization, where redundant information is potentially provided by ILDs, ITDs, and spectral cues (Keating et al. 2015). When relearning to localize sounds after unilateral hearing loss, mammalian listeners reweight sounds to reduce the influence of altered binaural cues and give greater importance to monaural spectral cues received by the unaffected ear (Hofman and Van Opstal 2002; Keating et al. 2016).

Auditory cortex is implicated as a key site for plasticity, for both binaural cues and restoring the perception of sound elevation after the alteration of spectral cues via ear molds (Trapeau and Schonwiesner 2018). For example, animals with auditory cortical lesions or inactivation are unable to relearn to localize sounds after the introduction of an earplug (Nodal et al. 2010; Bajo et al. 2019). However, a complex network of brain areas also interacts with auditory cortex to adapt to sensory processing. In particular, the cholinergic system is a key driver of plasticity, because lesioning the cholinergic basal forebrain also prevents animals from adapting to an earplug (Leach et al. 2013). Descending pathways are also critical for implementing plasticity – selectively ablating the descending connections from layer V in auditory cortex to the IC also prevents the recalibration of the auditory space (Bajo et al.

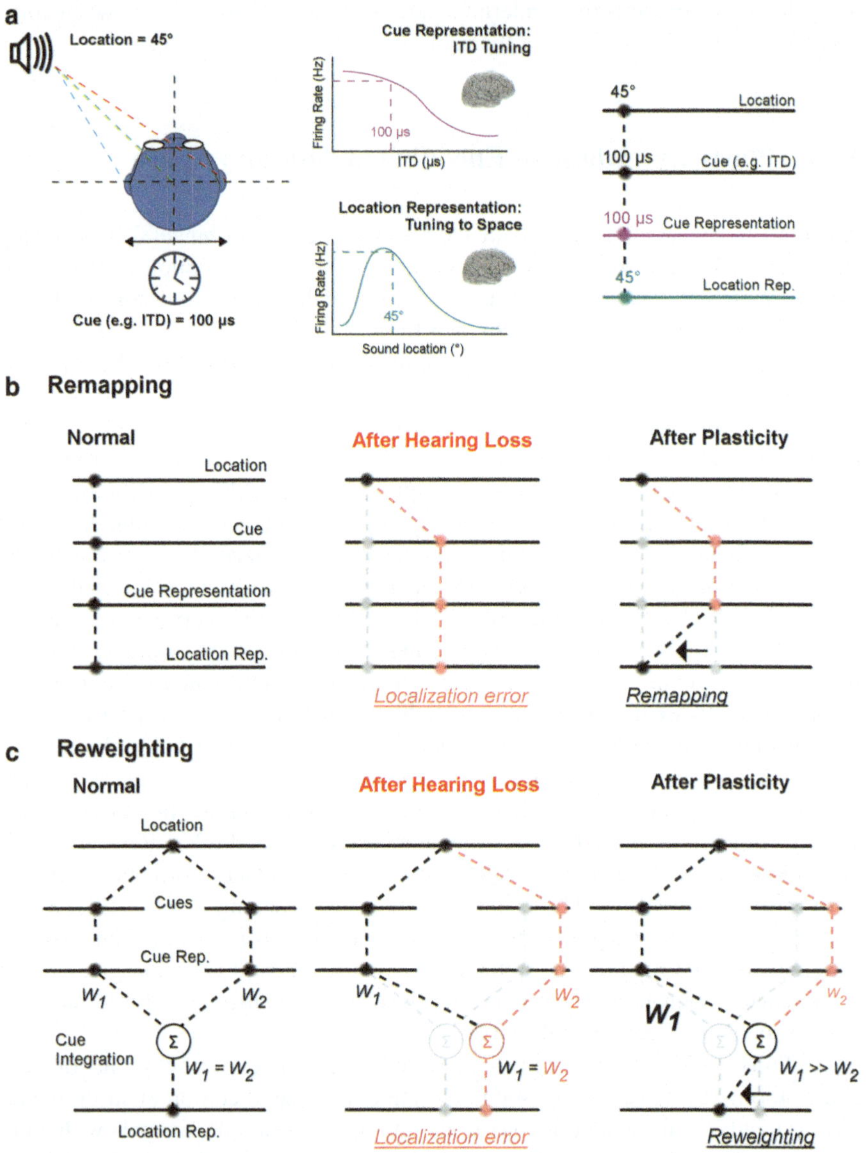

Fig. 12.3 Remapping and reweighting cues in cortical processing. (**a**) cartoon (*left*) illustrating how a sound source at a particular location generates acoustic cues such as ITDs. In the brain, neurons are tuned to particular ITDs (*top center*) such that the firing rate of a given neuron can indicate the ITD that was presented. Other neurons (*bottom center*) represent the spatial position itself, which is presumably computed from cue-sensitive neurons. At each step, the localization cue and firing rates of neurons can be schematically represented as values on a scale, with the relationship between values indicated by their relative position (*right*). (**b**) in remapping, hearing loss produces distorted acoustic cues that initially affect the representation of the location of the sound within the brain, leading to errors in localization. Plasticity allows the brain to compensate

2010). This suggests that adaptation of binaural processing requires adjustments in subcortical processing by top-down cortical control signals.

12.5 Unanswered Questions

12.5.1 Does the Auditory Cortex Represent Space or Acoustic Localization Cues?

An open question is to what extent spatial tuning in the auditory cortex reflects the encoding of acoustic cues for localization (i.e., ITDs, ILDs, and spectral cues) or the actual location of the sound source in the world. Although sound location is tightly linked with the values of acoustic cues, our perception of a location in space does not depend on any one cue. A cue-based representation indicates that a neuron is tuned to specific physical variables calculated from sounds arriving at the ears, whereas representations of actual location reflect the computation of an abstract property of a sound source across multiple acoustic cues. Dedicated brainstem nuclei are thought to compute ITDs and ILDs and thus encode specific acoustic cues (see Owrutsky, Benichoux, and Tollin, Chap. 5; Stecker, Bernstein, and Brown, Chap. 6); however, the nature of cortical representations remains unclear.

The duplex theory (Rayleigh 1907; see Hartmann, Chap. 2) posits that the use of ITDs and ILDs is frequency dependent, with locations of high- (>2.5-kHz) and low- (<1.5-kHz) frequency sounds determined primarily by ITDs or ILDs, respectively (Wightman and Kistler 1992; MacPherson and Middlebrooks 2002). In contrast, representations of actual location should integrate across cues to represent the sound position consistently across frequencies. Few studies have explicitly tested whether cortical neurons encode acoustic cues or actual sound location because stimuli in such experiments tend to be presented over headphones or with broadband energy (and thus multiple acoustic cues available).

One exception is a study of the primary auditory cortex in ferrets performing a relative localization task, where subjects reported whether sounds moved left or right in space (Wood et al. 2019). During behavior, the spatial tuning of neurons was mapped in response to free-field noise that was filtered below 1 kHz or above 3 kHz, so that ITDs or ILDs (but not both) were primarily available as acoustic cues. The authors hypothesized that neurons encoding the actual sound location should have

Fig. 12.3 (continued) for the altered cues by altering the relationship between neural representations of the acoustic cues and sound location (*arrow*). (**c**) In reweighting, multiple cues (e.g., ITDs and ILDs) are initially represented in the brain and will be integrated using specific weights for each cue (W_1 and W_2) to localize sounds. If hearing loss distorts one cue (e.g., ILD), this will lead to localization errors. However, if other cues remain reliable, the brain will learn to rely more heavily on the intact cues by changing the size of W_1 and W_2. Reweighting therefore reduces the contribution of distorted cues so that localization accuracy recovers after plasticity similar tuning curves when tested with either stimulus, whereas neurons tuned to specific acoustic cues should only be spatially tuned when tested with one stimulus type. Consistent with the encoding of the actual sound location, a subset of neurons was found to be tuned to the same locations across stimulus types, indicating that primary auditory cortex can indeed encode abstract information about the sound location across acoustic cues.

The representation of actual location rather than acoustic cues in ferret auditory cortex parallels findings in humans. For example, using an adaptor-probe design in MEG studies, Salminen et al. (2015) showed that auditory cortex is highly sensitive to changes in sound location regardless of whether those changes are conveyed by ITD or ILD. However, the limited spatial resolution of MEG makes it difficult to determine whether the same neuronal population responded to both binaural cue types. Using multivoxel pattern analysis of fMRI data, Higgins et al. (2017b) showed that varying ILDs or ITDs to position sounds in VAS activated overlapping regions of the caudal auditory cortex. Furthermore, support vector machine classifiers that were trained to decode ILDs from voxel responses to stimuli containing primarily ILDs. Without further training, these classifiers could also spontaneously decode ITDs from voxel responses to stimuli containing primarily ITDs. Likewise, decoders trained to extract ITDs could also recover ILDs without additional training. Both results suggest that the auditory cortex contains a redundant representation of location that brings together multiple acoustic cues.

12.5.2 Representing Sound Location in Multiple Coordinate Frames

Given the evidence that auditory cortex represents space as well as spatial cues, a yet-to-be addressed question is to what extent the representation in space is tied to head-centered coordinates or to another reference frame (see Takahashi, Kettler, Keller, and Bala, Chap. 4). Because the cues to sound localization are inherently head-centered, a head-centered or *craniocentric* representation of auditory space has (with little verification) been long assumed. However, it is difficult to reconcile head-centered representations with properties of spatial sound perception. For example, when a subject rotates their head in the presence of a static sound source, they perceive a stable source location, despite different populations of neurons firing over time in response to changes in the sound position relative to the head. This perceptual stability may be explained by the transformation of sound location into other coordinate systems. In animals that make significant eye movements, such as primates, spatial tuning could reflect neural selectivity for sound location relative to the position of the eyes (an eye-centered or *oculocentric* reference frame). Alternatively, spatial sensitivity could reflect tuning to a particular location in the local environment independent of the subject's head position or direction (a world-centered or *allocentric* representation).

To illustrate why the coordinates of space are important, consider a scenario in which a listener wants to check the caller on their phone while working on the computer (Fig. 12.4a). Initially, they may orient their eyes toward the phone while keeping their head stationary. In this case, the relevant coordinate frame for action (should they answer the call?) is defined by the eyes rather than the head (Fig. 12.4b, c) and so the brain must therefore transform head-centered representations of sound location into oculocentric coordinates.

Aligning our perception of sound location with visual input is a critical step in multisensory integration and requires conversion between the craniocentric and oculocentric coordinates. This computation is thought to be achieved by the level of the lateral intraparietal cortex where LIP neurons have similar spatial tuning to auditory and visual stimuli (Mazzoni et al. 1996; Mullette-Gillman et al. 2005). The coordinate frames of auditory representations in the LIP have been studied in delayed-saccade tasks in which monkeys are presented with a sound and required to make eye movements after a short time interval (Stricanne et al. 1996). In the interval between sound offset and eye movement, LIP neurons show activity that is selective for sound location in either head-centered (33%; 14/43 units) or eye-centered (44%; 19/43 units) coordinates, with the remainder of neurons (23%; 10/43 units) showing an intermediate representation. Coordinate frame shifts have also been demonstrated in sound-evoked responses, where neurons throughout the PPC (including lateral, medial, and ventral intraparietal areas) represent sounds in head-centered, eye-centered, and intermediate coordinate frames (Mullette-Gillman et al. 2005; Schlack et al. 2005). Together, these results indicate that information about sound location measured relative to the head through binaural cues is converted within the central auditory system into new spaces. Furthermore, they may explain how gaze direction affects sound localization and spatial stream segregation in human hearing (Maddox et al. 2014; Pomper and Chait 2017).

Although the parietal cortex may be critical for converting between reference frames, neural activity at earlier processing stages is also influenced by visual information. The effects of eye position are observed in the IC (Groh et al. 2001), and visual stimuli can impact on auditory spatial receptive fields, in some cases, sharpening tuning for auditory-visual stimuli relative to auditory-only conditions (Bizley and King 2008). Indeed, as many as 80% of recording sites in the auditory cortex show an impact of eye position, with a bias for matching eye position and sound source location (Fu et al. 2004). The effects of gaze are strongest in the superficial/supragranular layers and occur on a relatively slow timescale, suggesting that these signals are likely to be feedback signals, possibly originating in the parietal or prefrontal cortex.

Returning to the example of the ringing phone, if the listener is working on something important, they may simply want to silence the call without shifting our gaze; they might therefore reach for the phone and press the relevant button without looking at all. To do this, head-centered sound location must be related to the motor system and, specifically, to the coordinate systems of the hand, arm, and body (Fig. 12.4d). Cohen and Andersen (2000) extended the study of coordinate frame

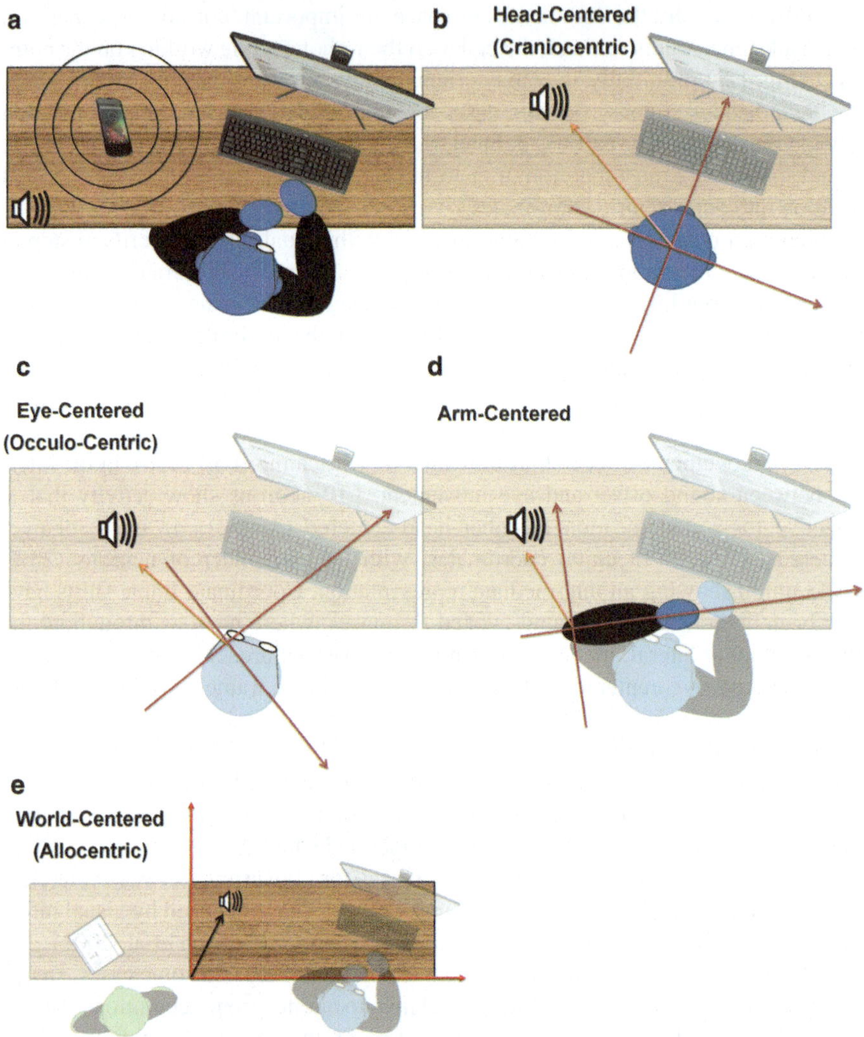

Fig. 12.4 Representing sound location in multiple coordinate frames. Cartoons show a scenario where a listener, seated at a desk, hears a ringing phone. (**a**) in such a situation, the sound can be represented in multiple coordinate systems relative to or independently from an observer. B-E: for each system, the coordinates are depicted by the axes, and the angle of the position of the phone is depicted by the *orange arrow*. (**b**) head-centered sound location that has been traditionally associated with binaural cue processing. (**c**) eye-centered location that has been studied in the inferior colliculus, auditory cortex, and parietal cortex. (**d**) arm-centered/hand-centered location that is relevant for neurons in the parietal reach region. (**e**) world-centered coordinate frames that have been observed in auditory cortex

representations to similar circumstances by asking monkeys to (1) press buttons depending on sound location while fixating gaze on a constant point or (2) make saccades to sound locations while maintaining a constant hand position. During task performance, neurons were recorded from the parietal reach region (PRR), an area of the PPC in which cells are active when planning reaches (Andersen and Buneo 2002). Within the PRR, neurons were found to be sensitive to sound location, and this spatial sensitivity was modulated by both hand and eye position (although more neurons were modulated by eye than hand position). Thus, binaural cues sampled at the ear are transformed into a range of spatial coordinate frames that are relevant for the motor system and behavioral output.

Finally, if the example listener is fortunate enough to have a helpful colleague willing to answer the phone, they could ask their colleague to take the call. If the listener wants to describe the location of the phone to this other person, however, it is unhelpful to refer to the location of the sound in a coordinate system defined by the listener's own head, eyes, or other body part because the colleague does not share the listener's egocentric reference frames. Instead, it is more useful to describe the location of the sound in a common reference frame, such as that defined by the outside world (i.e., an allocentric representation; Fig. 12.4e). Thus, one person might ask another to pick up the phone "from my coat pocket" rather than the phone "to my right, slightly above my head." Such descriptions indicate that listeners can represent sound location in their environment, and have motivated investigations of corresponding neural correlates. Using EEG, several groups have studied the coordinate frames underlying spatial mismatch negativity, an amplified neural response to changes in location of a repeated sound. Subjects were presented with a sequence of standard sounds (noise bursts) from a constant location before being tested with a deviant stimulus that changed position in either head or world coordinate systems. Changes in both world-centered and head-centered sound location elicited spatial mismatch negativity responses, indicating that the human auditory cortex encoded the location of standard sounds in both craniocentric and allocentric space (Altmann et al. 2009; Schechtman et al. 2012).

EEG studies are limited in their ability to localize the sources of neural activity within cortex; however, the coexistence of world-centered and head-centered representations has been confirmed using single-neuron recordings within the auditory cortex (Town et al. 2017). By recording neurons in freely moving subjects, sound location was varied in both head and world coordinate systems. This allowed the spatial tuning of neurons to be contrasted in the two spaces, with most units (80%) tuned to the head-centered location, and a smaller population (20%) representing sound position in the world across changes in the animals' head direction.

Across the central auditory system then, binaural cues sampled relative to the head are used to represent sound location in a variety of coordinate frames. Coordinate frame transformations allow listeners to organize behavioral responses using their body and to infer the spatial organization of an auditory scene in the environment beyond themselves. However, many open questions remain about how neural circuits within the brain perform these transformations and integrate signals from other sensory systems into auditory processing.

12.6 Chapter Summary

The work summarized in this chapter emphasizes that central processing of binaural cues involves large-scale cortical networks, including many brain regions. The flow of information within these networks appears to be complex because the neural processing of stimuli arriving at the ear occurs within the context of ongoing neural activity that is shaped by the subject's attentional state before sound onset. Thus, binaural cue processing does not simply proceed serially through the ascending auditory system but rather involves the integration of incoming sounds with simultaneous multidirectional signaling up (ascending) and down (descending) pathways. Disentangling these processes to understand the neural mechanisms of spatial sound perception is thus a significant challenge but one that is critical for treating sound localization deficits after the impairment of hearing or cognitive function.

To understand how neural circuits function, we need as many tools from neuroscience as possible. One of the challenges for spatial hearing is that most advanced techniques for studying brain activity (e.g., optogenetics, calcium imaging) have been developed in mice. However, it is not clear how binaural processing in mice, in which ILDs are the dominant localization cue, can be related to humans, other primates, or carnivores in which both ITDs and ILDs are combined for spatial perception. Thus, to understand our own spatial hearing, it may be that cutting edge approaches developed for mice must be expanded to other species, particularly those that share our sensitivity to low frequencies and our ability to localize sounds with high acuity.

As the principles of spatial auditory processing continue to evolve, it will be important to understand their implications for neuroplasticity and therapeutic strategies addressing hearing loss. For example, plastic changes in brain function after hearing impairment (e.g., earplugging) occur within the context of existing spatial processing. The implications (and therapeutic opportunities) of phenomena such as remapping and reweighting therefore must be considered within the context of binaural cue integration, when ILDs and ITDs are integrated into a representation of space across cues. Cue integration may enable more reliable estimates of sound location, but it also provides a specific locus for which reweighting of each cue would be possible after unilateral hearing loss. Thus, if an integrated signal of spatial location is the output from auditory cortex (rather than both binaural cues), then the opportunity to reweight cues later on in the network might be limited, and we would expect remapping to become the dominant form of plasticity in regions such as the prefrontal and parietal cortices. Such hypotheses remain to be tested when the effects of hearing loss on neural processing are studied beyond the auditory cortex.

Overall, decades of research have shown that the processing of binaural cues and sound location are significant components of auditory function across the cortex. However, we have only scratched the surface of how neural circuits operate, and future developments will likely require integration of knowledge across multiple cortical fields, between cortical and subcortical brain areas and from fields as diverse as cellular physiology, computational neuroscience, and clinical audiology to understand healthy and impaired spatial hearing. Furthermore, to understand how the central auditory system turns binaural cues into sound perception, it will be critical to study neural processing during active listening in behaving subjects where processes such as attention and dynamic decision making can be observed.

Acknowledgments This work was supported by Wellcome Trust/Royal Society Henry Dale Award 098418/Z/12/Z to Jennifer K. Bizley and Biotechnology and Biological Sciences Research Council (BBSRC) Project Grant BB/R004420/1 to Jennifer K. Bizley and Stephen M. Town. The authors are grateful to Dr Fred Dick for advice when preparing Fig. 12.1.

Compliance with Ethics Requirements

Stephen M. Town declares that he has no conflict of interest.

Jennifer K. Bizley declares that she has no conflict of interest.

References

Ahveninen J, Jaaskelainen IP, Raij T et al (2006) Task-modulated "what" and "where" pathways in human auditory cortex. Proc Natl Acad Sci U S A 103:14608–14613

Alain C, Arnott SR, Hevenor S et al (2001) "What" and "where" in the human auditory system. Proc Natl Acad Sci U S A 98:12301–12306

Alain C, Shen D, Yu H et al (2010) Dissociable memory- and response-related activity in parietal cortex during auditory spatial working memory. Front Psychol 1:202

Altmann CF, Wilczek E, Kaiser J (2009) Processing of auditory location changes after horizontal head rotation. J Neurosci 29:13074–13078

Andersen RA, Buneo CA (2002) Intentional maps in posterior parietal cortex. Annu Rev Neurosci 25:189–220

Arnott SR, Binns MA, Grady CL et al (2004) Assessing the auditory dual-pathway model in humans. NeuroImage 22:401–408

Azuma M, Suzuki H (1984) Properties and distribution of auditory neurons in the dorsolateral prefrontal cortex of the alert monkey. Brain Res 298:343–346

Bajo VM, Nodal FR, Bizley JK et al (2007) The ferret auditory cortex: descending projections to the inferior colliculus. Cereb Cortex 17:475–491

Bajo VM, Nodal FR, Moore DR et al (2010) The descending corticocollicular pathway mediates learning-induced auditory plasticity. Nat Neurosci 13:253–260

Bajo VM, Nodal FR, Korn C et al (2019) Seilencing cortical activity during sound-localization trinaing impairs auditory perceptual learning. Nat Commun 10:3075

Bauer RW, Matuzsa JL, Blackmer F et al (1966) Noise localization after unilateral attenuation. Journal of the Acoustical Society America 40:441–444

Behrens D, Klump GM (2016) Comparison of mouse minimum audible angle determined in prepulse inhibition and operant conditioning procedures. Hear Res 333:167–178

Belin P, Zatorre RJ (2000) 'What', 'where' and 'how' in auditory cortex. Nat Neurosci 3:965–966

Belliveau LA, Lyamzin DR, Lesica NA (2014) The neural representation of interaural time differences in gerbils is transformed from midbrain to cortex. J Neurosci 34:16796–16808
Benson DA, Hienz RD, Goldstein MH (1981) Single-unit activity in the auditory-cortex of monkeys actively localizing sound sources - spatial tuning and behavioral dependency. Brain Res 219:249–267
Bizley JK, Cohen YE (2013) The what, where and how of auditory-object perception. Nat Rev Neurosci 14:693–707
Bizley JK, King AJ (2008) Visual-auditory spatial processing in auditory cortical neurons. Brain Res 1242:24–36
Bizley JK, Nodal FR, Parsons CH et al (2007) Role of auditory cortex in sound localization in the midsagittal plane. J Neurophysiol 98:1763–1774
Bizley JK, Walker KM, Silverman BW et al (2009) Interdependent encoding of pitch, timbre, and spatial location in auditory cortex. J Neurosci 29:2064–2075
Blauert J (1996) Spatial hearing: the psychophysics of human sound localisation. MIT, Cambridge, MA
Bourquin NM, Murray MM, Clarke S (2013) Location-independent and location-linked representations of sound objects. NeuroImage 73:40–49
Brugge JF, Dubrovsky NA, Aitkin LM et al (1969) Sensitivity of single neurons in auditory cortex of cat to binaural tonal stimulation; effects of varying interaural time and intensity. J Neurophysiol 32:1005–1024
Bushara KO, Weeks RA, Ishii K et al (1999) Modality-specific frontal and parietal areas for auditory and visual spatial localization in humans. Nat Neurosci 2:759–766
Campbell RA, Doubell TP, Nodal FR et al (2006a) Interaural timing cues do not contribute to the map of space in the ferret superior colliculus: a virtual acoustic space study. J Neurophysiol 95:242–254
Campbell RA, Schnupp JW, Shial A, King AJ (2006b) Binaural-level functions in ferret auditory cortex: evidence for a continuous distribution of response properties. J Neurophysiol 95:3742–3755
Carlile S (1990) The auditory periphery of the ferret. I. Directional response properties and the pattern of interaural level differences. J Acoust Soc Am 88:2180–2195
Clarke S, Bellmann Thiran A, Maeder P et al (2002) What and where in human audition: selective deficits following focal hemispheric lesions. Exp Brain Res 147:8–15
Cohen YE, Knudsen EI (1995) Binaural tuning of auditory units in the forebrain archistriatal gaze fields of the barn owl - local-organization but no space map. J Neurosci 15:5152–5168
Cohen YE, Andersen RA (2000) Reaches to sounds encoded in an eye-centered reference frame. Neuron 27:647–652
Cohen YE, Russ BE, Gifford GW et al (2004) Selectivity for the spatial and nonspatial attributes of auditory stimuli in the ventrolateral prefrontal cortex. J Neurosci 24:11307–11316
Day ML, Delgutte B (2013) Decoding sound source location and separation using neural population activity patterns. J Neurosci 33:15837–15847
Divac I, Lavail JH, Rakic P et al (1977) Heterogeneous afferents to the inferior parietal lobule of the rhesus monkey revealed by the retrograde transport method. Brain Res 123:197–207
Eisenman LM (1974) Neural encoding of sound location: an electrophysiological study in auditory cortex (AI) of the cat using free-field stimuli. Brain Res 75:203–214
Fritz J, Shamma S, Elhilali M et al (2003) Rapid task-related plasticity of spectrotemporal receptive fields in primary auditory cortex. Nat Neurosci 6:1216–1223
Fritz JB, David SV, Radtke-Schuller S et al (2010) Adaptive, behaviorally gated, persistent encoding of task-relevant auditory information in ferret frontal cortex. Nat Neurosci 13:1011–1019
Fu KM, Shah AS, O'Connell MN et al (2004) Timing and laminar profile of eye-position effects on auditory responses in primate auditory cortex. J Neurophysiol 92:3522–3531
Gifford GW, Cohen YE (2004) Effect of a central fixation light on auditory spatial responses in area LIP. J Neurophysiol 91:2929–2933
Gifford GW, Cohen YE (2005) Spatial and non-spatial auditory processing in the lateral intraparietal area. Exp Brain Res 162:509–512

Graziano MS, Reiss LA, Gross CG (1999) A neuronal representation of the location of nearby sounds. Nature 397:428–430
Griffiths TD, Rees A, Witton C et al (1997) Spatial and temporal auditory processing deficits following right hemisphere infarction. A psychophysical study. Brain 120:785–794
Groh JM, Trause AS, Underhill AM et al (2001) Eye position influences auditory responses in primate inferior colliculus. Neuron 29:509–518
Grunewald A, Linden JF, Andersen RA (1999) Responses to auditory stimuli in macaque lateral intraparietal area I. Effects of training. J Neurophys 82:330–342
Hackett TA (2015) Anatomic organization of the auditory cortex. Handb Clin Neurol 129:27–53
Hackett TA, Preuss TM, Kaas JH (2001) Architectonic identification of the core region in auditory cortex of macaques, chimpanzees, and humans. J Comp Neurol 441:197–222
Hackett TA, Stepniewska I, Kaas JH (1998) Thalamocortical connections of the parabelt auditory cortex in macaque monkeys. J Comp Neurol 400:271–286
Hackett TA, Stepniewska I, Kaas JH (1999) Prefrontal connections of the parabelt auditory cortex in macaque monkeys. Brain Res 817:45–58
Harrington IA, Stecker GC, Macpherson EA et al (2008) Spatial sensitivity of neurons in the anterior, posterior, and primary fields of cat auditory cortex. Hear Res 240:22–41
Heffner HE, Heffner RS (1990) Effect of bilateral auditory cortex lesions on sound localization in Japanese macaques. J Neurophysiol 64:915–931
Higgins NC, McLaughlin SA, Da Costa S et al (2017a) Sensitivity to an illusion of sound location in human auditory cortex. Front Syst Neurosci 11:35
Higgins NC, McLaughlin SA, Rinne T et al (2017b) Evidence for cue-independent spatial representation in the human auditory cortex during active listening. Proc Natl Acad Sci U S A 114:E7602–E7611
Hofman PM, Van Opstal AJ (2002) Bayesian reconstruction of sound localization cues from responses to random spectra. Biol Cybern 86:305–316
Imig TJ, Adrian HO (1977) Binaural columns in the primary field (A1) of cat auditory cortex. Brain Res 138:241–257
Kaas JH, Hackett TA (2000) Subdivisions of auditory cortex and processing streams in primates. PNAS 97:11793–11799
Kacelnik O, Nodal FR, Parsons CH et al (2006) Training-induced plasticity of auditory localization in adult mammals. PLoS Biol 4:e71
Kaiser J, Birbaumer N, Lutzenberger W (2002) Magnetic oscillatory responses to lateralization changes of natural and artificial sounds in humans. Eur J Neurosci 15:345–354
Kaiser J, Lutzenberger W, Preissl H et al (2000) Right-hemisphere dominance for the processing of sound-source lateralization. J Neurosci 20:6631–6639
Keating P, Dahmen JC, King AJ (2015) Complementary adaptive processes contribute to the developmental plasticity of spatial hearing. Nat Neurosci 18:185–187
Keating P, King AJ (2013) Developmental plasticity of spatial hearing following asymmetric hearing loss: context-dependent cue integration and its clinical implications. Front Syst Neurosci 7:123
Keating P, Rosenior-Patten O, Dahmen JC et al (2016) Behavioral training promotes multiple adaptive processes following acute hearing loss. eLife 5:e12264
Knudsen EI, Esterly SD, Knudsen PF (1984) Monaural occlusion alters sound localization during a sensitive period in the barn owl. J Neurosci 4:1001–1011
Knudsen EI, Knudsen PF (1996) Contribution of the forebrain archistriatal gaze fields to auditory orienting behavior in the barn owl. Exp Brain Res 108:23–32
Leach ND, Nodal FR, Cordery PM et al (2013) Cortical cholinergic input is required for normal auditory perception and experience-dependent plasticity in adult ferrets. J Neurosci 33:6659–6671
Lee CC, Middlebrooks JC (2011) Auditory cortex spatial sensitivity sharpens during task performance. Nat Neurosci 14:108–114
Lee CC, Middlebrooks JC (2013) Specialization for sound localization in fields A1, DZ, and PAF of cat auditory cortex. J Assoc Res Otolaryngol 14:61–82

Lewald J, Wienemann M, Boroojerdi B (2004) Shift in sound localization induced by rTMS of the posterior parietal lobe. Neuropsychologia 42:1598–1607
Lewis JW, Van Essen DC (2000) Corticocortical connections of visual, sensorimotor, and multimodal processing areas in the parietal lobe of the macaque monkey. J Comp Neurol 428:112–137
Linden JF, Grunewald A, Andersen RA (1999) Responses to auditory stimuli in macaque lateral intraparietal area. II. Behavioral modulation. J Neurophysiol 82:343–358
Linden JF, Schreiner CE (2003) Columnar transformations in auditory cortex? A comparison to visual and somatosensory cortics. Cereb Cortex 13:83–89
Lomber SG, Malhotra S (2008) Double dissociation of 'what' and 'where' processing in auditory cortex. Nat Neurosci 11:609–616
Lutzenberger W, Ripper B, Busse L et al (2002) Dynamics of gamma-band activity during an audiospatial working memory task in humans. J Neurosci 22:5630–5638
Macpherson EA, Middlebrooks JC (2002) Listener weighting of cues for lateral angle: the duplex theory of sound localization revisited. J Acoust Soc Am 111:2219–2236
Maddox RK, Billimoria CP, Perrone BP et al (2012) Competing sound sources reveal spatial effects in cortical processing. PLoS Biol 10:e1001319
Maddox RK, Pospisil DA, Stecker GC et al (2014) Directing eye gaze enhances auditory spatial cue discrimination. Curr Biol 24:748–752
Maeder PP, Meuli RA, Adriani M et al (2001) Distinct pathways involved in sound recognition and localization: a human fMRI study. NeuroImage 14:802–816
Magezi DA, Krumbholz K (2010) Evidence for opponent-channel coding of interaural time differences in human auditory cortex. J Neurophysiol 104:1997–2007
Malhotra S, Hall AJ, Lomber SG (2004) Cortical control of sound localization in the cat: unilateral cooling deactivation of 19 cerebral areas. J Neurophysiol 92:1625–1643
Mazzoni P, Bracewell RM, Barash S et al (1996) Spatially tuned auditory responses in area LIP of macaques performing delayed memory saccades to acoustic targets. J Neurophysiol 75:1233–1241
Mesulam MM (1981) A cortical network for directed attention and unilateral neglect. Ann Neurol 10(4):309–325
Michalka SW, Rosen ML, Kong L et al (2016) Auditory spatial coding flexibly recruits anterior, but not posterior, visuotopic parietal cortex. Cereb Cortex 26:1302–1308
Middlebrooks JC, Bremen P (2013) Spatial stream segregation by auditory cortical neurons. J Neurosci 33:10986–11001
Middlebrooks JC, Onsan ZA (2012) Stream segregation with high spatial acuity. J Acoust Soc Am 132:3896–3911
Middlebrooks JC, Xu L, Eddins AC et al (1998) Codes for sound-source location in nontonotopic auditory cortex. J Neurophysiol 80:863–881
Middlebrooks JC (1999) Virtual localization improved by scaling nonindividualized external-ear transfer functions in frequency. J Acoust Soc Am 106:1493–1510
Middlebrooks JC, Makous JC, Green DM (1989) Directional sensitivity of sound-pressure levels in the human ear canal. J Acoust Soc Am 86:89–108
Miller LM, Recanzone GH (2009) Populations of auditory cortical neurons can accurately encode acoustic space across stimulus intensity. Proc Natl Acad Sci U S A 106:5931–5935
Milner AD, Goodale MA (2006) The visual brain in action, Oxford psychology series, vol 43, 2nd edn. Oxford University Press, Oxford, New York
Mullette-Gillman OA, Cohen YE, Groh JM (2005) Eye-centered, head-centered, and complex coding of visual and auditory targets in the intraparietal sulcus. J Neurophys 94:2331–2352
Nakamoto KT, Zhang J, Kitzes LM (2004) Response patterns along an isofrequency contour in cat primary auditory cortex (AI) to stimuli varying in average and interaural levels. J Neurophysiol 91:118–135
Nakamura K (1999) Auditory spatial discriminatory and mnemonic neurons in rat posterior parietal cortex. J Neurophys 82:2503–2517

Nodal FR, Kacelnik O, Bajo VM et al (2010) Lesions of the auditory cortex impair azimuthal sound localization and its recalibration in ferrets. J Neurophysiol 103:1209–1225

Ortiz-Rios M, Azevedo FAC, Kusmierek P et al (2017) Widespread and opponent fMRI signals represent sound location in macaque auditory cortex. Neuron 93:971–983

Pandya DN, Kuypers HG (1969) Cortico-cortical connections in the rhesus monkey. Brain Res 13:13–36

Panniello M, King AJ, Dahmen JC et al (2018) Local and global spatial organization of interaural level difference and frequency preferences in auditory cortex. Cereb Cortex 28:350–369

Pavani F, Meneghello F, Ladavas E (2001) Deficit of auditory space perception in patients with visuospatial neglect. Neuropsychologia 39:1401–1409

Petrides M, Pandya DN (2001) Comparative cytoarchitectonic analysis of the human and the macaque ventrolateral prefrontal cortex and corticocortical connection patters in the monkey. Eur J Neurosci 16:291–310

Pomper U, Chait M (2017) The impact of visual gaze direction on auditory object tracking. Sci Rep 7:e4640

Rauschecker JP, Scott SK (2009) Maps and streams in the auditory cortex: nonhuman primates illuminate human speech processing. Nat Neurosci 12:718–724

Rauschecker JP, Tian B (2000) Mechanisms and streams for processing of "what" and "where" in auditory cortex. Proc Natl Acad Sci U S A 97:11800–11806

Rayleigh L (1907) On our perception of sound direction. The London, Edinburgh, and Dublin. Philosophical Magazine and Journal of Science 13:214–232

Reale RA, Brugge JF (1990) Auditory cortical neurons are sensitive to static and continuously changing interaural phase cues. J Neurophysiol 64:1247–1260

Retsa C, Matusz PJ, Schnupp JWH et al (2018) What's what in auditory cortices? NeuroImage 176:29–40

Rodgers CC, DeWeese MR (2014) Neural correlates of task switching in prefrontal cortex and primary auditory cortex in a novel stimulus selection task for rodents. Neuron 82:1157–1170

Rohlfing T, Kroenke CD, Sullivan EV et al (2012) The INIA19 template and NeuroMaps atlas for primate brain image parcellation and spatial normalization. Front Neuroinform 6:27

Romanski LM, Tian B, Fritz J et al (1999) Dual streams of auditory afferents target multiple domains in the primate prefrontal cortex. Nat Neurosci 2:1131–1136

Rozzi S, Calzavara R, Belmalih A et al (2006) Cortical connections of the inferior parietal cortical convexity of the macaque monkey. Cereb Cortex 16:1389–1417

Ryugo DK, Fay RR, Popper AN (2011) Auditory and vestibular efferents. Springer, New York

Salminen NH, May PJ, Alku P et al (2009) A population rate code of auditory space in the human cortex. PLoS One 4:e7600

Salminen NH, Takanen M, Santala O et al (2015) Integrated processing of spatial cues in human auditory cortex. Hear Res 327:143–152

Salminen NH, Tiitinen H, Yrttiaho S et al (2010) The neural code for interaural time difference in human auditory cortex. J Acoust Soc Am 127:EL60–EL65

Schadwinkel S, Gutschalk A (2010) Activity associated with stream segregation in human auditory cortex is similar for spatial and pitch cues. Cereb Cortex 20:2863–2873

Schechtman E, Shrem T, Deouell LY (2012) Spatial localization of auditory stimuli in human auditory cortex is based on both head-independent and head-centered coordinate systems. J Neurosci 32:13501–13509

Schlack A, Sterbing-D'Angelo SJ, Hartung K et al (2005) Multisensory space representations in the macaque ventral intraparietal area. J Neurosci 25:4616–4625

Schmitt LI, Wimmer RD, Nakajima M et al (2017) Thalamic amplification of cortical connectivity sustains attentional control. Nature 545:219–223

Schofield BR, Cant NB (1999) Descending auditory pathways: projections from the inferior colliculus contact superior olivary cells that project bilaterally to the cochlear nuclei. J Comp Neurol 409:210–223

Scott BH, Malone BJ, Semple MN (2009) Representation of dynamic interaural phase difference in auditory cortex of awake rhesus macaques. J Neurophysiol 101:1781–1799
Shackleton TN, Skottun BC, Arnott RH, Palmer AR (2003) Interaural time difference discrimination thresholds for single neurons in the inferior colliculus of Guinea pigs. J Neurosci 23:716–724
Smith DV, Davis B, Niu K et al (2010) Spatial attention evokes similar activation patterns for visual and auditory stimuli. J Cogn Neurosci 22:347–361
Society for Neuroscience (2017) 3D Brain https://www.brainfacts.org/3D-Brain, Accessed 07 Aug 2019
Spierer L, Meuli R, Clarke S (2007) Extinction of auditory stimuli in hemineglect: space versus ear. Neuropsychologia 45:540–551
Stecker JKB, Middlebrooks SMT (2003) Distributed coding of sound locations in the auditory cortex. Biol Cybern 89:341–349
Stecker GC, Harrington IA, Macpherson EA et al (2005a) Spatial sensitivity in the dorsal zone (area DZ) of cat auditory cortex. J Neurophysiol 94:1267–1280
Stecker GC, Harrington IA, Middlebrooks JC (2005b) Location coding by opponent neural populations in the auditory cortex. PLoS Biol 3:e78
Stricanne B, Andersen RA, Mazzoni P (1996) Eye-centered, head-centered, and intermediate coding of remembered sound locations in area LIP. J Neurophysiol 76:2071–2076
Tanaka H, Hachisuka K, Ogata H (1999) Sound lateralisation in patients with left or right cerebral hemispheric lesions: relation with unilateral visuospatial neglect. J Neurol Neurosurg Psychiatry 67:481–486
Tardif E, Murray MM, Meylan R et al (2006) The spatio-temporal brain dynamics of processing and integrating sound localization cues in humans. Brain Res 1092:161–176
Tian B, Reser D, Durham A et al (2001) Functional specialization in rhesus monkey auditory cortex. Science 292:290–293
Town SM, Brimijoin WO, Bizley JK (2017) Egocentric and allocentric representations in auditory cortex. PLoS Biol 15:e2001878
Town SM, Wood KC, Bizley JK (2018) Sound identity is represented robustly in auditory cortex during perceptual constancy. Nat Commun 9:4786
Trapeau R, Schonwiesner M (2018) The encoding of sound source elevation in the human auditory cortex. J Neurosci 38:3252–3264
Ungerleider LG, Mishkin M (1982) Two cortical visual systems. In: Ingle DJ, Goodale MA, Mansfield RJ (eds) Analysis of visual behavior. MIT, Boston
Vaadia E, Benson DA, Hienz RD et al (1986) Unit study of monkey frontal-cortex - active localization of auditory and of visual-stimuli. J Neurophys 56:934–952
Vaadia E (1989) Single-unit activity related to active localization of acoustic and visual stimuli in the frontal cortex of the rhesus monkey. Brain Behav Evol 33:127–131
Wegener JG (1973) The sound-locating behavior of brain-damaged monkeys. J Auditory Res 13:191–219
Whitfield IC, Cranford NJ, Ravizza R et al (1972) Effects of unilateral ablation of auditory cortex in cat on complex sound localization. J Neurophysiol 35:718–731
Wightman FL, Kistler DJ (1989) Headphone simulation of free-field listening. II psychophysical validation. J Acoust Soc Am 85:868–878
Wightman FL, Kistler DJ (1992) The dominant role of low-frequence interaural time differences in sound localization. J Acoust Soc Am 91:1648–1661
Wimmer RD, Schmitt LI, Davidson TJ et al (2015) Thalamic control of sensory selection in divided attention. Nature 526:705–709
Winkowski DE, Knudsen EI (2006) Top-down gain control of the auditory space map by gaze control circuitry in the barn owl. Nature 439:336–339
Winkowski DE, Knudsen EI (2007) Top-down control of Multimodal sensitivity in the barn owl optic tectum. J Neurosci 27:13279–13291

Winkowski DE, Knudsen EI (2008) Distinct Mechanisms for top-down control of neural gain and sensitivity in the owl optic tectum. Neuron 60:698–708

Winkowski DE, Nagode DA, Donaldson KJ et al (2018) Orbitofrontal cortex neurons respond to sound and activate primary auditory cortex neurons. Cereb Cortex 28:868–879

Wood KC, Town SM, Atilgan H et al (2017) Acute inactivation of primary auditory cortex causes a sound localisation deficit in ferrets. PLoS One 12(1):e0170264. https://doi.org/10.1371/journal.pone.0170264

Wood KC, Town SM, Bizley J (2019) Primary auditory cortex represents the location of auditory objects in a cue-invariant manner. Nat Commun 10:3019

Woods TM, Lopez SE, Long JH et al (2006) Effects of stimulus azimuth and intensity on the single-neuron activity in the auditory cortex of the alert macaque monkey. J Neurophysiol 96:3323–3337

Yao JD, Bremen P, Middlebrooks JC (2015) Emergence of spatial stream segregation in the ascending auditory pathway. J Neurosci 35:16199–16212

Zatorre RJ, Bouffard M, Ahad P et al (2002) Where is 'where' in the human auditory cortex? Nat Neurosci 5:905–909

Zundorf IC, Karnath HO, Lewald J (2014) The effect of brain lesions on sound localization in complex acoustic environments. Brain 137:1410–1418

Chapter 13
Binaural Hearing with Devices

Todd Andrew Ricketts and Alan Kan

13.1 Hearing with Devices

13.1.1 Amplification of Sound with Hearing Aids

Hearing aids have gone through several major transformations. Acoustic devices (e.g., ear trumpets, hearing thrones), which used resonance to increase sound level, were first introduced in the seventeenth century. After electronic hearing aids were introduced in 1898, several advancements allowed for several significant improvements (Ricketts et al. 2019). Specifically, the amount of amplification possible and the amplified frequency range (audible frequency bandwidth) were increased while at the same time distortion and instrument size were reduced. These advancements were largely made possible by implementing transistors (1952) and electret microphones (1961) into hearing aid design as well as the development and miniaturization of the balanced armature driver. Throughout much of this history, from ear trumpets through body-style hearing aids, sound was usually received at a single location (a hearing aid case or external microphone) and then routed to one or both ears. Until about the 1950s, essentially all electric hearing aids were body-worn devices. Specifically, body aids were relatively large and included a boxlike case containing the microphone and amplification stages. A receiver worn in the ear was attached to the case by a cord. These cases could be worn in a variety of positions, including on a cord around the neck, in a shirt pocket, or attached to another part of the user's clothing.

T. A. Ricketts (✉)
Department of Hearing and Speech Sciences, Vanderbilt University Medical Center, Nashville, TN, USA
e-mail: todd.a.ricketts@vanderbilt.edu

A. Kan
School of Engineering, Macquarie University, Macquarie Park, NSW, Australia
e-mail: alan.kan@mq.edu.au

© Springer Nature Switzerland AG 2021

R. Y. Litovsky et al. (eds.), *Binaural Hearing*, Springer Handbook of Auditory Research 73, https://doi.org/10.1007/978-3-030-57100-9_13

Table 13.1 List of abbreviations

Abbreviation	Definition
A/D	Analog to digital
ANSI	American National Standards Institute
BTE	Behind the ear
CAMFIT	CAMbridge FITting formula
CIC	Completely in the canal
CIS	Continuous interleaved sampling
DI	Directivity index
DNR	Digital noise reduction
DSL v5	Desired sensation level v5
DSP	Digital signal processing
HRTFs	Head-related transfer functions
ILD	Interaural level difference
IPM	Interaural place of stimulation mismatch
ITC	In the canal
ITD	Interaural time difference
ITE	In the ear
NAL-NL2	National Acoustics Laboratory-nonlinear v2
Pps	Pulses per second
RIC	Receiver in the canal
SNR	Signal-to-noise ratio
SPL	Sound pressure level
TFS	Temporal fine structure
WDRC	Wide dynamic range compression

The first use of ear-level amplification was behind-the-ear-style (BTE; see Table 13.1 for a list of abbreviations) hearing aids, which are still in use today. In modern BTE hearing aids, incoming sound is directed to the microphone through opening(s) in the top of the case (microphone ports). Within the hearing aid case, the input is then transduced into a digital form for processing. After processing and amplification, the signals are transduced back into an acoustic form and the output of the receiver is routed from the case through an attached earhook that is used primarily for retention (keeping the device secure on the ear). The sound is then directed to the ear through tubing and a custom-made earmold (typically vented) that terminates in the ear canal. As advances in electronic circuitry continued, manufacturers were able to further reduce the size of the hearing aid, thereby introducing custom in-the-ear (ITE) and in-the-canal (ITC) style devices in the 1960s. In these instruments, all electronics are contained within a plastic case that is small enough to fit in the concha bowl and ear canal. Typically, an ear shell is manufactured using accurate impressions of the patient's ear so that the fit is customized to the individual. The microphone port(s) is in a faceplate, which is attached to the top of the ear shell and is typically the only part of the device that is visible when the hearing aid is placed correctly in the ear. Further miniaturization led to the introduction of the completely-in-canal (CIC) style that is fitted entirely inside the ear canal. Because the faceplate and microphone port openings are typically recessed slightly inside the ear canal in CIC instruments, this is the position in which sound is received.

The introduction and increasing popularity of mini-BTE products have greatly reduced the use of custom products. As the name suggests, these hearing aids are significantly smaller than traditional BTE styles, but the case still fits behind the ear. Like the traditional BTE style, the microphone receives sound on the top of the case. In addition to a smaller case, either a very thin tube is used to transmit amplified sound to the ear or a thin plastic covered wire runs from the hearing aid case to an external receiver placed in the ear canal. This latter style, commonly referred to as a receiver-in-canal (RIC; Fig. 13.1), is currently the most popular hearing aid style, accounting for about 66% of all hearing aids sold (Ricketts et al. 2019). Although current mini-BTE hearing aids can be fitted with custom "eartips," they also differ from other styles in that they are most often fitted by one of several sizes and configurations or noncustom eartips. The BTE form factor has 80% of the market (mostly mini-BTE styles). The small case and thin tube or wire of the mini-BTE style offer excellent cosmetics on the ear, whereas the use of noncustom eartips precludes the need for an ear impression, thereby reducing patient visits and improving efficiency.

Although diotic presentation (same sound presented to both ears) occurred even in the era of ear trumpets, it was not until ear-level devices became popular that amplified sound, sampled at the location of each ear, was delivered to the two ears individually (bilateral amplification). Even then, it took some time before research identifying the bilateral benefits informed clinical practice. Up to the early 1970s, it was still common for audiologists to recommend a single hearing aid. Indeed, some

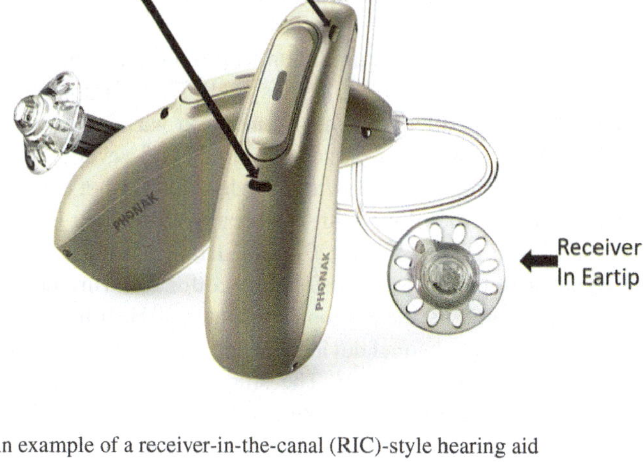

Fig. 13.1 An example of a receiver-in-the-canal (RIC)-style hearing aid

professionals argued that dispensing bilateral hearing aids was simply an attempt to double profit. A 1975 ruling by the Federal Trade Commission supported this belief by requiring professionals who wished to dispense bilateral amplification to disclose that there were no benefits associated with a second hearing aid. The importance for children to be fitted bilaterally, however, was beginning to be widely accepted about this time. However, it was not until the 1990s that the majority of people in the United States were fitted bilaterally. About 80% of the US fittings were bilateral in 2018.

It is important to distinguish the terms "unilateral and bilateral," which are used to describe wearing devices on one or both ears, from the terms "monaural and binaural," which are used to describe hearing with one or both ears. As described in this chapter, devices and/or processing can distort naturally occurring binaural cues. Furthermore, in the case of hearing aids, it is common for some binaural information to be audible when an individual is fitted with only one instrument (unilaterally). Even in cases where processing attempts to restore binaural cues, considerable distortion often remains. Therefore, unlike much of the rest of this book, this chapter focuses on bilateral versus unilateral performance with devices and, when appropriate, compares these outcomes with binaural hearing.

13.1.2 Recovery of Hearing with Cochlear Implants

Although hearing aids attempt to improve the access to sound for patients with some usable acoustic hearing, the motivation behind the creation of cochlear implants was to restore speech understanding to profoundly hearing-impaired patients. Cochlear implants bypass many of the peripheral components of the auditory system (outer ear, tympanic membrane, middle ear, and cochlea) and provide a sense of hearing through electrical stimulation of the auditory nerves near the cochlea. There are two main components to a cochlear implant: (1) an external sound processor and (2) an array of electrodes surgically implanted into the cochlea structure.

The feasibility of using electric stimulation to provide hearing to deaf patients was first demonstrated by Djourno and Eyriès in 1957. By implanting induction coils into a bilaterally deaf patient, a sense of audition was achieved whereby the patient was able to hear some environmental sounds and several words but was unable to understand speech. In the following two decades, further experiments were sparsely conducted around the globe with electrically stimulated hearing but with mixed success (see Eisen 2006; Wilson 2019 for detailed historical reviews). The legitimacy of cochlear implants as a possible device for restoring hearing came in 1975 with a study sponsored by the National Institutes of Health. Thirteen patients were implanted with a single-channel device and underwent extensive psychoacoustic, audiological, and vestibular testing. The study report concluded that single-channel devices could not support speech understanding but aided in lipreading and speech production and enhanced the quality of living for patients (Bilger et al. 1977).

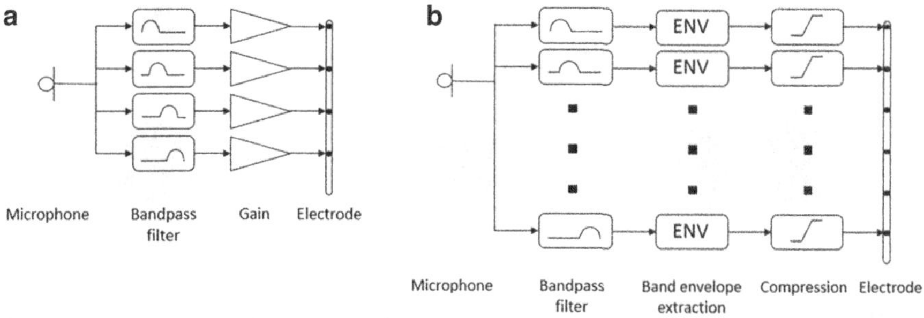

Fig. 13.2 Two general examples of cochlear implant processing strategies. The signal processing blocks for analog (**a**) and continuous interleaved sampling (CIS; **b**) processing is shown. ENV, envelope. Details of the differences between these strategies are described in the text

A major turning point for cochlear implants came in the 1980s, when open-set speech recognition was reported with multichannel devices (Clark et al. 1981). These devices used an array of electrodes to stimulate the auditory nerve fibers at different places in the cochlea to take advantage of the frequency of the cochlea to place mapping. The incoming signal was divided into different channels by passing it through a series of band-pass filters, and the output of each filter was sent to the different electrodes along the array (see Fig. 13.2A). The electrodes located from the base to the apex were stimulated with high- to low-frequency information, respectively. Subsequent reports showed significant improvements in speech understanding performance of multichannel over single-channel devices (Gantz et al. 1988; Cohen et al. 1993).

Unilateral use of multichannel cochlear implants have enabled many patients to recover usable speech understanding (>80% correct) without lipreading in quiet situations (Firszt et al. 2004; Wilson and Dorman 2007). However, there is a large variability in outcomes among the implanted population, and understanding speech in noise with only one cochlear implant is still very challenging. To improve outcomes in understanding speech in noisy environments, bilateral implantation has become more common since the early 2000s (Peters et al. 2010).

13.2 Bilateral Hearing with Devices

Several studies have demonstrated significantly better outcomes when listening with two devices rather than using one. This advantage, commonly termed bilateral benefit, has been demonstrated in a number of different domains, including speech recognition, listening effort, spatial/localization abilities, and sound quality. With regard to speech recognition, bilateral benefits may result from the head shadow effect, binaural squelch, and binaural redundancy (Fig. 13.3; see also Culling and Lavandier, Chap. 8). The head shadow effect results from a physical improvement in the signal-to-noise ratio (SNR) and is essentially a monaural phenomenon. That

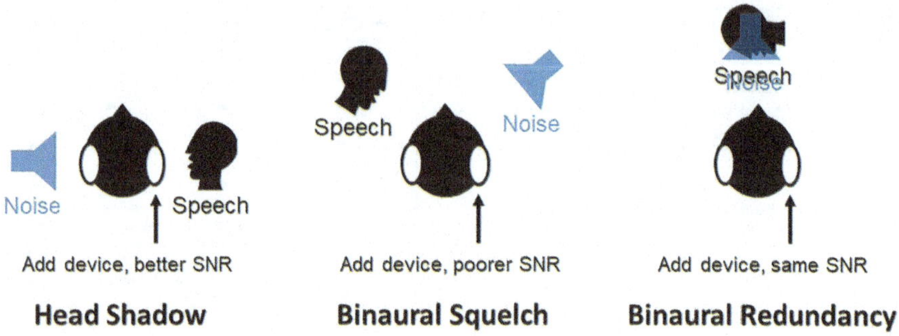

Fig. 13.3 Schematic representation of adding a second device to obtain benefits from the head shadow effect (*left*), binaural squelch (*center*), and binaural redundancy (*right*). Details regarding the potential benefits in each of these conditions are described in the text

is, improved performance due to a reduction in head shadow effects is primarily associated with conditions in which the SNR is different at the two ears. In contrast to the head shadow effect, diotic summation (also known as binaural redundancy) and binaural squelch are considered to be binaural effects that are based on complex neural processing (Hawley et al. 1999, 2004). Binaural squelch may occur when the interaural spectral or temporal differences of the target speech signal are different from those of the background noise (as occurs, e.g., when the target speech signal and background noise come from different spatial positions in the horizontal plane). Binaural auditory processing in this situation can result in an effective improvement in the SNR relative to the actual SNR measured at either ear (Zurek 1993). In contrast, diotic summation refers to the advantage that results from having redundant (Ching et al. 2006) or complimentary (Kokkinakis and Pak 2014) information at the two ears and can lead to improved speech recognition in quiet as well as in the presence of noise. These effects can combine to improve speech recognition by 3 dB or more when considering one versus two ears. However, less benefit has typically been observed in hearing-impaired listeners.

13.2.1 Bilateral Benefits with Hearing Aids

As described by Hartmann (Chap. 2), listeners with normal hearing use binaural cues to assist in localization, particularly for localization in the horizontal plane. External sounds may reach the two ears of a listener at slightly different times or levels depending on the angle of arrival. For example, if a loudspeaker is placed on the right side of the head (90° azimuth), sound will reach the right ear before the left ear (because the right ear is closer to the sound), and some of the sound will reach the right ear at a higher level than that at the left ear (because the head will block some of the high-frequency energy as the sound travels around to the left ear). These time and level differences are referred to as interaural time differences (ITDs) and interaural level differences (ILDs), respectively, and they provide crucial information for localizing sounds in the horizontal plane. Specifically, an ITD provides an important cue for the localization of lower frequency signals (<1500 Hz) and an ILD provides an important cue for the localization of higher frequency signals (>2000 Hz). These interaural differences are used in conjunction with monaural high-frequency spectral information (>5000 Hz), which is used for front-back and vertical resolution (Slattery and Middlebrooks 1994; Blauert 1997). The low-frequency cues are especially important to listeners with high-frequency hearing loss for whom the high-frequency ILDs and monaural spectral cues may be inaudible (Neher et al. 2009; Jones and Litovsky 2011).

Although hearing aids can improve audibility, they generally do not improve localization for listeners with hearing loss (Köbler and Rosenhall 2002; Van den Bogaert et al. 2006). In addition, one of the hallmarks of aided localization is high intersubject variability. Some listeners exhibit aided localization that is quite poor, whereas a small percentage achieve performance in the normal range. Average aided-localization performance, however, is typically significantly poorer than that found in listeners with normal hearing (Van den Bogaert et al. 2011). Figure 13.4 provides a comparison of localization performance between listening with binaural hearing and listening bilaterally with hearing aids. Although performance remains outside the normal range on average, bilateral fittings generally allow for better localization than unilateral fittings, as indicated by subjective reports of improved localization (Köbler et al. 2001; Boymans et al. 2009) and laboratory tests of horizontal auditory localization (Byrne et al. 1992; Boymans et al. 2008). For example, Köbler and Rosenhall (2002) tested experienced hearing aid users with their personal hearing aids. They presented speech from one of eight loudspeakers, and the other seven loudspeakers presented background noise with the same long-term spectrum as the speech signal at a + 4 dB SNR overall. Listeners were instructed to identify the loudspeaker with the speech signal and also to repeat the speech presented. Results with bilateral devices indicated that the localization ability improved approximately 10 percentage points compared with unilateral fittings and was returned to unaided performance levels.

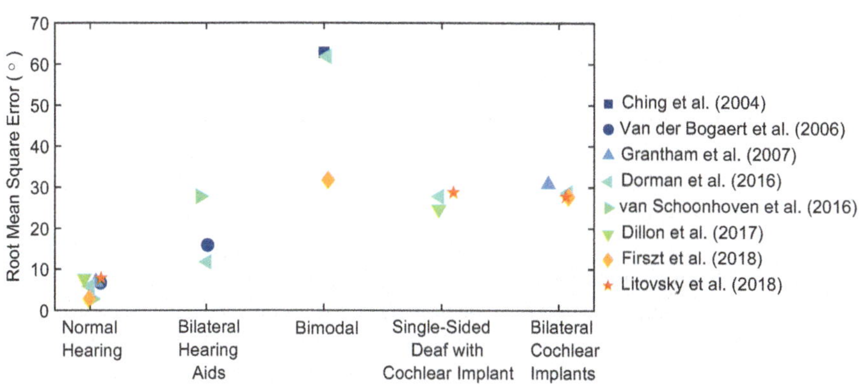

Fig. 13.4 The average sound localization performance across common bilateral device configurations and in normal-hearing listeners as reported by eight previous studies

For many listeners with bilateral hearing loss, there is some residual audibility for speech in the unaided ear. This audibility provides the potential for at least some binaural speech recognition benefits when fitted with unilateral amplification. However, adding a second hearing aid can still further improve speech recognition in specific listening situations, even when the SNR is similar at the two ears. Because there is the potential to take advantage of both binaural squelch and binaural redundancy, it is perhaps not surprising that consistent bilateral benefits for speech recognition are most often reported in studies that included speech and noise sources that were spatially separated (Hawkins and Yacullo 1984; Freyaldenhoven et al. 2006). For example, bilateral benefits of approximately 3 dB have been reported in hearing aid users by several investigators that had a speech source in front of the listener and uncorrelated noise sources surrounding or to the sides of the listener (Ricketts 2000a; Boymans et al. 2008). When speech and noise are colocated (i.e., the potential for binaural redundancy as shown in Fig. 13.3), a smaller percentage of listeners exhibit benefits. For example, Walden and Walden (2005) reported that only 11% of hearing aid listeners demonstrated significant bilateral benefits for sentences presented in colocated noise. A study has demonstrated that a bilateral hearing aid fitting is actually *worse* than a unilateral hearing aid fitting in specific listening conditions (Henkin et al. 2007). In that study, speech was presented from a loudspeaker in front of the listener and noise was presented from behind. Even when speech and noise are spatially separated, significant bilateral benefits are not always found. Indeed, several investigators have reported similar speech recognition performance for unilateral and bilateral hearing aid fittings (Hedgecock and Sheets 1958; Punch et al. 1991). Potential factors contributing to the variability in speech recognition outcomes are described in Sects. 13.3.4 through 13.3.6.

Another domain in which bilateral hearing with hearing aids may be beneficial is "listening effort," which is often described as the cognitive resources required for understanding speech (Fraser et al. 2010; Sugawara and Nikaido 2014). Listening effort has been shown to increase in adverse or complex listening situations (Murphy et al. 2000; Picou et al. 2013) and can be improved by hearing aids and the activation of some types of advanced sound processing (Sarampalis et al. 2009; Picou et al. 2017). Although factors that improve speech recognition also generally decrease the listening effort, speech recognition and listening effort are likely distinct constructs (Strand et al. 2018). Therefore, it is possible that even if the bilateral benefits for speech recognition are limited, bilateral hearing aid use could decrease the listening effort. Indeed, several researchers have reported subjective benefits of bilateral fittings on listening effort (Noble and Gatehouse 2006; Most et al. 2012), which can be present even when speech recognition is at the ceiling (Rennies and Kidd 2018). However, the potential bilateral benefits for objective listening effort have not yet been demonstrated.

Subjective bilateral benefits have been consistently reported in laboratory studies within the dimensions of sound quality, including clarity (i.e., clear versus muffled), loudness (i.e., soft versus loud), and balance (i.e., equal level at both ears versus unequal levels) (Balfour and Hawkins 1992; Naidoo and Hawkins 1997). Subjective preferences also generally favor bilateral fittings in real-world trials (Boymans et al. 2008; Cox et al. 2011). For example, the majority of respondents reported better speech recognition and overall sound quality with bilateral hearing aids in a survey of experienced hearing aid users (Köbler et al. 2001). In addition, most respondents reported that bilateral hearing aid use was beneficial when attending a lecture, in group conversations, while listening to music, and while watching television. Loudness is one sound-quality dimension for which bilateral fittings are sometimes not favored and, instead, are rated less comfortable than unilateral fittings or are rated too loud (Boymans et al. 2008; Cox et al. 2011).

Despite tendencies for higher subjective ratings when using two hearing aids compared with using just one, the number of patients who ultimately choose to be fit bilaterally is not consistently high. Researchers who have examined whether listeners prefer one versus two hearing aids have reported that preference for two hearing aids ranges from approximately 30–55% in field studies (Erdman and Sedge 1981; Vaughan-Jones et al. 1993) to approximately 70–95% in retrospective studies (Boymans et al. 2009; Bertoli et al. 2010). When listeners are fitted with their preferred fitting type (unilateral or bilateral), hearing aid outcomes have generally been shown to be similar on indices of use, satisfaction, benefit, and residual handicap (Walden and Walden 2004; Boymans et al. 2009), although some studies have shown improved outcomes on these dimensions in bilateral hearing aid users (Bertoli et al. 2010; Cox et al. 2011). These mixed preferences and outcomes are perhaps not surprising given the mixed benefits measured in the laboratory.

13.2.2 Bilateral Benefits with Cochlear Implants

Around 2000 when bilateral benefits were being observed with two hearing aids, some investigators began exploring whether bilateral implantation would improve outcomes (Tyler et al. 2002; van Hoesel 2004). Experiments comparing performance between unilateral and bilateral cochlear implantation have generally shown improved speech understanding in noise (Litovsky et al. 2009; Dunn et al. 2010) and greater sound localization abilities (Grantham et al. 2007; Dorman et al. 2016). Inspired by the results of bilateral implantation, some researchers have begun implanting the deaf ear of patients with some residual hearing or even normal hearing in the contralateral ear. For some of these patients, it is possible to wear a hearing aid in the impaired ear. The combination of electric and acoustic hearing in these bimodal hearing listeners have also provided some advantages in speech understanding in noise (Mok et al. 2006; Kokkinakis and Pak 2014) and sound localization abilities (Ching et al. 2004; Firszt et al. 2018). Furthermore, since around 2008, a growing number of single-sided deaf patients have been provided with a cochlear implant in the deaf ear as a treatment for tinnitus (Van de Heyning et al. 2008; Arts et al. 2012). Although it is unclear why cochlear implantation alleviates debilitating tinnitus, the addition of a cochlear implant appears to improve sound localization abilities similar to those in bilateral cochlear implant users (Dillon et al. 2017; Litovsky et al. 2018), but the benefits for listening to speech in noise are not as comparable (Bernstein et al. 2017; Döge et al. 2017).

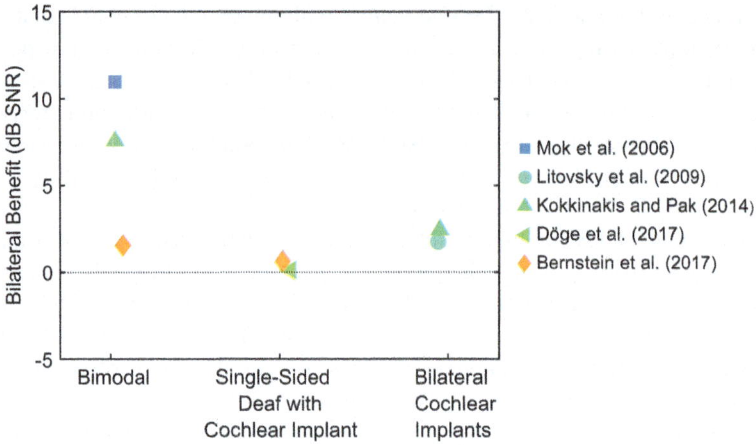

Fig. 13.5 Speech understanding benefit of adding a cochlear implant to enable bilateral hearing. Performance in both conditions was measured by the speech reception threshold for 50% performance (SRT-50). SNR, signal-to-noise ratio

Although bilateral listening with cochlear implants has improved outcomes, there is still a gap in performance between normal-hearing listeners and cochlear implant users. When listening to speech in noise, the benefit of adding a cochlear implant is usually due to the benefit of an acoustic head shadow (e.g., Litovsky et al. 2009; Gartrell et al. 2014). However, when trying to understand speech in the presence of noise or multiple talkers that surround the listener, the benefits of listening bilaterally with a cochlear implant can be quite small (see Fig. 13.5).

For sound localization, the gap in performance between normal-hearing listeners and cochlear implant users is due to a reliance on ILDs for locating sounds (see Fig. 13.4). This is in contrast to normal-hearing listeners who predominantly rely on ITDs for the precise localization of sounds (Wightman and Kistler 1992; Macpherson and Middlebrooks 2002). However, at least for bilateral cochlear implant users, the reliance on ILDs for sound localization (Grantham et al. 2007; Aronoff et al. 2010) is not due to a lack of sensitivity to ITDs with electrical stimulation. Psychophysical studies conducted using specialized research processors have found that bilateral cochlear implant users are sensitive to ITDs presented via electrical pulses (see Kan and Litovsky 2015; Laback et al. 2015 for detailed reviews). These studies have shown that bilateral cochlear implant users are sensitive to ITDs at low-pulse rates (van Hoesel et al. 2009; Litovsky et al. 2012), in amplitude-modulated high-rate pulse trains (Noel and Eddington 2013; Ihlefeld et al. 2014), or in aperiodic high-rate pulse trains (Laback and Majdak 2008; Srinivasan et al. 2018). However, sensitivity to ITDs is typically poorer than that in normal-hearing listeners. Median ITD just-noticeable differences for normal-hearing listeners presented pure tones (frequency range of 500–1000 Hz) is around 11.5 μs, whereas in bilateral cochlear implant users, median just-noticeable differences are around 144 μs for low-rate (≤100 pulses per second [pps]) electrical pulse trains (Laback et al. 2015). Although

ITD sensitivity with electrical stimulation has been observed in the laboratory, there are many factors that hinder access to usable ITDs when listening with modern cochlear implant processors. These factors are discussed in Sect. 13.3.

13.3 Factors Affecting Binaural Hearing with Devices

A hallmark of bilateral hearing with devices is the substantial variability in outcomes across and within studies. The reasons for the variability may be attributed to numerous factors including sound acquisition and delivery, signal processing, and individual variations in the degree of hearing loss and contralateral interference for speech. Although specific details of the design and signal processing of hearing aids and cochlear implants are far beyond the scope of this chapter, it is important to understand how these devices can intentionally or unintentionally modify signals resulting in changes to acoustic cues important for binaural processing. For the interested reader, detailed explanations of the design of hearing aids can be found in Kates (2008), Lyons (2010), and Ricketts et al. (2019) and of cochlear implants in Zeng et al. (2008).

13.3.1 Sound Acquisition and Delivery

In hearing devices, the acquisition of sound in the listener's environment is typically accomplished using a microphone. The electronic signal at the output of the microphone is intended to mimic the pressure changes of sounds in the environment at the microphone location. The electret or microelectromechanical system (MEMS) microphones used in modern devices are characterized by a relatively flat and broad frequency response. Because modern microphones are relatively transparent acoustically (i.e., flat-frequency response and nearly zero-added delay or distortion), distortion of natural binaural cues in sound acquisition is mainly related to the microphone position, which varies by style as described in Sect. 13.1.1. It follows that the nearer the microphones are to the natural position of sound acquisition (the tympanic membranes), the less distortion of the natural binaural cues.

Like hearing aids, the first ear-level cochlear implant processors were BTE styles with all the signal-processing hardware contained within the external processor. A transcutaneous radio frequency link, typically located above and behind the pinna, delivers power and communication of stimulus information to the implanted electrode array. The link is held in place by a magnet implanted just under the skin. Although BTE styles are still common, cochlear implant manufacturers have made a shift toward providing button-shaped processors that are magnetically held in place on the external side of the radio frequency link. These button-shaped processors shift the microphone location from behind the ear to a location that is above and behind the pinna, although the consequences of this change in location is currently unknown.

The specific location of the microphone port affects not only the frequency response of sounds arriving from specific angles but also the relative sensitivity to sound as a function of the angle of arrival. Changes in angular sensitivity to sound can have a direct effect on the SNR that a listener experiences in realistic listening environments. The relative sensitivity for sounds arriving from the front versus all other angles of arrival is quantified by the directivity index (DI). In environments in which the listener is surrounded by noise and speech arrives from the front, changes in speech recognition in noise performance are linearly related to changes in the DI after correcting for audibility (Ricketts et al. 2005). It is notable that many of the studies with hearing aids that demonstrated a lack of bilateral benefits for speech recognition are older (e.g., Jerger and Dirks 1961; Punch et al. 1991). As a result, all of these studies used traditional BTE hearing aids with omnidirectional microphones, often placed in a suboptimal location. Specifically, the microphone ports were often on the back or even the bottom of the BTE case. This location can result in a negative DI because the hearing aid will be more sensitive to sounds from behind than from the front because the pinna acts to provide some attenuation for higher frequency sounds arriving from in front of the listener (Ricketts et al. 2019). Therefore, although these BTE devices improved audibility, they also decreased the SNR compared with unaided listening. Audibility could be improved by a single hearing aid of this type. However, the addition of a second hearing aid does not improve audibility and further decreases the SNR. Modern hearing aids do not use this suboptimal location and often also include directional microphones or advanced microphone array technologies, which may account for the increased consistency of bilateral speech recognition benefits measured in newer studies.

Unfortunately, current hearing aids have not addressed another issue with BTE microphone port placement. Specifically, because the BTE microphone location is above the pinna, any advantages related to pinna diffraction effects (e.g., monaural spectral cues) will be eliminated. However, access to these cues for many listeners is likely limited regardless of hearing aid style. The amount of gain and subsequent audible bandwidth is limited by the hearing aid receiver. Audible bandwidth depends on many factors, including the power needed and the magnitude of hearing loss in the lowest and highest frequencies. Typical receivers are able to deliver amplified low frequencies down to 100–300 Hz. However, listeners with normal or near normal low-frequency hearing may be able to access even the lowest frequencies, unamplified, as long as there is adequate venting. In the high frequencies, even receivers that are considered "wideband" typically have a limited output available above 7–8 kHz. For listeners with severe-to-profound hearing loss, high-frequency audibility is often limited to only 4–5 kHz or lower. Instead of a receiver, the output stage may include a vibrating oscillator that stimulates the cochlea through bone conduction or a variety of other specialized transducers, including those associated with middle ear implants. In some cases, the bandwidth of these oscillators is considerably smaller than that delivered by traditional receivers, although the audible bandwidth is equivalent or even broader in some devices.

Limiting bandwidth reduces access to monaural spectral cues, which may be one reason why hearing aid wearers generally exhibit poorer than normal localization in the vertical plane, as described in Sect. 13.2.1. Access to extended high frequencies (from 5 to approximately 10 kHz in modern instruments) has also been shown to improve speech recognition for spatially separated speech in noise (Levy et al. 2015). Although benefits were larger in listeners with normal hearing (1.3–3.0 dB), they were still present in listeners with impaired hearing (0.5–1.3 dB). Therefore, improving access to this extended high-frequency information in hearing aid wearers still has the potential to slightly improve binaural outcomes.

In cochlear implants, one manufacturer (Advanced Bionics) provides the option of having an adapter that allows the microphone input to be located close to the entrance of the ear canal. A small handful of studies have shown some benefits for speech understanding by having the microphone located in this position compared with that behind the ear (Gifford and Revit 2010; Kolberg et al. 2015). However, the benefits of having microphones in the ears may still be limited. Jones et al. (2016) investigated the impact of microphone position on the horizontal-plane sound localization in the frontal hemisphere in cochlear implant users. By measuring individualized head-related transfer functions (HRTFs) of the bilateral cochlear implant for microphones positioned in the ear, behind the ear, and on the shoulders and then using the HRTFs to generate virtual stimuli Carlile (1996) found no significant improvement in localization performance with in-the-ear microphones. In addition, frequencies above 8 kHz are typically not available with cochlear implants, which would limit their ability to use high-frequency pinna-related spectral cues for front-back discrimination and vertical plane localization (Majdak et al. 2011). However, even if high-frequency spectral cues were available, it is likely that current spread would limit the sensitivity to spectral profile differences (Goupell et al. 2008).

A second factor related to sound acquisition that is important for hearing aids, but not for cochlear implants, is venting. All hearing aids or earmolds allow for at least a small amount of sound to leak out of and into the residual ear canal space. The space that allows sound to pass around the borders of the hearing aid/earmold in the concha and ear canal is referred to as a slit leak. Venting simply refers to the intentional process of increasing the amount of this leakage, usually by creating an additional sound channel. Because venting provides a pathway for external sounds to leak into the ear, these sounds may be audible in listeners with normal or near normal hearing in some frequency ranges. This may lead to audible natural binaural cues that could improve binaural outcomes. However, there is also the potential for incongruent cues between the same sound being accessed naturally and a delayed version of the sound after hearing aid processing and amplification, as described in Sect. 13.3.2.1. Access to natural acoustic sounds by hearing aid wearers will depend on the sound level, the magnitude of venting, and the degree of residual hearing. Most commonly, it will occur in hearing aid wearers that have little or no low-frequency hearing loss because they are also typically fitted with large vents.

13.3.2 *Signal Processing*

The goal of signal processing in hearing aids and cochlear implants has been to provide understandable speech information. However, the two device types are very different because of the mode of signal delivery to the patients. In hearing aids, signal processing aims to provide amplification at frequencies where a patient has difficulty hearing. In contrast, a cochlear implant has to convert the acoustic signal into an electrical code that the brain can understand as speech. Hence, there are different issues associated with how signal processing affects binaural hearing abilities with each device.

13.3.2.1 Hearing Aids

In analog hearing aids, the continuously varying voltage at the microphone output is filtered, amplified, and delivered to the receiver coil where it is transduced back into acoustic sound pressure. This process has a low latency, but signal modifications are quite limited. However, nearly all modern hearing aids are digital and the electrical output of the microphone is converted to a string of representative numbers. The process of obtaining a digital representation of sound through sampling and quantization is called analog-to-digital (A/D) conversion. Sampling is the process of measuring the signal amplitude at discrete points in time. Quantifying amplitude (or change in amplitude) through the assignment of numeric values at discrete sample intervals is a process referred to as quantization. Simple-to-complex signal modifications can then be achieved by applying mathematical functions to this digital representation of the input sound. In addition, the incoming signal can be analyzed and different functions can be applied for different inputs. Groups of mathematical functions with a combined goal are commonly referred to as digital algorithms or just algorithms. Together, these algorithms when applied to digital representations of sound are referred to as digital signal processing (DSP) or just processing. After applying the desired DSP, the digital signal is converted back into analog form for delivery to the hearing aid wearer.

Because patients do not typically have the same magnitude of hearing loss at all frequencies, it is often of interest to apply different processing in different frequency regions. Frequency-specific analysis and processing is typically completed for gain processing, digital noise reduction (DNR), wind noise reduction, and activation and control of many other special features. Consequently, it is typically necessary to break up the output of the A/D convertor into frequency ranges that are then analyzed in a variety of ways (e.g., amplitude, phase, changes compared with previous samples) to make processing decisions (see Ricketts et al. 2019 for further details).

Overall, the goal is for an accurate digital representation of sound, real-time signal analysis, and processing with limited delay, all while ensuring an adequate battery life (e.g., an 18 + −hour listening day in the case of rechargeable systems). To achieve these goals, a variety of complex and, in some cases, proprietary DSP methods are used by manufacturers but they are well beyond the scope of this chapter. Although modern hearing aids do many things very well, a continued challenge is delay. More accurate digital representation of the sound, greater numbers of algorithms, and more complex algorithms all increase processing time. It is important to clarify, however, that the amount of time that the sound is delayed by the hearing aid, referred to as total delay, is dependent on a number of factors in addition to processing speed. For example, both A/D and D/A conversions require some processing time. The type of filtering also has an effect on the delay. Specifically, the delay may be frequency independent (essentially the same delay at all frequencies as is the case for finite impulse response filters) or frequency dependent (typically more delay in the low frequencies falling to less delay in the higher frequencies as is the case for infinite impulse response filters). Furthermore, although more filters in a filter bank will provide a higher frequency resolution for processing, it will also increase the delay. Processing and/or analysis after obtaining a group of samples (e.g., block processing) is also necessary for some algorithms. However, although increasing the number of samples in a block is often desirable (particularly for improving the accuracy of analysis), it will also increase the delay. From 2005 to 2008, the total delay measured for some digital hearing aids was as large as 11–12 ms (unpublished data). Total delays are now typically between 2 and 8 ms (Alexander 2016). Delays as short as 5–6 ms may be noticed by listeners when fitted with hearing aids using highly vented eartips due, in part, to incongruent cues across natural acoustic and hearing aid-processed sound pathways (Stone and Moore 2005; Stone et al. 2008). Therefore, keeping group delay to a minimum is also a design concern when manufacturers introduce new algorithms.

The potential for incongruent cues in listeners with relatively normal low-frequency hearing was mentioned in Sect. 13.3.1. Specifically, because these individuals are often fitted with significant venting, there is no effective processing delay in the unamplified low frequencies but considerable processing delay in the amplified higher frequencies. A similar frequency-specific distortion of ITDs is also present in commercial digital hearing aids that have delay that differs greatly in the low and high frequencies. Although one might assume that this could be problematic, it is not because frequency-specific interaural differences are always constant. That is, even though there is an overall delay in the high frequencies, the magnitude of the ITD remains accurate as long as the same delay is present in both hearing aids. Therefore, it is perhaps not surprising that research to date has not found a performance detriment related to incongruent processing delay across frequencies (Byrne et al. 1996).

The majority of probe microphone verification systems used for fitting hearing aids implement two or three popular and validated prescriptive methods for the assignment of level- and frequency-specific gain in hearing aids. The current versions of these methods are the National Acoustics Laboratory-Nonlinear v2 (NAL-NL2; Keidser et al. 2011); the Desired Sensation Level v5 (DSLv5; Scollie et al. 2005); and the CAMFIT 2 (Moore et al. 2010). The primary goals of these three prescriptive gain methods relate to optimizing audibility and speech recognition for a wide range of speech inputs, avoiding loudness discomfort, and providing good sound quality. Although hearing thresholds increase with sensorineural hearing loss, there is not a concomitant increase in the thresholds of discomfort because of loudness recruitment (Hellman and Meiselman 1993). Recruitment refers to the abnormally rapid growth in the perceived loudness with increases in the sound level above the hearing threshold that is associated with damage to the inner hair cells of the cochlea. As a result, sensorineural hearing loss results in a decrease in a listener's residual dynamic range (the range from hearing threshold to threshold of discomfort). To offset the decrease in the residual dynamic range, modern prescriptive gain procedures prescribe decreasing gain with increasing signal input level. These level-dependent gain changes are typically achieved using compression. Compression reduces the dynamic range from the lowest to highest levels at the output of the hearing aid relative to this input range of levels. When combining gain and compression processing in a hearing aid, the resulting amplification is referred to as nonlinear gain. When applied appropriately, nonlinear gain can effectively provide audibility for even soft speech, ensure average speech inputs are comfortable, and, at the same time, prevent loudness discomfort for many listeners with sensorineural hearing loss. Nearly every hearing aid today uses at least one type of compression, and most use at least two different types (Ricketts et al. 2019). One common scheme is to apply multichannel wide dynamic-range compression (WDRC) in combination with compression limiting. Compression limiting is designed to greatly reduce gain when hearing aid output would otherwise exceed a criterion level that is very high (e.g., 115 dB sound pressure level [SPL]). That is, the goal is to ensure that high-level inputs do not exceed the patient's threshold of discomfort. In contrast, WDRC uses frequency-specific activation threshold levels (knee points). For input levels below the knee point, the gain is constant regardless of input; for input levels above the knee point, the gain is decreased with increasing input. In modern hearing aids, compression knee points are commonly lower than most speech input levels (e.g., 35–55 dB SPL). This ensures that most speech inputs (soft to loud) are compressed. When input signals are above the kneepoint in any given channel, any change in level will result in a change in gain based on a predefined ratio (compression ratio) and timescale. For example, if an input increases by 10 dB, a compression ratio of 2:1 will result in a change in output of only 5 dB. The timescale is defined by an "attack time" (the time it takes to decrease gain in response to an increase in signal level) and a "release time" (the time it takes to increase gain in response to a decrease in signal level). Compression attack and release times are defined electroacoustically, which involves nearly instantaneously increasing and then decreasing frequency-specific input signal levels from 55 to 90 back to 55 dB SPL and measuring the time it takes to reach stable output values within ±3 dB (American National Standards Institute [ANSI] 2014). Modern hearing aids have attack times varying from <1 ms to a few seconds. Release times exhibit an even greater range, from <30 ms to >5 s. Compression attack and release times are sometimes described relative to the length of the speech segment that is effectively compressed (e.g., phonemic compression, syllabic compression). In contrast, slow-acting compression will not compress the dynamic range of the speech input for a single talker. Instead, the range of average output levels over longer time windows is reduced. Compression is often referred to as automatic gain control because the gain of the hearing aid changes automatically as the input level changes. Even hearing aid models that do not use true compression still typically apply alternative methods to decrease gain with increasing input level.

Binaural listening increases the perceived sound loudness compared with listening monaurally, particularly for sound levels above threshold (i.e., Hall and Harvey 1985). Interestingly, the two popular prescriptive methods apply slightly different corrections for unilateral versus bilateral fittings to account for this suprathreshold loudness summation for speech. Specifically, DSLv50 includes an optional bilateral correction that reduces speech targets by 3 dB. Conversely, NAL-NL2 has a bilateral gain correction that increases with input level. It is 2 dB at low-input levels (40 dB SPL and below) and increases to 6 dB at high-input levels for symmetrical losses. Smaller corrections are used for asymmetrical losses.

Because compression distorts ILDs, it has the potential to limit binaural outcomes, particularly in complex environments in which ITD information may be distorted more than ILD information (see Zahorik, Chap. 9). Importantly, there are data suggesting that listeners will weigh either ILD or ITD cues more heavily depending on which cue is more accessible (e.g., Bibee and Stecker 2016). Because WDRC applies gain based on the input intensity in multiple channels, changes to the ILD are level dependent. Moreover, because all popular prescriptive gain procedures prescribe higher compression ratios in those frequency regions where an individual's thresholds are poorer (typically in the higher frequencies), typical ILDs will also be disrupted in a frequency-specific fashion. Similarly, DNR can affect the frequency- and level-specific ILD for nonspeech sounds. DNR generally acts to filter or reduce the gain when the frequency-specific input is deemed to be noise based on acoustic analyses. In most hearing aids, the magnitude of gain reduction will increase with increasing level and decreasing estimated SNR. Although the data to date suggest that compression, DNR, and unilateral beamforming have no more than a limited effect on binaural outcomes in experienced hearing aid wearers (Keidser et al. 2006), much of this work was completed in relatively simple environments. It may therefore be of interest to examine the bilateral benefits in more complex listening environments.

13.3.2.2 Cochlear Implants

Multichannel stimulation in cochlear implants presents a few interesting issues that can affect binaural hearing outcomes. Early multichannel implants used analog electric currents to provide speech understanding. These currents were presented on all channels simultaneously and led to interactions between channels because of summation of the electric fields of individual electrodes (White et al. 1984). These interactions cause cross talk between channels such that auditory nerve fibers in one location of the cochlea may respond to stimulation from multiple electrodes at once, thereby distorting the spectral representation of the acoustic signal. To overcome channel interaction, interleaved pulsatile stimulation was introduced. This approach, named continuous interleaved sampling (CIS), stimulated only one electrode at any time with a biphasic electric pulse (Wilson et al. 1991a). The amplitude of each pulse is derived from the envelope of the signal for that channel (see Fig. 13.2B). Studies comparing analog stimulation and CIS showed improvements in open-set speech recognition in most patients with the latter approach (Wilson et al. 1991b; Loizou et al. 2003). Hence, CIS-like approaches have become the common approach for delivering electrical stimulation with cochlear implants. Modern sound-coding strategies typically use high-stimulation rates (≥900 pps/channel) based on research showing that electrical stimulation should be at least four times the highest envelope frequency that is to be presented for envelope pitches to be unaffected by the carrier pitch (McKay et al. 1994) and that pulse rates around 1000 pps were typically better (although not always) for speech perception compared with lower rates (Kiefer et al. 2000; Loizou et al. 2000).

With multichannel stimulation, an important question is how many electrode channels are necessary to convey speech understanding with fidelity. The channel vocoder (Dudley 1939) has been an instrumental tool for answering this question in listeners with normal hearing. Like CIS, the channel vocoder band-pass filters an incoming acoustic signal into a number of channels and extracts the envelope of each channel (Loizou 2006). However, rather than using the amplitudes of the envelope to modulate electrical pulses, the envelope could be used to modulate sine tones or band-pass filtered noise so that they can be presented to listeners with normal hearing (Shannon et al. 1995). With the channel vocoder, the effect of varying the number of channels on speech understanding has been studied in normal-hearing listeners, with results showing that three to four channels are needed to achieve reasonable speech understanding in quiet and at least eight channels are needed for speech understanding in noise (Loizou et al. 1999; Shannon et al. 2004). Similar findings have also been reported in cochlear implant users (Friesen et al. 2001; Croghan et al. 2017).

Depending on the manufacturer, modern cochlear implants use between 12 and 22 intracochlear electrodes for encoding sounds into electrical stimulation. Although CIS is the common approach for delivering electrical stimulation, each cochlear implant manufacturer uses a slightly different sound-coding strategy to convert acoustic signals into electrical stimulation. The differences in strategy arise from the different choices made in the design of the internal implant as well as which features of the acoustic signal to encode with electrical pulses. Details of the different sound-coding strategies used by each manufacturer can be found in Verhaert et al. (2012), Wouters et al. (2015), and Zeng et al. (2015). Although each sound-coding strategy is different, many patients can recover usable speech understanding (>80% correct) without lipreading in quiet situations, irrespective of which manufacturer's device is being used. However, there can be a large variability in outcomes within the users of the same manufacturer's device (Firszt et al. 2004; Lazard et al. 2012).

The method by which acoustic information is converted into electrical stimulation arguably plays an important role in determining binaural hearing performance with cochlear implants because binaural cues in the acoustic signal need to be encoded in the electrical stimulation for the brain to be able to access them.

However, modern cochlear implants have not been designed for encoding binaural cues because the original motivation for cochlear implants was for restoring speech understanding unilaterally. Recall that the CIS method used in modern cochlear implant devices modulates electrical pulses by the signal envelope extracted from band-pass-filtered channels. Although the signal envelope is the minimum amount of information needed for speech understanding, the acoustic signal also contains temporal fine structure (TFS) information that has been shown to be important for the lateralization of ITDs (Smith et al. 2002; Dietz et al. 2013). The importance of TFS for sound localization was shown by Jones et al. (2014). By replacing the original TFS with different acoustic carriers using a channel vocoder, Jones et al. showed that sound localization performance in normal-hearing listeners became like that of bilateral cochlear implant users. However, even though TFS ITDs do not appear to be encoded in a way that is accessible by cochlear implant users, envelope ITDs should (in theory) still be encoded to some degree of fidelity (Kan et al. 2018). However, independent unilateral processing at the two ears can lead to envelope ITDs varying dynamically and unreliably (van Hoesel 2004; Litovsky et al. 2012).

Attempts to improve binaural hearing with cochlear implants have focused on encoding TFS ITDs using individual electrical pulses presented at low stimulation rates on the apical channels (van Hoesel and Tyler 2003; Arnoldner et al. 2007). This approach follows the assumption that normal-hearing listeners are most sensitive to ITDs at low frequencies (Wightman and Kistler 1992; Macpherson and Middlebrooks 2002), and hence TFS ITDs should be provided in the apical channels of the cochlear implant. However, this assumption may not be necessary in cochlear implant users, largely because ITD sensitivity can be measured throughout the length of the electrode array, with no place of best ITD sensitivity among bilateral cochlear implant users (van Hoesel et al. 2009; Litovsky et al. 2012). A few studies have shown that ITD information presented throughout the length of the electrode array can promote good ITD sensitivity with multielectrode stimulation (Kan et al. 2015a; Thakkar et al. 2018).

Bilateral implantation implies that an electrode array needs to be surgically inserted into each ear. Unfortunately, it is extremely difficult to place electrodes in the two ears at precisely the same insertion depth. Differences in insertion depth can be a problem because cochlear implants were designed to take advantage of the tonotopic organization of the cochlea, and insertion depth differences will likely lead to an interaural place-of-stimulation mismatch (IPM) for electrodes of the same number in the two ears. Unfortunately, current cochlear implant signal processing and audiological practices do not necessarily account for IPM because each sound processor is programmed independently for each ear. Hence, the range of frequencies assigned to the same numbered electrode can stimulate different cochlear locations in each ear (see Fig. 13.6). The impact of IPM on binaural hearing abilities in bilateral cochlear implant users have been studied using specialized research processors. IPM has been shown to decrease binaural sensitivity, although the resulting impact on ITD and ILD sensitivities may differ. In general, with increasing IPM, ITD thresholds double with approximately 3 mm of IPM, whereas ILD sensitivity remains much more consistent for an IPM up to about 6 mm (Poon

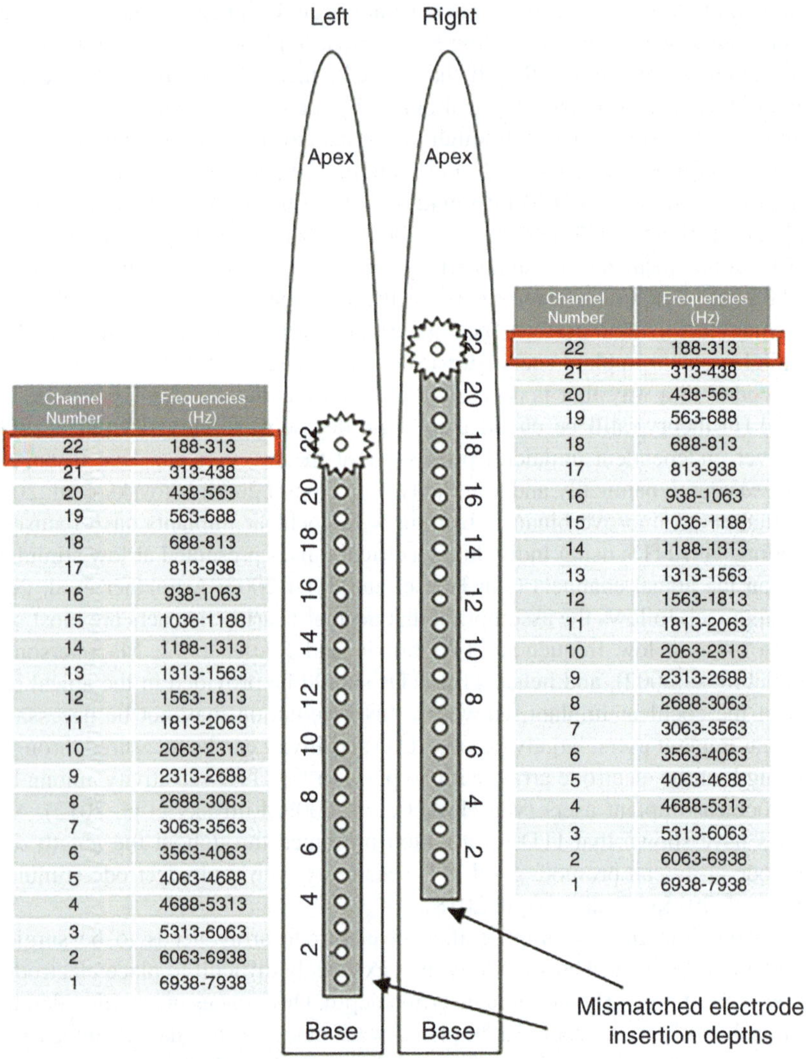

Fig. 13.6 Schematic of an interaural place-of-stimulation mismatch (IPM) in bilateral cochlear implant users. The figure shows an unrolled cochlea with an inserted electrode array. For a sound in a particular frequency range (e.g., 250 Hz; *red boxes*), a difference in electrode insertion depth will lead to different place of stimulation in the two ears

et al. 2009; Kan et al. 2015b). However, much of this work was conducted without loudness roving and a monaural confound may exist when the impact of the IPM on the ILD was measured. Although the impact of an IPM on spatial unmasking has not been directly measured in cochlear implant users, studies in normal-hearing listeners using vocoders have shown that IPM reduces the binaural benefits for speech understanding in noise (Yoon et al. 2011, 2013; Goupell et al. 2018b; see also Best, Goupell, and Colburn, Chap. 7).

Although bilateral cochlear implant users appear to solely rely on ILDs for sound localization when using clinical sound processors (Grantham et al. 2008; Aronoff et al. 2010), the fidelity with which ILDs are transmitted through the sound processor is likely compromised. This is because the available dynamic range in electric hearing is much smaller than that of naturally occurring sounds (Zeng and Shannon 1994) and the acoustic signal needs to be compressed to be encoded. Generally, the amount of compression applied to speech input levels in cochlear implants is much greater than that in hearing aids. After cochlear implant processing, ILDs of 15–17 dB can be reduced to 3–4 dB (Dorman et al. 2014), which may explain the poorer sound localization performance in bilateral cochlear implant users.

13.3.3 Beamforming Technology

To improve the SNR, beamforming technology was introduced in hearing aids and cochlear implants. Beamforming technologies are designed to have greater sensitivity for sounds arriving from in front of the listener, with relatively lower sensitivity for sounds arriving from behind and/or the side. Angle-specific sensitivity is commonly displayed via polar plots for easier visualization (e.g., Ricketts et al. 2019). The angle-specific sensitivity of these devices increases the effective SNR in environments for which the listener is facing the talker of interest and surrounded by noise (Ricketts 2000b). Consequently, in such environments, speech recognition is significantly improved by beamforming (Ricketts 2000a; Ricketts et al. 2005). Unilateral beamformers (directional microphones) and bilateral beamformers (higher order microphone arrays) have been shown to distort ILDs and ITDs and, in turn, affect sound localization. With regard to ITDs, unilateral beamformers generally have little-to-no effect; however, some bilateral beamformers can result in a severe distortion. For example, one commercial hearing aid provided an ITD of 0° azimuth regardless of the actual angle of arrival (Picou et al. 2014). With regard to ILDs, the intentional differences in sensitivity as a function of angle distort ILDs in both unilateral and bilateral beamformers (Fig. 13.7). In this example, adopted from the data presented by Picou et al. (2014), ILD distortion (i.e., a reduction in ILD compared to the unaided condition) for the unilateral beamformers is generally concentrated at more lateral angles. This is due to the fact that the angular sensitivity of the two unilateral beamformers are most different near 90°. In contrast, the specific bilateral beamformer investigated in that study distorted ILDs for all angles greater than 30°.

A few studies have found a decreased localization performance in the horizontal plane with unilateral beamformers (Keidser et al. 2006; Van den Bogaert et al. 2006). Consistent with the pattern of distortion of the ILDs, localization was disrupted near 90° but not distorted for sounds nearer to the midline. Even with this distortion, overall sound localization performance can be better using unilateral beamformers than omnidirectional processing due to a reduction in front-back confusions. Specifically, because beamformers are generally designed to provide the greatest reduction of level for sounds arriving from behind, they exaggerate the

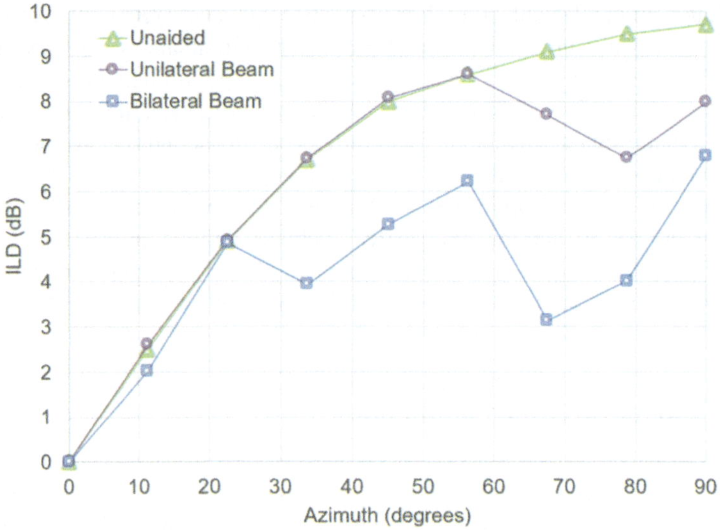

Fig. 13.7 The relative effects of commercially available examples of unilateral and bilateral beamformers on interaural level differences (ILDs) compared with those ILDs measured in the unaided ear. (Data from Picou et al. 2014)

normal level differences between front-arriving and rear-arriving sounds. This exaggeration of the front-to-back level difference can reduce the number of front-back confusions compared with omnidirectional microphone settings (Carette et al. 2014).

Bilateral beamformers can present a more severe distortion to interaural difference cues. In general, bilateral beamforming uses the output from all four microphones in bilaterally fitted hearing aids to generate a single output with a higher DI, which is then routed to both hearing aids simultaneously. With this configuration and limited venting, the resulting diotic presentation could eliminate all interaural differences with clear negative consequences (perceptually, all sounds would emanate from the center of the head). As a result, all modern commercial implementations of bilateral beamformers attempt to preserve or restore some of the naturally occurring interaural differences. In one technique, the frequency region over which the bilateral beamformer provides a diotic signal is band limited to the higher frequencies and a dichotic signal is presented in the lower frequencies via traditional directional processing. In a second and more common technique, the ITDs or ILDs are estimated at the input to the two hearing aids and partially reintroduced after beamformer processing through convolution with an average head-related transform, either in the low frequencies in some research designs (Best et al. 2017) or over a wider frequency range (Picou et al. 2014). In this second, commercially available example, the ILD is reintroduced across all frequencies (rather than just in the high frequencies as would occur in listeners with normal hearing). This provides a constant ILD across all frequencies and is done to offset the distortion to the ITD for low-frequency band limited sounds. Despite these corrections, bilateral beamformers still distort ILDs and ITDs (Brown et al. 2016) and the associated localization performance, particularly in the absence of visual cues (Van den Bogaert et al. 2008; Picou et al. 2014). However, at least one study has demonstrated that a relatively simple localization task (closed-set speaker identification with four possible sources in a background of cafeteria noise) was not significantly disrupted by a commercial bilateral beamformer in comparison to a unilateral beamformer if visual cues were also present. In addition, there was no significant preference difference between these two types of microphone processing during a short (approximately 20-minute) trial during which patients were asked to switch between processing types while walking around a noisy hospital (Picou et al. 2014).

13.3.4 Attempts to Limit Interaural Distortion

There is some evidence suggesting that providing some bilateral control over the previously independent changes in hearing aid gain may be beneficial. For example, a study has shown that some systems that control interaural phase across bilaterally fitted hearing aids can result in a higher proportion of individuals with normal or near normal localization (Drennan et al. 2005). There are also limited data suggesting that directional processing should be activated or deactivated simultaneously for both devices. For example, one study examined the effect of nonsynchronized microphones on localization for 12 hearing-impaired listeners (Keidser et al. 2006). Results showed that left/right localization error was largest when an omnidirectional microphone mode was used on one side and a directional processing one was used on the other. If the microphones settings matched, the localization error decreased by approximately 40%. There has also been at least one study examining the potential benefits of bilateral control that provided simultaneous changes in directional processing and the DNR in comparison to two hearing aids operating independently (Smith et al. 2008). In a crossover design, this study evaluated 20 participants using hearing aids in the real world via the Speech, Spatial, and Qualities of Hearing Scale. The outcomes revealed a trend for the bilaterally linked condition to be rated higher in 12 of 14 questions related to speech understanding and in 14 of 17 questions related to the localization or spatial qualities of sound. Despite these trends, however, significant preference for the bilaterally controlled condition only occurred for one item: *"You are sitting around a table or at a meeting with several people. You can see everyone. Can you tell where any person is as soon as they start speaking?"*

In addition to bilateral control, there have been efforts to enhance binaural cues, with the goal of improving localization and speech recognition performance in listeners with impaired spatial abilities. Unfortunately, these studies have generally found little or no benefits related to "binaural enhancement." For example, artificially exaggerating pinna cues as a way to limit front-back reversals and improve localization in the vertical plane in listeners with impaired hearing has been shown to have limited benefits, and the average minimum audible angle performance remained significantly worse than that exhibited by listeners with normal hearing (Rønne et al. 2016). It seems that despite the complexity of some of these processing algorithms, exaggerating interaural differences does not appear to be particularly beneficial for hearing aid users with reduced sensitivity to these same cues.

13.3.5 Patient Factors

Hearing abilities may not always be symmetrical across the two ears. The degree of hearing loss in each ear of an individual can arise from the same or different factors ranging from hereditary, acquired (such as through noise exposure and ototoxicity), and other unknown causes. Even if the factor(s) causing hearing loss in both ears is the same, the severity and progression of loss can still be different. With hearing loss, there can also be physiological differences. Within each ear, poor innervation of spiral ganglion cells can lead to decreased audibility and frequency selectivity with prolonged periods without hearing leading to a decrease in spiral ganglia cells (Moore 2007). If the extent of loss is different in each ear, this can lead to hearing asymmetries, which can affect binaural hearing abilities. As hearing loss increases, signals in an unaided ear become less audible, and hence binaural hearing benefits will be smaller (e.g., Durlach et al. 1981). Therefore, without bilateral amplification, signals may not be audible in both ears, limiting access to binaural cues. Consequently, it is not surprising that several investigators have reported greater and more consistent bilateral benefits for listeners with greater degrees of hearing loss on measures of speech recognition (Festen and Plomp 1986; McArdle et al. 2012), localization (Byrne et al. 1992), preference (Chung and Stephens 1986), and subjective ratings of benefit (Noble 2006; Boymans et al. 2008). van Schoonhoven et al. (2016) demonstrated that the magnitude of bilateral benefits from hearing aids generally increases with increasing hearing loss in multiple domains.

In cochlear implant users, there is increasing evidence that the early onset of hearing loss can affect binaural hearing outcomes. Early profound loss of hearing may lead to a less developed binaural hearing system that appears to affect sensitivity to ITDs with electrical stimulation (Litovsky et al. 2010). Laback et al. (2015) conducted a survey of the literature that measured ITD sensitivity with research processors and low-rate stimulation ($\leq$100 pps). Their review showed that patients who lost their hearing earlier in life were more likely to have poorer ITD sensitivity. The impact of the early onset of deafness is more pronounced in bilaterally implanted children (Litovsky and Gordon 2016). Children who have had no experience with acoustic hearing and use bilateral cochlear implants for listening were more likely to have no measurable ITD sensitivity. In contrast, all bilateral cochlear implant users appear to be able to use ILDs to judge left from right sounds at a constant presentation level (Gordon et al. 2014; Ehlers et al. 2017).

13.3.6 Contralateral Interference for Speech

"Contralateral interference for speech" refers to a dichotic deficit wherein binaural speech recognition performance with two ears is measurably worse than monaural performance during a situation where the reverse would be expected. That is, the additional auditory information from the second ear interferes with performance. Past literature has commonly referred to this phenomenon using the less descriptive and more general term "binaural interference." Although the prevalence of contralateral interference for speech in the general population may be small (between 5 and 18%; Allen et al. 2000; Mussoi and Bentler 2017), several authors have reported that contralateral interference for speech can be a predictor for unsuccessful bilateral hearing aid use (Jerger et al. 1993; Köbler et al. 2010). Contralateral interference for speech has also been reported in bilateral cochlear implant users (Goupell et al. 2016, 2018a).

13.4 Summary

Hearing devices have become relatively effective in aiding in, and even restoring, speech communication in quiet situations for hearing-impaired patients. In comparison to unilateral use, bilateral fitting of devices can improve speech recognition in noisy situations and sound localization. These benefits seem to also occur for patients who receive a cochlear implant in one ear while still having acoustic hearing in the other. However, despite these gains, a gap in performance still exists between device users and normal-hearing listeners. In this chapter, many of these factors that may be contributing to this gap were described, but there is still much work to be done to improve binaural hearing when listening with devices. Of immediate interest are the relative effects and trade-offs of processing that distort binaural cues (e.g., advanced beamforming technologies), particularly in complex reverberant listening situations for which the relative importance of ILDs and ITDs are less well understood. Another area of interest is the potential trade-offs for technologies when considering a target talker versus talkers in other locations (i.e., overhearing and group conversations) as a function of individual listener differences. Furthermore, a better understanding of how to encode ITDs is needed for cochlear implants.

Compliance with Ethics Requirements
Todd Ricketts declares that he has no conflict of interest.
Alan Kan owns stocks in Cochlear Ltd.

References

Alexander J (2016) Hearing aid delay and current drain in modern digital devices. Can Audiol 3

Allen RL, Schwab BM, Cranford JL, Carpenter MD (2000) Investigation of binaural interference in normal-hearing and hearing-impaired adults. J Am Acad Audiol 11:494–500

ANSI (2014) Specification of hearing aid characteristics (ANSI S3.22-2014). New York

Arnoldner C, Riss D, Brunner M et al (2007) Speech and music perception with the new fine structure speech coding strategy: preliminary results. Acta Otolaryngol 127:1298–1303. https://doi.org/10.1080/00016480701275261

Aronoff JM, Yoon Y, Freed DJ et al (2010) The use of interaural time and level difference cues by bilateral cochlear implant users. J Acoust Soc Am 127:EL87–EL92. https://doi.org/10.1121/1.3298451

Arts RAGJ, George ELJ, Stokroos RJ, Vermeire K (2012) Review: cochlear implants as a treatment of tinnitus in single-sided deafness. Curr Opin Otolaryngol Head Neck Surg 20:398–403. https://doi.org/10.1097/MOO.0b013e3283577b66

Balfour PB, Hawkins DB (1992) A comparison of sound quality judgments for monaural and binaural hearing aid processed stimuli. Ear Hear 13:331–339. https://doi.org/10.1097/00003446-199210000-00010

Bernstein JGW, Schuchman GI, Rivera AL (2017) Head shadow and binaural squelch for unilaterally deaf Cochlear Implantees. Otol Neurotol 38:e195–e202. https://doi.org/10.1097/MAO.0000000000001469

Bertoli S, Bodmer D, Probst R (2010) Survey on hearing aid outcome in Switzerland: associations with type of fitting (bilateral/unilateral), level of hearing aid signal processing, and hearing loss. Int J Audiol 49:333–346. https://doi.org/10.3109/14992020903473431

Best V, Roverud E, Mason CR, Kidd G (2017) Examination of a hybrid beamformer that preserves auditory spatial cues. J Acoust Soc Am 142:EL369. https://doi.org/10.1121/1.5007279

Bibee JM, Stecker GC (2016) Spectrotemporal weighting of binaural cues: effects of a diotic interferer on discrimination of dynamic interaural differences. J Acoust Soc Am 140:2584–2592. https://doi.org/10.1121/1.4964708

Bilger RC, Black FO, Hopkinson NT, Myers EN (1977) Implanted auditory prosthesis: an evaluation of subjects presently fitted with cochlear implants. Trans Am Acad Ophthalmol Otolaryngol 84:677–682

Blauert J (1997) Spatial hearing, 2nd edn. The MIT Press, Cambridge, MA

Boymans M, Goverts ST, Kramer SE et al (2008) A prospective multi-Centre study of the benefits of bilateral hearing aids. Ear Hear 29:930–941

Boymans M, Goverts ST, Kramer SE et al (2009) Candidacy for bilateral hearing aids: a retrospective multicenter study. J Speech Lang Hear Res 52:130–140. https://doi.org/10.1044/1092-4388(2008/07-0120)

Brown AD, Rodriguez FA, Portnuff CDF et al (2016) Time-varying distortions of binaural information by bilateral hearing aids. Trends Hear 20:233121651666830. https://doi.org/10.1177/2331216516668303

Byrne D, Noble W, Glauerdt B (1996) Effects of earmold type on ability to locate sounds when wearing hearing aids. Ear Hear 17:218–228. https://doi.org/10.1097/00003446-199606000-00005

Byrne D, Noble W, LePage B (1992) Effects of long-term bilateral and unilateral fitting of different hearing aid types on the ability to locate sounds. J Am Acad Audiol 3:369–382

Carette E, Van den Bogaert T, Laureyns M, Wouters J (2014) Left-right and front-back spatial hearing with multiple directional microphone configurations in modern hearing aids. J Am Acad Audiol 25:791–803. https://doi.org/10.3766/jaaa.25.9.2

Carlile S (1996) Virtual auditory space: generation and applications. Springer, Berlin Heidelberg, Berlin, Heidelberg

Ching TYC, Incerti P, Hill M (2004) Binaural benefits for adults who use hearing aids and Cochlear implants in opposite ears. Ear Hear 25:9–21. https://doi.org/10.1097/01.AUD.0000111261.84611.C8

Ching TYC, Incerti P, Hill M, van Wanrooy E (2006) An overview of binaural advantages for children and adults who use binaural/bimodal hearing devices. Audiol Neurotol 11:6–11. https://doi.org/10.1159/000095607

Chung SM, Stephens SD (1986) Factors influencing binaural hearing aid use. Br J Audiol 20:129–140. https://doi.org/10.3109/03005368609079006

Clark GM, Tong YC, Martin LF, Busby PA (1981) A multiple-channel cochlear implant. An evaluation using an open-set word test. Acta Otolaryngol 91:173–175

Cohen NL, Waltzman SB, Fisher SG (1993) A prospective, randomized study of Cochlear implants. N Engl J Med 328:233–237. https://doi.org/10.1056/NEJM199301283280403

Cox RM, Schwartz KS, Noe CM, Alexander GC (2011) Preference for one or two hearing aids among adult patients. Ear Hear 32:181–197. https://doi.org/10.1097/AUD.0b013e3181f8bf6c

Croghan NBH, Duran SI, Smith ZM (2017) Re-examining the relationship between number of cochlear implant channels and maximal speech intelligibility. J Acoust Soc Am 142:EL537–EL543. https://doi.org/10.1121/1.5016044

Dietz M, Marquardt T, Salminen NH, McAlpine D (2013) Emphasis of spatial cues in the temporal fine structure during the rising segments of amplitude-modulated sounds. Proc Natl Acad Sci 110:15151–15156. https://doi.org/10.1073/pnas.1309712110

Dillon MT, Buss E, Anderson ML et al (2017) Cochlear implantation in cases of unilateral hearing loss. Ear Hear 38:611–619. https://doi.org/10.1097/AUD.0000000000000430

Döge J, Baumann U, Weissgerber T, Rader T (2017) Single-sided deafness: impact of Cochlear implantation on speech perception in complex noise and on auditory localization accuracy. Otol Neurotol 38:e563–e569. https://doi.org/10.1097/MAO.0000000000001520

Dorman MF, Loiselle L, Stohl J et al (2014) Interaural level differences and sound source localization for bilateral cochlear implant patients. Ear Hear 35:633–640. https://doi.org/10.1097/AUD.0000000000000057

Dorman MF, Loiselle LH, Cook SJ et al (2016) Sound source localization by Normal-hearing listeners, hearing-impaired listeners and Cochlear implant listeners. Audiol Neurootol 21:127–131. https://doi.org/10.1159/000444740

Drennan WR, Gatehouse S, Howell P et al (2005) Localization and speech-identification ability of hearing-impaired listeners using phase-preserving amplification. Ear Hear 26:461–472. https://doi.org/10.1097/01.aud.0000179690.30137.21

Dudley HW (1939) The vocoder. Bell Labs Rec:122–126

Dunn CC, Noble W, Tyler RS et al (2010) Bilateral and unilateral cochlear implant users compared on speech perception in noise. Ear Hear 31:296–298. https://doi.org/10.1097/AUD.0b013e3181c12383

Durlach NI, Thompson CL, Colburn HS (1981) Binaural interaction in impaired listeners: a review of past research. Int J Audiol 20:181–211. https://doi.org/10.3109/00206098109072694

Ehlers E, Goupell MJ, Zheng Y et al (2017) Binaural sensitivity in children who use bilateral cochlear implants. J Acoust Soc Am 141:4264. https://doi.org/10.1121/1.4983824

Eisen MD (2006) History of the Cochlear implant. In: Cochlear implant. Thieme Medical Publishers Inc, New York, pp 1–10

Erdman SA, Sedge RK (1981) Subjective comparisons of binaural versus monaural amplification. Ear Hear 2:225–229. https://doi.org/10.1097/00003446-198109000-00009

Festen JM, Plomp R (1986) Speech-reception threshold in noise with one and two hearing aids. J Acoust Soc Am 79:465–471

Firszt JB, Holden LK, Skinner MW et al (2004) Recognition of speech presented at soft to loud levels by adult Cochlear implant recipients of three Cochlear implant systems. Ear Hear 25:375–387. https://doi.org/10.1097/01.AUD.0000134552.22205.EE

Firszt JB, Reeder RM, Holden LK, Dwyer NY (2018) Results in adult Cochlear implant recipients with varied asymmetric hearing. Ear Hear 39:845–862. https://doi.org/10.1097/AUD.0000000000000548

Fraser S, Gagne J-P, Alepins M, Dubois P (2010) Evaluating the effort expended to understand speech in noise using a dual-task paradigm: the effects of providing visual speech cues. J Speech Lang Hear Res 53:18. https://doi.org/10.1044/1092-4388(2009/08-0140)

Freyaldenhoven MC, Plyler PN, Thelin JW, Burchfield SB (2006) Acceptance of noise with monaural and binaural amplification. J Am Acad Audiol 17:659–666. https://doi.org/10.3766/jaaa.17.9.5

Friesen LM, Shannon RV, Baskent D, Wang X (2001) Speech recognition in noise as a function of the number of spectral channels: comparison of acoustic hearing and cochlear implants. J Acoust Soc Am 110:1150–1163. https://doi.org/10.1121/1.1381538

Gantz BJ, Tyler RS, Knutson JF et al (1988) Evaluation of five different cochlear implant designs: audiologic assessment and predictors of performance. Laryngoscope 98:1100–1106. https://doi.org/10.1288/00005537-198810000-00013

Gartrell BC, Jones HG, Kan A et al (2014) Investigating long-term effects of cochlear implantation in single-sided deafness: a best practice model for longitudinal assessment of spatial hearing abilities and tinnitus handicap. Otol Neurotol 35:1525–1532. https://doi.org/10.1097/MAO.0000000000000437

Gifford RH, Revit LJ (2010) Speech perception for adult cochlear implant recipients in a realistic background noise: effectiveness of preprocessing strategies and external options for improving speech recognition in noise. J Am Acad Audiol 21:441–451. https://doi.org/10.3766/jaaa.21.7.3

Gordon KA, Deighton MR, Abbasalipour P, Papsin BC (2014) Perception of binaural cues develops in children who are deaf through bilateral cochlear implantation. PLoS One 9:e114841. https://doi.org/10.1371/journal.pone.0114841

Goupell MJ, Kan A, Litovsky RY (2016) Spatial attention in bilateral cochlear-implant users. J Acoust Soc Am 140:1652. https://doi.org/10.1121/1.4962378

Goupell MJ, Laback B, Majdak P, Baumgartner W-D (2008) Current-level discrimination and spectral profile analysis in multi-channel electrical stimulation. J Acoust Soc Am 124:3142–3157. https://doi.org/10.1121/1.2981638

Goupell MJ, Stakhovskaya OA, Bernstein JGW (2018a) Contralateral interference caused by binaurally presented competing speech in adult bilateral cochlear-implant users. Ear Hear 39:110–123. https://doi.org/10.1097/AUD.0000000000000470

Goupell MJ, Stoelb CA, Kan A, Litovsky RY (2018b) The effect of simulated interaural frequency mismatch on speech understanding and spatial release from masking. Ear Hear. https://doi.org/10.1097/AUD.0000000000000541

Grantham DW, Ashmead DH, Ricketts TA et al (2007) Horizontal-plane localization of noise and speech signals by postlingually deafened adults fitted with bilateral cochlear implants. Ear Hear 28:524–541. https://doi.org/10.1097/AUD.0b013e31806dc21a

Grantham DW, Ashmead DH, Ricketts TA et al (2008) Interaural time and level difference thresholds for acoustically presented signals in post-lingually deafened adults fitted with bilateral cochlear implants using CIS+ processing. Ear Hear 29:33–44. https://doi.org/10.1097/AUD.0b013e31815d636f

Hall JW, Harvey AD (1985) Diotic loudness summation in normal and impaired hearing. J speech Lang hear res 28:445-448. doi.org/10.1044/jshr.2803.448

Hawkins DB, Yacullo WS (1984) Signal-to-noise ratio advantage of binaural hearing aids and directional microphones under different levels of reverberation. J Speech Hear Disord 49:278–286. https://doi.org/10.1044/jshd.4903.278

Hawley ML, Litovsky RY, Colburn HS (1999) Speech intelligibility and localization in a multi-source environment. J Acoust Soc Am 105:3436–3448. https://doi.org/10.1121/1.424670

Hawley ML, Litovsky RY, Culling JF (2004) The benefit of binaural hearing in a cocktail party: effect of location and type of interferer. J Acoust Soc Am 115:833–843. https://doi.org/10.1121/1.1639908

Hedgecock LD, Sheets BV (1958) A comparison of monaural and binaural hearing aids for listening to speech. AMA Arch Otolaryngol 68:624–629

Hellman RP, Meiselman CH (1993) Rate of loudness growth for pure tones in normal and impaired hearing. J Acoust Soc Am 93:966

Henkin Y, Waldman A, Kishon-Rabin L (2007) The benefits of bilateral versus unilateral amplification for the elderly: are two always better than one? J Basic Clin Physiol Pharmacol 18:201–216

Ihlefeld A, Kan A, Litovsky RY (2014) Across-frequency combination of interaural time difference in bilateral cochlear implant listeners. Front Syst Neurosci 8:22. https://doi.org/10.3389/fnsys.2014.00022

Jerger J, Dirks D (1961) Binaural hearing aids. An Enigma J Acoust Soc Am 33:537–538. https://doi.org/10.1121/1.1908711

Jerger J, Silman S, Lew HL, Chmiel R (1993) Case studies in binaural interference: converging evidence from behavioral and electrophysiologic measures. J Am Acad Audiol 4:122–131

Jones GL, Litovsky RY (2011) A cocktail party model of spatial release from masking by both noise and speech interferers. J Acoust Soc Am 130:1463–1474. https://doi.org/10.1121/1.3613928

Jones H, Kan A, Litovsky RY (2014) Comparing sound localization deficits in bilateral cochlear-implant users and vocoder simulations with normal-hearing listeners. Trends Hear 18:233121651455457. https://doi.org/10.1177/2331216514554574

Jones HG, Kan A, Litovsky RY (2016) The effect of microphone placement on interaural level differences and sound localization across the horizontal plane in bilateral cochlear implant users. Ear Hear 37:e341–e345. https://doi.org/10.1097/AUD.0000000000000297

Kan A, Jones HG, Litovsky RY (2015a) Effect of multi-electrode configuration on sensitivity to interaural timing differences in bilateral cochlear-implant users. J Acoust Soc Am 138:3826–3833. https://doi.org/10.1121/1.4937754

Kan A, Litovsky RY (2015) Binaural hearing with electrical stimulation. Hear Res 322:127–137. https://doi.org/10.1016/j.heares.2014.08.005

Kan A, Litovsky RY, Goupell MJ (2015b) Effects of interaural pitch matching and auditory image centering on binaural sensitivity in cochlear implant users. Ear Hear 36:e62–e68. https://doi.org/10.1097/AUD.0000000000000135

Kan A, Peng Z, Moua K, Litovsky RY (2018) A systematic assessment of a cochlear implant processor's ability to encode interaural time differences. Asia-Pacific Signal and Information Processing Association Annual Summit and Conference. Honolulu, Hawaii, In

Kates JM (2008) Digital hearing aids. Plural Publishing, Inc, San Diego, CA

Keidser G, Dillon HR, Flax M et al (2011) The NAL-NL2 prescription procedure. Audiol Res 1:e24. https://doi.org/10.4081/audiores.2011.e24

Keidser G, Rohrseitz K, Dillon H et al (2006) The effect of multi-channel wide dynamic range compression, noise reduction, and the directional microphone on horizontal localization performance in hearing aid wearers. Int J Audiol 45:563–579. https://doi.org/10.1080/14992020600920804

Kiefer J, von Ilberg C, Hubner-Egner J et al (2000) Optimized speech understanding with the continuous interleaved sampling speech coding strategy in patients with Cochlear implants: effect of variations in stimulation rate and number of channels. Ann Otol Rhinol Laryngol 109:1009–1020. https://doi.org/10.1177/000348940010901105

Köbler S, Lindblad A-C, Olofsson Å, Hagerman B (2010) Successful and unsuccessful users of bilateral amplification: differences and similarities in binaural performance. Int J Audiol 49:613–627. https://doi.org/10.3109/14992027.2010.481774

Köbler S, Rosenhall U (2002) Horizontal localization and speech intelligibility with bilateral and unilateral hearing aid amplification. Int J Audiol 41:395–400. https://doi.org/10.3109/14992020209090416

Köbler S, Rosenhall U, Hansson H (2001) Bilateral hearing aids - effects and consequences from a user perspective. Scand Audiol 30:223–235. https://doi.org/10.1080/01050390152704742

Kokkinakis K, Pak N (2014) Binaural advantages in users of bimodal and bilateral cochlear implant devices. J Acoust Soc Am 135:EL47–EL53. https://doi.org/10.1121/1.4831955

Kolberg ER, Sheffield SW, Davis TJ et al (2015) Cochlear implant microphone location affects speech recognition in diffuse noise. J Am Acad Audiol 26:51–58. https://doi.org/10.3766/jaaa.26.1.6

Laback B, Egger K, Majdak P (2015) Perception and coding of interaural time differences with bilateral cochlear implants. Hear Res 322:138–150. https://doi.org/10.1016/j.heares.2014.10.004

Laback B, Majdak P (2008) Binaural jitter improves interaural time-difference sensitivity of cochlear implantees at high pulse rates. Proc Natl Acad Sci U S A 105:814–817. https://doi.org/10.1073/pnas.0709199105

Lazard DS, Vincent C, Venail F et al (2012) Pre-, per- and postoperative factors affecting performance of postlinguistically deaf adults using cochlear implants: a new conceptual model over time. PLoS One 7:e48739. https://doi.org/10.1371/journal.pone.0048739
Levy SC, Freed DJ, Nilsson M et al (2015) Extended high-frequency bandwidth improves speech reception in the presence of spatially separated masking speech. Ear Hear 36:e214–e224. https://doi.org/10.1097/AUD.0000000000000161
Litovsky RY, Gordon K (2016) Bilateral cochlear implants in children: effects of auditory experience and deprivation on auditory perception. Hear Res 338:76–87. https://doi.org/10.1016/j.heares.2016.01.003
Litovsky RY, Goupell MJ, Godar S et al (2012) Studies on bilateral cochlear implants at the University of Wisconsin's binaural hearing and speech laboratory. J Am Acad Audiol 23:476–494. https://doi.org/10.3766/jaaa.23.6.9
Litovsky RY, Jones GL, Agrawal S, van Hoesel R (2010) Effect of age at onset of deafness on binaural sensitivity in electric hearing in humans. J Acoust Soc Am 127:400–414. https://doi.org/10.1121/1.3257546
Litovsky RY, Moua K, Godar S et al (2018) Restoration of spatial hearing in adult cochlear implant users with single-sided deafness. Hear Res 372:69–79. https://doi.org/10.1016/j.heares.2018.04.004
Litovsky RY, Parkinson A, Arcaroli J (2009) Spatial hearing and speech intelligibility in bilateral cochlear implant users. Ear Hear 30:419–431. https://doi.org/10.1097/AUD.0b013e3181a165be
Loizou PC (2006) Speech processing in vocoder-centric cochlear implants. In: Møller AR (ed) Cochlear and brainstem implants. S. Karger AG, pp 109–143
Loizou PC, Dorman M, Tu Z (1999) On the number of channels needed to understand speech. J Acoust Soc Am 106:2097–2103
Loizou PC, Poroy O, Dorman M (2000) The effect of parametric variations of cochlear implant processors on speech understanding. J Acoust Soc Am 108:790–802
Loizou PC, Stickney G, Mishra L, Assmann P (2003) Comparison of speech processing strategies used in the clarion implant processor. Ear Hear 24:12–19. https://doi.org/10.1097/01.AUD.0000052900.42380.50
Lyons RG (2010) Understanding digital signal processing, 3rd edn. Prentice-Hall, Inc., Englewood Cliffs, NJ
Macpherson EA, Middlebrooks JC (2002) Listener weighting of cues for lateral angle: the duplex theory of sound localization revisited. J Acoust Soc Am 111:2219–2236. https://doi.org/10.1121/1.1471898
Majdak P, Goupell MJ, Laback B (2011) Two-dimensional sound localization in Cochlear Implantees. Ear Hear 32:198. https://doi.org/10.1097/AUD.0b013e3181f4dfe9
McArdle RA, Killion M, Mennite MA, Chisolm TH (2012) Are two ears not better than one? J Am Acad Audiol 23:171–181. https://doi.org/10.3766/jaaa.23.3.4
McKay CM, McDermott HJ, Clark GM (1994) Pitch percepts associated with amplitude-modulated current pulse trains in cochlear implantees. J Acoust Soc Am 96:2664–2673. https://doi.org/10.1121/1.411377
Mok M, Grayden D, Dowell RC, Lawrence D (2006) Speech perception for adults who use hearing aids in conjunction with cochlear implants in opposite ears. J Speech Lang Hear Res 49:338. https://doi.org/10.1044/1092-4388(2006/027)
Moore B (2007) Cochlear hearing loss physiological, psychological and technical issues. Second. John Wiley & Sons, Inc., West Sussex
Moore BCJ, Glasberg BR, Stone MA (2010) Development of a new method for deriving initial fittings for hearing aids with multi-channel compression: CAMEQ2-HF. Int J Audiol 49:216–227. https://doi.org/10.3109/14992020903296746
Most T, Adi-Bensaid L, Shpak T et al (2012) Everyday hearing functioning in unilateral versus bilateral hearing aid users. Am J Otolaryngol 33:205–211. https://doi.org/10.1016/j.amjoto.2011.06.001
Murphy DR, Craik FIM, Li KZH, Schneider BA (2000) Comparing the effects of aging and background noise on short-term memory performance. Psychol Aging 15:323–334. https://doi.org/10.1037/0882-7974.15.2.323

Mussoi BSS, Bentler RA (2017) Binaural interference and the effects of age and hearing loss. J Am Acad Audiol 28:5–13. https://doi.org/10.3766/jaaa.15011
Naidoo SV, Hawkins DB (1997) Monaural/binaural preferences: effect of hearing aid circuit on speech intelligibility and sound quality. J Am Acad Audiol 8:188–202
Neher T, Behrens T, Carlile S et al (2009) Benefit from spatial separation of multiple talkers in bilateral hearing-aid users: effects of hearing loss, age, and cognition. Int J Audiol 48:758–774. https://doi.org/10.3109/14992020903079332
Noble W (2006) Bilateral hearing aids: a review of self-reports of benefit in comparison with unilateral fitting. Int J Audiol 45(Suppl 1):S63–S71. https://doi.org/10.1080/14992020600782873
Noble W, Gatehouse S (2006) Effects of bilateral versus unilateral hearing aid fitting on abilities measured by the speech, spatial, and qualities of hearing scale (SSQ). Int J Audiol 45:172–181. https://doi.org/10.1080/14992020500376933
Noel VA, Eddington DK (2013) Sensitivity of bilateral cochlear implant users to fine-structure and envelope interaural time differences. J Acoust Soc Am 133:2314–2328. https://doi.org/10.1121/1.4794372
Peters BR, Wyss J, Manrique M (2010) Worldwide trends in bilateral cochlear implantation. Laryngoscope 120(Suppl):S17–S44. https://doi.org/10.1002/lary.20859
Picou EM, Aspell E, Ricketts TA (2014) Potential benefits and limitations of three types of directional processing in hearing aids. Ear Hear 35:339–352. https://doi.org/10.1097/AUD.0000000000000004
Picou EM, Moore TM, Ricketts TA (2017) The effects of directional processing on objective and subjective listening effort. J Speech Lang Hear Res 60:199. https://doi.org/10.1044/2016_JSLHR-H-15-0416
Picou EM, Ricketts TA, Hornsby BWY (2013) How hearing aids, background noise, and visual cues influence objective listening effort. Ear Hear 34:e52–e64. https://doi.org/10.1097/AUD.0b013e31827f0431
Poon BB, Eddington DK, Noel V, Colburn HS (2009) Sensitivity to interaural time difference with bilateral cochlear implants: development over time and effect of interaural electrode spacing. J Acoust Soc Am 126:806–815. https://doi.org/10.1121/1.3158821
Punch JL, Jenison RL, Allan J, Durrant JD (1991) Evaluation of three strategies for fitting hearing aids binaurally. Ear Hear 12:205–215. https://doi.org/10.1097/00003446-199106000-00008
Rennies J, Kidd G (2018) Benefit of binaural listening as revealed by speech intelligibility and listening effort. J Acoust Soc Am 144:2147–2159. https://doi.org/10.1121/1.5057114
Ricketts T (2000a) The impact of head angle on monaural and binaural performance with directional and omnidirectional hearing aids. Ear Hear 21:318–328. https://doi.org/10.1097/00003446-200008000-00007
Ricketts T (2000b) Directivity quantification in hearing aids: fitting and measurement effects. Ear Hear 21:45–58. https://doi.org/10.1097/00003446-200002000-00008
Ricketts TA, Bentler RA, Mueller HG (2019) Essentials of modern hearing aids : selection, fitting, and verification. Plural Publishing, Inc, San Diego, CA
Ricketts TA, Henry PP, Hornsby BWY (2005) Application of frequency importance functions to directivity for prediction of benefit in uniform fields. Ear Hear 26:473–486. https://doi.org/10.1097/01.aud.0000179691.21547.01
Rønne FM, Laugesen S, Jensen NS, Pedersen JH (2016) Minimum audible angles measured with simulated normally-sized and oversized pinnas for normal-hearing and hearing- impaired test subjects. Adv Exp Med Biol 894:207–217. https://doi.org/10.1007/978-3-319-25474-6_22
Sarampalis A, Kalluri S, Edwards B, Hafter E (2009) Objective measures of listening effort: effects of background noise and noise reduction. J Speech Lang Hear Res 52:1230. https://doi.org/10.1044/1092-4388(2009/08-0111)
Scollie S, Seewald R, Cornelisse L et al (2005) The desired sensation level multistage input/output algorithm. Trends Amplif 9:159–197. https://doi.org/10.1177/108471380500900403
Shannon R, Fu Q-J, Galvin J III (2004) The number of spectral channels required for speech recognition depends on the difficulty of the listening situation. Acta Otolaryngol 124:50–54(5). https://doi.org/10.1080/03655230410017562

Shannon RV, Zeng F-GG, Kamath V et al (1995) Speech recognition with primarily temporal cues. Science 270:303–304
Slattery WH, Middlebrooks JC (1994) Monaural sound localization: acute versus chronic unilateral impairment. Hear Res 75:38–46. https://doi.org/10.1016/0378-5955(94)90053-1
Smith P, Davis A, Day J et al (2008) Real-world preferences for linked bilateral processing. Hear J 61:33–34. https://doi.org/10.1097/01.HJ.0000325657.80281.9c
Smith ZM, Delgutte B, Oxenham AJ (2002) Chimaeric sounds reveal dichotomies in auditory perception. Nature 416:87–90. https://doi.org/10.1038/416087a
Srinivasan S, Laback B, Majdak P, Delgutte B (2018) Introducing short Interpulse intervals in high-rate pulse trains enhances binaural timing sensitivity in electric hearing. J Assoc Res Otolaryngol. https://doi.org/10.1007/s10162-018-0659-7
Stone MA, Moore BCJ (2005) Tolerable hearing-aid delays: IV. Effects on subjective disturbance during speech production by hearing-impaired subjects. Ear Hear 26:225–235. https://doi.org/10.1097/00003446-200504000-00009
Stone MA, Moore BCJ, Meisenbacher K, Derleth RP (2008) Tolerable hearing aid delays. V. Estimation of limits for open canal fittings. Ear Hear 29:601–617. https://doi.org/10.1097/AUD.0b013e3181734ef2
Strand JF, Brown VA, Merchant MB et al (2018) Measuring listening effort: convergent validity, sensitivity, and links with cognitive and personality measures. Am J Speech-Language Pathol:1–24. https://doi.org/10.1044/2018_JSLHR-H-17-0257
Sugawara E, Nikaido H (2014) Properties of AdeABC and AdeIJK efflux systems of Acinetobacter baumannii compared with those of the AcrAB-TolC system of Escherichia coli. Antimicrob Agents Chemother 58:7250–7257. https://doi.org/10.1128/AAC.03728-14
Thakkar T, Kan A, Jones HG, Litovsky RY (2018) Mixed stimulation rates to improve sensitivity of interaural timing differences in bilateral cochlear implant listeners. J Acoust Soc Am 143:1428–1440. https://doi.org/10.1121/1.5026618
Tyler RS, Gantz BJ, Rubinstein JT et al (2002) Three-month results with bilateral cochlear implants. Ear Hear 23:80S–89S
Van de Heyning P, Vermeire K, Diebl M et al (2008) Incapacitating unilateral tinnitus in single-sided deafness treated by cochlear implantation. Ann Otol Rhinol Laryngol 117:645–652. https://doi.org/10.1177/000348940811700903
Van den Bogaert T, Carette E, Wouters J (2011) Sound source localization using hearing aids with microphones placed behind-the-ear, in-the-canal, and in-the-pinna. Int J Audiol 50:164–176. https://doi.org/10.3109/14992027.2010.537376
Van den Bogaert T, Doclo S, Wouters J, Moonen M (2008) The effect of multimicrophone noise reduction systems on sound source localization by users of binaural hearing aids. J Acoust Soc Am 124:484–497. https://doi.org/10.1121/1.2931962
Van den Bogaert T, Klasen TJ, Moonen M et al (2006) Horizontal localization with bilateral hearing aids: without is better than with. J Acoust Soc Am 119:515–526. https://doi.org/10.1121/1.2139653
van Hoesel RJM (2004) Exploring the benefits of bilateral cochlear implants. Audiol Neurotol 9:234–246. https://doi.org/10.1159/000078393
van Hoesel RJM, Jones GL, Litovsky RY (2009) Interaural time-delay sensitivity in bilateral cochlear implant users: effects of pulse rate, modulation rate, and place of stimulation. J Assoc Res Otolaryngol 10:557–567. https://doi.org/10.1007/s10162-009-0175-x
van Hoesel RJM, Tyler RS (2003) Speech perception, localization, and lateralization with bilateral cochlear implants. J Acoust Soc Am 113:1617–1630. https://doi.org/10.1121/1.1539520
van Schoonhoven J, Schulte M, Boymans M et al (2016) Selecting appropriate tests to assess the benefits of bilateral amplification with hearing aids. Trends Hear 20:233121651665823. https://doi.org/10.1177/2331216516658239
Vaughan-Jones RH, Padgham ND, Christmas HE et al (1993) One aid or two?—more visits please! J Laryngol Otol 107:329–332. https://doi.org/10.1017/S0022215100122947

Verhaert N, Lazard DS, Gnansia D et al (2012) Speech performance and sound localization abilities in Neurelec Digisonic® SP binaural cochlear implant users. Audiol Neurootol 17:256–266. https://doi.org/10.1159/000338472
Walden TC, Walden BE (2004) Predicting success with hearing aids in everyday living. J Am Acad Audiol 15:342–352. https://doi.org/10.3766/jaaa.15.5.2
Walden TC, Walden BE (2005) Unilateral versus bilateral amplification for adults with impaired hearing. J Am Acad Audiol 16:574–584. https://doi.org/10.3766/jaaa.16.8.6
White MW, Merzenich MM, Gardi JN (1984) Multichannel cochlear implants. Channel interactions and processor design. Arch Otolaryngol 110:493–501
Wightman FL, Kistler DJ (1992) The dominant role of low-frequency interaural time differences in sound localization. J Acoust Soc Am 91:1648–1661
Wilson BS (2019) The remarkable cochlear implant and possibilities for the next large step forward. Acoust Today 15:53–61
Wilson BS, Dorman MF (2007) The surprising performance of present-day cochlear implants. IEEE Trans Biomed Eng 54:969–972. https://doi.org/10.1109/TBME.2007.893505
Wilson BS, Finley CC, Lawson DT et al (1991a) Better speech recognition with cochlear implants. Nature 352:236–238. https://doi.org/10.1038/352236a0
Wilson BS, Lawson DT, Finley CC, Wolford RD (1991b) Coding strategies for multichannel cochlear prostheses. Am J Otol 12(Suppl):56–61
Wouters J, McDermott HJ, Francart T (2015) Sound coding in cochlear implants: from electric pulses to hearing. IEEE Signal Process Mag 32:67–80. https://doi.org/10.1109/MSP.2014.2371671
Yoon Y-S, Shin Y-R, Fu Q-J (2013) Binaural benefit with and without a bilateral spectral mismatch in acoustic simulations of cochlear implant processing. Ear Hear 34:273–279. https://doi.org/10.1097/AUD.0b013e31826709e8
Yoon Y, Liu A, Fu Q-J (2011) Binaural benefit for speech recognition with spectral mismatch across ears in simulated electric hearing. J Acoust Soc Am 130:EL94–E100. https://doi.org/10.1121/1.3606460
Zeng F-G, Rebscher S, Harrison W et al (2008) Cochlear implants: system design, integration, and evaluation. IEEE Rev Biomed Eng 1:115–142. https://doi.org/10.1109/RBME.2008.2008250
Zeng F-G, Rebscher SJ, Fu Q-J et al (2015) Development and evaluation of the Nurotron 26-electrode cochlear implant system. Hear Res 322:188–199. https://doi.org/10.1016/j.heares.2014.09.013
Zeng FG, Shannon RV (1994) Loudness-coding mechanisms inferred from electric stimulation of the human auditory system. Science 264:564–566
Zurek PM (1993) Binaural advantages and directional effects in speech intelligibility. Acoustical Factors Affecting Hearing Aid Performance, In, pp 255–276

Correction to: Computational Models of Binaural Processing

Mathias Dietz and Go Ashida

Correction to:
Chapter 10 in: R. Y. Litovsky et al. (eds.), *Binaural Hearing*, Springer Handbook of Auditory Research 73, https://doi.org/10.1007/978-3-030-57100-9_10

In the original version of this chapter, equation (10.1) present in page 286 was published without brackets.

The correct equation

$$\gamma(\tau) = \frac{\langle s(t+\tau)s^{*}(t)\rangle}{\langle |s(t)|^{2}\rangle} \tag{10.1}$$

has now been updated in this revised version of this chapter.

The updated online version of this chapter can be found at
https://doi.org/10.1007/978-3-030-57100-9_10

© Springer Nature Switzerland AG 2021

R. Y. Litovsky et al. (eds.), *Binaural Hearing*, Springer Handbook of Auditory Research 73, https://doi.org/10.1007/978-3-030-57100-9_14

The manufacturer's authorised representative in the EU is Springer Nature Customer Service Centre GmbH, Europaplatz 3, 69115 Heidelberg, Germany. If you have any concerns regarding our products, please contact ProductSafety@springernature.com

Printed and bound by CPI Group (UK) Ltd, Croydon, CR0 4YY
22/09/2026
02221048-0007